International Maritime and Port Technology and Development Conference & the 4th International Conference on Maritime Autonomous Surface Ships (MTEC/ICMASS 2022)

Journal of Physics: Conference Series Volume 2311

Singapore
5 - 7 April 2022

ISBN: 978-1-7138-6033-4
ISSN: 1742-6588

Printed from e-media with permission by:

Curran Associates, Inc.
57 Morehouse Lane
Red Hook, NY 12571

Some format issues inherent in the e-media version may also appear in this print version.

Institute of Physics
Dirac House, Temple Back
Bristol BS1 6BE UK

Phone: 44 1 17 929 7481
Fax: 44 1 17 920 0979

techtracking@iop.org

Additional copies of this publication are available from:

Curran Associates, Inc.
57 Morehouse Lane
Red Hook, NY 12571 USA
Phone: 845-758-0400
Fax: 845-758-2633
Email: curran@proceedings.com
Web: www.proceedings.com

TABLE OF CONTENTS

The International Maritime and Port Technology and Development Conference, and the 4th International Conference on Maritime Autonomous Surface Ships (MTEC/ICMASS 2022)

Journal of Physics: Conference Series **2311** (2022) 011001 doi:10.1088/1742-6596/2311/1/011001

The International Maritime and Port Technology and Development Conference, and the 4th International Conference on Maritime Autonomous Surface Ships (MTEC/ICMASS 2022)

Preface

This Special Issue presents peer-reviewed papers presented at the International Maritime and Port Technology and Development Conference (MTEC), and the 4th International Conference on Maritime Autonomous Surface Ships (ICMASS), held jointly as MTEC/ICMASS in Singapore on 6th-7th April 2022.

MTEC is arranged every second year and alternating between Singapore and Norway. MTEC is organised to enable sharing of new developments, concepts, practices and case studies in maritime and port technology. The first ICMASS was arranged in 2018 in Busan, South Korea, and has since been held in Norway and South Korea. ICMASS is organised to facilitate information and knowledge exchange of research and academic activities on Maritime Autonomous Surface Ships (MASS).

MTEC/ICMASS 2022 was jointly organised by the Maritime and Port Authority of Singapore (MPA) and The Research Council of Norway (RCN), in partnership with Singapore Maritime Institute (SMI) and SINTEF Ocean.

The main tracks of MTEC/ICMASS 2022 include:

A. Maritime Digitalisation, Connectivity and Cybersecurity
   - Optimisation of port operations with digitalisation and data analytics
   - Digital twin development to enhance design of ports and vessels, safe and secure ship-shore communications
   - Digital integration of various stakeholder systems to streamline processes
   - Big data analytics to support vessel/fleet management
   - Artificial Intelligence to support decision-making
   - Cybersecurity

B. Maritime Safety
   - Risk analysis, modelling and control tools
   - Advanced navigation systems
   - Smart safety equipment and systems
   - Emergency and oil spill response, incident analysis and management
   - Human factors in maritime safety
   - Immersive tools and technologies for training

C. Smart and Autonomous Shipping
   - New technologies and systems for design and operations of smart and autonomous vessels
   - Management of operations in a mixed environment (with smart, remote-controlled and autonomous vessels)

- On-shore support requirements for smart and autonomous fleet
- New skill sets and assessment standards for smart and autonomous fleet operators

D. Green Shipping and Ports
  - Maritime decarbonisation tools and approach
  - Cleaner power and energy options
  - Intelligent technologies and systems to optimise energy use and reduce emissions on-board vessels
  - Eco-friendly ship designs and operations/operating paradigms
  - Enhancement of terminal planning and operations to optimise energy demand
  - Circular economy and resource management for terminals

We would like to thank all that have contributed their time and effort to making this event successful. 73 presentations were delivered at MTEC/ICMASS 2022 and 37 full length papers have been included in the proceedings. We would like to thank the Steering Committee that have been responsible for the overall direction and management of the event, and the Editorial Committee and reviewers who supported the peer review process, which included the review of an abstract followed by the review of the full paper.

MTEC/ICMASS 2022 Steering Committee:

- Co-Chair Ng Yi Han, Maritime and Port Authority of Singapore
- Co-Chair Ørnulf Jan Rødseth, SINTEF Ocean
- Thomas Ting, Maritime and Port Authority of Singapore
- Tan Cheng Peng, Singapore Maritime Institute
- Sanjay Kuttan, Global Centre for Maritime Decarbonisation
- Toh Ah Cheong, Centre of Excellence in Maritime Safety, Singapore Polytechnic
- Lam Siu Lee Jasmine, Maritime Energy & Sustainable Development Centre of Excellence, Nanyang Technological University
- Chew Ek Peng, Centre of Excellence in Modelling and Simulation for Next Generation Ports, National University of Singapore
- Chua Kie Hian, Centre of Excellence for Autonomous & Remotely Operated Vessels, Technology Centre for Offshore and Marine, Singapore
- Daniel Zhang, Centre of Excellence in Maritime Safety, Singapore Polytechnic
- Kjell Røang, The Research Council of Norway
- Mary Ann Lundteigen, Norwegian University of Science and Technology
- Bjørn Egil Asbjørnslett, Norwegian University of Science and Technology
- Per Christer Lund, Innovation Norway

MTEC/ICMASS 2022 Editorial Committee:

- Co-Chair Ørnulf Jan Rødseth, SINTEF Ocean
- Co-Chair Allan Magee, Technology Centre for Offshore and Marine, Singapore
- Chew Ek Peng, Centre of Excellence in Modelling and Simulation for Next Generation Ports, National University of Singapore
- Lee Loo Hay, Centre of Excellence in Modelling and Simulation for Next Generation Ports, National University of Singapore
- Chai Kah Hin, Centre for Maritime Studies, National University of Singapore
- Ng Szu Hui, Centre for Maritime Studies, National University of Singapore
- Lam Siu Lee Jasmine, Maritime Energy & Sustainable Development Centre of Excellence, Nanyang Technological University
- Lau Hoong Chuin, Singapore Management University
- Kenneth Low, Singapore Institute of Technology
- Zhou Jianying, iTrust, Singapore University of Technology and Design
- Frederick James Francis, Singapore Maritime Academy, Singapore Polytechnic
- Brian Murray, SINTEF Ocean
- Svein Peder Berge, SINTEF Ocean

- Agus Hasan, Norwegian University of Science and Technology
- Bjørn Egil Asbjørnslett, Norwegian University of Science and Technology
- Ibrahim A. Hameed, Norwegian University of Science and Technology
- Karl Henning Halse, Norwegian University of Science and Technology
- Mary Ann Lundteigen, Norwegian University of Science and Technology
- Mehdi Zadeh, Norwegian University of Science and Technology

MTEC/ICMASS 2022 Working Committee:

- Ng Yi Han, Maritime and Port Authority of Singapore
- Thomas Ting, Maritime and Port Authority of Singapore
- Pauline Chua, Maritime and Port Authority of Singapore
- Tan Cheng Peng, Singapore Maritime Institute
- Murniyati Nordin, Singapore Maritime Institute

Journal of Physics: Conference Series **2311** (2022) 011002 doi:10.1088/1742-6596/2311/1/011002

Peer Review Statement

All papers published in this volume have been reviewed through processes administered by the Editors. Reviews were conducted by expert referees to the professional and scientific standards expected of a proceedings journal published by IOP Publishing.

- **Type of peer review:** Single Anonymous
- **Conference submission management system:** Morressier
- **Number of submissions received:** 46
- **Number of submissions sent for review:** 46
- **Number of submissions accepted:** 37
- **Acceptance Rate (Submissions Accepted / Submissions Received × 100):** 80.4
- **Average number of reviews per paper:** 2
- **Total number of reviewers involved:** 28
- **Contact person for queries:**
  Name: Murniyati Nordin
  Email: murni@maritimeinstitute.sg
  Affiliation: Singapore Maritime Institute

DigiShip - Digitalisation of ship operations

S P Yue, T Y Chua and J W Fan

ST Engineering, 16 Benoi Road, Singapore 629889

E-mail: siewpeng.yue@stengg.com

Abstract. Platform machineries and sensors onboard the vessel necessary for ship operations are connected to ST Engineering's NERVA Ship Management System and Sensemaking System (SMS^2) to remotely control and monitor platform machineries and sensors from the bridge, or monitored from an operator station on the shore. A case study on remote connection to the shore command centre for a tugboat established using local telco's 4.5G network with operations in the coastal waters of Singapore is discussed. A systematic approach for security assessment of the NERVA SMS^2 is also discussed. Threat modelling and risk assessment for major systems are done using STRIDE and HEAVENS methodologies respectively. With platform systems and sensors connected to NERVA, sensors data collected can be analyzed via the predictive maintenance function of NERVA SMS^2 to reduce the vessel's life-cycle cost. NERVA SMS^2 performs advanced analytics on the collected sensors data, providing predictive diagnostics information, which predicts potential machinery failure before it happen, which can greatly reduce unplanned downtime. In addition, there is decision support for maintenance, which are aligned with OEM recommendations, guiding the maintenance operator on part that need to be replaced or serviced based on the analysed symptoms reflected through the data.

1. Introduction

Digitalisation and digital transformation has started and accelerated very quickly across many industries in recent years. The challenging times of the COVID-19 pandemic also contributed as one of the driving forces of the accelerated transformation. At a different level, maritime digitalisation is also taking place at ports and terminals across the world. Maritime and Port Authority of Singapore (MPA) has a digital vision for maritime Singapore with its upcoming digital platforms and ecosystems digitalPORT@SG and digitalOCEANS. This paper shares some of the work on digitalisation of ship operations through the design concept of DigiShip and NERVA Ship Management System and Sensemaking System (SMS^2).

2. Concept of DigiShip design

The concept of DigiShip encompasses five areas of focus, as shown in Figure 1. The DigiShip shall be an "One Stop Data Shop", adopting an integrated data lake, aggregating data from both combat systems and platform systems to support the crew's operational and maintenance duties. With the use of advanced analytics to optimize maintenance efficiency, and equipping the crew with Digital Assistance tools to access various know-hows. The DigiShip uses automated systems to perform tasks with minimum human intervention, coupled with Human Factors Engineering (HFE) to support physical demands, to achieve a "No Hands On Deck" concept. Fully connectivity enabled by wired and wireless networks infrastructure on the DigiShip sets the foundation for the use of smart sensors and smart equipment, giving rise to an "Internet-of-Things (IoT) at Sea", allowing crew to communicate and perform tasks at their convenience. DigiShip is also a multi-purpose "Enhanced Mothership" that is able to carry out many types of missions and control many unmanned assets with a lean but resourceful crew, enabling true versatility.

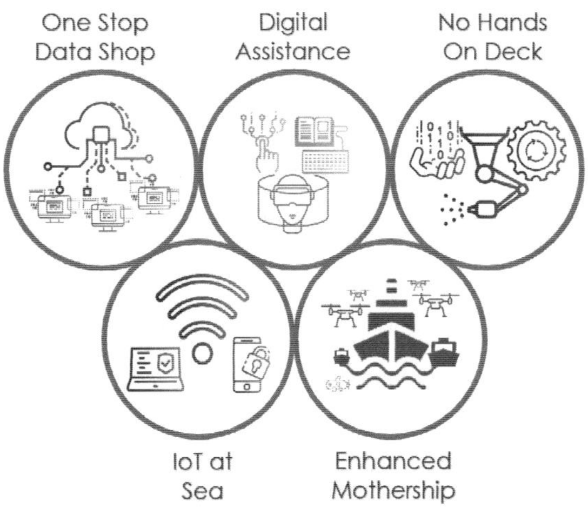

Figure 1. DigiShip focus areas

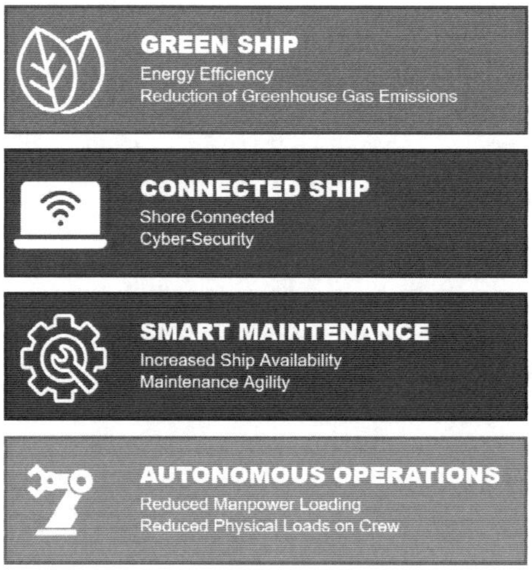

Figure 2. Pillars of DigiShip

The concept of DigiShip is supported by four pillars of technology enablers, as shown in Figure 2. This paper will focus on the Connected Ship Pillar and the Smart Maintenance Pillar of the DigiShip. These two pillars are addressed and underpinned by the NERVA Ship Management System and Sensemaking System (SMS^2). The NERVA SMS^2, designed and developed by ST Engineering, provides centralised control and monitoring of platform sensors and systems, as well as equipment health monitoring and predictive diagnostics. By providing ship operators with real-time, onboard and remote access, NERVA guarantees secure and efficient operations across a variety of vessels.

Based on an open architecture design, NERVA integrates commercial off-the-shelf equipment and enables remote monitoring on tablets and other mobile devices. Its modular design enables customised solutions that suit the operational needs of ship operators, and a centralised control system provides a clear view of the ship's operations. These include propulsion, electrical functions, auxiliaries and damage control. Leveraging on the data collected from platform systems and sensors on board, NERVA is incorporated with data analytics techniques to perform equipment health monitoring and predictive diagnostics. NERVA is further enhanced with a decision support engine which provides recommendations of required preventive actions to the operator. This allows for a condition-based maintenance approach which will reduce unplanned downtime and maintenance costs.

3. Connected ship using 4.5G network

Requirements and considerations for continuous and reliable network connectivity between ship and shore is heavily dependent on the operation of the vessel. The options available for a tugboat that operates mainly in the coastal waters of Singapore versus and a container vessel that travels across oceans and continents would also be different.

Satellite communication is a common solution for network connectivity for maritime applications. The advantage of satellite communication is global network coverage across oceans and continents. This comprehensive coverage of satellite communication for network connectivity comes with a high cost of usage, which will increase the operating cost of the vessel. For vessels that operates along Singapore coastal waters, network connectivity with a shore command centre could be achieved through 4.5G private APN network with the local telco, as shown in Figure 3.

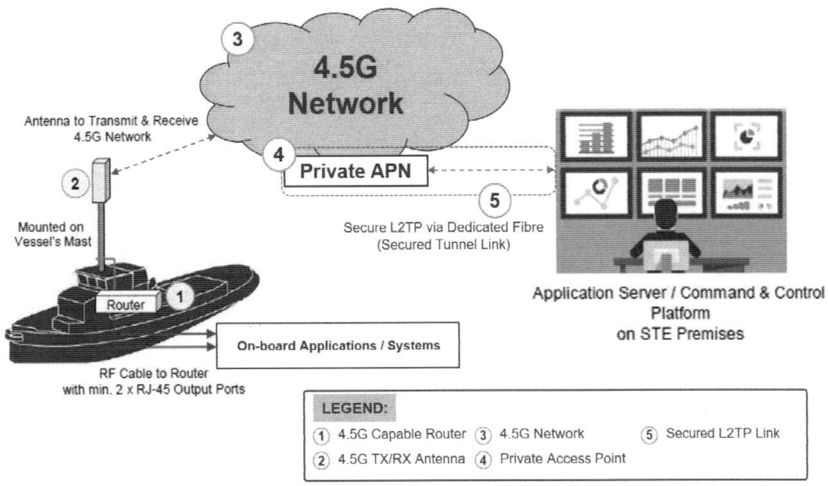

Figure 3. Shore connectivity on 4.5G network

The concept of operations for ship-to-shore connectivity could include a remote command centre of unmanned surface vessel regularly receiving ship navigation status, platform system status and video streaming footages for situation awareness. It could also include a fleet management centre at the shore, which receives data piped back from a fleet of vessels. Data collected from these vessels will also be used to derive a predictive analytics maintenance model through machine learning techniques.

A project on an autonomous tugboat, the Smart Maritime Autonomous Vessel (SMAV), adopted the above architecture to maintain its connectivity with a shore command centre. Remote control of the SMAV from the shore command centre requires a robust and reliable communication link. Beyond Line-of-Sight (BLOS) telemetry is critical for maritime autonomous vessel applications. Various communications means were also considered for the SMAV. Radio mesh network was deemed unfeasible due to the high cost as well as the need for communication nodes. Radio nodes would have to be placed midway between the shore command centre and the trial area, requiring additional administrative approval and installation work. TV whitespace (TVWS) was also considered, but previous trials conducted by MPA did not yield satisfactory results for autonomous vessel applications. SATCOM technology was not employed for the SMAV due to bandwidth limitations. Therefore, 4.5G network connectivity from local telco was selected as it provided BLOS communications.

The trial site was selected to be at Raffles Reserved Anchorage. Some optimization by the local telco is required to enhanced the coverage at the trial site. With the mounted antenna on the mast of the tugboat, and the optimized coverage to the trial site, the network connectivity and performance at the designated trial site was good. It was observed that there were interruption to the network connectivity when large vessels goes in between the local telco transmission base station at Raffles Lighthouse and the vessel itself. It was also observed that the 4.5G coverage en route from shipyard to trial site was poor. Through this project, it was noted that there are scenarios that 4.5G network from local telco may have its limitations for maritime applications.

4. Cyber security assessment

Cyber security has been of concern lately with increasing reported cyber incidents from across various industries. Ship classification societies have also formalized recommended practices, guidelines and rules on cyber security for marine vessels. One of the key systems onboard a vessel that could be of interest to malicious hackers, and vulnerable to cyber security attacks is the system that controls the platform systems onboard the vessel. ST Engineering's Cyber Security team and the NERVA SMS^2 development team worked together to conduct a cyber security assessment for NERVA SMS^2 to systematically identify attack surfaces and possible threats and vulnerabilities. This assessment identifies the gaps that need to be closed for NERVA SMS^2 to be protected against such attacks.

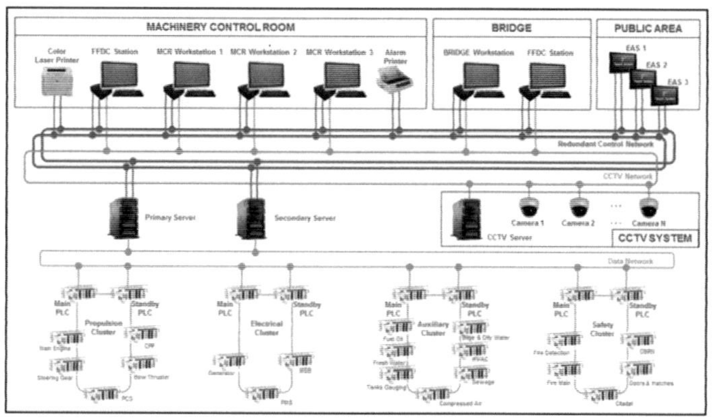

Figure 5. NERVA SMS^2 Architecture

Figure 4. Methodology for security assessment for NERVA SMS^2

The methodology used for this security assessment includes system review, threats analysis, risk assessment and security control recommendations, as shown in Figure 4. The cyber security assessment framework [2] is adopted for this security assessment. STRIDE and HEAVENS methodologies are applied for threat modelling and risk assessment respectively. The network architecture of NERVA SMS^2 used in the security assessment is shown in Figure 5.

The NERVA SMS^2 system is then broken down into various elements, and represented visually using data flow diagram (DFD), as shown in Figure 6, to facilitate subsequent threat modelling and analysis. From the DFD, it can easily identify the attack surfaces. It is also critical to take note of the type of interfaces between elements, as well as the data flowing between the elements. Microsoft's STRIDE methodology [1] for threat modelling is adopted to perform threat analysis on NERVA SMS^2. STRIDE classifies and characterizes threats based on the attacker motivation or exploit types. STRIDE was initially introduced for the identification of threats in a software system, however in recent years, the applicability of STRIDE has also been extended to cyber-physical systems. The six threat categories of STRIDE are Spoofing of identity (S), Tampering

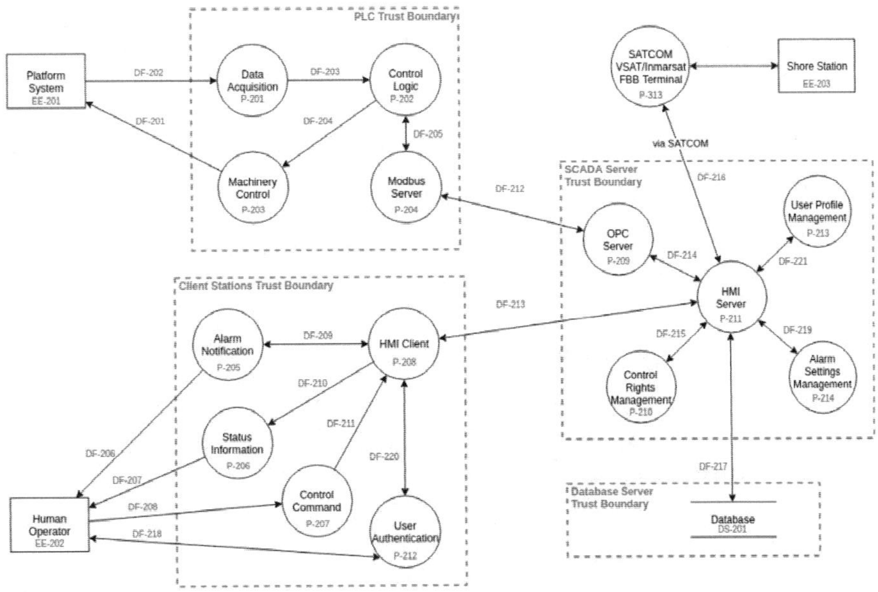

Figure 6. NERVA SMS^2 Data Flow Diagram (DFD)

of data (T), Repudiation (R), Information disclosure (I), Denial of service (D), and Elevation of privilege (E). For each attack surface identified, each threats category was considered in order to derive a list of threats for each attack surface.

After identification and analysis of threats, risk assessment using HEAVENS methodology [3] is applied. Although HEAVENS was developed for automotive systems, it is aligned with established industrial standards. HEAVENS also uses a detailed representation of security levels that associates security levels to each attack surface - threat pair. There are three steps in performing the risk assessment. First step is to determine the threat level (TL) of the identified list of threats. The threat level is dependent on potential, as well as opportunity to perform the attack, which is the equivalence of "likelihood" component of risk. The next step is to determine the impact level (IL) of the identified list of threats, in terms of safety, privacy, operation and financial aspects. The Security Level (SL) is then derived from the TL and IL, using the SL matrix as shown in Figure 7. The SL is derived for each attack surface - threat pair. The five levels of security are Quality Management (QM), Low, Medium, High and Critical. Quality Management indicates that no security measure is required.

		Impact Level (IL)				
		0	1	2	3	4
Threat Level (TL)	0	QM	QM	QM	QM	Low
	1	QM	Low	Low	Low	Medium
	2	QM	Low	Medium	Medium	High
	3	QM	Low	Medium	High	High
	4	Low	Medium	High	High	Critical

Figure 7. Security Level (SL) matrix

The results for threats analysis and risk assessment is shown in Figure 8. There are no attack surface threats at critical level, 25 (4%) at high level, 235 (37.5%) at medium level, and 366 (58.4%) at low level. This result is expected since this security assessment is done on the target system without consideration of any security measures in place, in order to identify its worst case scenario security gap. Addressing of the identified attack surfaces – threats pair will be done through implementation of security control measures and incorporating security requirements in the system development process, with the aim to reduce the attack surface, as threats continue to evolve along with the cyber threat landscape over time.

Security Level	Attack Surface Threats
Critical	0
High	25
Medium	235
Low	366
Quality Management	0

Figure 8. Security assessment results

5. Condition-based maintenance through NERVA SMS²

The Sensemaking System of NERVA SMS² supports the operator in adopting a condition-based maintenance (CBM) regime, by using algorithmic methods to forecast potential failures with critical equipment. These predictive algorithms are developed and trained using data analytics and machine learning techniques, as well as marine technology domain expertise, before being implemented on board. The Sensemaking System is deployed on an autonomous tugboat project, the Smart Maritime Autonomous Vessel (SMAV). Real-time data from selected platform systems and sensors – 2 x Main Engines and 2 x Diesel Generators – are connected to and processed by the Sensemaking System. The selection of parameters to be used are based on a physics-based understanding of the machinery's working behaviour, as well as a study of the typical operating profile of the machinery.

The first methodology involves mathematical processing of the machinery data. These processed results are compared against ideal performance curves of a healthy machine. Deviations from the ideal performance curves, past a certain threshold, indicate worsening of the machinery health. The processed results are also extrapolated, to predict the trend of its own value for a duration of time. The quantity of deviation from the predicted value allows the system to quantify the health of the machine and its degradation over time.

The second methodology involves tracking the usage characteristics of the machinery. An algorithm uses this to quantify the machinery's effective usage over time. By comparing the effective usage against the planned maintenance regime recommended by the OEM, the Sensemaking System is able to offer a countdown timer to the next OEM-defined maintenance event (e.g. engine overhaul). This time-to-maintenance metric allows the operator, operations manager or fleet manager to adjust and optimise the maintenance cycle of the SMAV accordingly.

In order to present the data-driven results of the aforementioned performance analyses, the Sensemaking System acts as a decision support system (DSS), where the operator is prompted to the results of both methodologies via visual and audible means. Where possible, OEM-recommended remedy actions are prompted for operator follow-up. Figure 9 shows a sample of the consolidated mimic page shown on board SMAV, where the diagnosis, operator on-site recommended action and maintenance crew's recommended remedy action is shown.

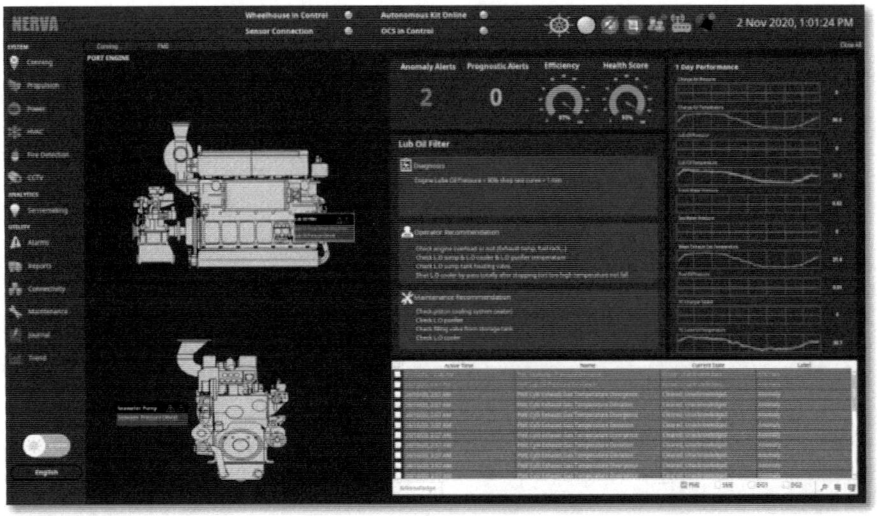

Figure 9. Screenshot of anomaly detected for Port side Main Engine

The Sensemaking System of NERVA SMS^2 is able to offer warnings towards potential early failure of the machinery, and prompt further investigation, which would otherwise have been missed. It would also prompt for a maintenance event either earlier or later than the OEM-recommended interval, for vessel and fleet planning and adjustment. Forecasting a potential failure before it occurs helps to prevent unplanned downtime of the machine and the vessel, especially when applied to critical systems on board. Conversely, by providing statistical analyses that the maintenance regime may be possibly extended, this enables cost optimisation and potential savings for the vessel. CBM provides a means for vessel operators and fleet managers to reduce the vessel's total life-cycle cost, while maintaining maximum operational availability.

6. Conclusions

This paper discusses the envisioned benefits of maritime digitalisation through the concept of DigiShip, and share on some of the works done in the area of maritime connectivity for vessels operating in coastal waters, cyber security assessment using STRIDE and HEAVENS for NERVA SMS^2, and condition-based maintenance using data analytics.

Acknowledgements

This work was partially funded by Maritime and Port Authority of Singapore (MPA). The Authors would like to thank MPA for their support. This work is also done in collaboration with M1 Limited, and ST Engineering Cyber Security team.

References

[1] Microsoft (2009) *The STRIDE Threat Model*, Retrieved January 30, 2022 from `https://msdn.microsoft.com/en-us/library/ee823878(v=cs.20).aspx`

[2] Singapore Standards Council (2019) *Technical Reference Autonomous Vehicles Part 3: Cybersecurity principles and assessment framework*, Enterprise Singapore.

[3] The HEAVENS Consortium (2016) *Security Models*, Retrieved January 30, 2022 from https://autosec.se/wp-content/uploads/2018/03/HEAVENS_D2_v2.0.pdf

Guidelines for cyber risk management in shipboard operational technology systems

Priyanga Rajaram, Mark Goh, Jianying Zhou

iTrust, Centre for Research in Cyber Security
Singapore University of Technology and Design (SUTD)

Email: priyanga@sutd.edu.sg, mark_goh@sutd.edu.sg, jianying_zhou@sutd.edu.sg

Abstract. Over the past few years, we have seen several cyber incidents being reported, where some of the primary causes were the lack of proper security controls onboard the ship and crew awareness on cybersecurity. In response to the growing cyber threat landscape in the maritime sector, we have developed a set of guidelines for maritime cyber risk management, focusing on four major shipboard Operational Technology (OT) systems that are crucial for the day-to-day operation of ships. These four OT systems are: Communication Systems, Propulsion, Machinery and Power Control Systems, Navigation Systems and Cargo Management Systems. The guidelines identify the cyber risks in each of the OT systems and recommend the necessary actions that can be taken to manage risks in each shipboard OT system. In this paper, we introduce the new guidelines, which include cyber risks, mitigation measures, cyber risk assessment, and a checklist to help shipowners and maritime authorities assess and enhance cyber hygiene of their vessels. Our guidelines have been disseminated by the Maritime and Port Authority of Singapore (MPA) to owners and operators of the Singapore Registry of Ships for their reference and use.

1. Introduction

Cybersecurity has become a vital subject for the maritime industry to consider, due to the increasing adoption of digital advancements that help enhance maritime operations. Despite their many benefits, digitalisation of shipboard operations can pose a significant threat to the safety of vessels and its crew, if there is a lack of awareness on cybersecurity among the crew and cyber hygiene habits are not put into practice. All the components in a vessel, such as OT systems, IT systems including routers, switches and network cables, etc., need to be safeguarded from cyber threats, since a weak cybersecurity posture can have a major impact on the safety of the crew, vessel and cargo. As shipboard OT systems are crucial for day-to-day operation of vessels, managing the cyber risks associated with those systems must take top priority.

We have developed a new set of guidelines [1] to address the impacts of cyber risks in shipboard OT systems and the necessary security controls to be enforced in those systems in order to maintain cyber hygiene in vessels. The guidelines are framed in such a way that they are feasible and cost-effective for enforcement by ship owners and maritime authorities. The guidelines are centered on four shipboard OT systems: Communication Systems, Propulsion, Machinery and Power Control Systems, Navigation Systems, and Cargo Management Systems. The International Maritime Organisation (IMO) has also highlighted these OT systems as the most vulnerable to cyberattacks [2].

In our new guidelines, the attack surfaces and cyber risks associated with the shipboard OT systems are listed, with a description of potential attack scenarios. We also described mitigation measures for those cyber risks identified. To assess the severity and likelihood of those cyber risks, a cyber risk assessment methodology is proposed. Finally, a checklist with a tiered security approach is included to guide maritime authorities in determining the cyber hygiene of vessels. It can also be referenced by those in charge of ensuring and enhancing cybersecurity of vessels, such as engineers and IT/OT system specialists. During each stage of developing the guidelines, webinars were hosted to exhibit the work done and to receive feedback from the stakeholders, which helped in fine-tuning the guidelines further.

The rest of this paper is organised as follows: Section 2 reviews the related guidelines that have been published by various maritime organisations. Section 3 investigates the cyber risks and attack surfaces associated with the shipboard OT systems. Section 4 highlights the features of our guidelines, especially the cyber risk assessment approach and the checklist with tiered security. Finally, Section 5 concludes the paper.

2. Related Guidelines

The International Maritime Organization (IMO) resolution MSC.428(98) *Maritime Cyber Risk Management in Safety Management Systems* states that a ship's Safety Management System (SMS) must include cyber risk management based on the objectives and functional requirements of the International Safety Management (ISM) code [3]. Before the first annual verification of company's Document of Compliance on 1 January 2021, IMO encourages administrators to address the cyber risks in SMS. Correspondingly, this resolution highlights IMO's *Guidelines on Maritime Cyber Risk Management* (MSC-FAL.1/Circ.3), which contains high-level recommendations on maritime cyber risk management, to safeguard the maritime industry from the existing and emerging cyber threats [2]. This guideline recommends five functional elements of cyber risk management defined by the National Institute of Standards and Technology (NIST) cybersecurity framework, that is, identify, protect, detect, respond, and recover [4]. The guidelines also emphasize on the fact that cyber risk management should be initiated by the senior management level, and awareness on cybersecurity must be established at all levels of an organisation.

Besides references to ISO/IEC 27001 and NIST cybersecurity frameworks, the IMO guideline also refers the *Guidelines on Cyber Security Onboard Ships Version 4 issued in 2020,* published and supported by various maritime organisations such as the Baltic and International Maritime Council (BIMCO), Intercargo, Intermanager, INTERTANKO, Chamber of Shipping of America (CSA), Digital Containership Association (DCA), ICS, IUMI, OCIMF, Sybass and World Shipping Council [5]. This guideline focuses on creating awareness on cybersecurity onboard ships and guides in enhancing the safety and security of the ships and the crew. It also outlines the cyber risks and the right measures to be implemented to ensure cybersecurity onboard vessels.

Det Norske Veritas (DNV) recommended a set of cybersecurity measures with the publication of *Cyber Security Resilience Management for Ships and Mobile Offshore Units in Operation in 2016,* to guide the personnel in charge of cybersecurity in their respective organisations [6]. This guideline is designed based on well recognised guidelines issued by organisations such as IMO and BIMCO. The guidelines spotlight various cyber threat factors targeting the maritime sector and focuses on crucial improvements and validation mechanisms necessary for cybersecurity resilience in ships. The 'cyber secure' guideline for class notation (October 2020 edition) provided by DNV is centred on the cybersecurity of a ship's primary functions and the operational needs of ship owners [7]. The notations, Cyber Secure, Cyber Secure (Essential) and Cyber Secure (Advanced) entail ten primary functions such as navigation, propulsion, steering, power generation, etc., and are suitable where cyber risk management system is set up and enforced into prevailing procedures and systems. Apart from this, the "+" notation can be used to add additional functions for a notation.

The European Network and Information Security Agency (ENISA) published two reports regarding maritime cybersecurity. In 2011, ENISA published a report on the *Analysis of Cyber Security Aspects in the Maritime Sector,* to highlight the cybersecurity challenges in the maritime industry and provide suggestions for the same [8]. Another related report published in 2019 is on *Good Practices for Cybersecurity in the Maritime sector,* focusing on port cybersecurity, primarily aiming at those responsible for OT and IT security in the port ecosystem [9]. That report includes good practices for ensuring port cybersecurity, related to both IT systems and OT systems.

Similarly, the Institution of Engineering and Technology (IET) published a *Code of Practice Cyber Security for Ships* in 2017, to recommend good practices on developing cybersecurity assessment plans, devising mitigation measures, and managing security breaches and incidents [10]. The guideline considers the importance of implementing cyber security measures, with respect to the technological aspects in maritime systems. On the other hand, the American Bureau of Shipping (ABS) published the *Guide for Cybersecurity Implementation for the Marine and Offshore Industries, ABS Cybersafety Volume 2,* in February 2021, providing cyber security recommendations and requirements for cyber-enabled systems [11]. ABS awards vessels with four notations CS-System, CS-Ready, CS-1 and CS-2, based on compliance with the requirements given in this guide.

Based on the review of the guidelines discussed above, while they do discuss maritime cyber risk management, their intended audiences are those at the management levels. On the contrary, our guidelines focus on managing risks in shipboard OT systems, with the target audience being those responsible for implementing cybersecurity controls in their respective organisations, and they include IT/OT system specialists and engineers. Our guidelines include a list of actionable mitigation measures organised as a checklist that can serve as a guide for maritime authorities and ship owners to check and ensure the cyber hygiene of the vessels.

3. Cyber Risks in Shipboard OT Systems

The shipboard OT systems that are fundamental for everyday vessel operations include communication systems, propulsion, machinery and power control systems, navigation systems and cargo management systems. These systems are vulnerable to cyber attacks due to various weaknesses in the systems or in the network in which they are connected. While enhancing operational efficiency with new technologies is important, cyber risks loom large in the background. For example, shipboard OT systems need to be connected to the Internet for operations, updates, communication, etc., and such interconnectivity between OT and IT network could be an entry point for attackers to compromise the ship's systems. The attack surfaces and corresponding cyber risks associated with each OT sub-system are provided in Tables 1, 2, 3, 4. For more details, please refer to our published journal article on cyber risks in shipboard operational technology (OT) systems in [12].

3.1. Communication Systems

Reliable communication is crucial during events such as alerting, sending and receiving maritime safety information. Cyberattacks targeting essential communication systems such as satellite communication system (SATCOM) and VoIP (Voice Over Internet Protocol) phones can disrupt communication among ships and between ship and shore.

3.2. Propulsion, Machinery and Power Control Systems

The ship's propulsion, power generation and management are vital for its safe passage. The performance parameters of machineries related to propulsion and power control are mostly monitored and controlled through a computer console. At times, these systems may use the Internet for various operational purposes, and remote updates, thus they can fall victim to cyberattacks. Compromising these systems could result in issues such as engine failure, fuel tank overflow, alarm failure and disrupted power supply in the vessel.

3.3. Navigation Systems

Navigation systems play a major role in determining a ship's speed, position and heading, which enable safe navigation. If cyberattacks are targeted on such systems, it may result in lack of navigation data, which in turn may result in severe consequences such as ship sinking, collision, etc.

3.4. Cargo Management Systems

Cargo management systems are a crucial part of vessel operations including assisting in the tracking and management of cargo, and in maintaining vessel's stability in varying environmental conditions with the use of ballast water system. With vast reliability on the Internet for managing cargo operations and due to human threat factors, these onboard systems can fall victim to cyberattacks. Attackers might take advantage of the weaknesses in these systems to compromise them, which could disrupt cargo operations and trigger vessel instability.

Table 1. Cyber risks in Communication Systems.

OT sub-system	Attack surface	Cyber risk
Satellite Communication System (SATCOM) Integrated Communication System (ICS)	Phishing emails, VSAT system/modem	Malware attack via Phishing emails, exploiting vulnerabilities in outdated VSAT software, unauthorised administrative access of vessel network, cross-site scripting, eavesdropping
Voice Over Internet Protocol (VoIP)	VOIP protocol (Session Initiation Protocol)	Denial of Service (DoS) attack, eavesdropping, vishing
Wireless Local Area Network (WLAN)	Access point (router)	Denial of Service (DoS) attack, access point tampering, eavesdropping/session hijacking

Table 2. Cyber risks in Propulsion, Machinery and Power Control Systems.

OT sub-system	Attack surface	Cyber risk
Engine Governor System Fuel Oil System Alarm Monitoring & Control System	USB ports in the machinery monitoring systems, serial communication network	Malware attack via USB ports, Man-in-the-middle (MITM) attack
Power Management System Emergency Generators and Batteries	USB ports in the power management system console, serial communication network	Malware attack via USB ports, Man-in-the-middle (MITM) attack

Table 3. Cyber risks in Navigation Systems.

OT sub-system	Attack surface	Cyber risk
Electronic Chart Display and Information System (ECDIS)	USB ports, serial communication network, outdated software	Malware attack via USB ports, spoofing, Denial-of-Service (DoS) attack
Radio Detection and Ranging (RADAR)	Ethernet switch, SMB (Server Message Block) service	Malware intrusion, Man-in-the-middle (MITM) attack
Automatic Identification System (AIS)	AIS software, AIS messages (VHF radio communication)	Spoofing, Replay attack, Frequency hopping attack
Global Positioning System (GPS)	GPS/GNSS receiver	GPS spoofing, GPS jamming
Dynamic Positioning System (DPS)	DP system software, GNSS Receiver	Denial-of-Service (DoS) attack, Backdoor attack, Spoofing

OT sub-system	Attack surface	Cyber risk
Global Maritime Distress and Safety System (GMDSS)	Very High Frequency (VHF) radio communication, poor/outdated software	Eavesdropping, spoofing, Denial-of-Service (DoS) attack
Voyage Data Recorder (VDR)	USB ports, VDR system	Man-in-the-middle (MITM) attack, Remote code execution
Integrated Navigation System (INS)	SMB (Server Message Block) service, Remote desktop protocol	Man-in-the-middle (MITM) attack, Remote code execution

Table 4. Cyber risks in Cargo Management systems.

OT sub-system	Attack surface	Cyber risk
Cargo Control Room (CCR)	Phishing emails, USB ports	Ransomware, Malware attack
Ballast Water System (BWS)	Phishing emails, USB ports	Phishing emails, Malware attack

4. Features of New Guidelines

While the existing guidelines mainly focus on maritime cyber risk management at the management level, our guidelines are framed in such a way that they can be readily adopted by those responsible for the cybersecurity of shipboard OT systems such as engineers, IT/OT system specialists and vessel inspectors. Here we highlight the features of our new guidelines.

4.1. Mitigation Measures

With heightening demands for cargo transportation by sea, if vital shipboard systems are subject to cyber attacks, the company might need to face severe repercussions such as economic and reputational loss. Ship operators, ship owners, engineers, IT/OT specialists, etc., must be aware of the potential cyber risks and the necessary mitigating actions to be taken. Our guidelines include a list of specific actionable mitigation measures to each of the cyber risks in shipboard OT systems (discussed in Section 3) that can be promptly implemented on those systems.

For example, security controls can be helpful when shipboard OT systems operate in the serial network. However, they often connect to the Internet for remote updates and operational purposes. Thus, technical and procedural security measures such as firewall, Virtual Private Network (VPN), antivirus software, email security mechanisms, regular SATCOM software updates, password security mechanisms, logging, access control lists, multi-factor authentication, network monitoring, and crew awareness programs are necessary for safe operation of the Internet-facing systems, as they are one of the common entry points for various cyberattacks. On the other hand, physical security measures are needed in cases where machineries need to be handled with care, for example, engine system, power generators, etc., as mishandling can cause harm to the crew and the ship. Likewise, it is also very important to secure USB ports in shipboard systems by blocking unused ports, to avoid insider attacks. The mitigation measures in our guidelines include a mixture of technical, procedural and physical security measures that can help safeguard shipboard OT systems from potential cyber attacks.

4.2. Cyber Risk Assessment Approach

In order to assess the cyber risks, a risk assessment approach is also detailed in our guidelines. Risk assessment helps in determining the impact and likelihood of risks, and this in turn can help to identify the places where security controls need to be improvised or freshly implemented. As the shipboard system conditions evolve due to upgrades, installation of new software, etc., it is good to perform risk assessment frequently.

In our guidelines, a four-step scale definition is defined to assess the severity and likelihood of a cyber risk. A 4x4 risk score matrix, as shown below in Figure 1, can be used to calculate the risk score for each cyber risk in an OT system, where the risk score is the product of severity score and likelihood score. The likelihood score of a risk can be decided based on the attack complexity and attacker's ease of access to necessary resources and attack surface. On the other hand, impact or severity score of a risk can be determined by the extent of environmental harm, financial loss and loss of confidentiality, integrity and availability that can result from a cyber attack. Tables 5 and 6 depict the definitions for evaluating the likelihood and severity scores respectively.

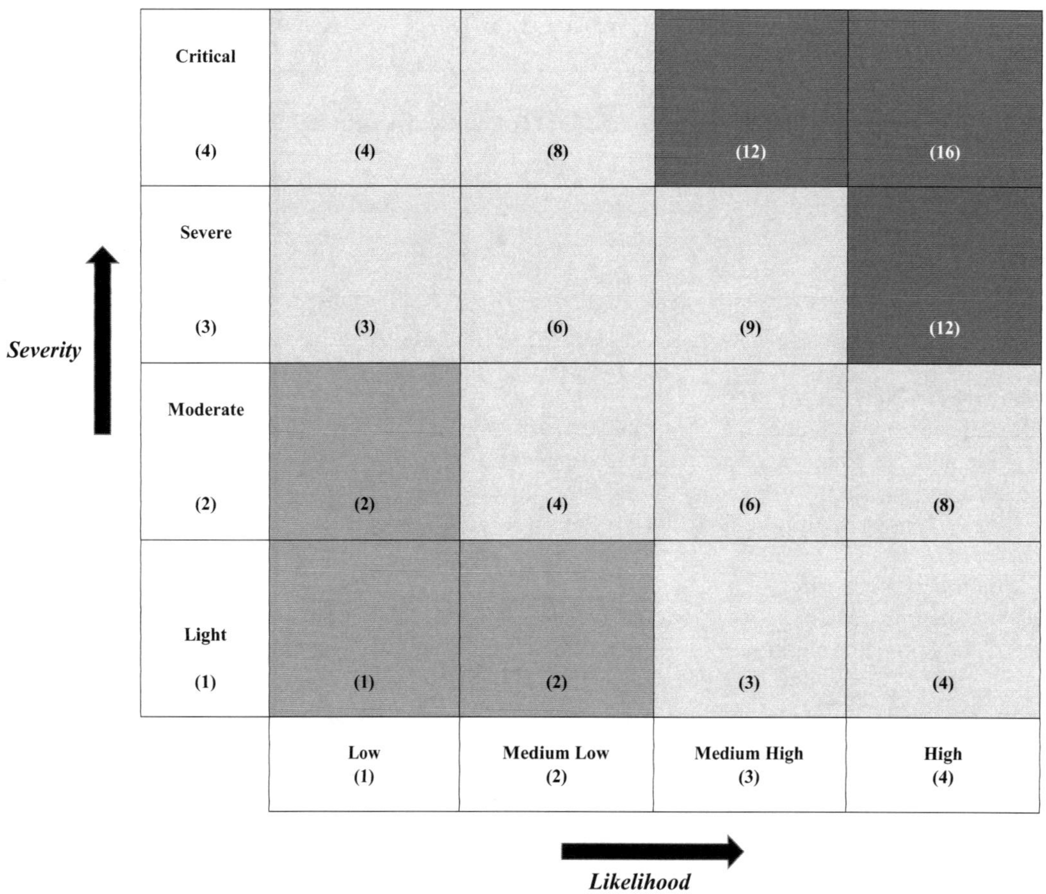

Figure 1. Cyber risk assessment matrix.

Table 5. Likelihood Definition.

Score	Label	Likelihood Definition
4	**High**	- Attack can be performed remotely or with physical access to open ports and systems in the vessel. - Can be performed with very minimal or no technical knowledge and with publicly available resources (e.g., Shodan, ExploitDB). - Can be attacked from external network.
3	**Medium High**	- Attack can be performed with basic technical knowledge.

Score	Label	Likelihood Definition
		- Can be performed with no change in exploits published online. - Possible if the attacker is in either an internal or external network.
2	**Medium Low**	- Attack can be performed with moderate technical knowledge. - Can be performed with minor changes in exploits published online. - Possible if the attacker is in either an internal or external network.
1	**Low**	- Attack can be performed with advanced technical knowledge. - Requires chaining of multiple exploits. - Requires physical/remote access to the OT system where there is restricted access.

Table 6. Severity Definition.

Score	Label	Severity Definition
4	**Critical**	Consequences of a cyberattack on the vessel operations, crew, and the enterprise, such as: - All operation systems, data, and resources unavailable, impacting safe operations. - Impact on the lives of crew members e.g., collision, explosion, sinking of ship, imbalance of ship, loss of life. - Economic loss to the enterprise.
3	**Severe**	Cyberattacks that lead to: - Unauthorised access of the vessel network, system, data, and other resources that affect day-to-day vessel operations such as navigation, communication, propulsion, generators, cargo. - Disruption of vessel network. - Disruption of ship-to-shore communication link.
2	**Moderate**	Cyberattack prevents or impairs the normal authorised functionality of networks, systems, or applications by exhausting resources. - Lack of availability of systems, data. - Misleading communication between OT systems, causing damage to the ship, cargo and disrupting its operations.
1	**Light**	Suspected or potential unauthorised access to the vessel systems which leads to data breach (sensitive/non-sensitive information).

Based on the risk scores determined with the 4x4 risk score matrix, as shown in Figure 1, the OT sub-systems and their associated cyber risks are classified into three categories based on their risk scores. The three categories are defined as follows: *High risk category (Risk Score: 12–16)*, *Medium risk category (Risk score: 3–9)* and *Low risk category (Risk score: 1–2)*. Such categorization facilitated us to define mitigation measures under each category in the form of a checklist.

4.3. Checklist with Tiered Security

Often, unsafe practices such as usage of default or weak passwords, outdated software or operating systems, etc., are some common causes that lead to cyber attacks. Hence, the most basic and crucial security controls must be implemented onboard the ships. Although there is no guarantee that a ship can be fully safeguarded from cyber threats, requisite actions can be taken as first line of defence against potential cyber threats.

In our guidelines we have provided a checklist, which is a list of mitigation actions that can be enforced in shipboard OT systems to manage the associated risks. To help maritime authorities and ship owners determine the cyber hygiene of their vessels, we introduce the concept of security tiers for listing the checklist. It consists of three tiers (Tier 1, Tier 2, Tier 3), where a tier of security refers to the *urgency of cyber risks* of a ship that ship owner needs to mitigate or manage.

The three security tiers are defined as follows:

- **Tier 1:** Includes cybersecurity measures for managing *high risk* cyber threats, which means that the measures stated under this tier are *highly recommended* to implement on-board vessels.
- **Tier 2:** Includes cybersecurity measures for managing *medium risk* cyber threats, which means that the security controls mentioned under this tier are *recommended* to have on-board.
- **Tier 3:** Includes cybersecurity measures for managing *low risk* cyber threats, which means that the measures listed under this tier are *good to implement on-board*, even though the risk level is low.

The checklist is designed to help ship owners and maritime authorities assess and improve the cyber hygiene of their vessels. The checklist can be easily enforced since the security measures are cost-effective and actionable.

5. Conclusions

Although adoption of emerging technologies in the maritime industry has become increasingly common, it opens up many attack surfaces for hackers to exploit. As OT systems play an important role in ship operations, it is crucial to protect them from the emerging cyber threats. In view of this, we developed guidelines that highlight the cyber risks and the impacts associated with OT systems, along with mitigation measures to manage those risks. The cyber risk assessment approach defined in the guidelines will aid in assessing the risks, whereas the checklist will guide in implementing appropriate security controls in shipboard OT systems. Thus, the guidelines will serve as a practicable guide for managing the cyber risks in shipboard OT systems and also ensure cyber hygiene in vessels.

Our guidelines have been disseminated by the Maritime and Port Authority of Singapore (MPA) to owners and operators of the Singapore Registry of Ships for their reference and use. An information paper "Voluntary cyber risk management guidelines for shipboard operational technology (OT) systems" has also been submitted by MPA to the 105th session of the IMO's Maritime Safety Committee, to make our guidelines freely available to all entities that may find them useful.

6. Acknowledgements

This work is carried out by iTrust, the Centre for Research in Cyber Security in Singapore University of Technology and Design (SUTD) and funded by the Singapore Maritime Institute (SMI) under the grant number SMI-2020-MA-04. The team would like to thank the Maritime and Port Authority of Singapore (MPA) for the support throughout this project. We would also like to extend our thanks to American Bureau of Shipping (ABS), Klynveld Peat Marwick Goerdeler (KPMG), Centre of Excellence in Maritime Safety (CEMS) in Singapore Polytechnic (SP), and the Singapore Shipping Association (SSA) for their valuable feedback on the draft version guidelines.

7. References

[1] iTrust Centre for Research in Cyber Security 2022 *Guidelines for cyber risk management in shipboard operational technology systems* https://itrust.sutd.edu.sg/research/projects/maritime-cyber/
[2] International Maritime Organization 2017 *Guidelines on Maritime Cyber Risk Management*
[3] International Maritime Organization 2017 *Maritime Cyber Risk Management in Safety Management Systems*
[4] National Institute of Standards and Technology 2018 *Framework for Improving Critical Infrastructure Cybersecurity*
[5] BIMCO et al., 2020 *Guidelines on Cyber Security Onboard Ships*
[6] Det Norske Veritas 2016 *Cyber Security Resilience Management for Ships and Mobile Offshore Units in Operation*
[7] Det Norske Veritas 2020 *Class Guideline Cyber Secure*
[8] European Network and Information Security Agency 2011 *Analysis of Cyber Security Aspects in the Maritime Sector*
[9] European Network and Information Security Agency 2019 *Port Cybersecurity - Good practices for cybersecurity in the maritime sector*
[10] Boyes H and Isbell R 2017 *Code of Practice Cyber Security for Ships*
[11] American Bureau of Shipping 2021 *Guide for Cybersecurity Implementation for the Marine and Offshore Industries ABS Cybersafety* Vol 2
[12] Rajaram P, Dumbala R, Goh M and Zhou J 2021 A Cyber Risk Study in Shipboard OT Systems *Society of Naval Architects and Marine Engineers Singapore 40th Annual Journal 2020/2021*

Deriving tidal currents from AIS data

Lucas J. YIEW, Allan R. MAGEE

Technology Centre for Offshore and Marine, Singapore (TCOMS),
12 Prince George's Park, Singapore 118411

Email: lucas_yiew@tcoms.sg

Abstract. In this study, a method is proposed to derive ocean surface current conditions using only information available through AIS – mainly speed over ground (SOG), course over ground (COG) and vessel heading. This method allows the mapping of current conditions over large areas, in doing so gives seafarers better awareness and advance warning of local environmental conditions. This also helps in route planning and management of voyage-related resources. In the proposed method, current speeds and directions are solved through a system of equations containing the resolved SOG, COG and heading of multiple ships. This method was tested using mathematical manoeuvring models, and then subsequently using AIS data collected from vessel traffic in the Western Singapore Straits. Current predictions using this method are validated against physics-based depth-averaged hydrodynamic models driven by astronomical tidal forcing.

1. Introduction

Seagoing vessels need to be aware of their surrounding environment in order achieve safe passage, avoid navigational hazards, and maximise fuel efficiency, among other reasons. Advanced knowledge of environmental conditions allows a captain to decide on the most economical route, or take the necessary corrective actions to reach his intended destination in a safe manner. Tidal currents are one such environmental phenomena that seafarers have to be acutely aware of. These periodic flows can have an appreciable effect on a vessel's manoeuvrability and trajectory, as well as its voyage in the longer term.

Conventional methods for determining tidal currents include consulting tide tables, atlases or referencing nautical charts for historical current speeds and directions. However, these resources generally report such information at a limited number of locations. For areas with high variability in current speeds and directions e.g. winding channels, areas of variable bathymetries, or around constrictions from land masses, it is important to know the current conditions on a smaller spatial scale that is sufficient for navigational control. Hydrodynamic models can be used to resolve the current conditions to such scales, but they require computational resources that are often unavailable onboard.

Another way of obtaining live current information is via a speed log. Vessels equipped with speed logs will have information on their speed through water (STW). Resolving the vectors for speed over ground (SOG) from GPS, and STW gives the current speed and direction at a ship's particular location. If collated across multiple ships, this data could be harnessed to aid in the navigation and control of ships plying the same waters. However, this data is not transmitted/shared through standard channels such as the Automatic Identification System (AIS), and thus, there is no means available for vessels to tap into the STW data of other vessels.

In this paper, a method is proposed to derive current information using only information available through AIS – mainly SOG, course over ground (COG) and vessel heading. This method allows the mapping of current conditions over large areas, in doing so gives seafarers better awareness and advance warning of local environmental conditions. This also helps in route planning and management of voyage-related resources.

In the proposed method, current speeds and directions are solved through a system of equations containing the resolved SOG, COG and heading of multiple ships. In the following sections, we test this method using (i) mathematical ship manoeuvring models, and (ii) using AIS data collected from vessel traffic in the Western Singapore Straits. In the latter, current predictions are validated against physics-based depth-averaged hydrodynamic models driven by astronomical tidal forcing.

2. Vessel motion in current

Figure 1 shows the velocity vectors of a ship when moving in current of speed, U_c, and direction, ϕ. Its effective global motion is represented by its speed and course over ground (SOG and COG, otherwise denoted as U_g and θ, respectively). When taking reference to the water body (i.e. moving frame of reference), a vessel's motion is represented by its speed through water (STW or U_w) and its heading (ψ). In Figure 1, the terms highlighted in red are the unknown variables which we will solve for. The remaining terms i.e. SOG, COG and heading can be obtained from AIS data.

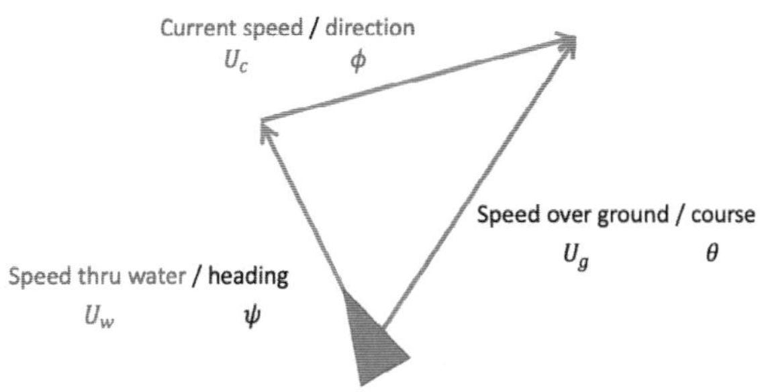

Figure 1: Decomposition of vessel and current velocity vectors.

The above velocity vectors can be resolved into global lateral and longitudinal components, according to the following equations:

$$U_w sin\,\psi + U_c\,sin\,\phi = U_g\,sin\,\theta$$
$$U_w \cos\psi + U_c \cos\phi = U_g \cos\theta.$$

To solve for the unknown terms, U_c, U_w and ϕ, velocity vectors of two vessels are needed. When assembled into a system of equations,

$$\begin{bmatrix} 1 & 0 & \sin\psi^1 & 0 \\ 0 & 1 & \cos\psi^1 & 0 \\ 1 & 0 & 0 & \sin\psi^2 \\ 0 & 1 & 0 & \cos\psi^2 \end{bmatrix} \begin{bmatrix} U_c \sin\phi \\ U_c \cos\phi \\ U_w^1 \\ U_w^2 \end{bmatrix} = \begin{bmatrix} U_g^1 \sin\theta^1 \\ U_g^1 \cos\theta^1 \\ U_g^2 \sin\theta^2 \\ U_g^2 \cos\theta^2 \end{bmatrix}$$

In the above, the superscripts 1 and 2 refer to ships 1 and 2. We assume both ships are affected by the same current conditions, U_c and ϕ. The current speed and direction can then be derived by solving the below:

$$\begin{bmatrix} U_c \sin \phi \\ U_c \cos \phi \\ U_w^1 \\ U_w^2 \end{bmatrix} = \begin{bmatrix} 1 & 0 & \sin \psi^1 & 0 \\ 0 & 1 & \cos \psi^1 & 0 \\ 1 & 0 & 0 & \sin \psi^2 \\ 0 & 1 & 0 & \cos \psi^2 \end{bmatrix}^{-1} \begin{bmatrix} U_g^1 \sin \theta^1 \\ U_g^1 \cos \theta^1 \\ U_g^2 \sin \theta^2 \\ U_g^2 \cos \theta^2 \end{bmatrix}$$

A similar method is proposed by Le Goff et al. (2021), but with the matrix expanded to include $2n$ equations from n AIS datapoints.

3. Validation using predictions from a manoeuvring model

Ship manoeuvring models utilise mathematical equations of motion to simulate the trajectories of a vessel for a given command action. For a 3 degrees-of-freedom (3-DOF) model, the surge, sway and yaw motions of a vessel are expressed (in an abridged form) as

$$m(\dot{u} - vr) = F_x$$
$$m(\dot{v} - ur) = F_y$$
$$I_{zG}\dot{r} = M_z$$

where u, v and r are the surge, sway and yaw velocities, and $\dot{u}, \dot{v}$ and $\dot{r}$ are their respective accelerations. I_{zG} represents the yaw moments of inertia. On the right hand side, F_x, F_y and M_z include the hydrodynamic damping forces (from the hull) and excitation forces (from propulsion and steering systems). For the full system of equations, readers should refer to Yasukawa & Yoshimura (2015).

Environmental disturbances can also be modelled in the above as external forces. For currents, this disturbance is modelled by superimposing the additional water speed (u_c, v_c) into the above equations of motion, i.e.

$$u = u_s + u_c$$
$$v = v_s + v_c$$

where (u_s, v_s) is the ship's local velocity. The following is applied to resolve the global current components $U_c = (C_x, C_y)$ into the ship-local reference frame:

$$u_c = C_x \cos \psi + C_y \sin \psi$$
$$v_c = -C_x \sin \psi + C_y \cos \psi$$

The KVLCC2 manoeuvring model of Yasukawa & Yoshimura (2015) is used for the purpose of validating the proposed current prediction method. Figure 2 shows the simulated trajectories of the vessel in a 3 knot, $\phi = 85°$ current. Two initial conditions are considered, $\psi^1 = 0°$ and $\psi^2 = 45°$. In both conditions, the ship speed is set to a propulsion point of 10 knots in calm water. At the final time step, the SOG, COG and headings are (5.51, 6.41) m/s, (16.2°, 53.9°) and (0°, 45°). Substituting these values into the matrix equation in Section 2 yields a predicted current speed and direction of 3 knots and 85°.

It is important to note that the proposed method is valid only when a vessel's rudder action, and correspondingly, yaw rate (or rate of turn) is zero. At non-zero rudder angles, steering action results in dynamic interactions between a vessel's SOG, COG and its heading, hence a stable equilibrium with current interactions cannot be achieved.

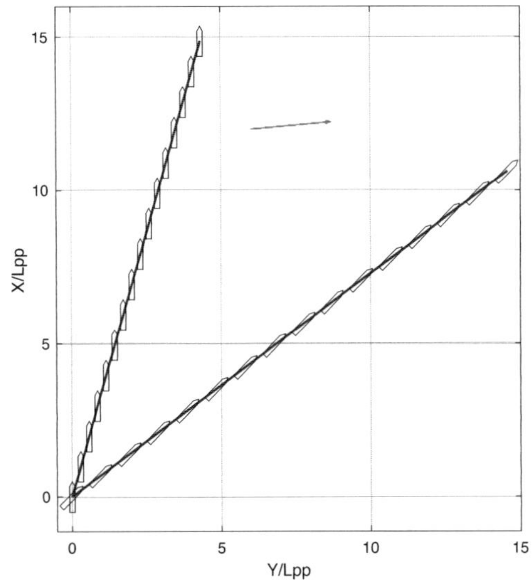

Figure 2: Simulated vessel trajectories in 3 knot current, ϕ= 85°.

Further, there are several other instances which lead to an ill-conditioned problem. For example when $\psi^1 = \psi^2$ or $|\psi^1 - \psi^2| = 180°$, the matrix equation in Section 2 becomes indeterminate. Another such condition is when the current direction coincides with the COG, or in other words when $\psi = \theta$ or $|\psi - \theta| = 180°$. In this instance there is no distinction between changes in velocity due to current effects or ship speed alone.

4. Current predictions using AIS data

AIS data from an open repository (AIS Logs, 2013), hosted by the National University of Singapore's School of Computing, was used to validate the proposed current prediction method. This data, which was obtained via an AIS receiver, covers the Western Singapore Straits, as shown in Figure 3.

Data over a 24-hour period on 3rd of November 2020 (0000H to 2359H) was extracted from the repository. Figure 4 shows the positions of each AIS datapoint over this interval. Colours denote the different classes of vessels, according to the 2 digit ship type code from AIS ship static information.

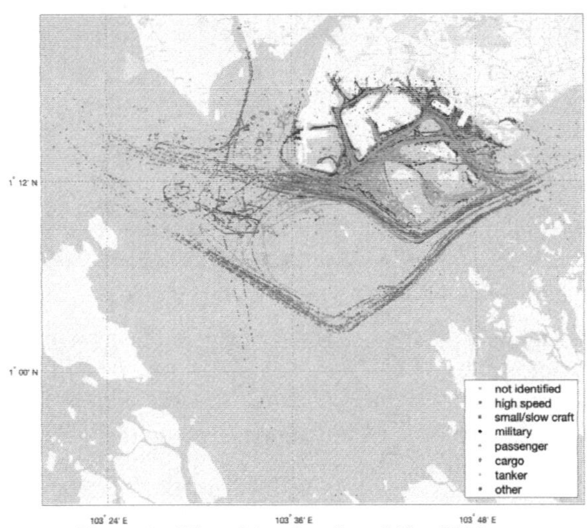

Figure 3: AIS position data from 3 Nov 2020

A 2-by-2 km grid, and a 1 hour moving-average window are used to segregate datapoints on a spatial and temporal scale. Current predictions are derived from the mean of ${}^{n}C_2$ combinations of AIS datapoints over this spatial and temporal interval. The superscript n refers to the number of AIS datapoints over these intervals.

To improve on the quality of data, the following filtering conditions are applied:

(i) SOG $\in [5, 15]$ kns
(ii) Rate of turn $\in (0, 5]$ deg/min
(iii) n^1, n^2 from different ships
(iv) $|\psi^1 - \psi^2| \geq 5$ deg
(v) $|\psi^1 - \psi^2| \notin [177.5, 182.5]$ deg

Condition (i) is implemented to filter out vessels that may be anchored or moored. Condition (ii) filters out the effects of steering action. Condition (iii) allows the filtering of ill-conditioned variables brought about by repetition of ship states, and is enforced by taking into consideration the MMSI number (unique ship identifier) tagged to each AIS datapoint. Conditions (iv) and (v) filters out ill-conditioned variables due to coincident COG angles.

Figure 4 shows a snapshot of current predictions from AIS data over 0000H-0100H on 3 Nov 2020. Red dots represent the unfiltered datapoints. Vectors represent the mean magnitude and direction of predicted currents for each 2x2 km grid. Appended to each vector are the number of AIS datapoints in each grid, n (after filtering), and the mean and standard deviation of current speed and directions. These values gives an indication of the statistical uncertainty associated with each prediction.

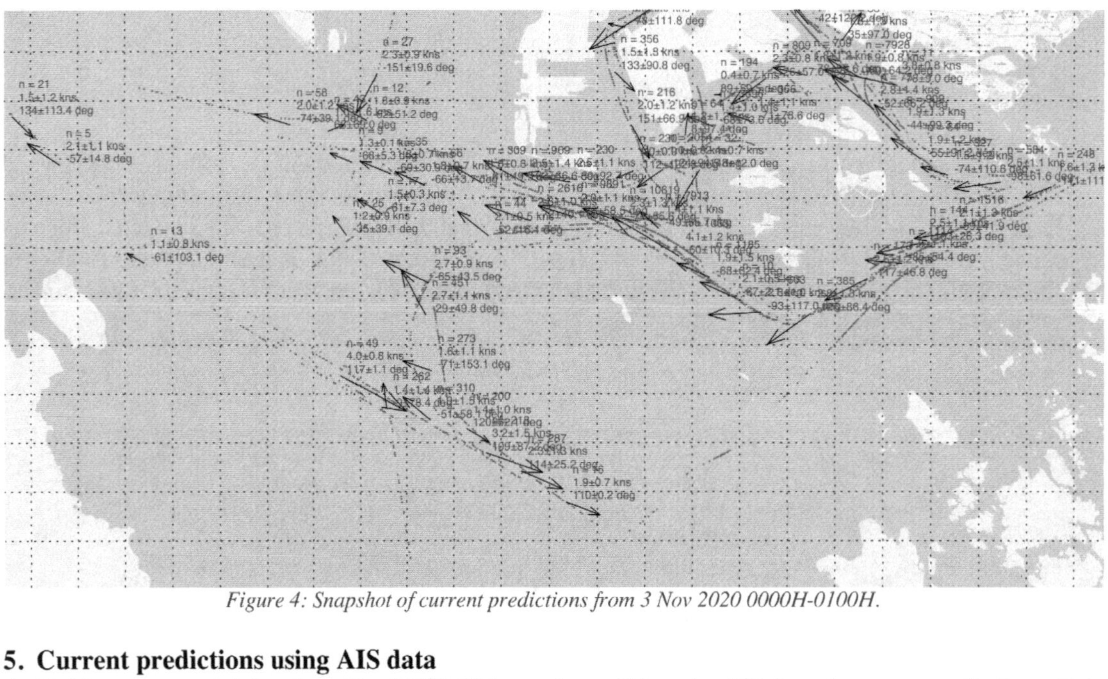

Figure 4: Snapshot of current predictions from 3 Nov 2020 0000H-0100H.

5. Current predictions using AIS data

A tidal forecast model developed by TCOMS is used to validate the AIS-based current predictions. This forecast model applies depth-averaged hydrodynamic models driven by astronomical tidal forcing to predict current conditions over a predefined mesh grid (see Figure 5). This numerical model is validated against observations from tidal stations.

Figure 5: Velocity vectors of the current forecast

For direct comparisons with AIS current predictions, the current forecast model is remeshed into the same 2x2 km grid, and the mean current velocity within each grid is used for comparisons with AIS results.

AIS-based current predictions from 1 hour moving-average windows are compared against current forecasts at 30 minute intervals. The mean errors over a 24 hour period are represented in Figure 6, for each 2x2 km grid. Results are delineated into current speeds and current directions. Mean errors for current speeds are reported in terms of percentage differences, while those for current directions are given in terms of absolute values (degrees).

Three observations can be deduced from the results:

1. Generally good predictions are obtained in locations away from port. Some examples are in the north-westerly/south-easterly and north/south sea lanes on the top- and bottom-left areas of the map in Figure 6.
2. Larger errors are observed closer to port. This could be due to the increased variability in current conditions due to the influence of land masses and bathymetry changes. The averaging over the 2x2 km grid may lead to resolution errors. However this grid size is needed for a large enough sample size, due to spareness of AIS data.
3. The largest errors are concentrated along the bottom-right of the map, in the north-easterly/ south-westerly sea lanes (otherwise known as the Traffic Separation Scheme, TSS). The direction of vessel traffic in this channel coincides with the general directions of currents during ebbing and flooding tides. As a result, this creates an ill-conditioned problem when solving for currents, as discussed in Section 2.

6. Discussions and Conclusions

AIS data, specifically SOG, COG and vessel headings from multiple ships, are used to predict current conditions in the absence of local current measurements or forecasts. This method of deriving current speeds and directions was tested using ship manoeuvring models and historical AIS data.

Barring certain conditions which lead to ill-conditioned problems, the method is shown to give accurate predictions. Some of these conditions, which were also highlighted in complementary studies (e.g. Le Goff et al., 2021), include having vessel headings that are roughly collinear. This occurs when vessels traverse sea lanes, for example. Another source of error pertains to inaccuracies with AIS data due to the quality of onboard sensors, or fluctuations in dynamic data due to inclement weather, as highlighted by Le Goff et al.. The effect of such errors can be minimised by averaging predictions over multiple data samples, as was done in this study, however this requires a sufficiently large sample size for statistical confidence.

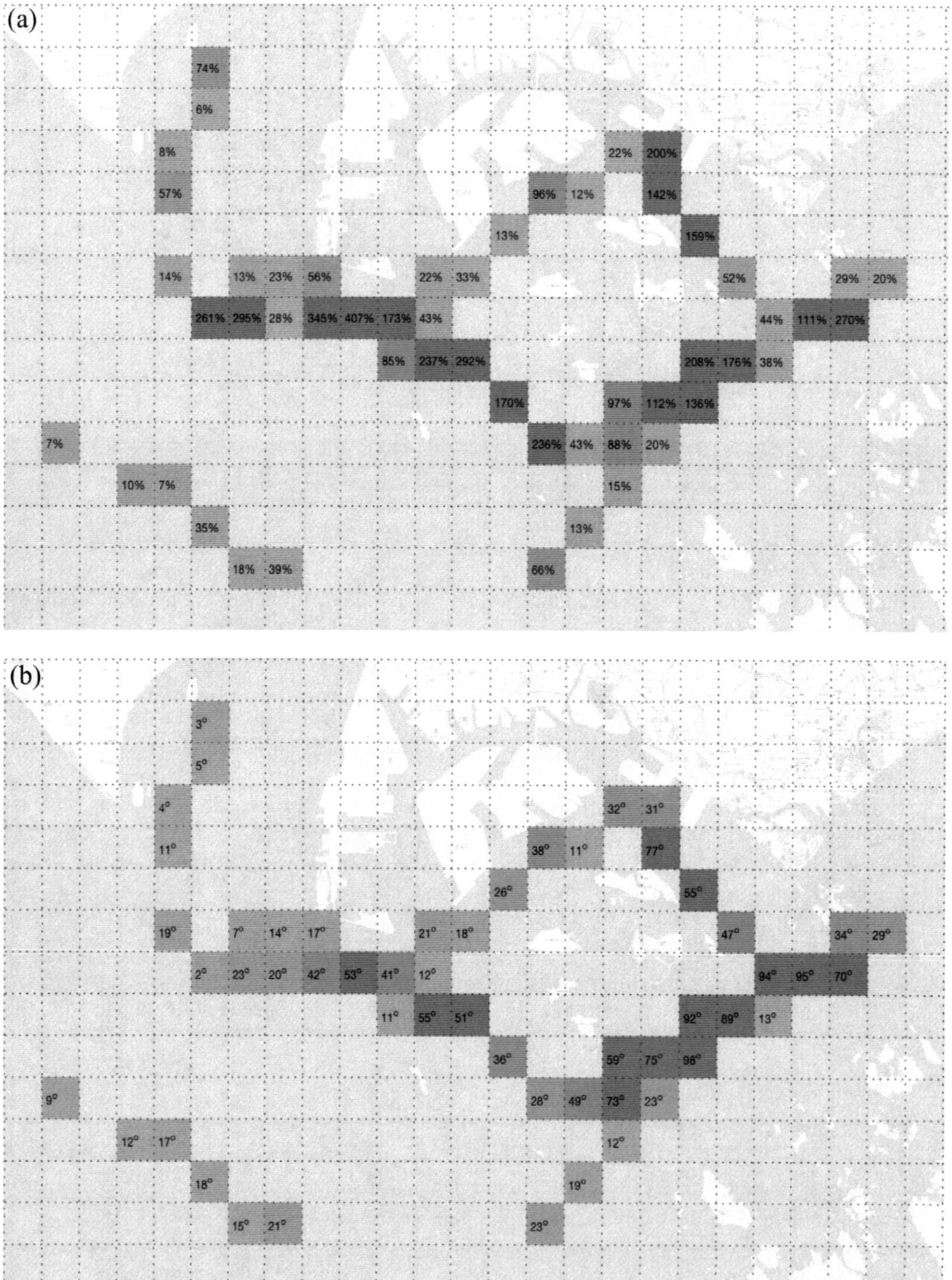

Figure 6: (a) Mean percentage error in current speed. (b) Mean absolute error in current direction.

This study also identified several other unique factors that could affect the accuracy of current predictions:

i. The influence of rudder action and dynamic ship motion affects the quality of current predictions. To account for this, we filtered out data where vessel rate of turns exceeded 5 degrees/min.

ii. Coincidental current direction and vessel course over ground will lead to an ill-conditioned problem. This occurs in constricted waters such as the Singapore Straits.

iii. High variability in current conditions close to shore, due to bathymetry changes and the influence of land masses. These result in larger errors in predictions. A higher spatio-temporal resolution should improve predictions.

As a case study, predictions of surface current conditions were derived using AIS data around the Western Singapore Strait. These were validated using a tidal forecast model. To date, methods of predicting currents from AIS data have only been validated in more open waters (e.g. Benaichouche et al., 2021 and Le Goff et al., 2021). For constricted waters such as the Singapore Straits, two problematic areas were identified: regions close to port, and regions where current directions and vessel traffic are approximately collinear. If we disregard these two regions, there is a fairly good correlation between AIS-based current predictions and forecasts. For future studies, we propose to validate this method against AIS data from vessel traffic further out in the open ocean.

Acknowledgements
This research is supported by the National Research Foundation, Singapore and the Singapore Maritime Institute under its Maritime Transformation Programme White Space Funding Support for the Centre of Excellence for Autonomous and Remotely Operated Vessels (CEAOPS) (SMI-2019-MTP-01). The authors thank Jeng Hei Chow and Kie Hian Chua for their technical contributions to the paper.

References

AIS Logs (2013). Data@Liánchéng. https://data.liancheng.science/ais_logs.html.

Benaichouche, S., Le Goff, C., Guichoux, Y., Rousseau, F., & Fablet, R. (2021). Unsupervised reconstruction of sea surface currents from ais maritime traffic data using learnable variational models. Remote Sensing, 13(3162). https://doi.org/10.1109/ICASSP39728.2021.9415038

Le Goff, C., Boussidi, B., Mironov, A., Guichoux, Y., Zhen, Y., Tandeo, P., Gueguen, S., Chapron, B. (2021). Monitoring the Greater Agulhas Current With AIS Data Information. Journal of Geophysical Research: Oceans, 126(5). https://doi.org/10.1029/2021JC017228

Yasukawa, H., & Yoshimura, Y. (2015). Intro-duction of MMG standard method for ship maneuvering predictions. Journal of Marine Science and Technology, 20, 37–52. https://doi.org/10.1007/s00773-014-0293-y

Numerical Forecasting of Wind-Waves for Singapore Region

Jeng Hei CHOW,[1] Kaushik SASMAL ,[1] Pavel TKALICH. [1,2]

[1]Technology Centre for Offshore and Marine Singapore (TCOMS), 12 Prince George's Park, Singapore 118411,
[2]Tropical Marine Science Institute (TMSI), S2S Building, 18 Kent Ridge Road, Singapore 119227
Corresponding author: Jeng Hei CHOW; e-mail: chowjh@tcoms.sg;
Tel.: 65 65165953

Abstract. A numerical wave model has been set up and tuned for the Singapore region extending to the parts of the South-China Sea and Malacca Strait. The MIKE 21 spectral wave model was set up using ERA5/ECMWF wind and waves reanalysis data to compute the phase-averaged significant wave heights. The outputs of the calibrated model validates well with the ECMWF data within the domain. The model increases the resolution of available wave data in narrow channelled regions like in the Singapore Straits. When supplied with the wind forecast data the resulting model allows for high resolution forecasting of phase-averaged wave spectra in the Singapore region. The study enables setting up of a wind-wave forecast dissemination framework on a visualization and data-sharing platform or the maritime community.

Keywords: spectral wind-waves, forecasting model, prediction framework, visualization platform

1. Introduction

With the global development of capabilities in maritime autonomy, as well as the transformation of national port waters towards smarter and more connected systems, safe and efficient maritime traffic operation is required. Currently, vessels rely on the experience and expert knowledge of mariners, as well as the Vessel Traffic Services (VTS) operators to navigate the busy waters around Singapore, which includes the congested Phillip Channel known for strong currents and waves that affects vessel maneuverability.

An accurate forecasting of the vessel's operating wave environment enhances the predictability of the vessel's responses, which can be harnessed for autonomous vessel navigation and control. Advanced data-sharing and visualisation platform can promote dissemination of accurate metocean information (namely waves, currents and wind) to maritime stakeholders. By incorporating the sea state forecast into navigation and control systems, the performance of Maritime Autonomous Surface Ship (MASS) and other smart maritime solutions can be further enhanced. Furthermore, the prediction of the hydrodynamic environment can also be used to improve the coordination of multi-vessel interactions at the VTS level, given that the maneuverability of each vessel will be affected by the surrounding environmental forces.

The wind-waves in a region can be represented by the phase-averaged wave variables defined from the wave spectrum:

- Significant wave heights
- Peak periods
- Mean directions
- Directional standard deviations

Overall, the variables define the wave energy in an area, which can be generated, dissipated, or reflected, following the energy conservation and balance assumptions.

The phase-averaged spectral wind-wave model have been used to simulate the growth, decay and transformation of wind waves and swell waves in offshore and coastal areas. The SWAN (Simulating WAves Nearshore) model [1] and the Spectral Wave Module in the MIKE by DHI software [2] are examples of simulation tools used for analyses. For the Spectral Wave Module, the basic governing equation is based on the conservation of Wave Action Density, N (m^2s^2/rad):

$$\frac{\partial N}{\partial t} + \nabla.(\vec{v}N) = \frac{S}{\sigma} \quad (1)$$

$$S = S_{in} + S_{nl} + S_{ds} + S_{bot} + S_{surf} \quad (2)$$

t represents the time (s), $\vec{v}$ is the propagation velocity vector of a wave group (m/s), and σ is the relative angular frequency (s^{-1}). The right-hand side of Equation (1) describes the physical phenomena of the sources, S (m^2s/rad), which can be further separated into the components in Equation (2):

- Wave growth by action of wind
- Non-linear wave-wave interaction
- Dissipation due to white-capping
- Dissipation due to bottom friction
- Dissipation due to depth-induced wave breaking

Other physics such as refraction and shoaling due to depth variations, wave-current interaction and the effects of time-varying water depth and flooding and drying are represented within the transport of wave action density by the propagation velocity in the left-hand side of Equation (1).

In recent decades, numerical models have been used in the assessment of wave climates in offshore and coast. Several investigations were done in the South China Sea (SCS) region [3]–[5] using the WAVEWATCH III wave model [6] developed by the National Centers for Environmental Prediction (NOAA/NCEP). The studies involve predictions of wind wave characteristics over a larger scale domain. A similar investigation [7]was performed in the Singapore Strait using the SWAN software package [1]. The study was focused more on resolving the wind-waves in the Straits channel without taking into account the wave action energy balance from the winds and waves generated in the South-China Sea. The study applied nesting of a finer grid within a coarser grid as a balance between computational cost and model resolution in the domain. Modern wind-wave numerical model codes [1], [2] allow the use of an unstructured grid that allows greater flexibility in refining regions of interests in the domain, that can be used, for example, in ship control algorithms.
The current study aims to setup a local model of the region extended to include the South-China sea region. The validated model can be used to further develop a high-resolution forecasting system of the local domain that takes the forecasted wind data and outputs the wind-wave conditions of the region.

The paper presents the methodology in setting up the numerical model with the domain and forcing data. The model settings used in the setup and the parameters for tuning are discussed. The results are presented and compared with reanalysis data. Finally, the future works on using the current model to develop a forecasting visualisation platform are described.

2. Methodology

The numerical tool used in the study for the wind-wave assessment is the spectral wave module in the MIKE by DHI software suite package [2]. The module is developed for unstructured grids and the discretization is performed using cell-centered finite volume method for the geographical and spectral spaces. The fully spectral formulation [8], [9] is chosen for the prediction runs.

2.1. Domain setup

The model computational domain encompasses the straits of Malacca to the Northwest (101.8° E, 2.5° N) and includes the South-China Sea to the east, stretching all the way to west Kalimantan (109.3° E, 2.2° S). A flexible mesh is set up using the mesh generator in the MIKE suite. The resulting mesh comprises of quadrangular and triangular cells (Figure 1).

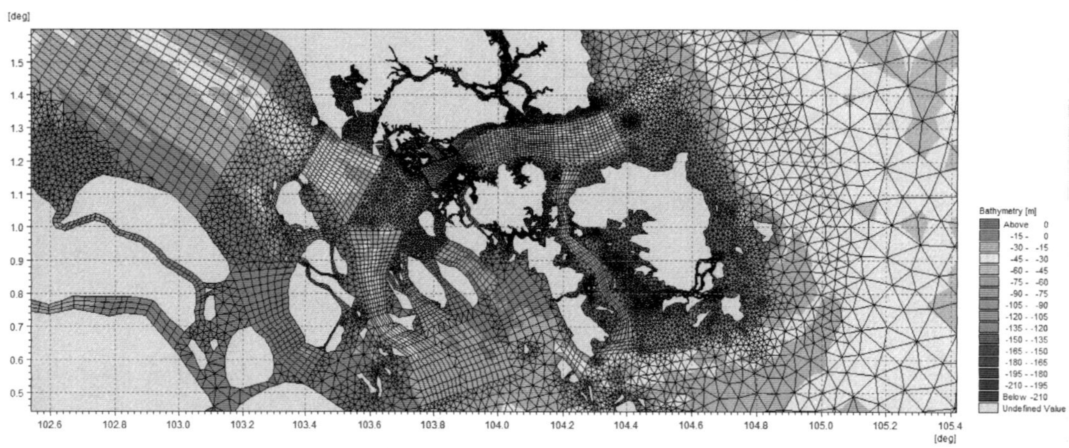

Figure 1. Computational domain zoomed in to the region of interest, the mesh in the Straits of Singapore comprises of triangular or quadrilateral elements based on directional or uni-directional flow

Quadrangular cells are preferred for long straits or channels where the flow is expected to follow the streams whereas triangular cells are used to fill up the rest of the domain to accurately capture the geometry around the land masses. The bathymetry of the region is prepared using scattered C-MAP data [10] interpolated using the natural neighbor of each computational cell. Five open boundaries were specified in the domain for forcing data as shown in (Figure 7).

2.2. Forcing data

To compute wind-waves over a very large region, the wind velocities applied over the domain usually are sufficient to resolve both the shorter period wind-sea and longer period swells. This is usually the procedure for analysis performed with WAVEWATCH III used over a larger domain. However, for smaller domains there is a higher possibility for swells generated outside the domain to be transferred in and are not attributed to the wind forcing data specified over the domain. As such, it is advantageous to specify the forcing data of wind-sea and swell phase-averaged parameters at the open boundaries. Alternatively, the wave energy spectrum can be specified which will include both the wind-sea and swell.

The hindcast exercises in this study are driven by the wind and wave data from European Centre for Medium-Range Weather Forecasts (ECMWF) ERA5 [11]. For the wind data, the velocities of the wind 10 m above the sea surface are required in the two horizontal directions x and y (u10, v10). For the wave data, the phase-averaged parameters – significant wave heights, peak periods, mean wave directions and directional standard deviations for the combined wind-sea and swells. The wave spectral directional width in the dataset can be converted into directional standard deviation (DSD) using Equation (3).

$$Directional\ width = \frac{\pi}{180} \times DSD \qquad (3)$$

The hourly wind velocities from ERA5 can be used directly after converting from the NetCDF format to the proprietary DFS2 format required by MIKE 21 Spectral Waves. The software will perform the necessary interpolations to obtain the velocities over each cell of the computational mesh. The phase-averaged wave parameters to be applied on the boundaries requires interpolation to the coordinates each point on the boundary lines. A script is written to perform bilinear interpolation and generate the hourly profile series of the five boundary lines. The Spectral Waves software will handle the time interpolation internally over the timesteps.

2.3. Model settings and Parameter tuning

The Spectral Waves software offers several settings to set up the simulations that best describes the phenomenon of wind-waves specific to a region.

The fully spectral formulation based on the conservation of Wave Action Density Equation (1) is used in the current setup. The directional discretization is set up with a 360 directional rose divided up to 72 discrete directions. The default setting of 16 directions is found to be insufficient, where the solution only converges at the regions of interest with 72 directions. Based on the solver default settings, the frequency is discretized into 30 frequencies with a minimum of 0.035 Hz and a logarithmic frequency factor of 1.1. In order to separate the wind-sea and swell, a dynamic frequency and wind direction scheme based on wave-age was chosen with the maximum threshold of 0.60 Hz. The wind-sea and swell separation settings are based on the solver defaults.

For the wind forcing applied over the domain, the coupled air-sea interactions option is used. The dissipation of waves due to white capping is described by the source S_{ds} in Equation (2). The source term is described by a function that can be tuned by the two dispersion coefficients C_{ds} (Cdis) and δ (DELTAdis). The constants are expected to have a significant impact on the energy in the domain in controlling the rate of dissipation. The results from the preliminary runs are found to be acceptable (Section 3.) using the default C_{ds} and δ constants of 4.5 and 0.5 respectively. No further tuning of the constants is needed. This is likely due to the smaller size of domain where the resolved significant wave heights internally are affected by the boundary conditions and less on the dissipations within. Other tuneable options such as modifying the bottom friction, wave-breaking parameters and inclusion of water levels and current velocities are excluded in this study as they have limited impact on the generated results during the preliminary runs.

3. Results

The ERA5 wind and waves variables are applied onto the boundaries to drive a smaller domain model with a higher spatial and temporal resolution. The flexible mesh of the smaller MIKE 21 SW model enables a much higher resolution of data in coastal regions, especially in the area of interest of the current study: the Singapore Straits. From the simulated outputs of the model shown in Figure 2, the coasts of Batam and Bintan islands can experience waves of higher significant wind-wave heights of up to 80 cm. Due to the geographic location, the southern coasts of Singapore will only observe about 15 - 20 cm of significant wind-wave heights.

In order to validate the MIKE 21 SW model, the significant wave heights in the domain are expected to be comparable with the ERA5 data. The exceptions will likely be in the regions near the coasts where the ERA5 lacks in the resolutions. The results from the Spectral Waves simulations are compared with the ERA5 data interpolated onto the flexible mesh computational domain. A subtraction is performed to determine the differences between the original ERA5 and the MIKE 21 SW model over the time steps.

Figure 3 shows the validation between the MIKE 21 SW model and the original ERA5 dataset interpolated over the same computations mesh. The subtracted differences show a very good agreement for the predicted wind-sea waves. Only slight under-predictions can be observed at the north boundaries and very small over-predictions in the South-China Seas. The Singapore Straits region are under-predicted but that is to be expected due to the lower resolution of the ERA5 data near the coasts. For the swell, the waves are generally more over-predicted but the for the overall SCS region they do not exceed 15 cm. This results in the combined significant wave heights having the same trend.

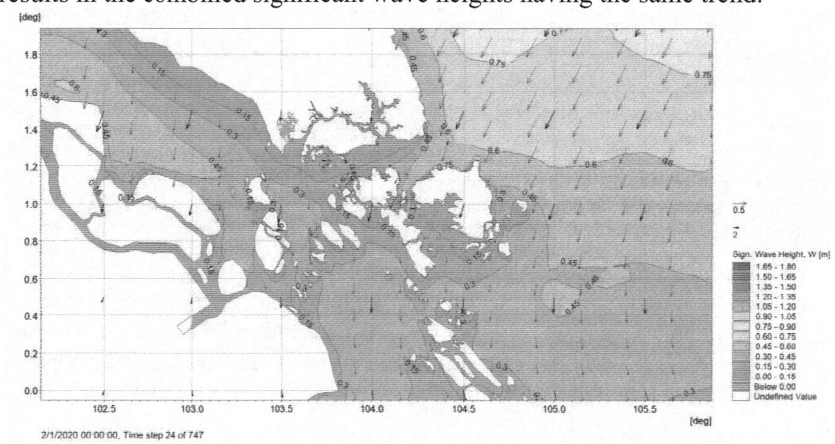

Figure 2. Contour plots of significant wind-wave heights in the Singapore Straits

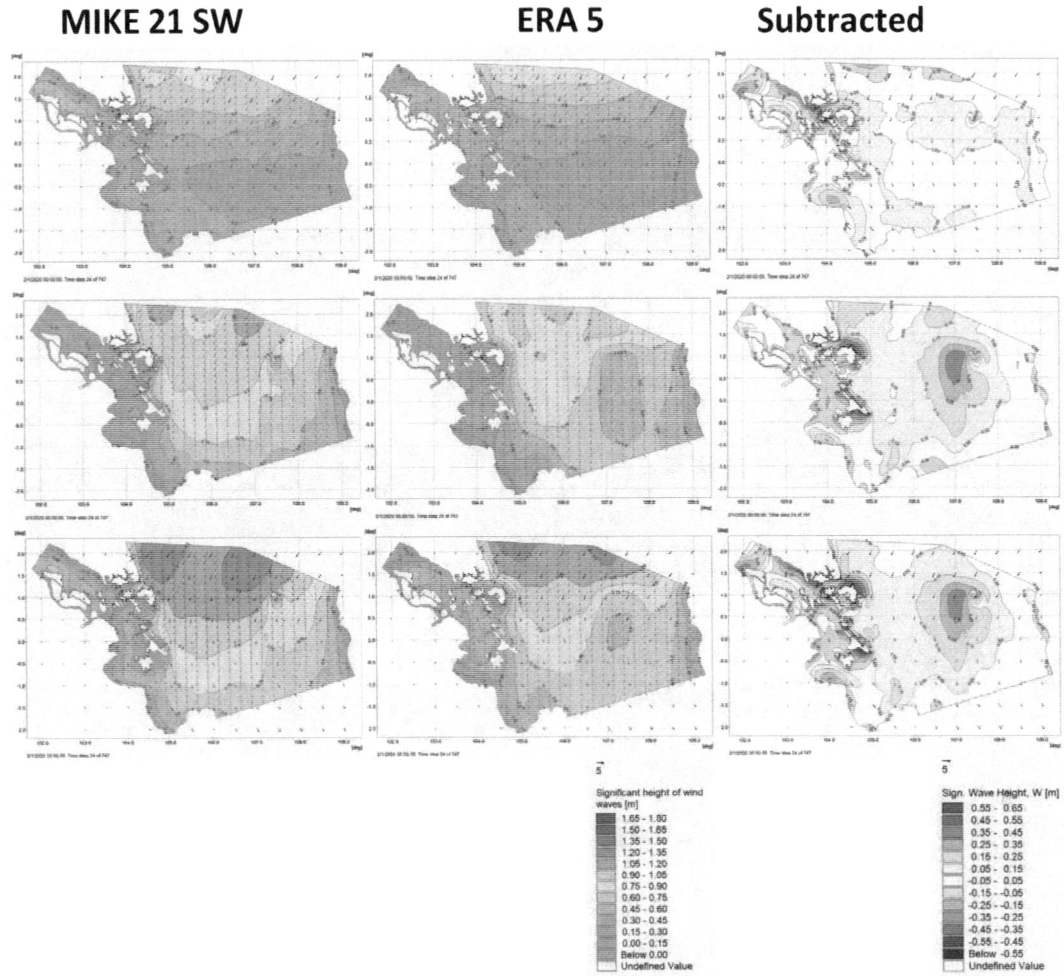

Figure 3. Contour plots of wind-sea (row 1), swell (row 2) and combined (row 3) significant wave heights. Plots show comparison between the simulated model (column 1) and the ERA5 (column 2) with the subtracted differences (column 3)

The time series of significant wave heights are extracted at three points in the domain. The points are chosen far away from the coasts in the South-China Sea region, as the coarse resolution of the ERA5 analyses will have low accuracies and missing data close to land. The points are located exactly on the grid points of the ERA5 dataset and thus no interpolation of the data is needed. (Figure 4) (Table 1)

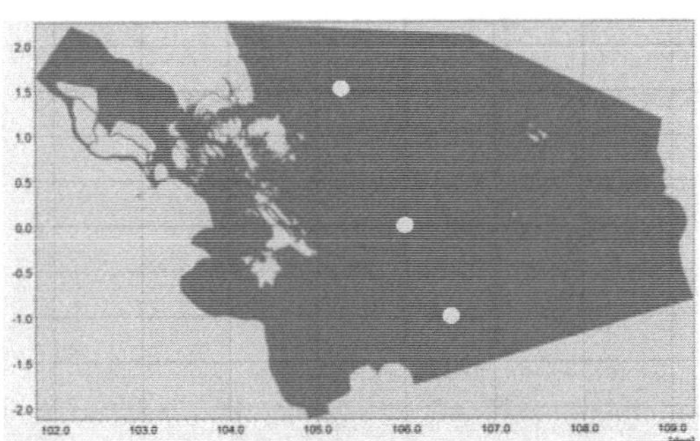

Figure 4. Locations of the comparison points

Table 1. Geographical coordinates of the comparison points

Point	Longitude, Latitude
1	105.25, 1.50
2	106.00, 0.00
3	106.50, -1.00

The time series of significant wave heights shows a very good agreement at the three locations, with the best results for the wind-sea waves. Slight over-predictions can be observed for the total swell and thus the combined significant wave heights (Figure 5). The time series scatter plots of wind-sea, swell and combined significant wave heights for the three points are shown in Figure 6 with the R-value and Root Mean Square Errors (RMSE). The same trend was observed, with the best agreement for the wind-sea waves. It can be seen that the simulated total swell deviates more from the ERA5 data but the overall combined significant wave height R-value are still above 0.944 with a maximum RMSE of 0.107.

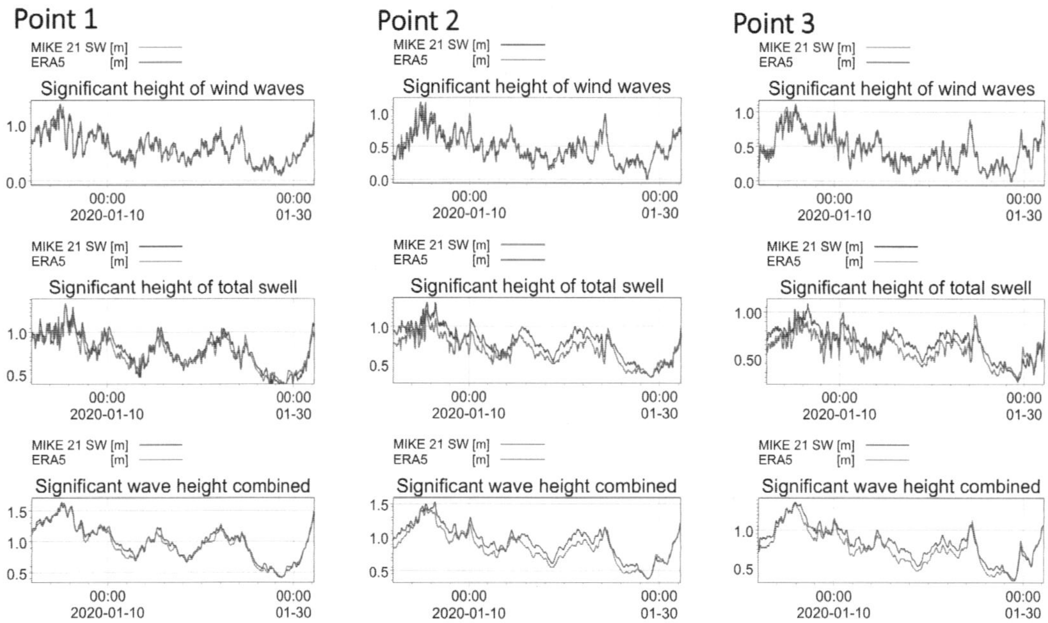

Figure 5. Time series plots of wind-sea (row 1), swell (row 2) and combined (row 3) significant wave heights, extracted at points 1, 2 and 3.
Comparison of MIKE 21 SW simulated data against ERA5 over January

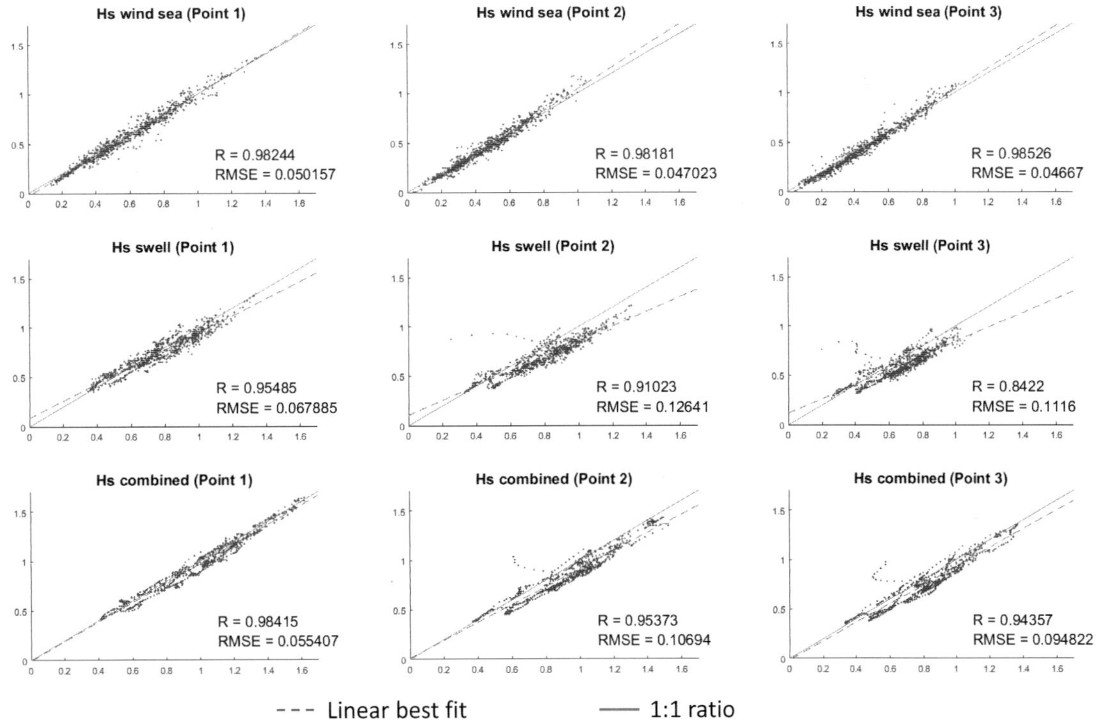

Figure 6. Time series scatter plots of wind-sea (row 1), swell (row 2) and combined (row 3) significant wave heights at points 1, 2 and 3

The overall wind and waves conditions in the region is shown in Figure 7. In the background, the significant wind-sea wave heights from the ERA5 global reanalysis are plotted with the unit vectors of the mean wave directions. In the local region bounded by the blue lines representing the boundaries, the MIKE 21 SW computed significant wind-sea wave heights are shown with the unit vectors of the mean wave directions.

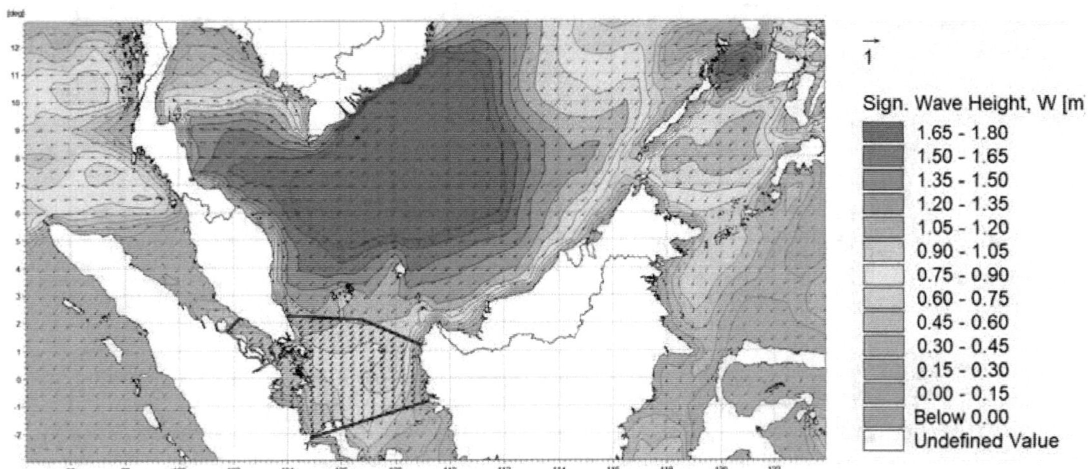

Figure 7. Computed wind-waves in the Singapore region (blue lines indicate the boundaries of the computational domain), overlapping the global reanalysis from ERA5 in the background, mean wave directions are represented by the unit vectors

4. Forecasting platform

The ERA5 With the hindcast validations demonstrated, the modelling capability can be expanded to a fortnightly forecasting framework with a visualization platform to disseminate the simulated phase-averaged wave data (Figure 8). In the platform, the ERA5 hindcast dataset from ECMWF is replaced with the forecasted wind data from National Oceanic and Atmospheric Administration (NOAA).The forecasted datasets are used to drive a regional scale wind-wave model like WAVEWATCH III. Following the methods in this study, the data outputs from WAVEWATCH III are used to setup a smaller local domain to generate forecasts with greater spatial and temporal resolutions.

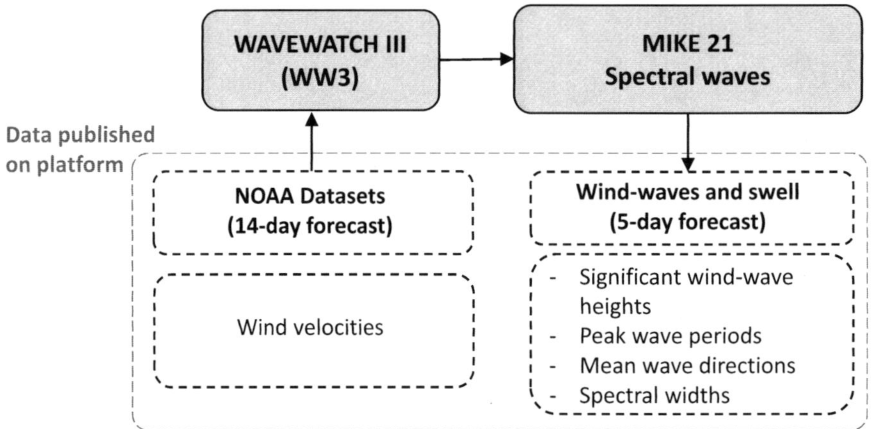

Figure 8. Flow chart of a forecasting framework to 1) process forecasted NOAA wind datasets 2) run the simulations with WAVEWATCH III, and 3) run MIKE 21 to generate the localized predictions

The forecasting of wind-wave is subject to uncertainties and there could be various sources of the uncertainties. For instance, inaccuracy in the forecasted wind data is a source of uncertainty. A 10% error in the wind speed may give rise to 20-30% error in the wave height [12]. The uncertainty in forecasted wave parameters e.g., significant wave height can be quantified by calculating the difference between the forecasted wave height and measured data at a specific location [13]. The mean and standard deviation of the difference in wave height can be used to estimate the uncertainty [13]. If the wave measurements are not available, which is the case in this study, alternatively the hindcast data can be used.

The forecast of wave heights of the swells and their arrival time is a major concern and is associated with uncertainties. The performance of the wave-forecast model is poor in terms of swell arrival time with an error of tens of hours both positive and negative [14]. This means swell predicted by wave forecast models can reach a location 10 hours earlier or later than the actual time. This is a major source of uncertainty and indicates that the wave model has limitations in resolving various mechanisms associated with swells [14].

5. Conclusions

The study aims to gain knowledge of and eventually set up a framework to predict wave environment in the Singapore region, as part of a wider effort to provide the maritime community with this information. A numerical model of the phase-averaged wave conditions of the region has been set up. Using the MIKE spectral wave (SW) model, the model was set up using hindcast ERA5 wind and waves data from ECMWF. The phase-averaged significant wave heights produced by the hindcast simulations of the domain has been well validated with the ECMWF data. The resulting model now provides high resolution phase-averaged wave data of the Singapore Straits region. The next phase of the research will focus on setting up a forecasting framework with a visualization and data sharing platform for dissemination of the wave information.

Acknowledgements
This research is supported by the National Research Foundation, Singapore and the Singapore Maritime Institute under its Maritime Transformation Programme White Space Funding Support for the Establishment and Operation of the Centre of Excellence for Autonomous and Remotely Operated Vessels (CEAOPS) (SMI-2019-MTP-01). The authors would also like to thank Dr Kie Hian CHUA for the advisory contributions to the research.

References

[1] N. Booij, L. H. Holthuijsen, and R. C. Ris, "THE SWAN WAVE MODEL FOR SHALLOW WATER," *Coastal Engineering Proceedings*, vol. 1, no. 25, Jan. 1996, doi: 10.9753/icce.v25.%p.

[2] DHI, "MIKE 21 Spectral Wave Module Scientific Documentation." DHI, 2021.

[3] Z. Wang *et al.*, "Wind wave characteristics and engineering environment of the South China Sea," *Journal of Ocean University of China*, vol. 13, pp. 893–900, Jan. 2014, doi: 10.1007/s11802-014-2331-0.

[4] C. Zheng, H. Zhuang, X. Li, and X. Li, "Wind energy and wave energy resources assessment in the East China Sea and South China Sea," *Science China Technological Sciences*, vol. 55, no. 1, pp. 163–173, 2012, doi: 10.1007/s11431-011-4646-z.

[5] A. Mirzaei, F. Tangang, and L. Juneng, "Wave energy potential assessment in the central and southern regions of the South China Sea," *Renewable Energy*, vol. 80, pp. 454–470, Aug. 2015, doi: 10.1016/J.RENENE.2015.02.005.

[6] The WAVEWATCH III Development Group WW3DG, "User manual and system documentation of WAVEWATCH III. Version 6.07. Tech. Note 333, NOAA/NWS/NCEP/MMAB, College Park, MD, USA," 2019.

[7] X. Zhu, P. Tkalich, and E. S. Chan, "Simulation of Typical Wind-Waves Characteristics in Singapore Strait," in *OCEANS 2006 - Asia Pacific*, 2006, pp. 1–5. doi: 10.1109/OCEANSAP.2006.4393949.

[8] I. R. Young, *Wind Generated Ocean Waves*, vol. 2. Elsevier Ocean Engineering Book Series, 1999.

[9] G. J. Komen, L. Cavaleri, M. Donelan, K. Hasselmann, S. Hasselmann, and P. A. E. M. Janssen, *Dynamics and Modelling of Ocean Waves*. Cambridge University Press, UK, 560pp, 1994. doi: 10.1017/CBO9780511628955.

[10] C-MAP, "C-MAP," 2020. https://www.c-map.com/ (accessed Jan. 10, 2022).

[11] H. Hersbach *et al.*, "ERA5 hourly data on single levels from 1979 to present.," *Copernicus Climate Change Service (C3S) Climate Data Store (CDS)*, 2018.

[12] E. Bouws *et al.*, *Guide to Wave analysis and forecasting*, World Met. Org., vol. WMO-No. 702. 1998.

[13] A. Natskår, T. Moan, and P. O. Alvær, "Uncertainty in forecasted environmental conditions for reliability analyses of marine operations," *Ocean Engineering*, vol. 108, pp. 636–647, Nov. 2015, doi: 10.1016/J.OCEANENG.2015.08.034.

[14] A. v Babanin *et al.*, "Waves and Swells in High Wind and Extreme Fetches, Measurements in the Southern Ocean," *Frontiers in Marine Science*, vol. 6, 2019, [Online]. Available: https://www.frontiersin.org/article/10.3389/fmars.2019.00361

Ship Navigation and Fuel Profiling based on Noon Report using Neural Network Generative Modeling

J Hadi[1], Z Y Tay[1], D Konovessis[2]

[1] Singapore Institute of Technology, 10 Dover Drive, Singapore 138683
[2] Strathclyde University, 16 Richmond Street, Glasgow, G11XQ, UK

Abstract. Harbor craft historical routes contain valuable information on how the experienced crews navigate around the known waters while performing jobs. The noon report logs each job timeframe which can be used to segregate the time-series positional data as routes. Other information from the noon report such as fuel consumption could be associated with a particular job as well. This paper offers a solution to encompass crew navigational experience into neural network models. The variational autoencoder, which is a generative model, can capture the routes into a knowledge base model. The same variational autoencoder is also able to train other neural networks to make predictions of route and fuel consumption based on job metadata (I.e., job duration, activity area, and route classification). The predicted routes could be used as a cost map for pathfinding algorithms such as A* or Dijkstra.

1. Introduction

One of the earliest pieces of evidence of marine activity is from 709,000 years ago when the ancestors of today's modern humans, made the sea crossings to and around what is now the Southeast Asian archipelagos[1]. Fast forward to the age of exploration, the marine navigation had been relying heavily on experienced crews or accounts from experienced crews for at least many centuries. Not only is the reliance on experienced crews due to the adverse and extreme conditions offshore, but also to the lack of assisting technology. It was not until the last century or so, that much of the technology and digitalization efforts have been put into marine applications. The application has gained even more traction into the 21st century[2]. One of the benefits of digitalization in the context of the current global climate crisis is the opportunity for achieving ship energy optimization and efficiency for the purpose of decarbonization[3]. In addition to achieving ship energy optimization and efficiency, decarbonization also mitigates a negative health effect on individuals[4].

The historical vessel routes from maritime navigation are a valuable source of information on how experienced crews navigate around known waters. It is especially true for harbor crafts that usually move about the area it serves frequently. This paper aims to capture a gold standard which is the crew's experience as a knowledge base of the historical routes, in the form of a trained neural network[5]. The knowledge base could be consumed by pathfinding algorithms to generate a suggested route[6]. The algorithm for pathfinding is essential in an autonomous or semi-autonomous system[7]. The most popular pathfinding algorithm such as A* or Dijkstra requires an area insight in form of a map[8,9]. The knowledge base generated from the solution presented in this paper could become a cost map. Not only for the said pathfinding algorithms, but also potentially for other pathfinding algorithms[10].

Nonetheless, in the case of less sophisticated smaller harbor vessels (i.e., tugboat, ferry, cargo, etc.), the data from onboard automatic digital logging system is not always available. Instead, the data that is regularly available is in form of daily manual logs such as noon reports that hold a considerable amount of information[11]. Given these scenarios, this paper aims to utilize the noon report combined with positional (navigation) and weather data from external sources to achieve its aim.

The subject vessel discussed in this paper is POSH Grace. The noon report data collected from POSH Grace, combined with externally sourced data such as positional and weather data for seven months (April to October 2020), make up the dataset. POSH Grace is a tugboat that is owned by PACC Offshore Services Holdings (POSH) as shown in Figure 1. The POSH Grace vessel specifications are given in Table 1. The main activities of the tugboat consists of anchoring, assisting the large vessels in docking, and piloting around the southern sea of Singapore. The purpose of collecting the operational data from the subject vessel was to conduct a research study in predicting the fuel rate to achieve fuel efficiency and better planning by data analytics and machine learning[12,13].

Table 1. Vessel specifications.

Main particulars	Value
Length overall (LOA)	29 m
Displacement	665 tons
Maximum speed	12 knots
Main engines	NIIGATA 6L26HLX
Number of engines	2
Total BHP	4000 BHP
Type of propulsion	Azimuth pod
Number of propulsors	2

Figure 1. The subject vessel, POSH Grace tugboat.

2. Methodology

There are four steps to achieve the solution offered in this paper as shown in Figure 2. All four steps use machine learning toolkit from computer science. Machine learning is a branch of artificial intelligence that uses data and algorithms to mimic the way human learns artificially using a digital tool (i.e., computer)[14]. A model (or network) must first be designed, which is akin to a pipeline to which the data is fed at both ends. Data that is fed at both ends while training the network are often called training X and y inputs. It is to let the algorithm iterate over the data, to adjust values in the network to get y value given the X value (backpropagation). The values in the network are function coefficients and are stored in the form of tensors. The function is often referred to as an activation function. Every iteration is evaluated by a loss function to let the algorithm know how far off it is from the desired value. A well-trained network can produce a close value of y (the prediction) to the expected value of y (the evidence/observation).

The final objective is to generate two outputs of Route Image (H) and Fuel Prediction (I) by step 4, from Job & Noon Report Metadata. Inputs (A) and (B) must first be prepared to complete all four steps.

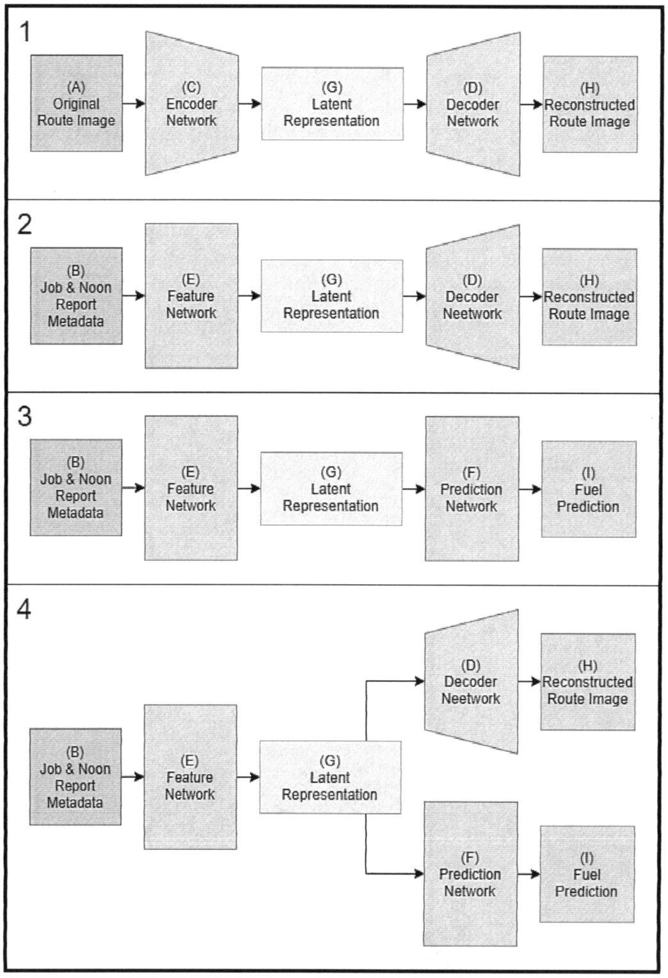

Figure 2. Block Diagram

2.1. Data Preparation

The Data Preparation sub-sections explain the methodologies to convert raw data into a dataset prior to proceeding to any steps shown in Figure 2.

2.1.1. Route Image Creation

A route of the vessel is comprised of a series of historical positions in form of GPS latitude-longitude coordinates between a timeframe from the noon report. The historical position of the vessel could be obtained from an AIS (Automatic Identification Systems) aggregator service by Marine Traffic. This is one of the external data discussed in the introduction. This work has been partially supported by Marine Traffic (www.marinetraffic.com).

The series of route positions is resampled to an interval (i.e., one minute). The series of one-minute vessel position data points are approximated and binned to a cell in a grid. Each route eventually forms a greyscale route image as shown in Figure 3(b). The greyscale images are the Original Route Image (A) in Figure 3(a).

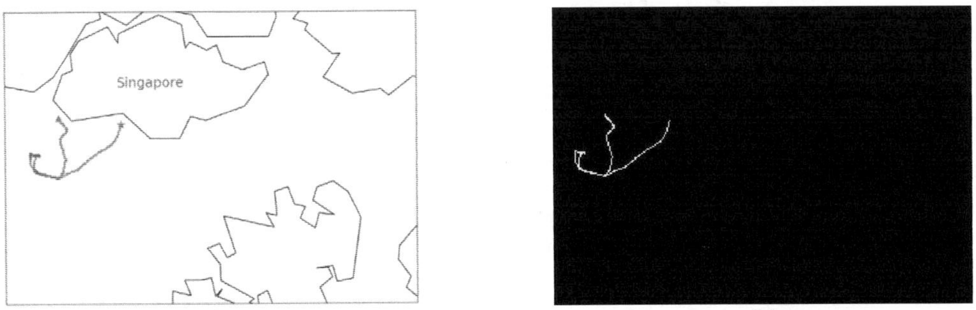

(a) (b)

Figure 3. The transformation of route information into a greyscale image.
(a) Original route over an illustrated map. Green triangle is origin, green star is destination.
(b) The route image (308 x 212 pixels).

In the noon report between April to October 2020, there are 405 jobs. Consequently, there are 405 route images, a few of which is shown in Figure 4. Routes from Figure 4 are used as a test dataset and excluded during training. The 400 images alone are insufficient to train a neural network in Figure 2 step 1. Hence, each image is duplicated randomly by 30 to 50 times. The total number of images for training the neural network is set to be around 15,000. The duplication of the images will unlikely be redundant as it will eventually generate non-identical samples of latent representation, which is explained in Appendix A (the reparameterization trick).

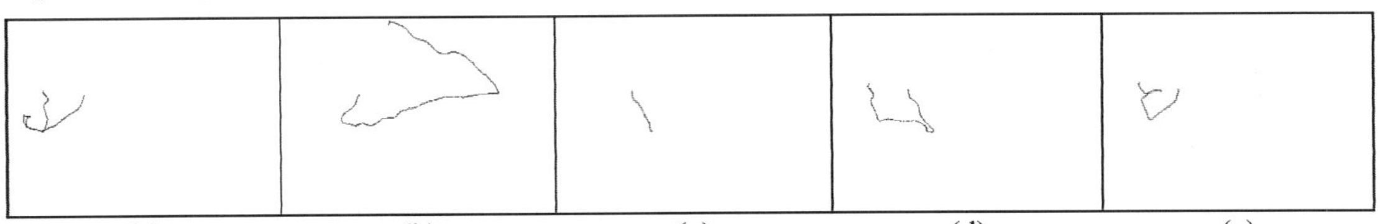

(a) (b) (c) (d) (e)

Figure 4. Sample of Original Route Images. The image greyscale is inverted for better visual.
Route number: (a) 22, (b) 264, (c) 288, (d) 383, (e) 387.

2.1.2. Route Classification

The vessel routes from jobs are classified based on the route similarity. Nonetheless, however similar two or more routes are, they are not identical in terms of vessel positional coordinate at a particular time. A sample case is shown in Figure 5.

(a) (b)

Figure 5. The route similarity comparison.
Route number: (a) 387 versus (b) 379.

For each job, the time-series positional coordinates are broken down into latitude and longitude elements. Latitude from each job is considered to measure the similarity distance using a dynamic time-warping algorithm[15]. The same method is applied to the longitude. Figure 6 shows the point-by-point associations for distance measurement. The cumulative absolute distance of latitude and longitude difference is combined by calculating a hypothenuse of the two. It is due to that the latitude and longitude are assumed to be orthogonal, thus ignoring the curvature of the earth.

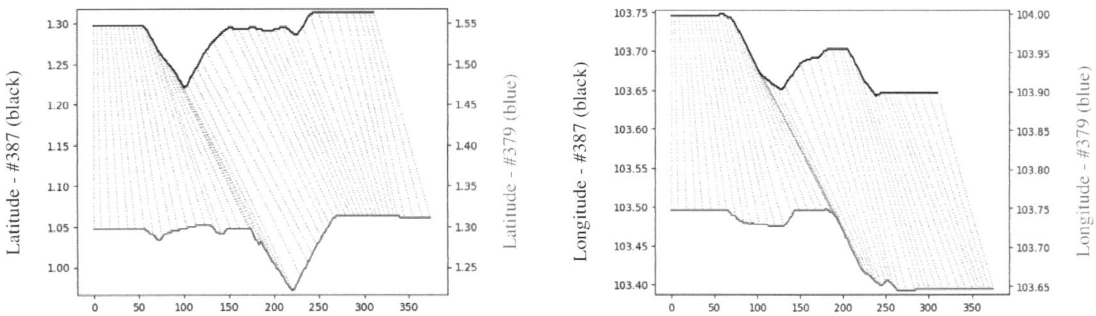

(a) (b)

Figure 6. Dynamic time-warping correlation on #387 and #379. (a) Latitudes, (b) Longitudes.

The pair similarity distance for every pair combination is stored as a distance matrix[16]. The visualized version of the distance matrix is shown in Figure 7. Each route's distance with another is binned in a cell of a matrix (table). Hence, it is mirrored diagonally. The dark areas are the regions where the distance value is low indicating there is a close similarity with other members. In contrast, the bright areas (lines) are the part where the distance value is high. From Figure 7, it can be observed that there are members (routes) that do not have similarities with other members (the routes with white lines).

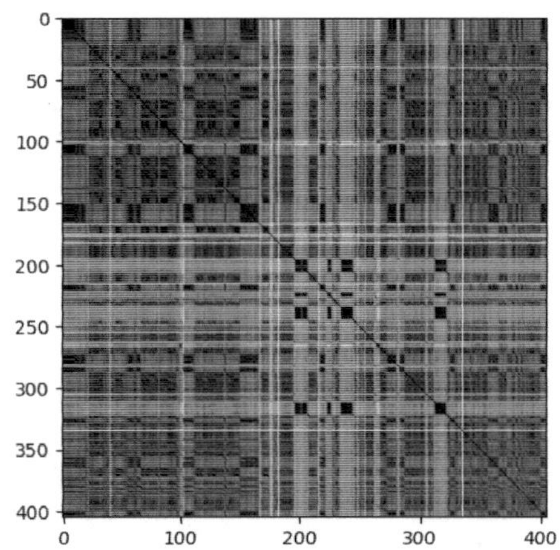

Figure 7. Visualized distance matrix.

The agglomerative hierarchical clustering is used to convert the distance matrix to class labels[17]. Agglomerative clustering is a hierarchical clustering, an unsupervised machine learning algorithm, that is a bottom-up approach. It starts by pairing up the individual/primary nodes (or data points) as secondary nodes. The pairing up criterion is based on the closest distance. It then joins (i.e., averages) the secondary nodes to become new primary nodes. It then again progressively pairs up and joins the now primary nodes as new secondary nodes based on the same criterion. The hypothetical shape of the clustering mimics the structure of a root of a tree. The clustering stops until either the distance threshold or the number of clusters is achieved.

Similar routes have the same class label indicating the cluster association. Figure 8 shows the other cluster members in the same cluster as route #387. The class labels (or cluster labels) are further one-hot encoded[18].

(a) (b) (c)

Figure 8. Other routes that are similar to route #387. All images are inverted greyscale image. (a) Route #98, (b) #228 & (c) #400.

2.1.3. Jobs & Noon Report Metadata

A set of secondary data that describes the primary data is known as metadata. Figure 9 shows the activity area from the position data, which forms a set of parameters (lat0, lat1, lon0, lon1). It is to denote the geographical area in which the job takes place. The activity area is a metadata as it describes the primary data.

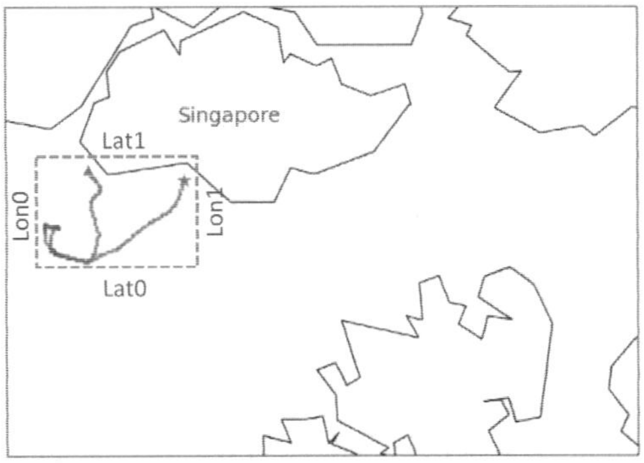

Figure 9. Activity area boundary is indicated by a dotted blue rectangle, producing a boundary of four parameters or features.

In addition to the activity area, the job duration from the noon report and the route classification classes are also metadata and are used as a complete set of Job & Noon Report Metadata (B). The total number of parameters for Jobs & Noon Report Metadata (B) depends on the number of route classification classes. Finally, fuel consumption is obtained from the noon report. The fuel consumption is used for training the prediction network (F).

2.2. Generative Modeling

Generative modeling refers to Step 1 in Figure 2. Generative modeling uses a variational autoencoder (VAE) neural network[19]. The VAE is comprised of two sub-networks: (C) Encoder Network, and (B) Decoder Network according to Figure 2. Figure 10(a) and Figure 10(b) show the encoder and decoder architectures. Several layers and layer types are used to construct both networks, which are readily available for implementation using TensorFlow framework[20]. VAE uses a combination of convolutional 2D (Conv2D) and fully connected (Dense) layers, as well as tensor shape transform layers (Flatten and Reshape)[21,22]. In the final layer of Encoder Network, a custom sampling layer (Sampling) is used. The sampling layer is discussed in section 3.

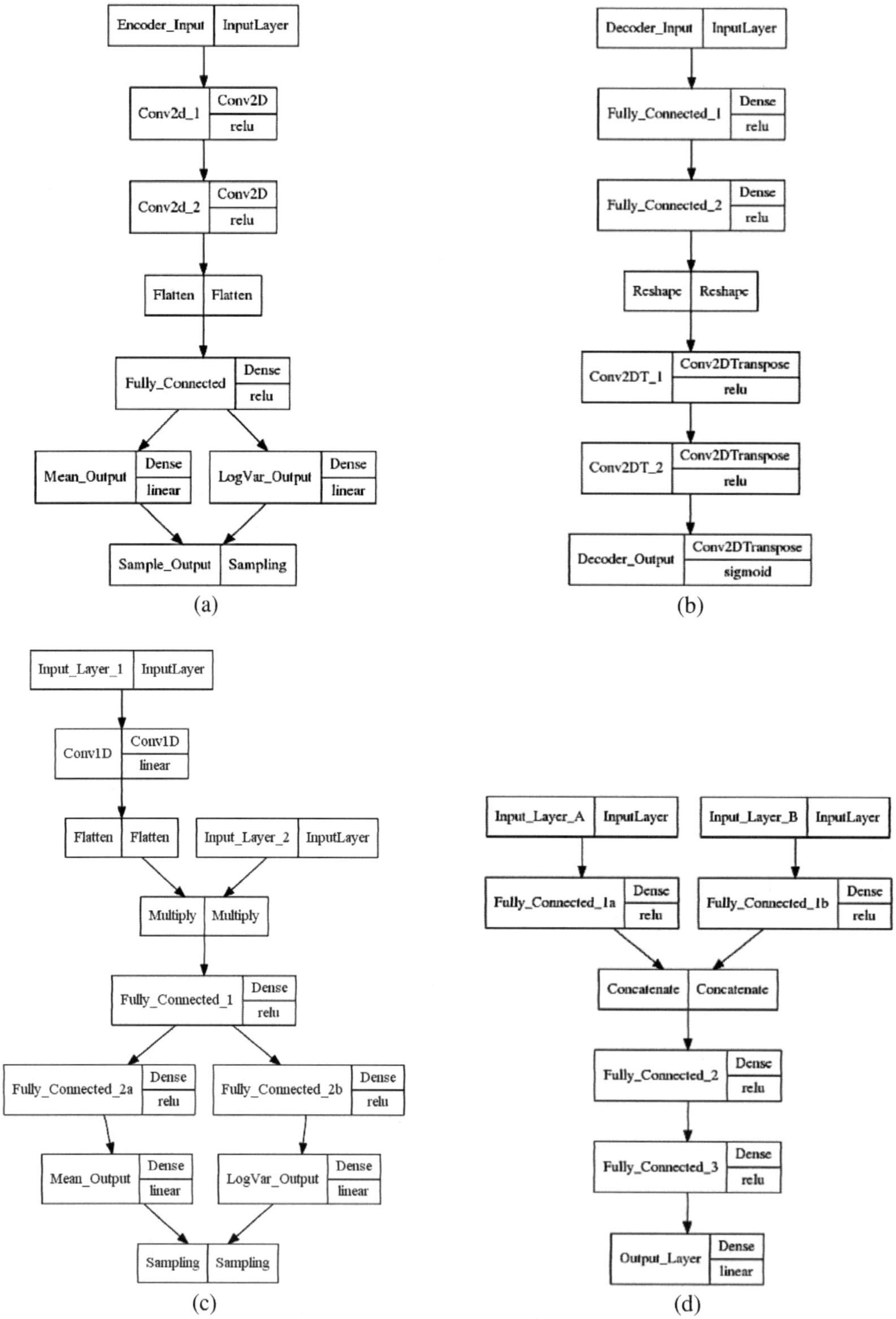

Figure 10. (a) Encoder network architecture, (b) decoder network architecture, (c) feature network architecture, and (d) prediction network architecture.

The aim of the Encoder Network is to reduce the original input (route images) to a representation in a lower dimensions space, also known as latent representation in latent space. The Latent Representation (G) is a set of probability distributions comprising mean (μ), and log-variance (σ). The number of sets of probability distribution (number of pairs of mean and log-variance) determines the number of the latent space dimension.

The aim of the Encoder Network in the solution presented in this paper is to convert the input of Original Route Image (A) into its Latent Representation (G) values. The Latent Representation will later be used to train the other two networks (E) and (F), which will be explained later. Eventually, the Encoder Network is removed from the final structure as shown in Figure 2 step 4.

Figure 11 shows outputs of a 2-dimensional slice of 7-dimensional latent space from a trained VAE network. The outputs are tiled in a 2-dimensional latent space. The VAE manages to fit the route images into functions in the latent hyperspaces of μ and σ.

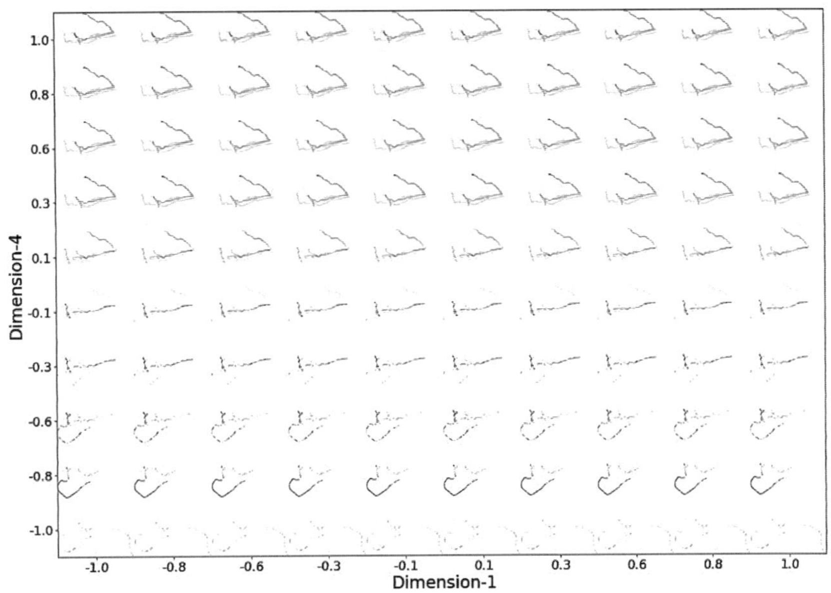

Figure 11. The visualized Latent Representation. The image is an inverted greyscale.

While training, the Decoder Network (D) accepts a sample of the Latent Representation (G) from the encoder network in attempt to reconstruct the original route image. The purpose of sampling in Encoder Network (C) is to sample out the probability distribution for training. It is due to that the backpropagation could not take place through a probabilistic function. Due to the probabilistic nature, it is highly unlikely that the sampling of probability distribution will produce identical values for a set of identically duplicated inputs as mentioned in section 2.1.1. The sampling (reparameterization trick) is discussed further in section 3.

Figure 12 shows the comparison of original and reconstructed route images from a well-trained VAE. While training VAE, a special loss function is used for the Encoder Network (C), the KL_{loss} also known as the Kullback–Leibler loss function[23].

$$KL_{loss} = -\frac{1}{2}(1 + \sigma - \mu^2 - e^{\sigma}) \tag{1}$$

KL_{loss} is a function of the probability distribution. σ is the log-variance, and μ is the mean, while e is Euler constant. The Decoder Network (D) uses binary cross-entropy as a loss function[24].

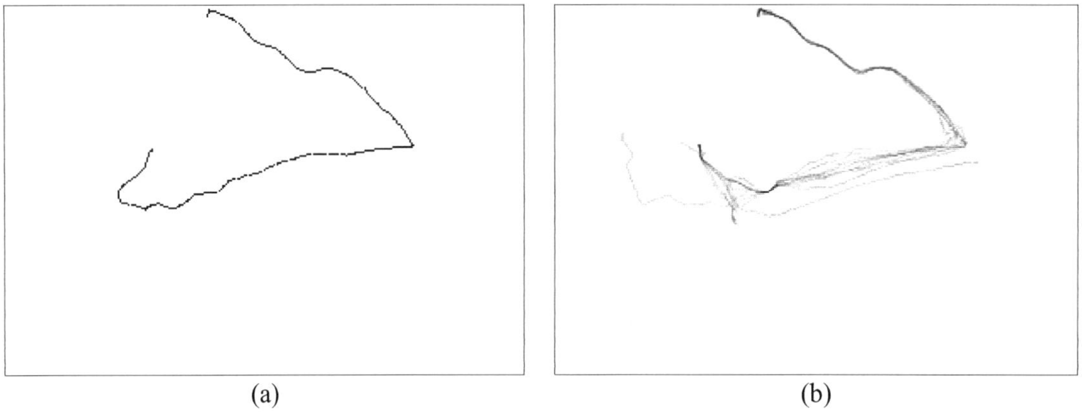

Figure 12. (a) Original of route #264 versus (b) reconstructed route image comparison from step 1. Both greyscale images are inverted for better visual.

2.3. Feature Modeling

Feature modeling refers to the second step in Figure 2. The architecture of the network is shown in Figure 10(c). It is a separate neural network replacing the encoder. The Feature Network (E) accepts Job & Noon Report Metadata (B) as training X, instead of route images for Encoder Network (C). For training y, it uses Latent Representation values (the μ and σ) generated by a trained Encoder Network (C). The route classification is fed into *Input_Layer_1*, while the job duration and job activity area are fed into *Input_Layer_2*. The Feature Network uses the same TensorFlow framework as VAE.

2.4. Prediction Modeling

Fuel prediction refers to step 3 in Figure 2. It uses fuel consumption information from the noon report as training y. As training X, the Prediction Network (F) uses Latent Representation (G) generated by Encoder Network (c). Figure 10(d) shows the architecture of the Prediction Network (F). *Input_Layer_A* takes in the mean (μ), while *Input_Layer_B* takes in the log-variance (σ). The prediction network uses exclusively fully connected (Dense) layers. Both feature and fuel prediction networks use mean absolute error as a loss function[25].

2.5. Navigation & Fuel Prediction Modeling

The Navigation & Fuel Prediction Modeling uses three trained networks as shown in Figure 2 step 4. In this final step, training is not essential, although it is possible to do so (also called fine-tuning)[26]. The final step 4 produces two deliverables: the Reconstructed Route Image (H), and Fuel Prediction (I) by taking in only the Job & Noon Report Metadata (B) as input. The Original Route Image (A) is no longer needed. The decoder, feature and prediction networks make up the final model.

2.5.1. Pathfinding Algorithm

Fuel prediction is evaluated by the percentage of difference of the fuel consumption in the noon report (evidence/observation) and prediction of fuel by the Prediction Network. The Reconstructed Route Image evaluation, however, is not as straightforward and requires an application of pathfinding algorithms such as A* and Dijkstra. Hence, the evaluation of Reconstructed Route Image (H) requires deliberation and is discussed in Section 3.

3. Results & Discussion

This section separately discusses the results from step 4 in Figure 2. Section 3.1 and 3.2 independently discuss two sets of results of Reconstructed Route Images (H) and the Fuel Prediction (I) respectively.

3.1. Route Image Reconstruction

The five Reconstructed Route Images from Figure 4 are presented in Figure 13. Compare them with the original images in Figure 4. Out of the five test samples, all appear to be visually acceptable except #22 and #264, the Figure 13(a)(b). The reconstructed #22 from Figure 13(a) is considerably different from the original image Figure 4(a) due to being a unique route. In other words, #22 is a one-member cluster and the final model is not trained with any information similar to #22. Nonetheless, the final model still manages to produce result that is not extremely different from the original.

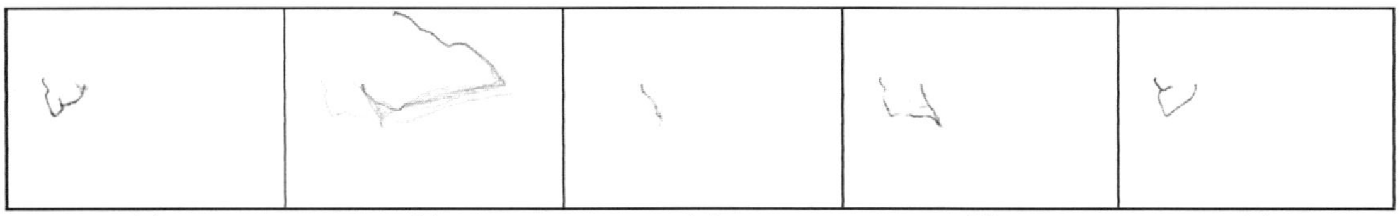

(a) (b) (c) (d) (e)

Figure 13. The reconstructed route images by final model. All greyscale images are inverted. Route number: (a) 22, (b) 264, (c) 288, (d) 383, (e) 387.

Figure 14(a) discusses the result of route #264. Despite there being less white area (areas which the vessel can visit), the result is still acceptable to become a cost map for the pathfinding algorithm. Figure 14(b), the inverted version of Figure 14(b), shows that there are grey areas which is the information from the past visits. The idea is to make the frequently visited areas lower cost to visit. In greyscale image, the darker shades are represented by lower values. In 8-bit representation, value 0 is black, value 255 is white (if normalized, 0 is black and 1 is white).

The transformed image as a cost map for the pathfinding algorithm is shown in Figure 14(c). The no-go zone or area is represented as value 0 as output by the final model. It is due to no evidence in the training data (400 images) that it has ever been visited. The no-go zone representation is a discontinuity and appears as red in Figure 14(c). In contrast, the frequently visited area appear in lighter shades by Figure 14(a), darker shades by Figure 14(b).

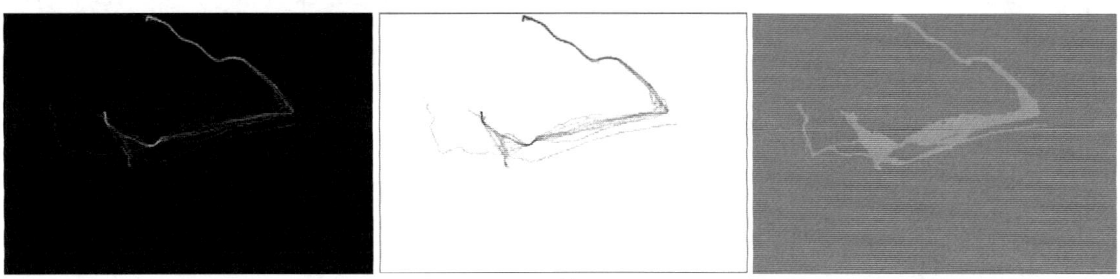

(a) (b) (c)

Figure 14. (a) reconstructed greyscale image as output by final model, (b) the invert of reconstructed greyscale image, (c) the visualization of cost map after transformation.

The comparison of the pathfinding algorithms using the cost map versus the observation (original route) is shown in Figure 15. Both pathfinding algorithms suggest that there are shorter paths. Provided with this information, the crew could have been advised of a shorter route. Ultimately, the crew still could have had the absolute discretion to determine the actual path. It may have been due to the adverse sea conditions that the crew may have experienced during route #264, or a certain objective to visit a certain location prior to the final location.

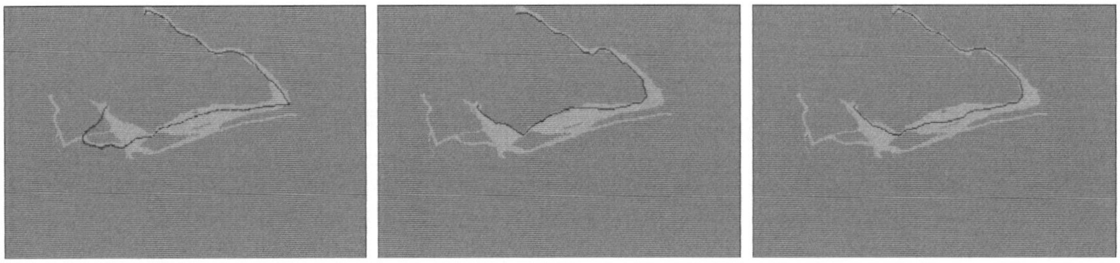

(a) (b) (c)

Figure 15. Application of pathfinding algorithm on reconstructed route image.
(a) Observation, (b) A*, (c) Dijkstra.

In addition to the one-member class issue discussed earlier, the solution presented in this paper is not able to make up for the shortcoming related to intermediate or multiple destinations as shown in Figure 16. If there is no explicit source of information (i.e., from the noon report) detailing or splitting the multiple trips in one job to separate trips, the solution presented in this paper will not be precise.

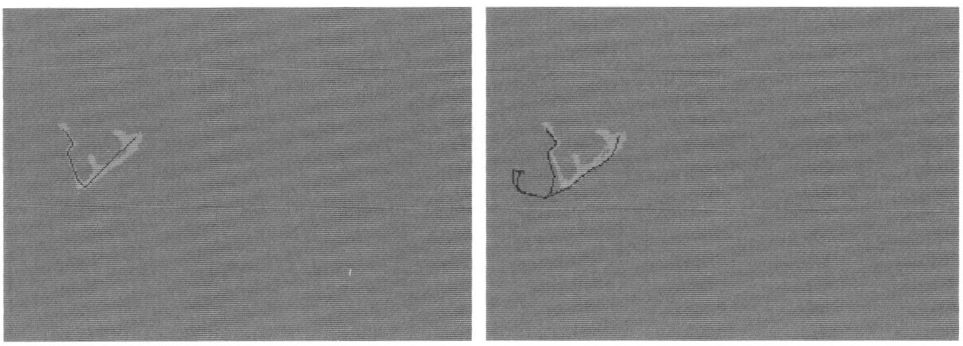

(a) (b)

Figure 16. Comparison of the route with multiple trips.
(a) Suggested path, (b) observation.

It is possible to customize the pathfinding algorithm to account for external factors (i.e., wind or wave situation), by making the cost of travel to a certain direction to be more expensive. Figure 17 shows the comparison of results with and without accounting for the weather. Figure 17(b) shows the effect of a southerly wind (downward) applied to Figure 17(a). The weather data is another source of external data mentioned in the introduction and is provided by www.meteoblue.com.

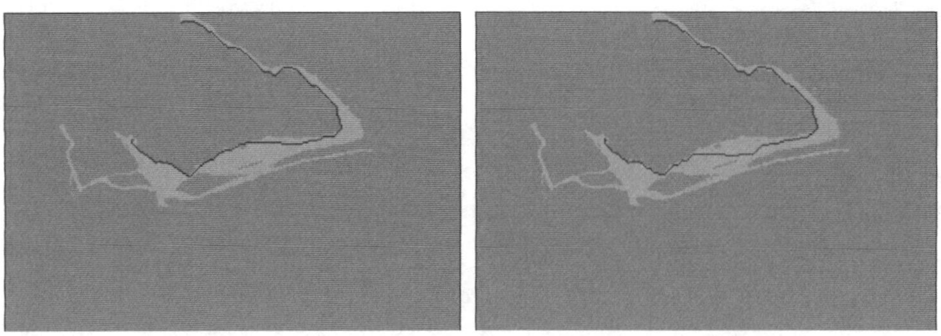

(a) (b)

Figure 17. Application of weather data in pathfinding algorithm.
(a) With weather data effect, (b) Without weather data effect.

3.2. Fuel Prediction

Table 2 shows the comparison of the predicted fuel consumption versus the observation. Route #22 is under-predicted as the number of training data for the similar route pattern is scarce. The result for route #22 is expected due to being a unique route and excluded during training. The final model is unable to predict what it had never learned before, and forced to guess. The best prediction result from route #288 and the worst prediction result from route #387 are discussed further in Table 3.

Table 3 lists down the training data from routes similar to #288 for comparison. It is labelled as class #0. The final model manages to use the job duration and activity area of route #288 to approximate to a very close prediction to the actual record in the noon report (the observation).

Table 2. Fuel prediction versus observation comparison.

Route #	Prediction[a]	Observation[a]	% Difference
22	1309	1688	-22.45
264	2426	2662	-8.88
288	454	450	0.78
383	1419	1310	8.28
387	1094	1245	-12.12

[a] In liters

Table 3 also lists down the training data from routes similar to #387 for comparison. It is labelled as class #1. There is a total of four data points by class #1 (only 3 for training: #98, #228, #400). Using three training data points, the final model still manages to approximate fuel prediction to a reasonable figure of 1094 liters, -12.12% off from the actual 1245 liters.

Table 3. Comparison of routes #288 and #387 with members from their own classes.

Class #	Route #	Job Duration[a]	Observation Fuel[b]
0	**288**	**7200**	**450**
0	124	9000	474
0	232	9600	605
0	259	18000	1025
0	303	7200	515
0	348	4500	275
1	**387**	**19800**	**1245**
1	98	18000	1050
1	228	15600	985
1	400	28800	1816

[a] In seconds
[b] In liters

During deployment, the user of the final model needs to input the space and time information in which the vessel is to operate in addition to selection of a route pattern (from route classification). The activity area and duration represent the space and time information the user needs to input. Figure 18 discusses the other features (geographical features) used to produce results from Table 3. Figure 18(a)(c)(e) show the case of route #288, compared with its own class members.

Figure 18(e) shows the correlation of fuel and duration as explained by Table 3, while Figure 18(a)(c) gives the geographical perspective of the activity areas. As route #288 is not an outlying member (in terms of both fuel-distance correlation and activity area), the final model manages to produce a very good fuel prediction result. It is in addition that there are more training data points compared to route

#387. The adversity of activity areas and good duration-fuel fitting make up a good set of training features for #288.

In contrast, route #387 does not have as many data points as route #288. The activity areas are not as adverse as route #288 either, as shown in Figure 18(d). Despite the lack of adversity, these feature combinations are still able to produce a reasonable ballpark prediction of 1094 versus the actual 1245 (observation). It is shown in Figure 18(f) as #387 that it could still fit relatively well between #98 and #400. In addition, the final model manages to generate a good route image as well, as shown in Figure 13(e). Hence, this presents an opportunity to increase the number and quality of job/route metadata to increase the quality of the features.

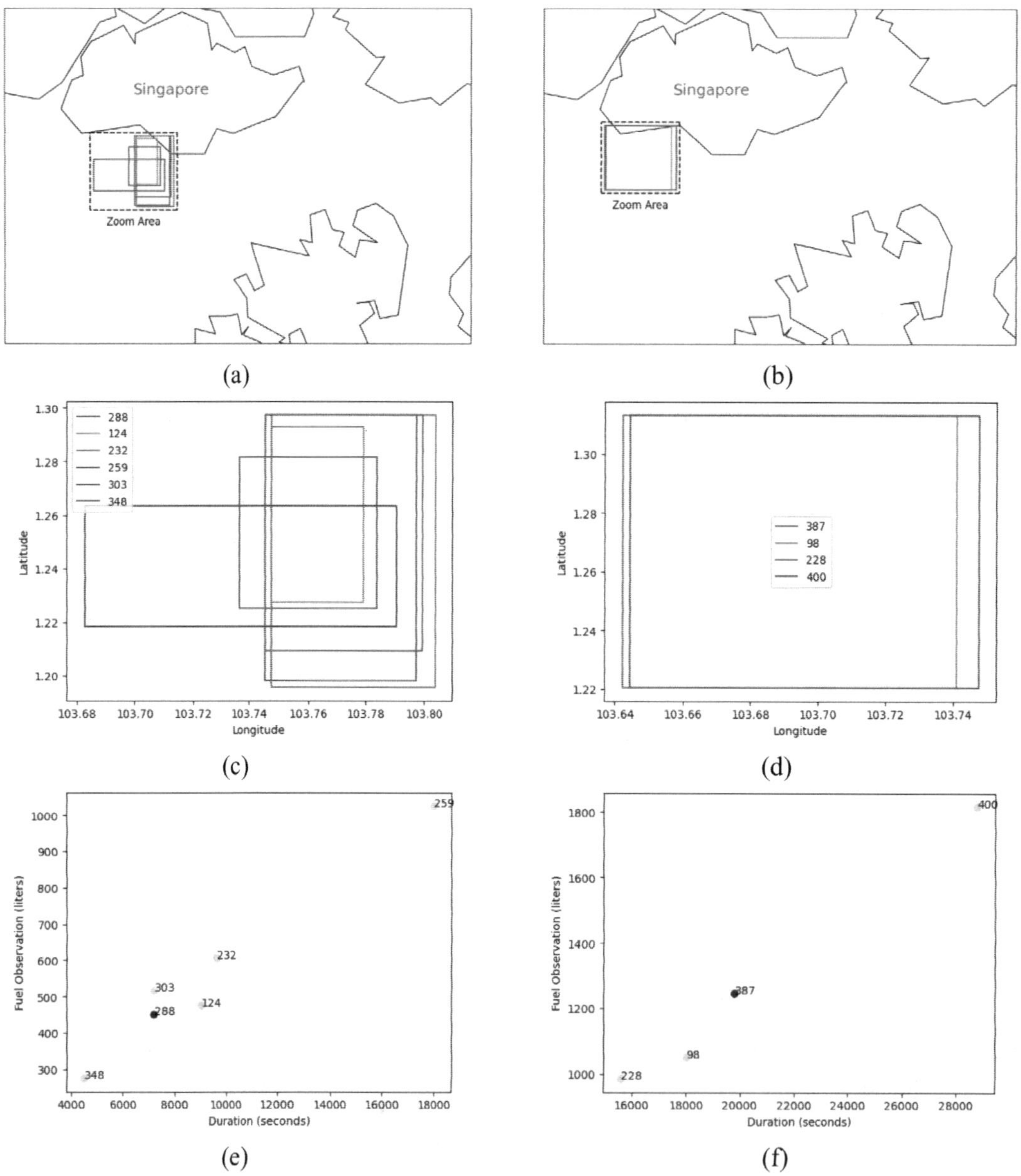

Figure 18. Comparison of features between route #288 versus route #387. Activity areas: (a) Route #288, (b) Route #387. Zoom-in (close-up) of activity areas: (c) Route #288, (d) Route #387. Fuel versus duration: (e) Route #288, (f) Route #387

4. Conclusion

Arguably, the very popular neural network applications are for dimensionality reduction purposes such as image classification. An image of certain pixel size that contains so much information is reduced to a tag for object A, B, etc. The neural network for regression takes a step up by attempting to find higher dimensional representations of lower dimension samples (points). Generative modeling takes it even further by creating what did not exist before, such as the reconstructed route image in the attempt to recreate the original. The reconstructed image is not the exact copy of the original, it is the slightly different version of the original.

The application of the solution presented in this paper attempts to utilize generative modeling to encompass valuable past human experiences in maritime navigation. The crew's experience is deemed as a gold standard from which the model is learning. The results presented in this paper demonstrate that not only is it possible to profile the navigation route and fuel consumption, but also able to perform well. The solution presented in this paper can become a tool on which other methods are applied, such as various pathfinding algorithms and its derivations/customizations.

A harbor tug movement is driven by the type of job and the number of jobs that the crew has on hand. For the harbor tug with a certain size and jobs, there is a rule-based expert system used to plan the job assignment depending on the time and location of the tug and type of job. The solution presented in this paper may be integrated with such a rule-based system to make up a semi-autonomous system. The implementation of the solution presented in this paper in such a system would have alleviated some load from the crew and staff, putting them into the supervisory role rather than the executor role. Finally, the other crews or other non-technical users can also benefit from the knowledge base in the form of a trained neural network, not only for route planning but also for fuel estimation.

Acknowledgement

The Authors wished to acknowledge the resources supported by SMI (R-SMI-A403-0001) and MOE (R-MOE-A403-C002/MOE2018-TIF-1-G-008).

Appendix A

Appendix A explains the Network Configurations of Figure 10. The configuration of the VAE network is as shown in Figure A1 for every layer in Figure 10(a)(b) for both Encoder Network and Decoder Network. Encoder Network accepts greyscale Original Route Image (A) of 308 x 212 pixels. It produces two outputs: Mean Output (μ) and LogVar (σ). The Sample Output (z) is derived from sampling the probability distribution (2) the sampling layer, where ε is the random value of a normal distribution (mean = 0, standard deviation = 1)[27]. The z is necessary to train the Decoder Network (D). This sampling technique is also known as reparameterization trick[28].

$$z = \mu + e^{\sigma/2} \times \varepsilon \tag{2}$$

In Figure A2(a), the configuration of the last 3 layers of Feature Network (E) output aligns with Encoder Network (C) output. It is to be trained with Job & Noon Report Metadata (B), and to be able to predict the latent representation values. The Feature Network (E) accepts 90 input parameters. The first set is the route classification input (Input_Layer_1) with 85 classes (85 route patterns). The second set is 5 job/route metadata (job duration and activity area) as an input to Input_Layer_2. Multiply layer performs element-wise multiplication.

Figure A2(b) is the Prediction Network (F) configuration. The output is one parameter of Fuel Prediction (I) output trained with consumed fuel information from the noon report. The input is a set of probability distribution (μ and σ) from the Feature Network (E). The latent representation is multi-dimensional. The network configuration used in this paper uses 7 dimensions of μ and σ.

Figure 10 also states "relu" in various locations. Relu (or ReLU) refers to rectified linear unit activation function used in the corresponding layers[29]. The function of ReLU is described in (3).

$$f(x) = \max(0, x) \tag{3}$$

Model: "Encoder"

Layer (type)	Output Shape	Param #	Connected to
Encoder_Input (InputLayer)	[(None, 212, 308, 1)]	0	[]
Conv2d_1 (Conv2D)	(None, 106, 154, 8)	520	['Encoder_Input[0][0]']
Conv2d_2 (Conv2D)	(None, 53, 77, 16)	8208	['Conv2d_1[0][0]']
Flatten (Flatten)	(None, 65296)	0	['Conv2d_2[0][0]']
Fully_Connected (Dense)	(None, 16)	1044752	['Flatten[0][0]']
Mean_Output (Dense)	(None, 7)	119	['Fully_Connected[0][0]']
LogVar_Output (Dense)	(None, 7)	119	['Fully_Connected[0][0]']
Sample_Output (Sampling)	(None, 7)	0	['Mean_Output[0][0]', 'LogVar_Output[0][0]']

Total params: 1,053,718
Trainable params: 1,053,718
Non-trainable params: 0

(a)

Model: "Decoder"

Layer (type)	Output Shape	Param #
Decoder_Input (InputLayer)	[(None, 7)]	0
Fully_Connected_1 (Dense)	(None, 16)	128
Fully_Connected_2 (Dense)	(None, 65296)	1110032
Reshape (Reshape)	(None, 53, 77, 16)	0
Conv2DT_1 (Conv2DTranspose)	(None, 106, 154, 16)	16400
Conv2DT_2 (Conv2DTranspose)	(None, 212, 308, 8)	8200
Decoder_Output (Conv2DTranspose)	(None, 212, 308, 1)	73

Total params: 1,134,833
Trainable params: 1,134,833
Non-trainable params: 0

(b)

Figure A1. VAE network parameters.
(a) Encoder, (b) Decoder.

Model: "Feature"

Layer (type)	Output Shape	Param #	Connected to
Input_Layer_1 (InputLayer)	[(None, 90, 1)]	0	[]
Conv1D (Conv1D)	(None, 1, 5)	455	['Input_Layer_1[0][0]']
Flatten (Flatten)	(None, 5)	0	['Conv1D[0][0]']
Input_Layer_2 (InputLayer)	[(None, 5)]	0	[]
Multiply (Multiply)	(None, 5)	0	['Flatten[0][0]', 'Input_Layer_2[0][0]']
Fully_Connected_1 (Dense)	(None, 1024)	6144	['Multiply[0][0]']
Fully_Connected_2a (Dense)	(None, 256)	262400	['Fully_Connected_1[0][0]']
Fully_Connected_2b (Dense)	(None, 256)	262400	['Fully_Connected_1[0][0]']
Mean_Output (Dense)	(None, 7)	1799	['Fully_Connected_2a[0][0]']
LogVar_Output (Dense)	(None, 7)	1799	['Fully_Connected_2b[0][0]']
Sampling (Sampling)	(None, 7)	0	['Mean_Output[0][0]', 'LogVar_Output[0][0]']

Total params: 534,997
Trainable params: 534,997
Non-trainable params: 0

(a)

Model: "Predictor"

Layer (type)	Output Shape	Param #	Connected to
Input_Layer_A (InputLayer)	[(None, 7)]	0	[]
Input_Layer_B (InputLayer)	[(None, 7)]	0	[]
Fully_Connected_1a (Dense)	(None, 128)	1024	['Input_Layer_A[0][0]']
Fully_Connected_1b (Dense)	(None, 128)	1024	['Input_Layer_B[0][0]']
Concatenate (Concatenate)	(None, 256)	0	['Fully_Connected_1a[0][0]', 'Fully_Connected_1b[0][0]']
Fully_Connected_2 (Dense)	(None, 512)	131584	['Concatenate[0][0]']
Fully_Connected_3 (Dense)	(None, 64)	32832	['Fully_Connected_2[0][0]']
Output_Layer (Dense)	(None, 1)	65	['Fully_Connected_3[0][0]']

Total params: 166,529
Trainable params: 166,529
Non-trainable params: 0

(b)

Figure A2. Feature and Prediction/Predictor network parameters.
(a) Feature, (b) Predictor.

References

[1] Ingicco T, van den Bergh G D, Jago-on C, Bahain J-J, Chacón M G, Amano N, Forestier H, King C, Manalo K, Nomade S, Pereira A, Reyes M C, Sémah A-M, Shao Q, Voinchet P, Falguères C, Albers P C H, Lising M, Lyras G, Yurnaldi D, Rochette P, Bautista A and de Vos J 2018 Earliest known hominin activity in the Philippines by 709 thousand years ago *Nature* **557** 233–7

[2] Tijan E, Jović M, Aksentijević S and Pucihar A 2021 Digital transformation in the maritime transport sector *Technological Forecasting and Social Change* **170** 120879

[3] Agarwala P, Chhabra S and Agarwala N 2021 Using digitalisation to achieve decarbonisation in the shipping industry *Journal of International Maritime Safety, Environmental Affairs, and Shipping* **5** 161–74

[4] Rosenbaum A, Hartley S and Holder C 2011 Analysis of diesel particulate matter health risk disparities in selected US harbor areas *Am J Public Health* **101 Suppl 1** S217–23

[5] Jain A K, Mao J and Mohiuddin K M 1996 Artificial neural networks: a tutorial *Computer (Long Beach Calif)* **29** 31–44

[6] Grifoll M, Martorell L, Castells M and de Osés F X M 2018 Ship weather routing using pathfinding algorithms: the case of Barcelona – Palma de Mallorca *Transportation Research Procedia* **33** 299–306

[7] Gu Y, Goez J C, Guajardo M and Wallace S W 2021 Autonomous vessels: state of the art and potential opportunities in logistics *International Transactions in Operational Research* **28** 1706–39

[8] Foead D, Ghifari A, Kusuma M B, Hanafiah N and Gunawan E 2021 A Systematic Literature Review of A* Pathfinding *Procedia Computer Science* **179** 507–14

[9] Shu-Xi W 2012 The Improved Dijkstra's Shortest Path Algorithm and Its Application *Procedia Engineering* **29** 1186–90

[10] Patle B K, Babu L G, Pandey A, Parhi D R K and Jagadeesh A 2019 A review: On path planning strategies for navigation of mobile robot *Defence Technology* **15** 582–606

[11] Bialystocki N and Konovessis D 2016 On the estimation of ship's fuel consumption and speed curve: A statistical approach *Journal of Ocean Engineering and Science* **1** 157–66

[12] Tay Z Y, Hadi J, Konovessis D, Loh D J, Tan D K H and Chen X 2021 Efficient Harbor Craft Monitoring System: Time-Series Data Analytics and Machine Learning Tools to Achieve Fuel Efficiency by Operational Scoring System *Volume 6: Ocean Engineering* (American Society of Mechanical Engineers)

[13] Tay Z Y, Hadi J, Chow F, Loh D J and Konovessis D 2021 Big Data Analytics and Machine Learning of Harbour Craft Vessels to Achieve Fuel Efficiency: A Review *Journal of Marine Science and Engineering* **9**

[14] Sarker I H 2021 Machine Learning: Algorithms, Real-World Applications and Research Directions *SN Computer Science* **2** 160

[15] Giorgino T 2009 Computing and Visualizing Dynamic Time Warping Alignments in R: The dtw Package *Journal of Statistical Software* **31** 1–24

[16] Weyenberg G and Yoshida R 2015 Chapter 12 - Reconstructing the Phylogeny: Computational Methods *Algebraic and Discrete Mathematical Methods for Modern Biology* ed R S Robeva (Boston: Academic Press) pp 293–319

[17] Davidson I and Ravi S S 2005 Agglomerative Hierarchical Clustering with Constraints: Theoretical and Empirical Results *Knowledge Discovery in Databases: PKDD 2005* ed A M Jorge, L Torgo, P Brazdil, R Camacho and J Gama (Berlin, Heidelberg: Springer Berlin Heidelberg) pp 59–70

[18] Harris S L and Harris D M 2016 3 - Sequential Logic Design *Digital Design and Computer Architecture* ed S L Harris and D M Harris (Boston: Morgan Kaufmann) pp 108–71

[19] Kingma D P and Welling M 2014 Auto-Encoding Variational Bayes

[20] Abadi M, Agarwal A, Barham P, Brevdo E, Chen Z, Citro C, Corrado G S, Davis A, Dean J, Devin M, Ghemawat S, Goodfellow I J, Harp A, Irving G, Isard M, Jia Y, Józefowicz R, Kaiser L, Kudlur M, Levenberg J, Mané D, Monga R, Moore S, Murray D G, Olah C, Schuster M, Shlens J, Steiner B, Sutskever I, Talwar K, Tucker P A, Vanhoucke V, Vasudevan V, Viégas F B, Vinyals O, Warden P, Wattenberg M, Wicke M, Yu Y and Zheng X 2016 TensorFlow: Large-Scale Machine Learning on Heterogeneous Distributed Systems *CoRR* **abs/1603.04467**

[21] Bengio Y and Lecun Y 1997 Convolutional Networks for Images, Speech, and Time-Series

[22] Sainath T N, Vinyals O, Senior A and Sak H 2015 Convolutional, Long Short-Term Memory, fully connected Deep Neural Networks *2015 IEEE International Conference on Acoustics, Speech and Signal Processing (ICASSP)* pp 4580–4

[23] Asperti A and Trentin M 2020 Balancing reconstruction error and Kullback-Leibler divergence in Variational Autoencoders *CoRR* **abs/2002.07514**

[24] Ramos D, Franco-Pedroso J, Lozano-Diez A and Gonzalez-Rodriguez J 2018 Deconstructing Cross-Entropy for Probabilistic Binary Classifiers *Entropy* **20**

[25] Chai T and Draxler R 2014 Root mean square error (RMSE) or mean absolute error (MAE)? *Geosci. Model Dev.* **7**

[26] Zhou Z, Shin J, Zhang L, Gurudu S, Gotway M and Liang J 2017 Fine-tuning Convolutional Neural Networks for Biomedical Image Analysis: Actively and Incrementally *Proc IEEE Comput Soc Conf Comput Vis Pattern Recognit* **2017** 4761–72

[27] Rezende D J, Mohamed S and Wierstra D 2014 Stochastic Backpropagation and Approximate Inference in Deep Generative Models

[28] Kingma D P, Salimans T and Welling M 2015 Variational Dropout and the Local Reparameterization Trick

[29] Xu B, Wang N, Chen T and Li M 2015 Empirical Evaluation of Rectified Activations in Convolutional Network

Containerized immutable maritime data sharing utilizing Distributed Ledger Technologies

Knut Erik Knutsen[1], Qin Liang[1], Nikita Karandikar[1], Imran Halimi Bin Ibrahim[2], Xiao Gang Tony Tong[2], Jacob Jun Hui Tam[2]

[1]DNV Norway, Veritasveien 1, 1363 Høvik, Norway
[2]DNV Singapore, 16 Science Park Dr, Singapore 118227

E-mail: knut.erik.knutsen@dnv.com

Abstract. In recent years, distributed ledger technologies (DLTs), especially blockchain technology, has become a popular topic and is being implemented in different industries. Ships are expected to be equipped with more advanced equipment and less crew onboard. The communication and information sharing between different maritime stakeholders such as business partners, port authorities, tax authorities, class societies, flag states, IMO and EU become more important than ever. Trust and authenticity are key properties when utilizing data for safety or business critical applications, and will only increase in importance as automation levels increase towards fully autonomous ships and trading systems. The implementation of DLTs as close as possible to the data source helps provide a reliable, trustworthy way to share, process and verify the data. The objective of this paper is to explore how tamperfree maritime data sharing can enable automated transactions between vessels and shore-based entities. Singapores TradeTrust platform has shown that digitalization of international trade has great potential. In this paper we explore how further enhancements may be implemented and how the distributed ledger VeChain, as compared to Ethereum, have some distinct advantages that may be beneficial for continued automation and efficiency gains. The electronic Bill of Lading was taken as an example to demonstrate the concept. Reporting requirements from the other maritime data sharing scenarios are also expected to require a way of ensuring tamperfree data.

1. Introduction

With the development of digital technologies, e.g., Internet of Things, edge computing devices, data and sensors, different industries start to transform their business with data-driven services. Maritime industry is one of the industries that benefits the most from this transformation. On the shore side, many countries are developing and applying smart port technologies to improve the efficiency and automation of port operation. For the offshore side, ships are expected to be equipped with more advanced equipment and sensors. From it, more operation and monitoring data becomes available. In addition, low earth orbit satellites and 5G networks are expected to significantly improve ship to shore connectivity. This makes data sharing between ships and ports possible and reliable. Currently, many maritime services require data sharing between ships and ports/shore. For example, Bill of Lading process, Ship Registration Process (SRP), port clearance and bunkering. Plenty of manual work, frequent email communications and complex identity and access control bring higher operation costs, slow processing speed, manual mistakes, and potential risk of fraud. The Bill of Lading is an important binding document between shipper and carrier that provides the details (e.g., cargo type, quantity and destination) needed to process the freight shipment and invoice it correctly. For many years, the bill of lading process involve paper communication between carriers, sellers, traders, banks, authorities and buyers which makes the process complex and inefficient. Port clearance is a time-intensive process which requires the stakeholders to make sure all the statutory and mandatory certificates are valid and send to the port authority to apply the Port Clearance Certificate (PCC) within a specific time window. From these two examples, it is obvious that the maritime data sharing is becoming increasingly important. The data sharing can be more challenging with a variety of influencing factors in practical maritime scenarios. If this can be optimized to a standardized and streamlined architecture with more automatization and security, the efficiency and trustworthiness can be significantly improved.

In recent years, Distributed Ledger Technologies (DLTs) are becoming more widely used and are being implemented in different industries. A central feature of DLTs is that the ledger is copied to the actors involved in a network and based on a consensus algorithm collaboratively provide trust in the network as a whole, rejecting bad actors within the network and providing a more resilient surface to attempts at manipulating the ledger. This greatly improves the tamperproofing capability of data. On the other hand, limitations of popular DLTs are becoming increasingly apparent; for example, with increasing size of blockchain networks it takes longer for individual transactions to be processed, and over time as the cryptographic puzzle required to create the next block becomes more difficult, competition between miners results in relatively high energy usage per transaction [1], which pulls into question the sustainability of this technology. Presently, novel consensus algorithms are or have been developed, aiming at reducing the consumption of resources and driving down cost. In addition, there are alternatives to the Blockchain technology which aims to reduce the energy usage and cost of DLTs significantly. In this paper, we argue that VeChain, as compared with Ethereum, has some benefits at least for certain applications, in particular electronic Bill of Lading and potential extensions to further automate international trade has been explored in the following. By comparing the working principles, performance and application scenarios, corresponding recommendations will be proposed. Future maritime applications can use this paper for reference.

In this paper, a conceptual implementation of an infrastructure based on DLTs and Docker for maritime data sharing will be introduced. The Bill of lading process is taken as the conceptual example for illustration. As of today, different maritime stakeholders collecting, storing and analyzing data needs to install and maintain proprietary hardware and software for this purpose, and each needs to arrange for access to vessel communications systems for ship-to-shore data transfer. Such solutions result in high complexity and low reliability and resource efficiency. The utilization of Docker containers and orchestration as part of a hosting environment to implement lifecycle management of DLT-assured Docker applications will contribute to the standardization of data sharing in maritime industries. If a standardized trusted infrastructure can be established, maritime stakeholders, e.g. ship owners and port authorities, can develop and deploy their services easily.

2. Docker based data sharing infrastructure and the Unified Hosting Service

The data sharing infrastructure that serves as our experimental platform to establish tamperfree maritime data sharing consists of a Docker orchestrator for docker deployment, MQTT serving as an event bus, and a full fledge Connectivity platform using 4G Mobile Broad Band (MBB) and VSAT connectivity to shore in combination, managed by a communications broker. The platform, called Unified Hosting Service, is delivered by Telenor Maritime, and may include a Wifi service for providing connectivity to people and equipment on the vessel. Configuration has been done with security in mind, full segregation of networks, firewall implementation on Layer 2-7 in the OSI 7 Layer model. All traffic is handled by a carrier aggregation, splitting traffic down and sending the data on different paths to shore. This is collected on a shore site, then reassembled and sent to their respective customers. The system is designed for redundancy wrt. storage and analysis as well as connectivity, but may be deployed also as a single server and modem setup.

A challenge arising when sharing data in the maritime industry is the lack of standard protocols and formats. Creating standardized exchange formats such as Maritime Context [2] and ISO19848 [3] compliant data and establishing appropriate APIs would allow microservices in the form of Docker containers to easily exchange data and thus use multiple data sources for a particular application. The UHS platform includes two APIs supporting the exchange formats above, and will be tested going forward. Docker is an open-source technology and a container file format for developing, shipping, and running applications [4] Docker allows users to separate applications in the infrastructure to form smaller particles (microservices), thereby increasing the speed of software delivery. The running environment of Docker images are standardized and can be deployed across different platforms. While Docker provides an open standard for packaging and distributing containerized applications, the potential complexities can accumulate fast. This makes it difficult to coordinate and schedule when many containers are running. The introduction of Docker Orchestration is to solve this problem. Docker Orchestration is a series of technologies and practices for the life cycle management of Docker containers at large scale. As applications are being designed as microservices with many containers running together, the orchestration process becomes necessary and important. Currently, there are two main technologies for Docker Orchestration: Kubernetes and Docker Swarm.

The MQTT broker will be used by Docker containers for publish/subscribe of sensor data as well as analysed/aggregated data in order to make it simple for e.g. data analysts to deploy SW to the platform and be provided access to the required data from the data owner. Crucially for the tamperproofing concept, the MQTT broker will also be used to share data hashes and ledger IDs for a DNV Microservice which publishes hashes to a Distributed Ledger (e.g. Blockchain), as seen illustrated in Figure 1.

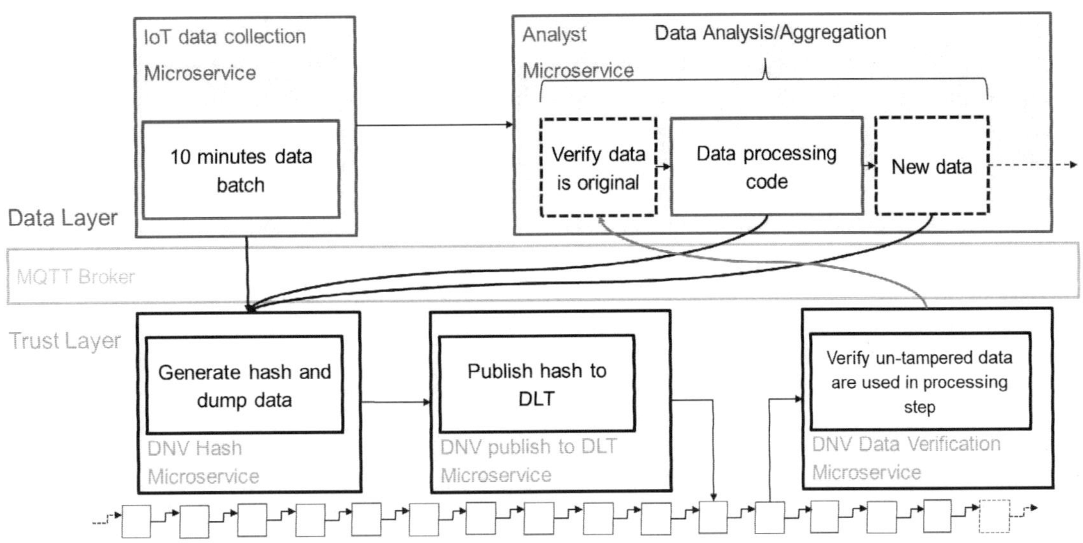

Figure 1. Conceptual illustration of tamperproofing methodology under development. By storing hashes of newly collected raw data as well as code or SW used for analysis/aggregation and any new data sets created, it is possible backtrace all changes to the data and to document the full data lineage. Any future consumers of raw or analyzed data can themselves verify that data remains tamperfree.

These hashes are publicly available in the DLT and can be used by data consumers to verify that data shared with them are original and unchanged from when they were collected, i.e. tamperfree, by hashing the same data again and comparing with the hash in the DLT. In the next sections we present a case study, and a proposed way for how to use the methodology proposed herein in the particular case.

Note that while our experimental platform is the Unified Hosting Service, the DNV tamperproofing methodology described herein is designed for deployment on any Docker based infrastructure, either directly or by some customization.

3. Case study: electronic Bill of Lading and Singapore MPA TradeTrust solution

The potential applications of blockchain in the maritime sector can be better understood with a concrete case study of one of the key players in this space. As the top bunkering port, including a growing LNG bunkering business, with the second highest container throughput in the world, Singapore is an important maritime state [5]. Their focus on and legislative support for digital transformation for the martime sector has led them to be ranked first in 2022 edition of the Leading Maritime Cities published by DNV [6]. International trade stands to gain significantly from moving from paper-based transaction to digital transaction, as emphasized by a statement at the April 2021 G7 meeting in London: "Paper-based transactions, which still dominate international trade, are a source of cost, delay, inefficiency, fraud, error and environmental impact". Singapore addresses this through the Electronic Transactions Act (ETA) amendments which allows electronic Bills of Lading (eBLs) to be legally equivalent to the traditional paper-based versions. Singapore can thus be regarded as a suitable focus for this study.

Singapore has been one of the leading nations embracing new technologies in the region as well as in the world. International trade forms the backbone of Singapore's open and trade-friendly economy. Singapore's institutions, infrastructure, policies and services, are designed to encourage the free flow of goods, to ensure its status as a major trading hub. With the opportunities provided by new digital technologies, Singapore aims to work with various countries to accelerate digital innovation in global trade.

The Maritime and Port Authority of Singapore (MPA) has been working with various agencies, industry associations and partners, both local and overseas, to accelerate adoption of electronic Bills of Lading (eBLs) for cross-border trade, including the Infocomm Media Development Authority of Singapore (IMDA), Singapore Customs, Singapore Shipping Association and Digital Container Shipping Association (DCSA). Conventionally, due to the nature of a Bill of Lading (BL) as a document of title, the original copy(s) of BL is sent around, exchanging hands from carrier to shipper, to financier (as collateral), traded by third parties before finally reaching the buyer, where cargo is released after payment. The processes are inefficient as cargo collection may be delayed with additional cargo holding costs if the BL arrives later than the shipment [7]. Apart from that, manual processing of BLs poses the risks of human errors and fraud with potential financial losses. In addition, about 20% of the total shipping costs are generally incurred by documentation processing [8]. Looking at the current market, only 0.1% of BLs are issued electronically today in limited or closed systems. It was estimated that about US$4bn savings can be made annually if 50% of today's container shipping lines adopt eBLs [9]. However, there were challenges to the adoption of eBL prior to TradeTrust's existence, including there were lack of legal & insurance certainty (e.g. not recognised by banks for trade financing), lack of interoperability as well as differences in documentation format.

To overcome the pain points of conventional BL as well as the challenges of eBLs, Singapore has amended her ElectronicTransactions Act (ETA) and adopted the Model Law on Electronic Transferable Records (MLETR) by the United To To overcome the pain points of conventional BL as well as the challenges of eBLs, Singapore has amended her ElectronicTransactions Act (ETA) and adopted the Model Law on Electronic Transferable Records (MLETR) by the United

Nations Commission on International Trade Law (UNCITRAL) to provide legal certainty to transactions involving electronic transferable title documents/instruments [10]. Besides, the IMDA has launched TradeTrust, a digital utility comprising a set of globally accepted standards and framework to enable trusted interoperability of electronic documents, including eBLs that meets the requirements of the MLETR.

First announced in 2019, TradeTrust, is spearheaded by the Infocomm Media Development Authority (IMDA) with the Maritime and Port Authority of Singapore, and is supported by Enterprise Singapore [11]. Advocates include the International Chamber of Commerce (ICC), the world's biggest business organization; the government of Singapore; and 17 major companies, including DBS Bank, Mastercard, Mitsubishi, Sumitomo, and others [12]. TradeTrust aims to reduce inefficiencies and complexities of cross-border trade arising from the current paper-based documentation, such as the Bill of Lading. This lowers operating costs for businesses and the risk of fraud while accelerating the digitalisation of cross-border trade processes thereby promoting more growth in trade. TradeTrust design principles include

(i) Public and Permissionless, in which no central governance authority
(ii) Data off-chain where this preserves data confidentiality
(iii) Payload agnostic where there is no data format or standard restriction
(iv) Open-source, enabling full transparency for faster adoption
(v) MLETR-Compliant, where it meets the requirements of the law.

From the technical perspective, TradeTrust technology stack comprises four different layers [13] (from bottom to top), namely:

(i) Legal harmonisation & standards development
(ii) Blockchain layer
(iii) Services layer
(iv) Application layer

The first layer enables legal recognition of cross-border digital trade documents and document exchange through standards. The blockchain layer enables blockchain network participation with certain requirements. Ethereum, Hyperledger, R3, Corda are the examples of networks that are supported in TradeTrust. The services layer consists of services that support the commercial applications and connection to other technologies, for instances, document verification, smart contracts, etc. The application layer enables commercial implementations of TradeTrust encompassing commercial applications, platforms and ecosystems.

In the trading context, a negotiable instrument is a signed document that promises a sum of payment to a specified person or the assignee. Such documents have been used extensively in international trade. The BL in maritime trade is one of the examples. The working concept of eBL in TradeTrade starts with the transport of electronic documents containing BL data from one party to another via platform, email, portal, USB, etc., followed by the reading and updating of Title Ownership by respective parties via the distributed ledger, which is the public blockchain (TradeTrust network). TradeTrust does also offer the flexibility in supporting situation where some participants, such as government authority or custom, do not require the change of title, but only need to read and verify the BL data. In 2021, MPA announced the successful completion of an eBL trial which demonstrated the interoperability between different digital trade platforms. In January 2021, Singapore and the Netherlands completed a shipment where an eBL "shadowed" a live shipment from Qui Nhon, Vietnam to Rotterdam, the Netherlands via transhipment in Singapore. The trial showed that there were significant time savings in the BL documentation process from an average of six to ten days when using a hardcopy to less than 24 hours when using an eBL. This successful trial also marks one of the first title transfer, across different digital platforms, which was facilitated by TradeTrust, an open-standard digital utility to enable interoperability [10].

Utilizing the Docker based infrastructure described in Section 2 can be integrated with platforms such as TradeTrust and opens for richer information flow between stakeholders including both trade documents between seller, transporter and buyer, as well as technical, regulatory and safety data, and as such represent an enabling technology to deploy the applications needed on a vessel to exchange any type of data and documents between the vessel and shore by combining on-chain and off-chain data to ensure immutability as well as privacy. The infrastructure is intended to be available for any stakeholder the shipowner provides access to for various analysis and documentation purposes. This includes the possibility of utilizing sensor data generated on the ship, such as container temperature and humidity, as well as trade documents which both may require immutability and privacy. The next section describes how our proposed DLT enabled on-board data infrastructure could be integrated with the current TradeTrust solution and how VeChain features specifically may improve certain properties of the solution as compared to Ethereum.

4. Conceptual implementation

Digitalization of the maritime value chain has the potential to improve efficiency and offer a host of services to the stakeholders. At the most basic level, this relates to sensing devices on machinery that monitor that state of the key components and environmental conditions, as well as actuators that perform a task in response to an input signal. These devices, if IoT enabled can allow the machines to communicate with each other and collaboratively offer services, allowing us to think of them as service providers rather than discrete objects.

Servitization of assets, as has been explored in the MIDAS project [14], refers to a transition by Original Equipment Manufacturers (OEM) from a product-based to an advanced service-based portfolio. This transition hinges on the basic idea that OEMs, rather than the shipowner, are better placed to monitor the health and incrementally improve the assets that they have designed. Under this paradigm, the OEMs would sell uptime instead of physical hardware to the shipowner. Further, machine to machine (M2M) communication can give rise to a scenario where machines owned by different stakeholders offer services to each other and conduct micropayment transactions in a machine to machine economy.

A M2M economy as delineated in Figure 2 shows a layered architecture with services from the lower layer flowing upwards to a higher layer, while payments flow from the top down. Thus, the data produced by the sensors is sold to the machine, which sells uptime or work to the vessel, which in turn abstracts away the lower services and sells the transport service, to the cargo owner. This chain produces valuable data about the services and the payments, in addition to the actual monetary transaction. This data is not only required for bookkeeping, but can be used to enable additional services.

However, the possibility of undetected tampering of the data can severely undermine trust in the data and services relying on it. Distributed ledger technologies (DLT), commonly known as Blockchain, can be used to bring verifiable trust to the data exchange value chain and is therefore an important component of the TradeTrust framework. TradeTrust [13] is a framework for interoperability of electronic trade documents exchanged between transacting parties and can be considered as the state of the art in this problem space.

In order to trust the electronic trade documents presented by a party, it must be possible to trace the provenance and origin of the document as well as verify that they have not been tampered. The TradeTrust framework uses the Ethereum blockchain to record issuance of certificates as well as register the ownership of transferable documents such as the bill of lading, that provide value to its holder. We note here that, Ethereum has released plans to upgrade to

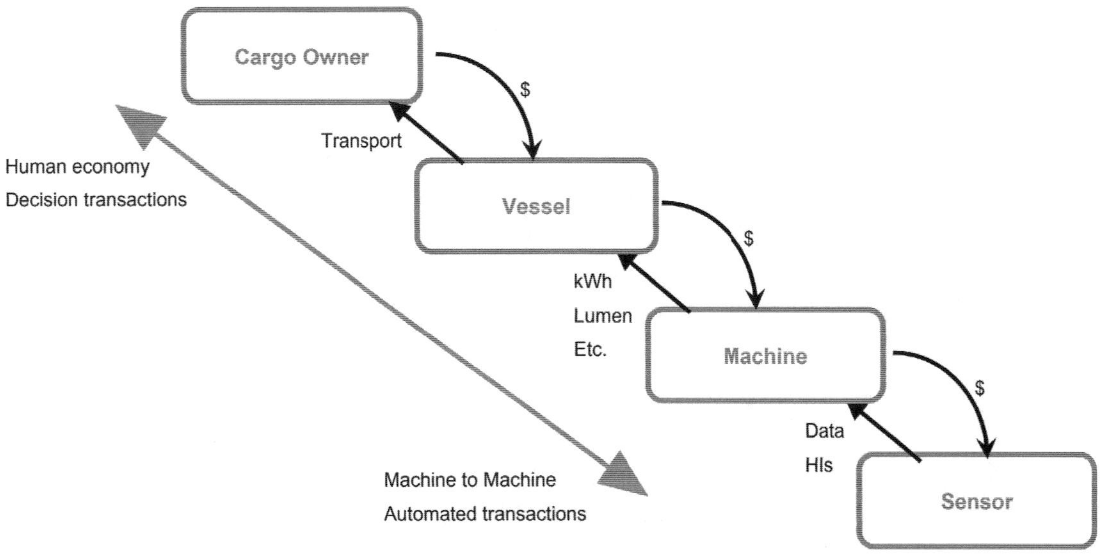

Figure 2. M2M Economy with flow of services flowing up from sensors to the cargo owner and payments flowing down the value chain.

a Proof of Stake based consensus in 2022 [15]. As the specifics of this upgrade are not clear, we do not speculate on whether this change could address some of the limitations discussed in this study.

Ethereum [16] is a public blockchain that allows anyone to propose transactions and create smart contracts that execute when predefined conditions are met, thus enabling decentralized applications. As this network is decentralized, a mechanism is needed to achieve consensus among the member nodes about the state of the ledger. Ethereum uses a consensus mechanism called Proof of Work in which, nodes called miners compete to verify transactions and add blocks to the ledger and earn a miner fee in 'ethers' (ETH), the cryptocurrency that drives the Ethereum network. Proof of Work is a computationally intensive brute force mechanism that expends tremendous amounts of electricity and requires specialized equipment and thus has a high operating cost. Users are charged fees for transaction processing and not only are these fees quite high, but as they are tied to the valuation of ETH, a publicly traded cryptocurrency, extremely prone to fluctuations [17]. A high and unstable transaction cost, that may appear to be a minor inconvenience at the proof-of-concept stages could limit the utility of a system relying on it for businesses having to contend with a high degree of uncertainty of managing their daily business operations. Separating the speculative publicly traded asset from the cost of running transactions on the blockchain would reduce the cost for practitioner businesses and promote the adoption of TradeTrust.

Such a two token system is an essential building block for VeChain, a public blockchain purpose built for enterprise solutions. The VeChain Token (VET), acts as a store and transfer of value, while the VeThor (VTHO) is used to pay for transactions on the VeChain. For comparison, the average transaction cost on VeChain was 1.10 VTHO which converts to \$0.0065 while the average Ethereum transaction of the same size was 0.00012 ETH which is around \$0.27. This difference in price is further put into context by considering use cases beyond BL ownership records. Consider the case of regular hashing of sensor data as highlighted in section 2. If we assume that a transaction is sent from a vessel every 10 minutes that would be 6 transactions per hour. For a system implemented in Ethereum, the cost would be around \$15061 per year, a much higher cost compared to approximately \$358 per year in case of a VeChain based system.

As the cost of VTHO is correlated with the utilization of blockchain resources, the cost is to a large extent predictable with an analysis of VTHO supply and demand for which data is readily available. Additionally, the VeChain foundation has measures in place to further stabilize the cost.

(i) Any VET holder address generates VTHO at a fixed velocity of $5 * 10^{-8}$ per VET per block, thus controlling the supply of VTHO in the system

(ii) Of the VTHO paid in each transaction, 70% is destroyed and the rest is awarded to authority master nodes, the consortium responsible for processing blocks

(iii) Moreover, when needed, the parameters such as price and velocity of the VTHO are adjusted to maintain the equilibrium of supply and demand. Such adjustments are carried out by a decentralized consensus

A predictable and lower cost for transaction processing could open several other avenues for application of the TradeTrust framework. The trust and traceability advantages offered by a blockchain based solution for a document such as the BL can be considered a starting point. With the possibility of frequent and cost effective ledger updates, more granular trust and traceability in the supply chain can be offered. VeChain offers native support for using RFID/NFC chips or QR codes to tag high value articles and tracking them with a unique nonreplicable VeChain ID. Pilot studies have already shown the efficacy of this approach for anti-counterfeiting [18] in retail value chains. A similar approach could be considered for tracking shipments between buyers and sellers where the QR code (or RFID chip) is scanned at predetermined intervals through the journey from point to point.

In the current TradeTrust framework, payments based on the electronic trade documents occur off chain, So, keeping with our example of eBL, any financial transactions taking place between the buyer and seller or shipper and receiver would be off chain as the solution registers and tracks the ownership of the document itself. However, the framework, if extended, can help converge physical, financial and document chains, thus making it possible to automate processes spanning all these chains by means of self enforcing smart contracts. Figure 3 shows how VeChain can enable an end to end solution for asset and value exchange between two parties with self enforcing smart contracts. Including the exchange of funds will allow us to capture more aspects of the trading relationship in an immutable digitized solution, flattening the value chain by reducing the number of actors and moving closer to machine to machine economy with micropayments.

The immutability of transactions is one of the biggest benefits of blockchain technology. However, due to the decentralized and asynchronous nature of the technology, implementations have to be wary of ledger forks which can call into question the immutability. Ledger Forks occur when different participants in a network have a different accepted order to transactions. The possibility of ledger forks is tied very closely with the consensus mechanism used by member nodes to agree on the status of the ledger. Ethereum relies on a probabilistic consensus algorithm that eventually achieve consistency to a high degree of probability. However, as past forks have shown, these algorithms are susceptible to divergent ledgers or forks. VeChain, uses a Proof of Authority based consensus which is based on an algorithm called SURFACE [19], an acronym for Secure, Use-case adaptive, and Relatively Fork-free Approach of Chain Extension. The name highlights the key features of SURFACE and its focus on relatively fork free ledgers for practical enterprise scale networks beyond proof of concept stages. The salient aspect of this is that instead of a miner, a committee would validate the proposed transaction blocks. The committee member chosen to process a given block is selected with a verifiable random function and other committee members selected in a pseudo random fashion serve as endorsers. Thus, for each block, a different set of committee members would be involved. This reduces the likelihood of forks as roles of committee members during a block creation stipulate their responsibilities.

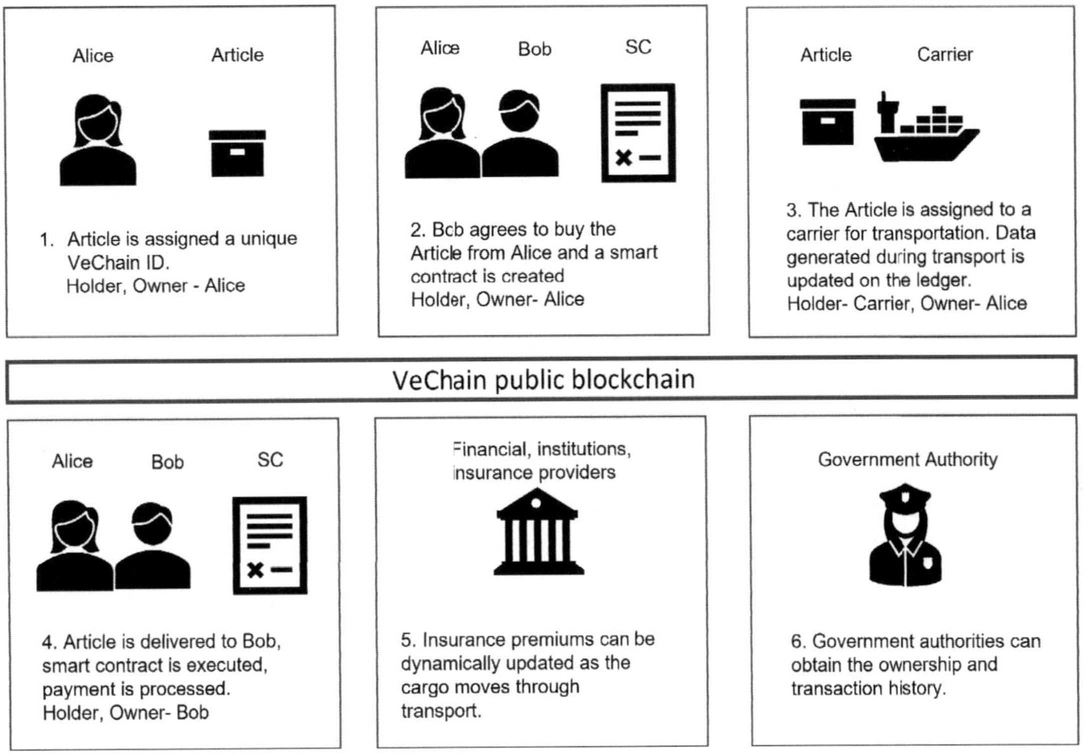

Figure 3. End to end value and asset exchange scenario

The transaction issuer and endorser committee are chosen with a pseudo random algorithm, so attacking the chain would require collusion among a high proportion of consortium nodes. Thus, VeChain reduces the attack surfaces by preventatively sacrificing some degree of decentralization for increased security.

TradeTrust uses the OpenAttestation, an open-source framework for tamper evident documents by providing methods to format data before hashing and updating to the ledger. An important component of this mechanism is the reliance on DNS addresses to identify issuers of documents. When viewing a document issued under this framework, a user will see text similar to "Document issued by `example.openattestation.com`". It is important to keep in mind that OpenAttestation does not endorse any issuers, but only verifies that the issuing party of the document has provided some sort of proof that it is the same party as claimed - for example, proving ownership over a domain by the ability to create a DNS record. However, for business-critical operations that require a higher level of identity assurance, additional guardrails may be necessary. VeChain Authority Master nodes on the VeChain work by tying the digital identity to an established real-world identity thus adding in reputational safeguards in the identity management process.

As miners in Ethereum can choose whether to process a given transaction, there is a possibility that a given proposed transaction may be ignored by the miners. In order to circumvent this, an additional tip can be added to the transaction to incentivize miners to process it, which adds to the costs but does not guarantee transaction processing. Thus, if a transaction is not chosen for processing the issues must simply wait. This is addressed in VeChain by including an expiration field in the transaction. The issuer can set the transaction to expire if not chosen, reassuring them that their transaction will never become stuck in the system or cause problems if an unscrupulous party tries to replay it. Moreover, with the rising popularity of VeChain and adoption into various supply chain ecosystems, the need for separate systems of record such as private blockchains is reduced, thus reducing complexity in the value chain. Additionally, if the interface to the blockchain is containerized as we propose, this can also contribute to the interoperatibility of the framework. If different transacting parties maintain separate systems of record in databases or private blockchains, they could still consider this framework for transactions requiring higher visibility beyond the private blockchain implementation. For instance, for an international shipment, there may be many stakeholders such as business partners, port authorities, tax authorities, class societies and flag states among others. Transactions such as these require a visibility spanning the value chain.

One of the key drivers of Ethereum is complete decentralization. However, a modern maritime setting is not really decentralized, but rather a consortium of authorities such as classification societies, port states, financial institutions, insurance providers. Thus, a consortium of world leaders in some of these arenas would represent the network better than a completely decentralized solution with unknown actors. While the integration of blockchain technology in the maritime sector is gaining traction, it has to contend with several barriers. Zhou et al. [20] identified cost of adoption, complexity of the technology and risk aversion to an untested new technology as the major barriers. VeChain has a lower transaction cost compared to Ethereum and a less complex structure as compared to directed acyclic graph ledgers such as IOTA. Pilot projects conducted with stakeholders have shown its suitability for a wide variety of use cases in supply chain management, food safety and digital assurance [21].

5. Conclusions and further research

We have shown how an on-vessel microservices infrastructure in combination with a DLT-based tamperproofing methodology can enable highly automated transactions of trusted data and documentation. Further we have argued that VeChain properties such as pseudo-deterministic consensus and proof of authority as well as the dual token solution makes VeChain particularly suited for several applications. Using this on-vessel docker based infrastructure together with the tamperproofing methodology is moving us closer to a machine-to-machine economy as will become increasingly necessary for autonomous shipping to become a reality. Although the above describes how the solution may fit into the cutting edge TradeTrust scheme specifically and possibly help expand on the functionality, we believe the same methodology can generalize to other applications such as automating the monetary transactions by moving also these to the public blockchain and strengthen the digital underpinnings of international trade. Another possible application may be in regards to alternative fuels such as LNG and ammonia bunkering. Today, DNV offers a service called FuelBoss [22] which is a fully digital collaborative tool for LNG fuel suppliers and ship owners. If fuel orders and bunker delivery notes could be moved on-chain, it would open the possibility to also move the monetary transactions on-chain and thus increase automation of bunkering procedures as well as enabling increased trust when fuel consumed is required for emissions reporting according to the MRV/DCS requirements. Similarly, for emissions related to the upstream production of fuel being bunkered can be linked to this scheme and taken into account for the reported emissions. Considering a likely future when reported ship emissions are subject to taxation, the increased trust and possibility of automating tax payments may be attractive with regards to both transparency as well as efficiency of a global emissions tax procedure.

Acknowledgements

This work was funded by the Research Council of Norway through the VidaMeco project and has been done in collaboration with Telenor Maritime. We also acknowledge the work done by DNV summerstudents Linnea F. Gustavsen and Ziyuan Zhang during the summer of 2021.

References

[1] Digiconomist, retrieved Jan 2022 from:
`https://digiconomist.net/bitcoin-energy-consumption`
[2] Maritime Context platform, retrieved Jan 2022 from:
`https://www.maritimecontext.org/about/`
[3] ISO 19848 Standard data for shipboard machinery and equipment, retrieved Jan 2022 from:
`https://www.iso.org/standard/66406.html`
[4] Azure Documentation, retrieved Jan 2022 from:
`https://azure.microsoft.com/en-us/topic/kubernetes-vs-docker/`
[5] MPA n.d., retrieved Jan 2022 from:
`https://www.mpa.gov.sg/web/portal/home/maritime-singapore/introduction-to-maritime-singapore/facts-and-trivia`
[6] DNV 2022, retrieved Jan 2022 from:
`https://www.dnv.com/news/leading-maritime-cities-report-2022-singapore-still-in-top-spot-followed-by-rotterdam-and-london-215099`
[7] MPA 2021, retrieved Jan 2022 from:
`https://www.mpa.gov.sg/web/wcm/connect/www/a5c13e4f-0f57-488f-b895-972ff1aa5ee6/Call-for-Proposal+for+eBL+Solutions+%2816+April%29.pdf?MOD=AJPERES&id=1621581460196`
[8] Biazetti, A. 2019. retrieved Jan 2022 from:
`https://www.tradelens.com/post/what-does-a-digitized-supply-chain-look-like`
[9] MPA 2021, retrieved Jan 2022 from:
`https://www.mpa.gov.sg/web/wcm/connect/www/017ed33a-d72d-473e-8339-5415ecd13d4c/Briefing+Slides+and+FAQs+%2820+May%29.pdf?MOD=AJPERES&id=1621582602869`
[10] MPA 2021, retrieved Jan 2022 from:
`https://www.mpa.gov.sg/web/portal/home/media-centre/news-releases/detail/0089937b-6453-446b-93c0-d660172d3d3b`
[11] Abdullah, A. Z. 2020. retrieved Jan 2022 from:
`https://www.channelnewsasia.com/singapore/singapore-agreement-icc-digital-trade-commerce-wef-786201`
[12] Takahashi, D. 2020. retrieved Jan 2022 from:
`https://venturebeat.com/2020/01/22/tradetrust-singapore-and-16-companies-back-perlin-blockchain-for-global-trade-platform/`
[13] TradeTrust, n.d. retrieved Jan 2022 from:
`https://www.tradetrust.io/static/images/webinar/tradetrust-tech-webinar-1-overview.pdf`
[14] MIDAS n.d., retrieved Jan 2022 from:
`https://www.dnv.com/research/review2019/featured-projects/modularity-in-design-for-asset-servitization.html`
[15] Ethereum n.d., retrieved Jan 2022 from:
`https://ethereum.org/en/upgrades/merge/`
[16] Buterin, 2014. A next-generation smart contract and decentralized application platform. white paper, 3(37).
[17] Donmez, 2022. Donmez, Anil, and Alexander Karaivanov. "Transaction fee economics in the Ethereum blockchain." Economic Inquiry 60.1 (2022): 265-29
[18] VeChain n.d., retrieved Jan 2022 from:
`https://www.vechain.com/solution/retail`
[19] Ren, 2020 Ren, Z. and Zhou, Z., 2020. SURFACE: A Practical Blockchain Consensus Algorithm for Real-World Networks. arXiv preprint arXiv:2002.07565.
[20] Zhou, 2020. Zhou, Y., Soh, Y.S., Loh, H.S. and Yuen, K.F., 2020. The key challenges and critical success factors of blockchain implementation: Policy implications for Singapore's maritime industry. Marine Policy, 122, p.104265.
[21] MyStory, retrieved Jan 2022 from:
`https://www.dnv.com/mystory/index.html`
[22] DNV, retrieved Jan 2022 from:
`https://www.dnv.com/maritime/fuelboss/`

Addressing digitalization though out a prototyping framework for agile smart services development: the case of Livorno Port

A Tardo[1], P Pagano[1], S Antonelli[1], S Rao[2]

[1]National Inter-University Consortium for Telecommunications, St. Moruzzi n. 1, Research Area CNR, 56124, Pisa, Italy
[2]BMT Singapore Pte Ltd, 456 Alexandra Road 15-01 Fragrance Empire Building, 119962, Singapore

alexandr.tardo@cnit.it, paolo.pagano@cnit.it, silvia.antonelli@cnit.it, shivaprakash.rao@bmtglobal.com

Abstract. One of the crucial challenges of maritime transport is digitalization through the implementation of proper IT solutions. Development and deployment of such solutions allows seaports to increase their overall efficiency, to lower costs and improve performance times of the intra-terminal operations, to improve information flow and decision-making and to decrease the environmental footprint by addressing the sustainability. Nevertheless, the digitalization levels are heterogeneous and depend on different factors, such as seaport size, cargo volumes and type, stakeholders' needs, business models and investments. We propose a prototyping framework for agile smart service development relying on a set of standard and open-source technologies. The proposed framework is validated at The Port Network Authority of the Northern Tyrrhenian Sea (Port of Livorno) and the transferability of research results is assessed against the Port of Singapore, the most important hub of the Asian region.

1. Introduction

One of the crucial challenges of maritime transport is digitalization through the implementation of proper IT solutions.

Development and deployment of such solutions allows seaports to increase their overall efficiency, to lower costs and improve performance times of the intra-terminal operations, to improve information flow and decision-making and to decrease the environmental footprint by addressing the sustainability. The logistics and port sector is facing enormous changes caused by the penetration of information and communication technologies into operational and management processes. Nevertheless, the digitalization levels are heterogeneous and depend on different factors, such as seaport size, cargo volumes and type, stakeholders' needs, business models and investments. Although different studies are still ongoing to come up with a common and comprehensive way to assess the level of digitalization and innovation of the seaport with the aim of meeting the requirements of the next-generation seaport (namely the Port of the Future). Erroneously, ports are historically considered static realities, with little inclination to change and whose activities are considered accessory and without added value. Port logistics is a strategic sector in which there is creation and diffusion of value in terms of optimization of logistics processes and in terms of environment and collective well-being. Therefore, anticipating the demand without being caught unprepared by disrupting innovation is essential. Ports have been involved in the digitalization era, where everything is connected in a smart way, businesses and their industrial processes have been reengineered according to the so called 4.0 paradigm. Smartness usually refers to the principles of automatic computing. However, a globally accepted definition of smart port has not been well defined in the literature, as well as an internationally accepted and standard definition for the word smart does not exist in the context of port industries and maritime scenario [1]. The Ports of the Future, increasingly smart and interconnected, are turning into real hubs for the creation of value and knowledge, based on technology transfer. Information and communication technologies enable a set of new digital services (user-centric, innovative and based on standard architectures, networks and data) that allow the continuous improvement of operational and environmental performances. In this context, we propose a prototyping framework for agile smart service development relying on a set of standard and open-source technologies. The proposed framework is validated at The Northern Tyrrhenian Sea Port Authority System (Port of Livorno) and the transferability of research results is assessed against the Port of Singapore, the most important hub of the Asian region. As a matter of fact, the Port of Singapore has positioned itself to be competitive to other ports and has relentlessly driven comparative advantage. There is immense scope to increase this advantage from digital transformation to improve operational efficiency and sustainability by positioning as smart and agile port. We propose technological solutions and strategies to accelerate maritime digitization, connectivity and cybersecurity in Singapore and the regional based on experience and lessons learnt in the Port of Livorno. Our approach is based on ICT state-of-the-art analysis for the Port of Singapore by considering some key factors driving port digitization and sustainability. According to this approach, we come up with the most effective strategy to accelerate and benefit from maritime digitization, connectivity and cybersecurity by addressing the operational efficiency and sustainability though out a prototyping ICT framework for agile services development as an effective tool to address the Port Community needs and requirements.

2. State-of-the-art analysis

Singapore has redefined the container logistic processes and operations through the adoption of different digital technologies such as Artificial Intelligence, Blockchain [2], Cloud Computing and Internet of Things [3].

Digital innovation in the port of Singapore has been disruptive, and case study [4] has presented how the port leveraged the information technologies to stay competitive in the industry. The Port Authority of Singapore (MPA) has employed mobile technology and wireless connectivity to improve efficiency, communications, and crew satisfaction in the Port of Singapore [5]. Multiple ports in this region already adopted the Integrated Port Management System (IPMS), which incorporates multiple ports in a single platform. A mega container port is being developed at Tuas in Singapore which would eventually consolidate the existing ports at Brani, Keppel, Tanjong Pagar and Pasir Panjang. This port is currently being developed the next generation port with digital twin of the physical systems and operations. The objective is to provide visualization, connectivity, analyzability, and granularity as depicted in Figure 1. Infocomm Media 2025 [6] in a 10-year plan released in August 2015 which aims to create a globally competitive infocomm media ecosystem that enables and complements Singapore's Smart Nation vision. This ecosystem supports people and enterprises in tapping on infocomm media to effect economic and social transformation and foster a common identity. To facilitate that objective to be achieved, a new Digital Industry Singapore, or DISG, office had been established. DISG will work with private companies to address issues on needs for securing talent and market access, building capabilities, and internationalizing. In this context, Singapore started building

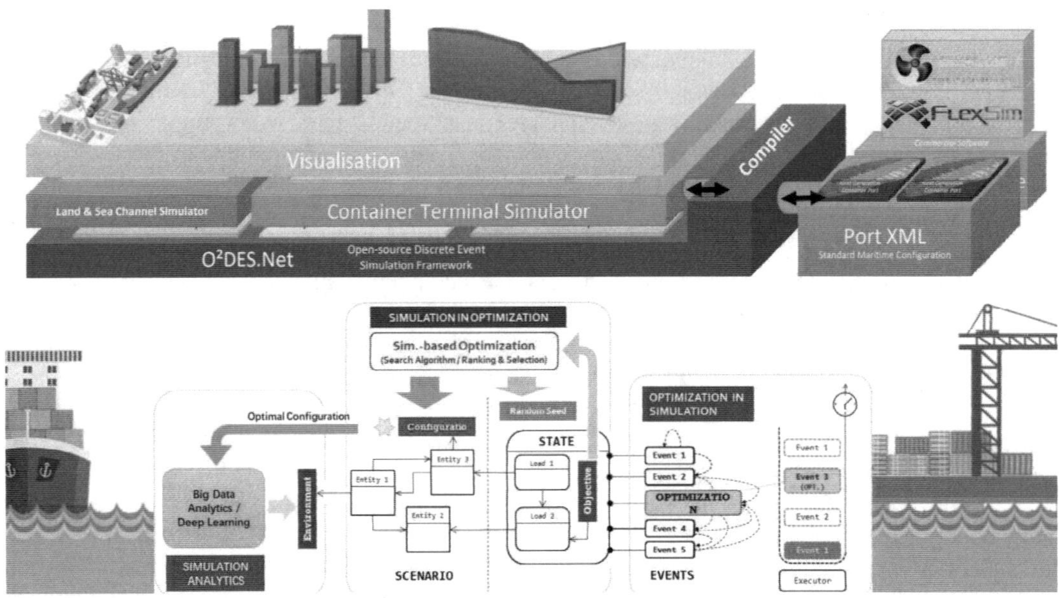

Figure 1. O2DES.Net framework for optimization in simulation.(NCSS, 2021).

on its existing R&D base, a vibrant maritime innovation cluster that is characterized by a diverse and tight-knit community of maritime enterprises (including start-ups and scale-ups), maritime technology solution providers, researchers, incubators, and venture capitalists, supported by an enabling environment. In June 2016, PSA, with the support of MPA and the Singapore Economic Development Board (EDB), committed to establish the PSA Living Lab, a living laboratory for the port and logistics industry [7]. The PSA Living Lab expects to enable technology solution providers to collaborate with PSA to develop and test-bed ideas in a live port environment, which is amongst the world's busiest container hubs. An example of a technology test-bedded at the PSA Living Lab is the Automated Guided Vehicle (AGV) system, which would be deployed for terminal operations at existing terminals, and eventually at Singapore's future Tuas mega port. Similarly, in March 2017, MPA announced the plan for a MPA Living Lab to provide a platform, with sufficient scale and real operating conditions, that technology providers and industry players can plug into for the co-development and piloting of innovative solutions by developing technological capabilities in the following four areas:

- Data analytics and intelligent systems;
- Smart and innovative infrastructure;
- Autonomous systems and robotics;
- Safety, security and environment.

As for the case of Livorno Port, the Port of Singapore relies on its own physical ICT infrastructure (with wireless connectivity, including commercial 4G and 5G technologies) to support two different types of digital systems and platforms for seaport operations management:

- Documentation Systems integrated with the Port Community System (e.g., billing, statistics and reporting operations);
- Planning and Operation Systems (e.g., berths, yard vehicles, ships and container gates management).

3. The ICT prototyping framework

The current information systems of the Livorno Port Authority are undergoing a deep digital transformation.

To address new requirements, the approach based on the usage of monolithic and technological solutions difficult to scale, has been descoped. For this reason, the Livorno Port Authority spent a lot of effort in the standardization of an ICT reference stack, named as MONICA Standard Platform [8], in order to support an agile development of innovative services for the actors from the supply chain and Livorno port community. This ICT stack is structured as a private cloud with full decoupling of the three canonical layers (data, business and application), and it is appointed by the Livorno Port Authority in current and future tenders and procurements (see Figure 2). The cloud has been deployed in a local laboratory within the Port of Livorno to allow testing and validation of the most recent functionalities, prior to the release time.

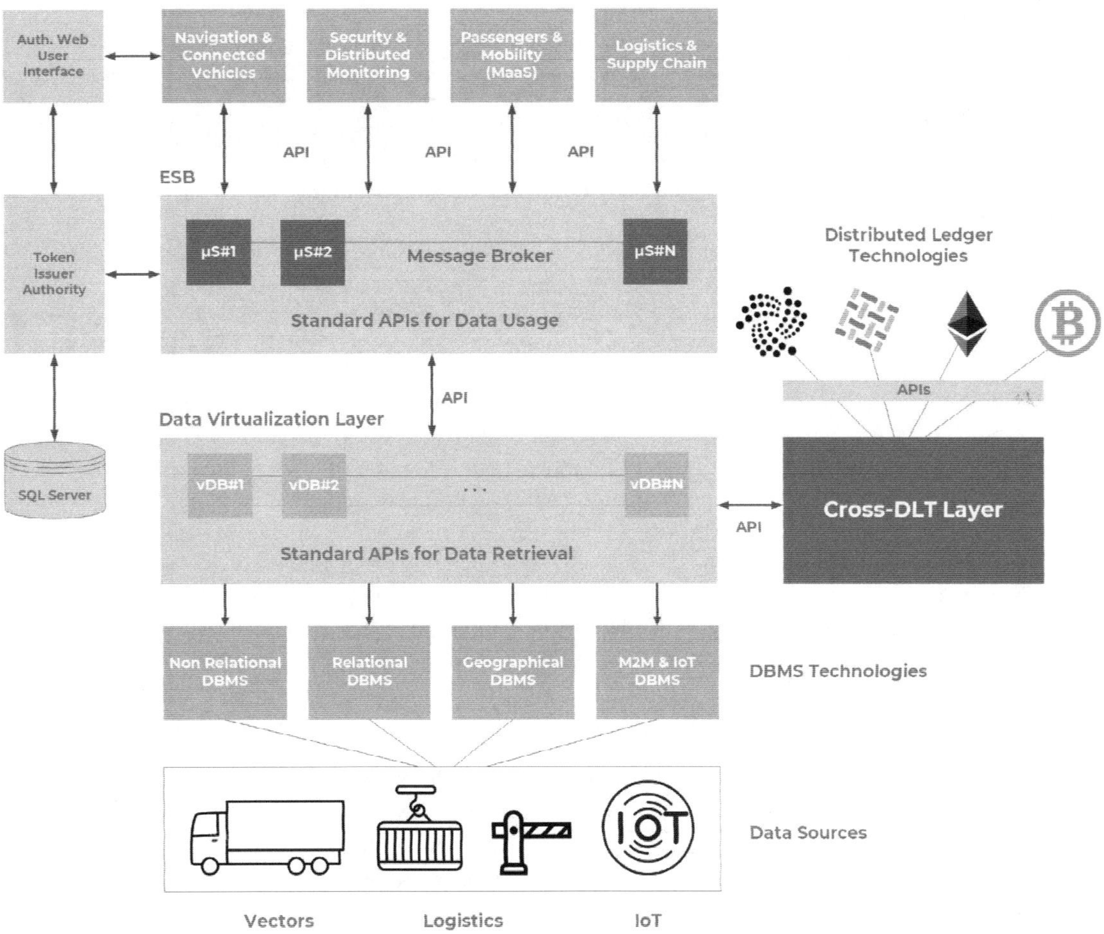

Figure 2. MONICA Standard Platform functional diagram.

The local cloud is configured as a staging environment for demo applications, which are turned into funding opportunities for industries allowing them to fill the gap from the prototype to the final product or service. All cloud resources are built on top of the local connectivity layer of the Port of Livorno, which is fully covered by a fiber optic backbone that capillary serves all terminals and gates in the port area, as depicted in Figure 3. On one hand, the fiber optic star-shaped network is allowing to connect all digital resources (i.e. sensors and actuators, network adapters, ordinary and industrial PCs) to a LAN. On the other hand, the network is mitigating the digital divide suffered by certain

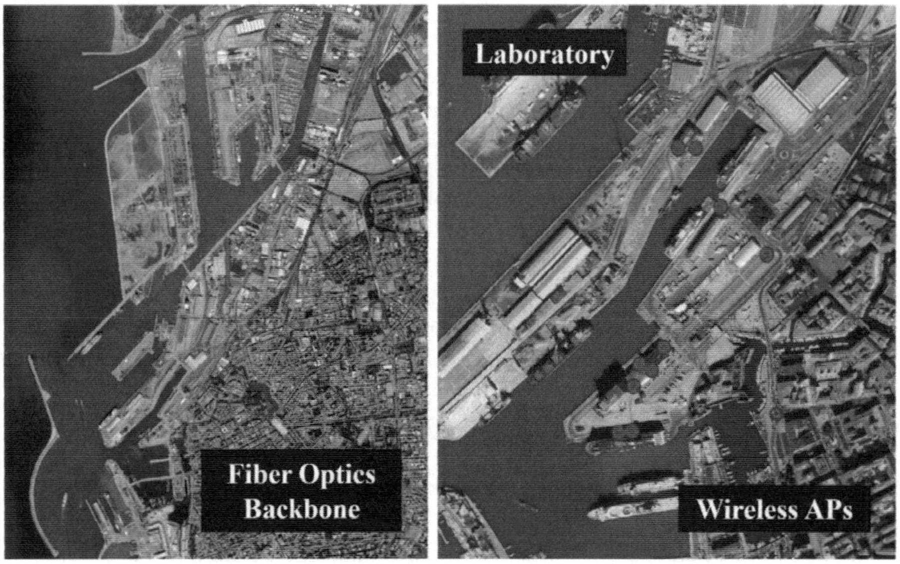

Figure 3. Staging environment and connectivity resources in the Port of Livorno.

areas, permitting to provide Internet access to government institutions (notably the Coast Guard, the State and Fiscal Police corps and Custom Agency), through the ISP selected by the Livorno Port Authority. In addition, a complementary wireless backbone is covering the maritime area with 100 Mbps Wi-Fi access points. The adoption of a standard and microservice-oriented architecture enables an agile integration of existing legacy systems into a single flexible and interoperable architecture and on the other side it enables a scalable approach for the development of new prototyping services. Furthermore, the introduction of open-source and software-based solutions such as Data Virtualization Layer (DVL) and Enterprise Service Bus (ESB) allow to define more efficient process control policies (e.g., services and data access, user management, APIs management, security, etc.) without compromising the scalability of the whole architecture. As depicted in Figure 2, MONICA Standard Platform is based on a full decoupling between the canonical layers of the three-tier architecture: data layer, business logic layer and application layer. In order to achieve this, we introduced two main software components into our ICT stack: Data Virtualization Layer and Enterprise Service Bus:

- *Data Virtualization Layer* (JBOSS Teiid [9]) is a cloud-native and open-source data virtualization platform enabling distributed databases, as well as multiple heterogeneous data sources, to be accessed by means of a common and standard set of APIs (e.g., JDBC, ODBC, REST, OData, SOAP, etc.). On one side, it allows aggregating data coming from disparate data sources, and on the other side, it permits to define a proper set of roles (according to create-read-updated-delete approach) allowing data consumers to use specific data sets. The current instance of the data virtualization layer is based on the WildFly application server [10], providing robust operations for transaction management, connection pooling, security configuration, resource management, and clustered deployment. The above-mentioned instance is deployed in the form of a docker container, running on a virtual machine with Ubuntu 20.04 Server operating system (64-bit architecture).
- *Enterprise Service Bus* (WSO2 API Gateway [11]) is an open-source, standardized, and componentized middleware platform that implements typical enterprise service bus functionalities by supporting microservices' logic development (SDK based on ASP.NET Core Framework) and APIs' life cycle management. It secures, protects, manages, and scales API calls by intercepting API requests and applying security policies. The WSO2 API

Gateway instance is deployed in a docker container, running in a virtual machine with Ubuntu 20.04 Server operating system (64-bit architecture).

As part of the overall ICT stack, a *cross-DLT layer* has been introduced to allow the interoperability with a set of most common Distributed Ledger Technologies (i.e. Bitcoin [12], Ethereum [13], IOTA [14]) as well as with commercial blockchain-based platforms (i.e. Tradelens [15]), so that data immutability can be guaranteed. Data privacy is also taken into account by implementing a pseudonymization function at DVL level in order to make sure that personal data are handled in respect to GDPR requirements [16]. Finally, the authentication and authorization aspects are properly addressed by the Authorization Server component of the ICT stack, based on industry-standard protocol OAuth2 [17]. According to this, we relied on OpenID Connect 1.0 Hybrid Flow [18] as the identity layer on top of the OAuth 2.0 protocol. This allows clients to verify the identity of the end-users based on the authentication performed by the Authorization Server, as well as to obtain basic profile information about such end-users in an interoperable and REST-like manner. The Authorization Server implements a single-sign-on (SSO) authentication scheme, and it is built upon the authorization token (JSON Web Token - JWT [19]) which is released to a specific user in order to invoke selected microservices. The JWT is emitted and validated by a token issuer authority, according to a given, as well as predefined, set of grants and access rules.

4. Validation of the proposed ICT stack

In this section, the design and development examples of three service prototypes that have been used to validate the considered ICT stack, are shortly presented and described.

The prototypes have been selected based on the needs and requirements from the Livorno Port Authority, according to the current Port Master Plan of the Livorno Port Authority [20]. This plan envisages the development of new digital innovative solution addressing (i) Rail–road and rail–sea integration (inter-modality) for people and goods, (ii) Navigation aid system for assistance to pilots and maritime authorities, as well as (iii) Integrated environmental monitoring and control of the port-logistics processes.

The first application consists in a *smart bathymetric service for seabed monitoring* in Livorno seaport. Dredging is essential for ports to remain in operation. It allows removing sediment and other materials from the sea beds. For this reason, seaports across the world dredge to enlarge and deepen access levels, ensuring safe navigation, competitive advantage by welcoming large vessels, remediation, and flood management. The seabed monitoring is based on a multibeam echo sounder probe installed on a monitoring boat from Coast Guard (see Figure 4).

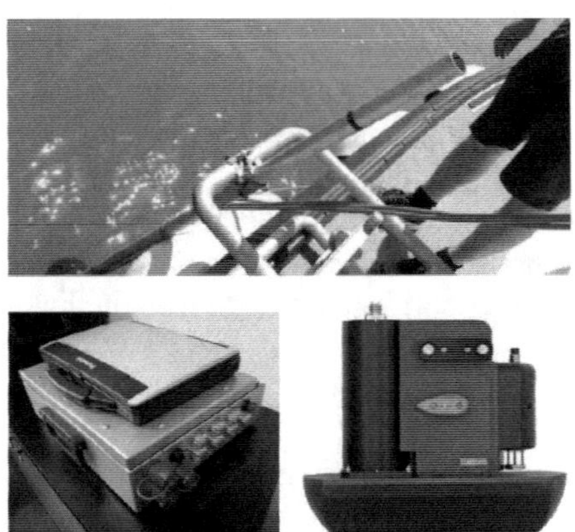

Figure 4. Bathymetric probe with dock station and its installation on the monitoring boat.

The digital twin of the Livorno Port, namely Port Monitoring System (PMS), allows to visualize collected data in a form of the isarithmic map that depicts the submerged topography and physiographic features of the considered seabed as depicted in Figure 5.

Figure 5. Bathymetric application user interface.

The second application, namely *yard vehicles management system*, allows a real-time monitoring of yard vehicles (e.g., forklifts, reach stackers, cranes, etc.) equipped with GPS trackers and supported by a low latency communication based on a private and prototyping 5G network deployed within the seaport. The application is based on general cargo management scenario at the Port of Livorno (a container terminal operator has been preliminary identified). From logistics perspective, it allows container terminal operators to monitor in real-time the status of the available forklifts within the container terminal storage area, by visualizing through a GUI (see Figure 6) the information related to each available forklift (ready or busy to perform cargo handling operations, forklift's unique identifier as well as its speed and GPS position) and the associated cargo (e.g., cargo unique identifier, width, height and length). The business logic of this application, allows to calculate the proper forklift choice according to a given algorithm in order to perform the forklift assignment (forklift-cargo association). Two parameters are taken into the account for this purpose: the distance between each forklift and the cargo to be handled, and the availability of each forklift. Finally, the third application performs trucks' and vessels' turn around time computation within the seaport of Livorno: Truck Turnaround Time (TTT) and Ship Stay Time (SST) respectively. The *Truck Turnaround Time application*, depicted in Figure 7, provides end users with information regarding the time spent by a given truck within the seaport area. Typically, when a truck arrives at the port entrance, the plate and container number (if present) is read and if the vehicle is authorized to enter the port facilities, the date and time of access are registered (gate-in event), and the barrier is lifted. After the loading/unloading operation, trucks leave the terminal and the port facilities: again, the plate and container number (if present) is read and if the vehicle is authorized to leave the port facilities, the date and time of exit are registered (gate-out event), and the barrier is lifted. The difference between the moment the truck enters the port facilities and the moment it exits the port facilities is called TTT. The knowledge of TTT can pave the way to different methods oriented to optimize the access to the port facilities and/or reduce the waiting time for vehicles at the port gate, leading to corresponding savings on direct costs for carriers.
The Ship Stay Time (SST) application, depicted in Figure 8 provides end users with information regarding the mooring time of a ship, meaning the time spent inside the port. The port information system registers the events of the ship entering and leaving the Port of Livorno, and the current infrastructure allows to seamlessly obtain and integrate data in the end-user application. Access to this kind of data allows to quickly identify inefficiencies in the current scenario, and act accordingly. Being able to spot crucial points in time when something goes wrong is a precious resource for domain experts that can take control of the situation and possibly take actions to mitigate the problem. Reducing the SST can positively affect the environmental impact of the seaport (a ship that docks for

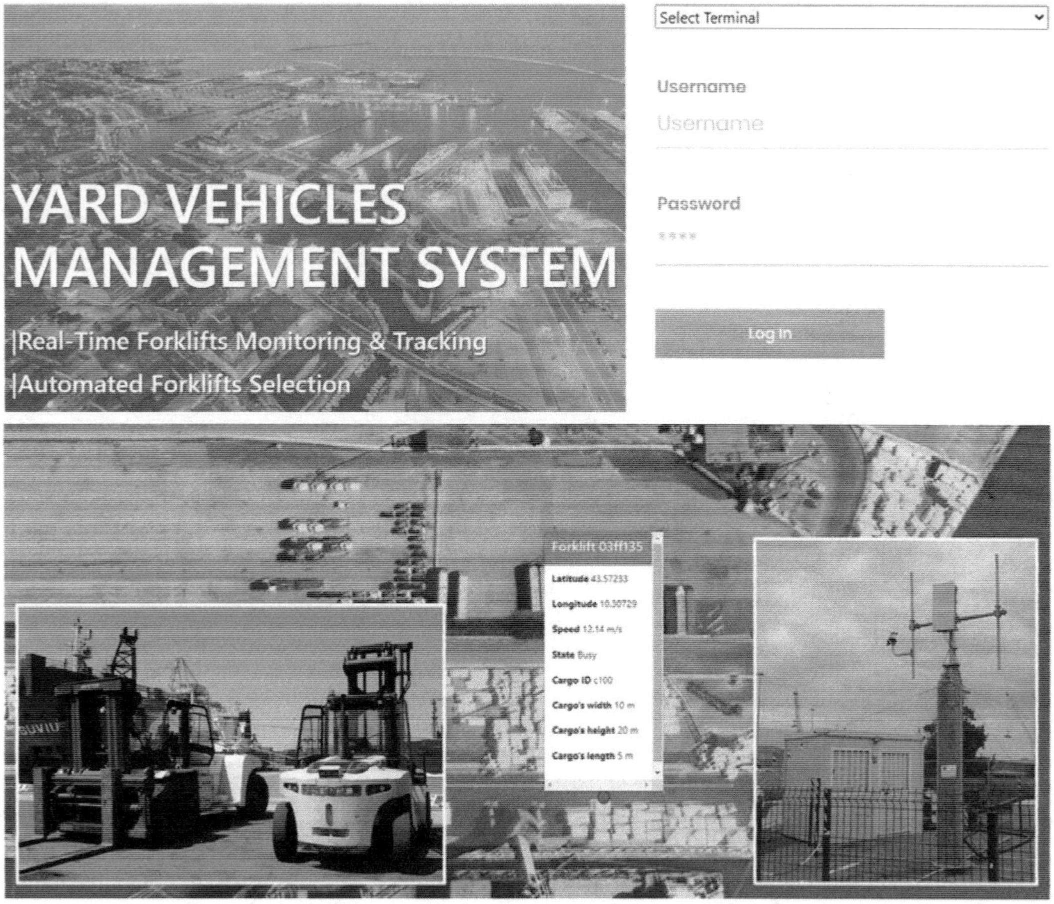

Figure 6. Yard vehicles management application and on-field equipment.

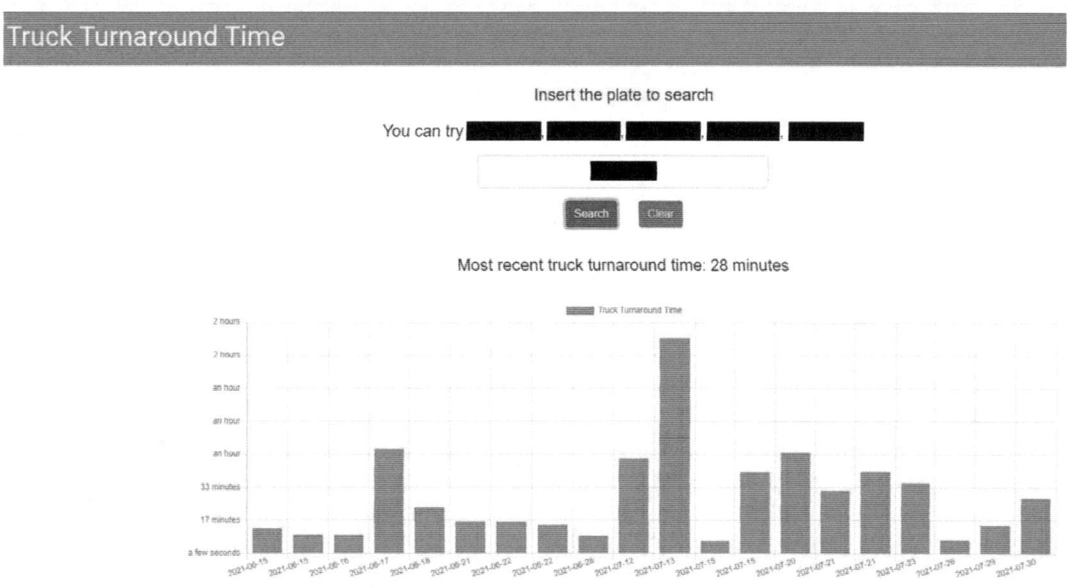

Figure 7. Trucks' turn around time computation – Graphic User Interface.

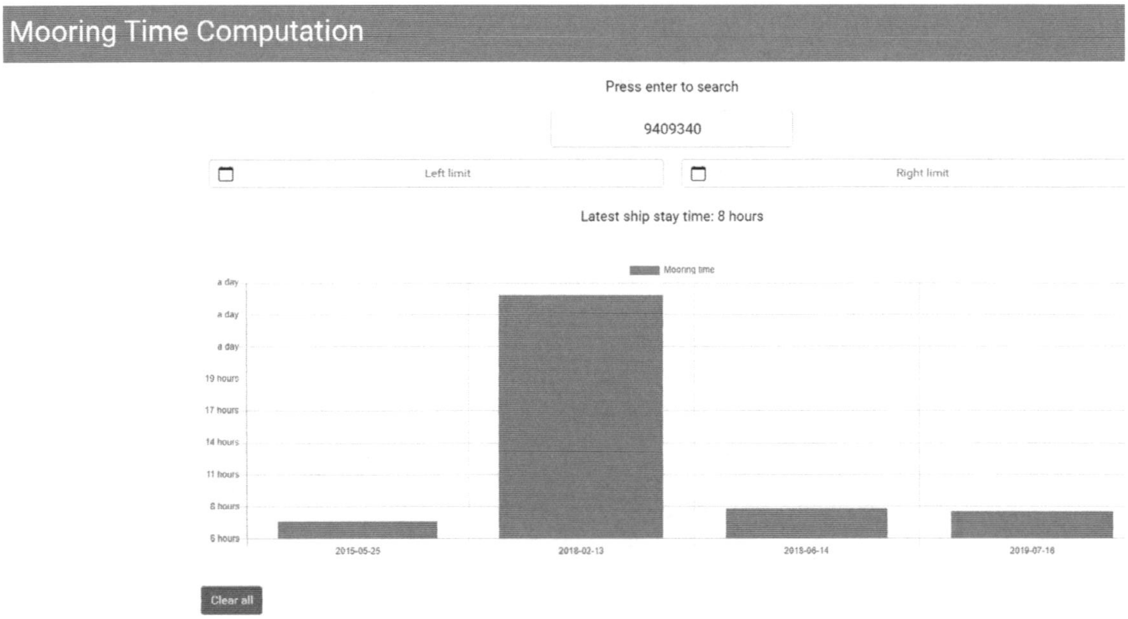

Figure 8. Vessels' turn around time computation – Graphic User Interface.

less time will likely pollute less) but also the costs (reducing the SST means being able to handle more ships). The business logic of the considered applications has been implemented in a form of a set of microservices by means of WSO2 API Gateway framework based on ASP.NET Core technology. For data access and retrieval the DVL component is used allowing the interaction with the relative data source. In case of the smart bathymetric service for seabed monitoring, raw data storage is supported by MongoDB (NoSQL and document-oriented database system [21]). As far as yard vehicles management system is concerned, we relied on Mobius [22] M2M platform (open source platform implementing ETSI OneM2M standard [23]) in order to store real-time data coming from GPS trackers. Finally, in the case of the TTT and SST computation applications, the database management systems of the existing platforms of the Livorno seaport are used as data sources (e.g., the RDBMS of the Port Community System). In all considered cases, the WSO2 API Gateway is used as the main framework for an agile microservices development while DVL is adopted as the unified interface for data lake management, data access and data retrieval. Furthermore, the communication between microservices is enabled by an open-source message broker (namely RabbitMQ [24]) that can be easily deployed in distributed and federated configurations to meet high-scale, high-availability requirements. Finally, the Authorization Server component manages the user authorization and authentication procedures, so that data access and APIs' invoking are properly handled.

5. Transferability and technological considerations

Transforming a traditional port into a smart port represents a true technological challenge in terms of increasing productivity, resource optimization and environmental and energy safeguarding.

In this respect, a significant investment in digital infrastructure is required. From an economic assessment point of view, ports of the future can achieve important impacts, or effects: firstly, in terms of microeconomic effects, since we have an increase in overall surplus (welfare) and, according to long-term growth, through the achievement of economies of scale and increased production specialization. According to the New Economic Geography Models [25], infrastructural upgrading can lead a company located in a specific area to have access to a wider market, as a result of lower transport costs. At the same time, other producers are encouraged to operate in that area, feeding an

increasingly attractive spatial concentration process and increasing the local market competitiveness [26]. In terms of macroeconomic effects, the Return on Investments in port infrastructure is expected to arise, as well as the long term GDP. This is a consequence of the high seaport infrastructures elasticity of GDP, which suggests that seaports infrastructures have an important indirect effect on economic growth. In order to validate the ports business models and to boost seaports quality of services (efficiency), any national economy must consider seaports as crucial business operating in a high competitive market, which require continuous development and specific infrastructures improvement and investments [27].
However, the seaport digital verticals can be classified into bundles, according to the C-Ports paradigm [28], which aim at offering modular and tailor-made solutions for the digitalization of port processes. C-Ports services will belong to three verticals and one horizontal, as depicted in Table 1:

- Vessel and Marine navigation (vertical);
- e-Freight & intermodal logistics (vertical);
- Passenger transport (vertical);
- Environmental sustainability (horizontal).

Table 1. R&D fields classification.

R&D field	Type	Regulator	Services
Vessel and Marine navigation	Vertical	ISO and IMO (mainly)	Vessel traffic management, vessel maneuvering in port waters, accident at sea, suspicious vessel/maneuver, berth allocation and docking.
E-freight and intermodal logistics	Vertical	UN/EDIFACT and not-standard proprietary solutions	Freight management and control, gates automation, in-port smart navigation, freight routing, accident at landside.
Passenger transport	Vertical	ISO, IEEE, MaaS Alliance	Infomobility and journey monitoring, integration with the traffic control centre (TCC), in-port smart and autonomous mobility (including safety).
Environmental sustainability	Horizontal	UN protocols	Pollution level (including COx and noise) reduction, road traffic level monitoring and control, dynamic pricing (all services) to vessels, terminals.

Services and product requirements are defined by the needs of the specific user-personas, representing port stakeholders. Thus, it is important to analyze the users' expectations and needs, to make them part of the process of development, by collecting feedback, deriving from their experience, with a continuous improvement approach. The level of involvement of the port communities can in fact be assessed through the definition of ad-hoc Key Performance Indicators (KPIs), as these indicators can provide important information about the validity of models and prototypes. Every smart port offers tailor-made services to its port community. This is achievable by using standard data interchange protocols and standard digital platforms, in order to let the systems interoperate with each other. The digital port services must follow a commercial exploitation plan, in order to extend the smart port concept to the other logistics hubs, which are part of the maritime network. In this sense, a feasibility study must be investigated, with the purpose to understand the services time-to-market depending on the specific business case. User requirements must be defined, in this sense a strategic partnership with the major ICT players is needed. Users adoption rate must be investigated, as well as the compliance to the standards and law requirements. Most specifically, according to the SWOT Analysis paradigm (Strengths, Weaknesses, Opportunities and Threats), the factors to keep in mind are both internal and external, as there are several aspects which can may have a positive or negative influence on the business strategy. While developing the new smart port features, the innovation lab has taken into account the most competitive and innovative business in the ICT and logistics sector, as well as telcos, insurance and shipping companies. As already mentioned, the transferability issue should be also assessed from a technological perspective. Technology is transforming maritime business models and value chains. Ships are becoming smarter and increasingly connected. Ports around the world are becoming more automated in the face of growing labour constraints. Maritime companies are leveraging developments in both hardware (e.g., advanced sensor systems, sea-to-shore connectivity, robotics, etc.) and software (e.g. data analytics, artificial intelligence, machine learning, etc.) to better optimize their operations. In this sense, seaports are at different starting positions and will adopt different degrees of digitalization. Some ports have taken pioneering steps to become first-movers in the adoption of advanced technologies for collaboration, synchronization, automation, and analytics. But, at many smaller ports, much of the information is not managed by digital means. Despite the user's needs and requirements are slightly different between the two considered seaports, a common and similar ICT infrastructure guarantees a certain level of freedom in development and testing of innovative services without additional costs. Moreover, the availability of digital platforms and monitoring applications (including legacy systems) in the Port of Singapore fulfils the minimum requirements as far as data availability is concerned. Moreover, an open and scalable ICT stack for data storage, access and retrieval as well as for smart services and applications business logic development is easily applicable in other maritime domains such as the one characterizing the Port of Singapore. The existing digital infrastructure, including the underlying communication technologies as well as the computational resources, enables the deployment of the proposed solution by providing an efficient and agile tool for the development of the innovative services driven by the Port of Singapore needs, requirements and challenges.

6. Conclusions

The potential offered by commercial ICT solutions is often limited by the lack of a standardized way to build new services on it.

We proposed and discussed a reference architecture, with the aim to provide a straightforward, easy-to-adopt solution to all the actors of the Port Community interested in the rapid development of service prototypes. In order to validate the proposed framework, we developed and presented some example applications to demonstrate how the considered approach can support an agile development as well as integration of new or already existing services at the Port of Livorno. Finally, we provided and discussed some considerations related to the ICT stack transferability with special focus to more challenging context such as the one of the Port of Singapore. Additionally, to motivate our approach, we expect to conduct a survey with some relevant members from the Port Community of Singapore, so that user needs and requirements can be used for the consolidation of the proposed solution.

Acknowledgements

The authors would like to thank the *Port Network Authority of the Northern Tyrrhenian Sea* for the appreciation of *CNIT* work in the last six years as well as all *Joint Lab Seaports* [29] members who contributed to the results discussed above. Moreover, we thank the *BMT Group* for the interest in light of ongoing joint activities

References

[1] A. Molavi, G. J. Lim and B. Race, *A Framework for Building a Smart Port and Smart Port Index*, International Journal of Sustainable Transportation, 2019.

[2] V. Campfens, C. Dekker, *Turning Rotterdam into the World's Smartest Port with IBM Cloud and IoT*, 2018.

[3] L. Heilig et al., *Digital transformation in maritime ports: analysis and a game theoretic framework. Netnomics: Economic research and electronic networking*, Presented at the Hawaii International Conference on System Sciences, 2017.

[4] L. Partridge et al., *Information technology management: the case of the Port of Singapore Authority*, 2000.

[5] *The Maritime Executive, Intellectual Capital For Leaders, Case Study* available at: https://www.maritime-executive.com/magazine/2015/3, 2015 Edition.

[6] Infocomm Media 2025, Ministry of Communications and Information, "*Creating Connections, Inspiring Innovation*", available at: www.mci.gov.sg/portfolios/infocomm-media/infocomm-media-2025, accessed on 01/02/2022.

[7] Singapore Economic Development Board (EDB), *PSA Living Lab - Port of the Future*, available at www.edb.gov.sg, accessed on 01/02/2022.

[8] MONI.C.A. Standard Platform, *Monitoring and Control Application*, available at: https://www.monicapmslivorno.eu/, accessed on 01/02/2022.

[9] Data Virtualization Layer – Teiid, *Cloud-native data virtualization*, available at https://github.com/teiid/teiid, accessed on 01/02/2022.

[10] WildFly Application Server, *Teiid based on WildFly application server*, available at https://teiid.github.io/teiid.io/teiid_wildfly, accessed on 01/02/2022.

[11] *Overview of the WSO2 API Gateway - API Manager Documentation 3.2.0*, available at https://apim.docs.wso2.com/en/3.2.0/learn/api-gateway/overview-of-the-api-gateway/, accessed on 01/02/2022.

[12] S. Nakamoto, *Bitcoin: A Peer-to-Peer Electronic Cash System*, 2008.

[13] V. Buterin, *A Next Generation Smart Contract & Decentralized Application Platform*, 2013.

[14] IOTA, *An Open, Feeless Data and Value Transfer Protocol*, available at https://www.iota.org/, accessed on 01/02/2022.

[15] Tradelens, *Transform container logistics by freeing from legacy data systems, manual document handling and poor visibility*, available at https://www.tradelens.com/, accessed on 01/02/2022.

[16] Regulation (EU) 2016/679 of the European Parliament and of the Council, *The protection of natural persons with regard to the processing of personal data and on the free movement of such data, and repealing Directive 95/46/EC (General Data Protection Regulation)*, available at https://eur-lex.europa.eu/legal-content/EN, accessed on 01/02/2022.

[17] OAuth 2.0, *Industry standard protocol for authorization*, available at https://oauth.net/2/, accessed on 01/02/2022.

[18] OpenID Connect 1.0, *RFC6749 protocol*, available at https://openid.net/specs/openid-connect-core-1_0.html, accessed on 01/02/2022.

[19] JSON Web Token, *Industry standard RFC 7519*, available at https://jwt.io/, accessed on 01/02/2022.

[20] Port Network Authority of the Northern Tyrrhenian Sea, *Archive of Port Authority three-year Master Plans*, available at https://www.portialtotirreno.it/pianificazione-e-opere/piano-operativo-triennale/, accessed on 01/02/2022.

[21] MongoDB, *Introducing Native Support for Time Series Data*, available at https://www.mongodb.com/, accessed on 01/02/2022.

[22] Mobius OneM2M, *oneM2M IoT Server Platform*, available at https://github.com/IoTKETI/Mobius, accessed on 01/02/2022.

[23] oneM2M Sets Standards for the Internet Of Things & M2M, *Global IoT Technical*

Specifications, available at https://www.onem2m.org/technical, accessed on 01/02/2022.
[24] RabbitMQ, "*Open source message broker*", available at https://www.rabbitmq.com/, accessed on 01/02/2022.
[25] P. Krugman, *Increasing Returns and Economic Geography*, Journal of Political Economy, vol. 99, no. 3, 1991.
[26] Banca D'Italia, *Le infrastrutture in Italia: dotazione, programmazione, realizzazione*, Roma, 2011.
[27] T. Jouili, *The Role of Seaports in the Process of Economic Growth*, International Knowledge Sharing Platform, 2016.
[28] P. Pagano, S. Antonelli and A. Tardo, *C-Ports: a proposal for a comprehensive standardization and implementation plan of digital services offered by the Port of the Future*, Elsevier, 2021.
[29] Joint Lab Seaports, *laboratory on advanced sensing and networking in sea ports*, available at https://jlab-ports.cnit.it/, accessed on 01/02/2022.

Ship-Shore Interaction: The Model in the Middle

Marianne Hagaseth[1], Jeppe Skovbakke Juhl[2], Ørnulf Jan Rødseth[1], Ben van Scherpenzeel[3], Andreas Maria van der Wurff[3]

[1]SINTEF Ocean, Trondheim, Norway, [2]BIMCO, Copenhagen, Denmark, [3]ITPCO, Rotterdam, Netherlands

Marianne.Hagaseth@sintef.no

Abstract. The complex picture of maritime transport involving more than 100 thousand ships, more than 8000 ports and a large number of stakeholders fulfilling different roles and being responsible for different processes means that providing up-to-date and timely information during port calls is a complex task. In addition, there are requirements to reduce both the environmental footprint and costs of the maritime transport. From 2019, the International Maritime Organization (IMO) made it mandatory to accept electronic clearance of ships entering foreign ports, meaning that the digitalization of ship-port interfaces to support the port call processes has become an important topic. Further, this means that the interoperability between a large number of ICT systems to support the end-users related to a port call, must be ensured. This paper describes the model in the middle, by which we mean a "least common multiple" regarding data elements and data models to support a consistent exchange of information between preferably all involved stakeholders during a port call.

1. Objective

The last year's disruption of the sea transportation has shown the importance of the role that the ports have in the wider supply chain. The complex picture of maritime transport involving more than 100 thousand ships [1], more than 8000 ports [4] and a large number of stakeholders fulfilling different roles and being responsible for different processes means that providing up-to-date and timely information during port calls is a complex task. In this regard, the digitalization of the interface between ship and shore and the related processes will become increasingly important. It is important to improve the quality and availability of data to be exchanged before, during and after a port call and to harmonize the standards to allow reliable and efficient information exchange between ship and shore systems. There is an overlap of information needs, both in the mandatory reporting to authorities and in the information needed to ensure efficient Just-in-Time arrival, timely execution of port services, safe navigation to the port, and automation of cargo operations and the mooring process.

This paper gives an overview of the relevant actors, processes and information elements related to ship-shore interaction during port calls with focus on the commonalities between operational services, authority processes and nautical operations. Focus will be on the "model in the middle" by which we mean the commonalities regarding information requirements as seen by the different actors and processes related to describing the timing and locations for a port call to be as efficient as possible.

Chapter 2 gives an overview of the methodology, while Chapter 3 describes the background and structure of the IMO Reference Model that is the main topic of this paper. The technical implementation of the reference model can be done by ISO 28005 on Electronic Port Clearance (Chapter 4), while other initiatives and related work are described in Chapter 5. Three sub-models in the reference model, namely for Location, Timestamp, and Maritime Service are described in Chapter 6, 7, 8 and 9. The top-level processes are described in Chapter 10 and the related actors are listed in Chapter 11.

2. Overview of Methodology

The paper defines relevant actors related to the ship operation, port and terminal services. Further, a description of the high-level processes is given. Various initiatives regarding standardization of the information is described. This will focus on information to describe the timing of the port arrival, departure and service execution and also information to describe locations related to the port call.

The goal is to outline the requirements to a harmonized description of timing and locations to ensure efficient port calls, both related to safe berth-to-berth navigation, berth planning done by the terminals, efficient execution of port services and cargo operations, improved information exchange with hinterland transport, timely execution of port and terminal services, and Just-in-Time arrival of ships which are closely linked to voyage optimization. Even if the information is available at different points in time and with different granularity for each use case, it is important to describe the commonalities that still exist, to ensure efficient interoperability among ship-side and shore-side ICT systems, and thus ease the operation both for the ships and the ports.

3. The IMO Reference Model

An important contribution to the "model in the middle" comes from the work done by the International Maritime Organization (IMO) in cooperation with the World Customs Organization (WCO), the United Nations Economic Commission for Europe (UNECE) and the International Organization for Standardization (ISO) through their establishment of a common reference model [2]. The original purpose of the IMO Reference Model was to ensure interoperability between government-to-government data exchanges (WCO), international trade data exchanges (UNECE) and ship operational data exchanges (ISO) in the scope defined by the Convention on Facilitation of International Maritime Traffic (FAL Convention). This is also often referred to as the scope of the "Maritime Single Window" (MSW).

More generally, the "model in the middle" contains a set of data elements that are given a harmonized description and a minimum representation requirement (data formats, code lists and business rules). This is done to ensure that each data element has a unique interpretation and thus can be used in various contexts with a well-defined meaning. This is important to support the "report once"-principle and at the same time ensure a consistent and clear information exchange between actors during a port call. One example of a data element that needs a clear definition, is the arrival and departure time to/from the port. Here, both the type of the timestamp must be defined (whether it is an estimated, actual, planned or requested time) and a specific location in the port (in some cases it is enough to specify the port with no further details). In addition comes the definitions of arrival and departure, for instance related to whether it is defined by the first/last line onshore/onboard, when the ship is ready for boarding etc. Also, the actor responsible for each of the timestamps and the related processes where this data is exchanged, must be defined. More generally, the data elements defined in the reference model will be relevant for one or more domains, which is shown as FAL Forms, JIT (Just-in-Time) Arrival, Certificates, Ship Reporting etc. in Figure 1. From the top, the picture in Figure 1 shows the following as described in the following sections:

3.1. *Controlling documents*

These are the legal basis for the reporting requirements and the data exchange between actors engaged in the port call.

3.2. *IMO Data Set*

This is the list of common data elements defined across all domains relevant for a port call and the voyage.

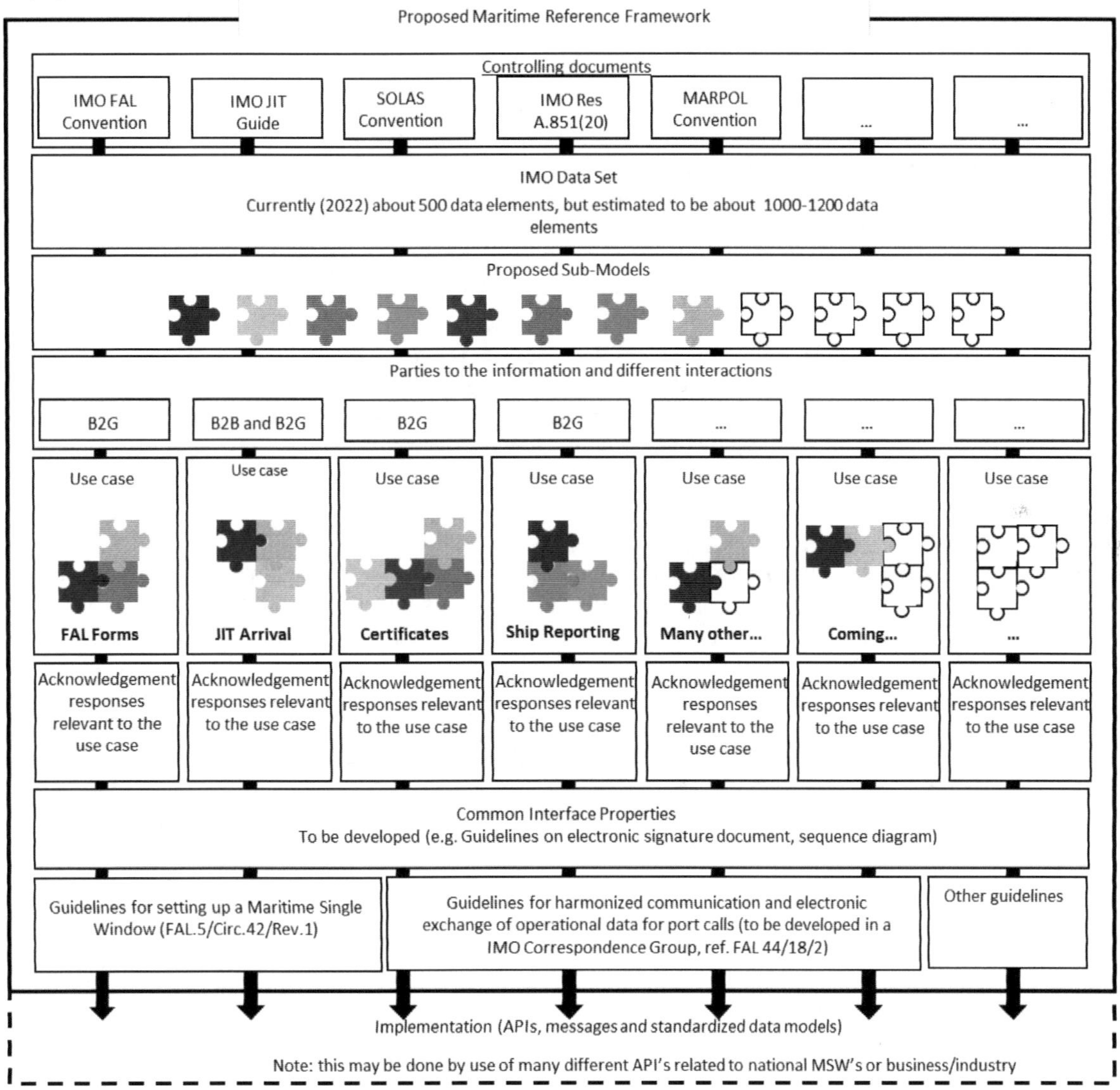

Figure 1. Sketch of a Maritime Reference Framework as proposed to IMO in EGDH 6/9

3.3. *Proposed Sub-Models*

The Reference Framework can be structured in sub-models covering specific topics, for instance static information about the ship, cargo, dangerous cargo, health, maritime services, waste, crew, passengers, timestamps, and the port call itself, among others.

3.4. Parties to the information and different interactions

This is the various types of actors which is relevant for the exchange of information described in the proposed maritime reference framework. Roles and actors are further described in Section 11.

3.5. Use cases

These are the different domains that the data elements and sub-models may be used for. Examples here are for reporting on the FAL forms, Just-in-Time arrivals, certificate reporting, mandatory ship reporting, etc. For each use case, a description of the relevant acknowledgement responses must be included.

3.6. Common Interface Properties

Across each of the domains and systems, common guidelines on electronic signatures, meta data on the message exchange, sequencing, and similar properties to ensure safe and secure data exchange, must be described.

3.7. Guidelines

This includes guidelines for setting up each of the types of systems relevant for the maritime framework, for instance guidelines for setting up Maritime Single Windows (MSWs), guidelines for the exchange of operational data needed for a port call, guidelines for how to use the proposed maritime reference framework to interface terminal operation systems, among others.

3.8. Implementation

The implementation-specific tasks are shown at the bottom of the figure. This includes the development of technical standards, message formats and APIs. Already existing examples here are the mappings that are described from the IMO reference model to the UNECE, WCO and ISO 28005 data models.

3.9. Example

An example on how the bricks in Figure 1 can be understood is the following: The dark blue brick at the "Proposed Sub-Models"-level can represent for instance the sub-model for the timestamp data set (see Section 7). Then, at the use case level, Figure 1 shows that this sub-model is used in the FAL forms use case, in the Just-in-Time Arrival use case and in the Ship Reporting use case. Similarly, the red brick can indicate that for instance the crew sub-model is used only in the FAL form use case.

The IMO Reference Data Model and the IMO Data Set support the semantic harmonization between the various reporting requirements (Controlling documents, top in Figure 1) and also the coordination between different technical implementations as shown at the bottom of Figure 1. Currently, the IMO Reference Data Model and Data Set are mapped to three different data models:

- The customs domain (the World Customs Organization (WCO) data model)
- The trade domain (the United Nations Economic Commission for Europe (UNECE/UNCEFACT) Core Component library)
- Ship-port-specific domain: The standard for electronic port clearance (ISO 28005-2 data model [3]).

The main concepts from the IMO Reference Model that are important for port call optimization are the operational data covering Location, Timestamps and Maritime Services. These concepts are now being implemented in a technical standard through the work done together with several maritime stakeholders through ISO TC8, both industry partners, authorities, and standardization organizations, to extend the 28005-2 standard. 28005-2 is further described in the next section.

4. ISO 28005 Electronic Port Clearance

The ISO 28005-series of standards maintained by ISO TC8 contains data elements covering the requirements for ship-shore reporting for authority information as defined in the following:

- Most required information sets as defined in the FAL Convention to be sent at arrival or departure:
    - General Declaration (FAL Form 1)
    - Cargo Declaration (FAL Form 2)
    - Ship's Stores Declaration (FAL Form 3)
    - Crew's Effects Declaration (FAL Form 4)
    - Crew List (FAL Form 5)
    - Passenger List (FAL Form 6)
    - Dangerous Goods Manifest (FAL Form 7)
- The document required under the Universal Postal Convention for mail (a reference to the physical or electronic document)
- Maritime Declaration of Health as based on the Maritime Declaration of Health (MDH) from WHO, 58th World Health Assembly, WHA58.3.
- Security-related information as required under SOLAS regulation XI-2/9.2.2 (ISPS code).
- Advanced electronic cargo information for customs risk assessment purposes
- Advanced Notification Form for Waste Delivery to Port Reception Facilities, based on the recommended reporting on ship-generated waste as defined in MEPC 644, which is mandatory within the European Union, as described in EU/2000/59.
- Required reporting as defined in the bulk loading and unloading code IMO Resolution A.862.
- Mandatory ship reporting system (MRS) requirements as defined in IMO Resolution A.851.
- ETA reporting to pilot station as defined in IMO Resolution A.960.

The new version (2021) of the ISO 28005 standard was updated to cover the data element list and the data model described in the IMO Reference Data Model, as described in the updated IMO FAL Compendium in FAL.5/Circ.43 from 23 October 2020 (IMO Compendium, 2021).

The information in 28005 is described as XML types in an XSD and also as classes in UML diagrams. The ISO 28005 standard also describes messages and the protocol on how to exchange these messages, including clearance, update, cancellation, receipt and acknowledgement. Most important, 28005 contains a mapping to the IMO Reference Model that ensures that systems implementing this standard is compatible with the semantic model agreed on by all partners involved in the development of the IMO Reference data model and data set.

Further, work has started in ISO TC8 to extend the 28005-standard with operational data related to the port call, especially the timestamps, identification of locations, and the definition of services. The aim of this work is to give a technical description of the data exchange between ship and shore for custom declarations, arrival times and completion times of cargo services, among others, including the synchronisation between each actor, and based on the semantic reference model agreed on by IMO and all representatives participating in the EGDH (Expert Group on Data Harmonization) working group.

5. Other Initiatives and Related Work

5.1. International Task Force Port Call Optimization

ITPCO [5] is an industry initiative to optimize the port call process in a trade agnostic manner from the contracting phase where a ship is chartered either for bulk or container and all the way through the port call and departure from the port. This process focuses on covering both bulk and container cargo and also to ensure that the processes handle the fact that decisions regarding the ship operation is made by the captain, including the timing and ordering of services in the port and terminal. The ITPCO process map gives the AS-IS-situation, except for the usage of the various time stamps, since today, only the usage of estimated times is common.

5.2. PortCDM/S-211

PortCDM focuses on the actual port call optimization and does not cover the processes before the port call as the contracting phase and how this influence the port approach from the pilot boarding place to the berth. The PortCDM/S-211 process model is an attempt to have an integrated view of all plans related to a port call, that means, plans for the port arrival and departure, berth management and for the various services to be provided to the ship, and to considered this as a whole. PortCDM/S-211 describes each of the port call process steps and the related sub-processes. For each step, a list of events is described. The PortCDM initiative is standardized by IEC in S-211, and this describes the port call message with the necessary information elements to cover the timestamp, location states and events to support optimization of the port call and Just-in-Time arrival of the ship to the port. This model covers data elements needed to be exchanged between the ship and port authority to cover the planning, execution and completion of the port arrival, port stay and port departure and also the synchronization of port and terminal services both in the planning phase, execution phase and completion phase.

5.3. DCSA

DCSA is an organization consisting of nine container shipping lines that has defined a Just-in-Time concept for port call optimization based on the work done by ITPCO. DCSA defines five high level port call processes as follows:

- *Berth Arrival Planning:* The carrier and terminal operator align on the short-term plan and the approximate time of availability of the berth.
- *Pilot Boarding Place Arrival Planning:* The carrier (represented either by the captain of the ship, the operation centre or the agent) and the port align on the arrival plan. A planned time at the Pilot Boarding Place (PBP) is agreed.
- *PBP and Berth Arrival/Start Cargo Operations:* The ship arrives at the PBP and is manoeuvred to the berth where cargo operations will start.
- *Cargo Completion & Port Departure Planning:* The terminal operator shares the planned time of cargo operation completion with the carrier, which enables departure planning with the port.
- *Cargo Completion & Berth Departure:* Cargo operations are completed, port clearance is granted, and the ship departs for the next port from the berth.

6. Operational Information for Port Calls

Since several different processes, actors and domains are involved in a ship visiting a port, the standardization of information to be exchanged should be held at a minimum. Several stakeholders (IMO, ISO, ITPCO, DCSA, S-131, IHO) have defined the following three: Location, Timestamp and Maritime Services as shown in Figure 2. This figure shows basic operational information needed before, during and after a port call.

The following sections will describe this information to ensure consistent exchange of data ensuring timing of the port arrival, departure and service execution and also information to describe locations related to the port call.

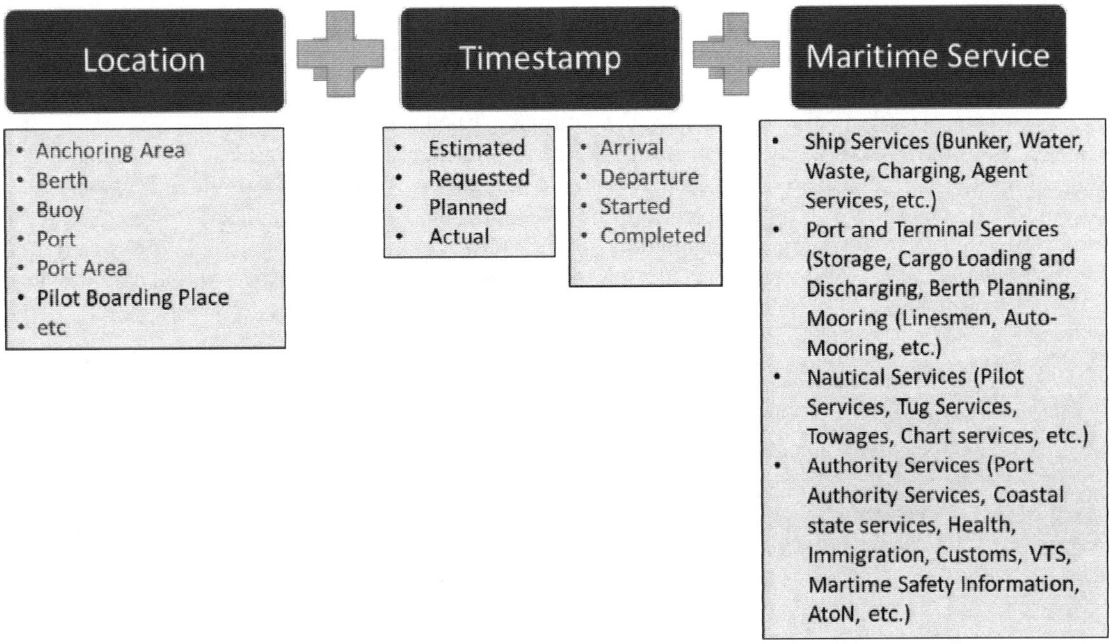

Figure 2. Operational Information for Port Calls

7. Timestamps

The IMO Reference Model gives a semantic definition of timestamps for port arrival, port departure and for service start and end as shown in Figure 3. ITPCO has described the processes using these timestamps, and the corresponding sequence diagram details are shown in Figure 4.

8. Locations

To ensure easy interoperability between the various systems and actors involved in a port call, it is important to be able to identify locations in the port for different usages. Several initiatives are relevant here, both from the sea side and the shore side. IHO is maintaining the S-57/S-100 standard covering ECDIS onboard the ships, while national representatives handle the mapping in the ports. In addition to this comes the different ways to identify locations for the different modes; road, rail and sea, and also for the port and terminal infrastructure. The IMO Reference data model has currently (2022) included a simple definition of locations. The port is identified by a UNLOCODE (or with the name if a UNLOCODE does not exist), and a location in port is defined as "the name of the berth, terminal or station where the ship is located in the referenced port". In addition, to handle the Just-inTime concepts, the concepts as listed in Figure 5 are defined, similar to what is done by IHO, ITPCO and others. The IMO Reference data model does not describe the data format of these fields other than being free text (or LAT/LON for positions). For the technical data model as for instance ISO 28005, this must be defined as GLN-codes from GS1, MRN-numbers from IALA or by other means.

9. Maritime Services

Figure 6 shows the definition of maritime services from the IMO Reference model. In ITPCO, the services are divided into:

- *Nautical services:* Services related to the safe passage and berthing of the vessel.

- *Vessel services:* Services related to the vessel, for instance bunkers, lube oil, potable water, provisions, stores, waste per IMO/MARPOL class, repairs, vetting, flag survey, periodic maintenance and others.
- *Cargo services:* Services related to the cargo, for instance cargo handling, cargo survey, lashing and others. The definition of maritime services is important to be able to link the reference model to the IHO S-100 framework for e-navigation [6] and its Common Maritime Data Structure [7][8]. Also, a clear definition of maritime services is needed to be used in the exchange of operational information before, during and after a port call to ensure efficient booking of services, follow up of service execution during the port call, and the completion of the services.

Time stamp	Definition
Date and time of arrival – actual	The date and time the ship arrives at a specified location, ATA.
Date and time of arrival - estimated	The date and time the ship is estimated to arrive at a specified location, ETA.
Date and time of departure - actual	The date and time the ship departs from a specified location, ATD.
Date and time of departure - estimated	The date and time the ship is estimated to depart from a specified location, ETD.
Date and time of arrival - planned	Date and time the ship plans to arrive at a specific location, PTA
Date and time of arrival - requested	The date and time the ship is requested to arrive at a specified location, RTA
Date and time of departure - planned	Date and time the ship plans to depart from a specific location, PTD
Date and time of departure - requested	The date and time the ship is requested to depart from a specified location, RTD
Date and time of service start - estimated	The date and time a service provider estimates a specified maritime service will start, ETS.
Date and time of service start - requested	The date and time a ship requests a service provider to start a specified maritime service, RTS.
Date and time of service start - planned	The date and time a service provider plans to start a specified service, PTS.
Date and time of service start - actual	The date and time a maritime service provider starts a specified service, ATS.
Date and time of service completion - estimated	The date and time a service provider estimates a specified maritime service will be completed, ETC.
Date and time of service completion - requested	The date and time a ship requests a service provider to complete a specified maritime service, RTC.
Date and time of service completion - planned	The date and time a service provider plans to complete a specified maritime service, PTC.
Date and time of service completion - actual	The date and time a service provider completes a specified maritime service, ATC.

Figure 3. IMO Reference Data Model timestamps

10. Ship-Shore Processes

As part of the work to describe autonomous ships calling at small ports and especially the mother/daughter vessel concept in AEGIS, an overview of processes related to ship operation, port and terminal operation and cargo operation were described, Figure 7. The processes cover both the phases Marketing, sales and alignment, Planning, Execution, and Completion, with the main focus on Planning and Execution.

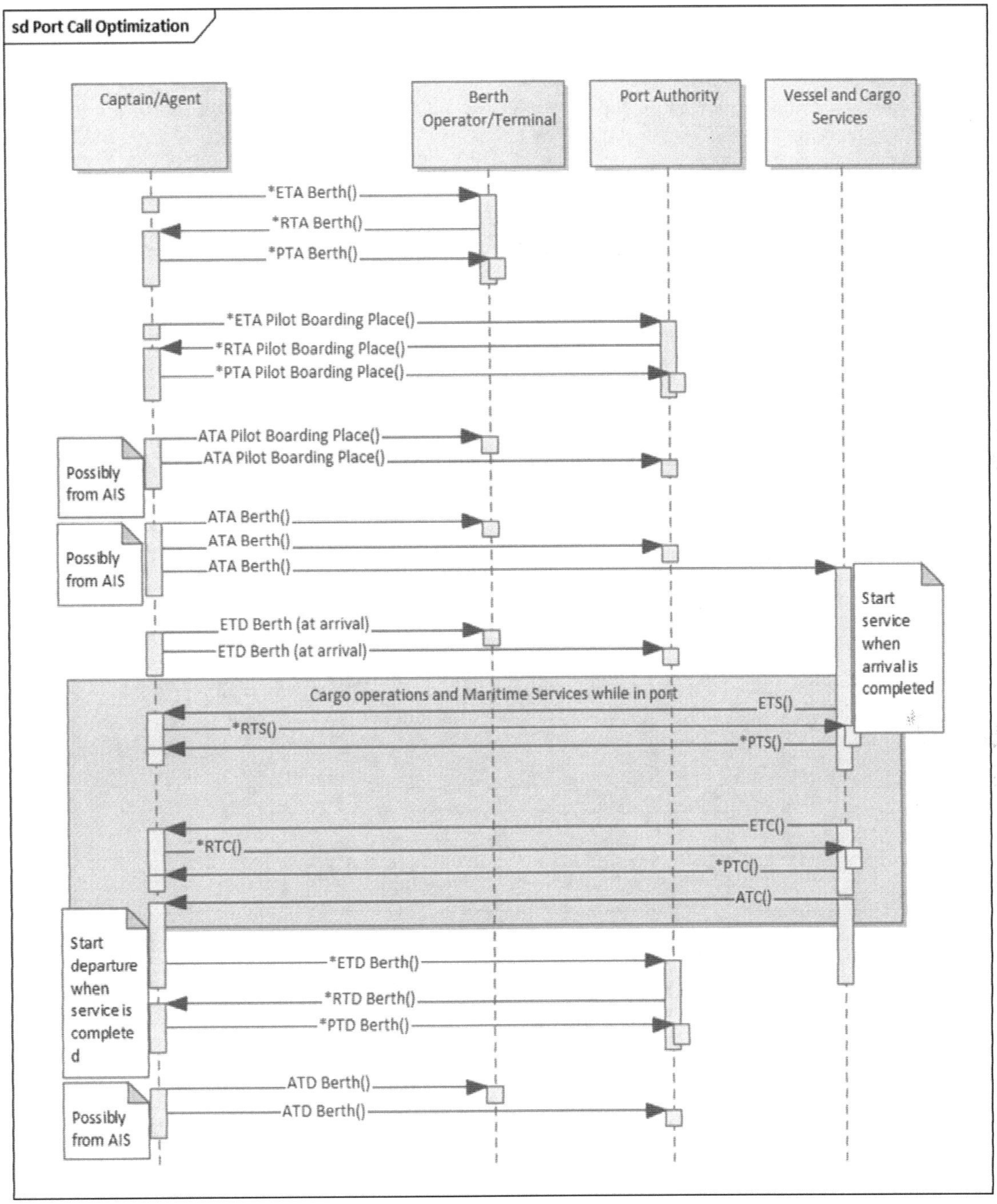

Figure 4. Sequence Diagram for timestamps

10.1. Voyage Planning

Based on ship details and the booking and cargo information, the ship representative (ship operation or captain) sets up the stowage plan (bay plan) (done by ship operation or captain) and the loading and discharge plan for each port call (done by ship agent or ship operation). This includes planning of the equipment that has to be used during the loading and discharge operation and the timing of the operation.

Included here is also the plan for the port call when several terminals are to be visited and several port and terminal services are needed in addition to berth and mooring, for instance for onshore power supply, bunkers, waste collection or fresh-water supply.

The shipping company may also need a collection plan and discharge plan describing interaction with hinterland transport, that is, how and when cargo will be delivered to the ship and fetched from the quay after discharging.

If the Requested Time of Arrival (RTA) to the pilot boarding place and to the berth could be sent directly to the (ECDIS), for instance using the S-421 route exchange format, this would provide the ship with better information to adjust their speed to arrive at the port in a timely manner.

Anchorage	An area in which ships anchor or may anchor (IHO Hydrographic Dictionary S-32)
Terminal	A number of berths grouped together and provided with facilities for handling cargo (IHO Hydrographic Dictionary S-32)
Pilot Boarding Place	At sea, the meeting place to which the pilot comes out.
Berth	The space assigned to, or taken up by, a ship when anchored or when lying alongside a quay, wharf, jetty, or other structure.
Berth Position	The position along the line of a berth, specified by one point (e.g. bollard, manifold or ramp number), allowing the ship to berth in the correct position along the berth.

Figure 5. IMO Reference Data Model for JIT Locations

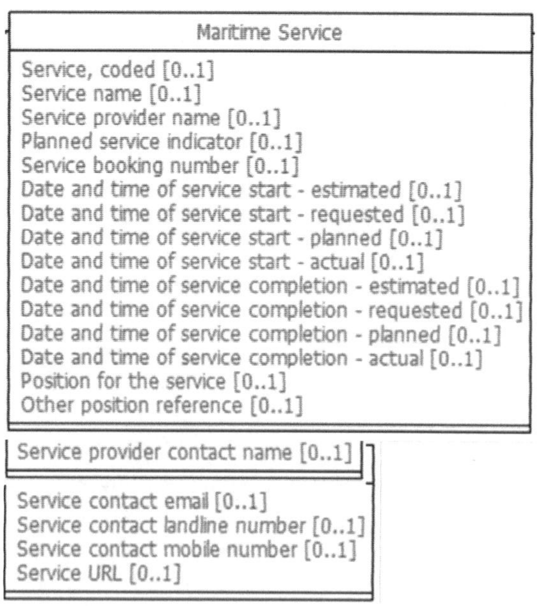

Figure 6. IMO Reference Model Maritime Service Data Elements

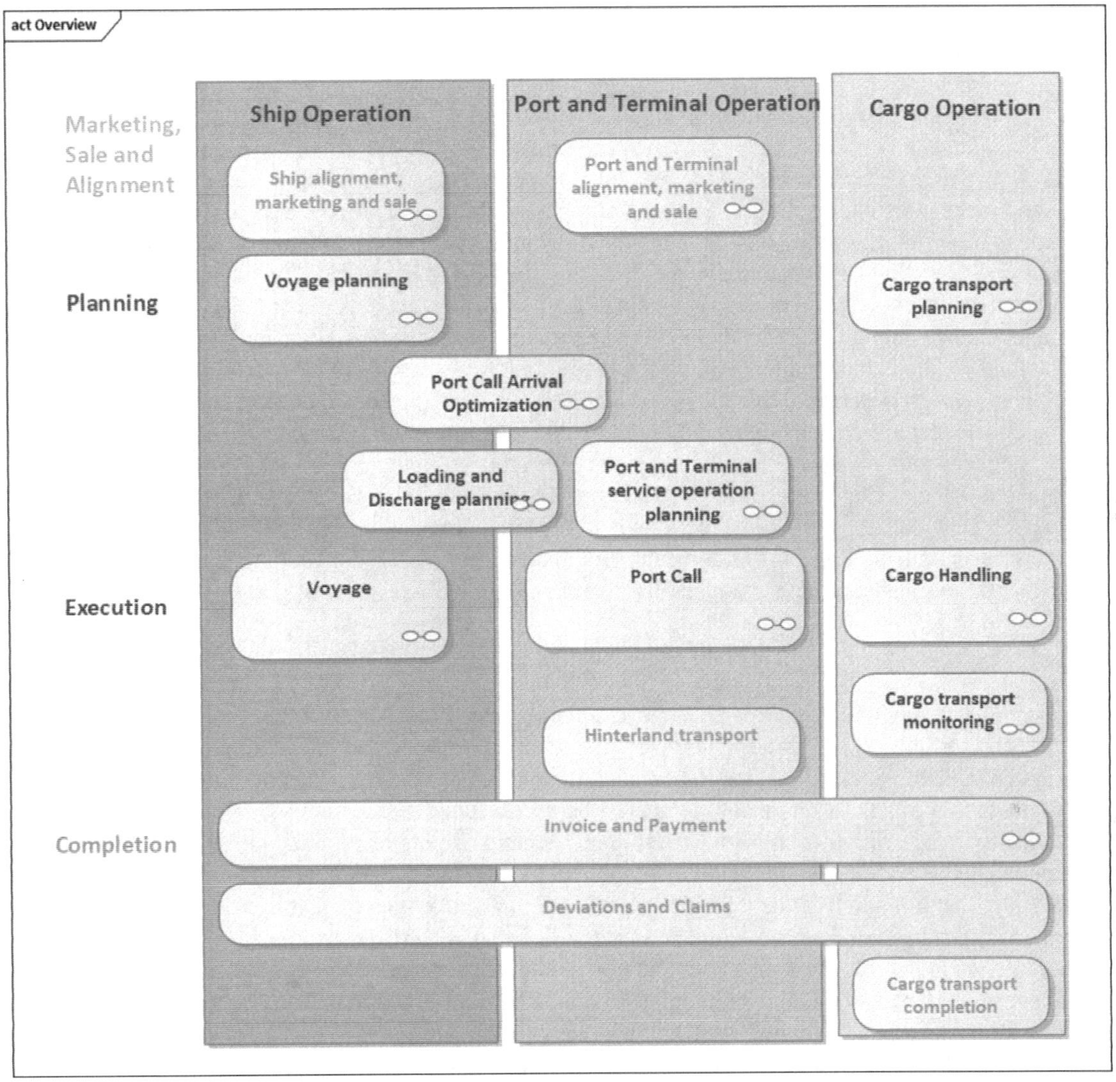

Figure 7. Port Call Processes [from AEGIS]

10.2. Port Call Arrival Optimization

ITPCO has given a very detailed description of the port call arrival optimization [5]. The basic principle is that the ETA has three stages, the ETA proposed by the captain or ship agent, the RTA (requested time of arrival) proposed by the terminal based on the ETA from the captain and the planned arrival time confirmed by the captain (PTA). To improve the terminal service operations, updates of the various time stamps must be given to all involved actors throughout this process. The planning of the arrival time may have several iterations, where the ship agent informs about changes in the ETA and ETD every day, which means that the loop of ETA-> RTA->PTA can be repeated several times.

Berth planning for a ship at the arrival port is usually organized by the terminal operator of the berth. The terminal operator tells which ship will be moored at what time, at which berthing position and which side should be alongside.

By giving the ship a better estimate of the arrival time to the berth, the ship can optimize its speed towards the port. This will have significant impact on the environment, since reduced speed will give large reductions in overall emissions. Also, improved berth planning will improve the hinterland transport planning when the cargo is transferred to another transport mode. The same will be the case for trans-shipment to for instance a mother or daughter ship.

10.3. Loading and Discharge Planning

Loading and discharge planning is done as part of the port call. As the loading and discharge approach, more and more information about the cargo will be available to the shipping company and the terminal operators. The shipping company sets up the handling instructions for cargo, also called shipping instructions, based on cargo information either from the booking they have received and confirmed, or from the transport contract that they have agreed on (typically for bulk cargo). Based on the handling instructions, the shipping company creates the sea waybill including a description of possible dangerous goods. Further, the shipping agent sets up the loading plan and discharge plan based on the sea waybill. These plans are used as input to the stowage plan (bay plan) of the ship. The stowage plan is created by the captain. The terminal plans the loading and discharge operations based on both the loading and discharge plans and the stowage plan (bay plan). Also related to the cargo flow, the shipping company and the terminal exchange information about empty containers, that is, the shipping company can request a specific container or just some containers, and this information is provided by the terminal. Optimalization can be done in the terminal, for instance by using the loading lists prepared by the shipping operator: The terminal can use this to order the containers according to this plan.

10.4. Port and Terminal Service Operation Planning

The port and terminals are the main actors in this process, consisting of the planning of operational activities according to the contract between port and terminal, and the shipping company. Also, interaction with various service providers and infrastructure providers in the port and terminal is part of this process. The result of this planning should be a clear definition of how the port call is performed regarding exchange of information, which ship movements are expected in the port, which ship services are needed and the timing of this. This information may be noted in an execution plan for the port call. In addition, the port and terminal service operation planning consists of each service and infrastructure provider having to plan their work and resource utilization with respect to each of the port calls. This includes planning of the traffic flow in the port and ensuring that safety aspects are maintained. For services that may be autonomous, it includes a clear definition of the requirements to be fulfilled before the operation can start and the responsibilities during the operation.

Resource allocation is a key activity in this process which contains identification and allocation of terminal and port resources needed during a port call. Security aspects and needs related to public inspections have to be taken into consideration in the planning. It must be possible for the regulators to execute inspections of goods e.g., customs clearance and health inspections. In cases of dangerous goods, safety and security are issues which must be planned for in accordance with regulators requirements and needs. Agreeing on the start and end times for the services is an important part of the resource allocation.

Booking is the process confirming that resources are made available for the required service based on the resource allocation process. This is based on the agreed timing of the services, the cargo information and the handling instructions.

Port planning for a ship arriving in a port is usually organized by the port authority, based on the berth planning done by the terminal located in the port. The port authority approves the request for a ship to enter the port and gives information about the pilot boarding place, if relevant. Port authorities often share their port planning with Vessel Traffic Services (VTS), for instance through VHF communication or a MSW, or as in Norway, through SSNN.

Ship services are usually planned by the ship agent together with the captain. The role of the ship agent is to coordinate with the ship service providers in the terminal, for instance for services as bunkers, provisions and waste collection. Cargo services are usually planned by the terminal in collaboration with the captain, and this requires that the terminal has a good overview of the operation of all the different service providers in the terminal to be able to set up a good plan for the ship stay. The terminal needs to know the various cargo services well in addition to be aware of services that requires clearance from the port authorities or the availability of crew on deck.

10.5. Cargo Transport Planning

The cargo transport planning coves the processes where the transport user (cargo handler) prepares customs and other import/export documents which require knowledge about public laws and regulations at national and international level, prepare special handling requirements which require information and knowledge about regulations and laws in different countries e.g. how to handle dangerous goods, environmental requirements related to the transport of cargo, and safety and security requirements that apply.

10.6. Voyage

The voyage consists of the sailing, the ship approaching the port, the port call and the ship departing from the port. Sailing includes the safe navigation of the ship using various e-Navigation tools, weather forecasts etc., and also noon-reporting to the ship operator and MRS and VTS reporting to public authorities. For border crossing, customs reporting may be needed. When it comes to cargo, it may include cargo monitoring to ensure that transport companies can give information to the transport users about status of the cargo. Other cargo reporting may include reporting of cargo information to other public authorities according to international or regional laws and regulations. Updates about the next port call may be received from the port or the ship operator. AIS and possibly other sensor information is made available for shore actors and other ships during the voyage. For the mother-daughter concept, position and status updates between the participants are important.

Ship departing from port includes departure reporting to the port and completion of the statement of facts that summarize all activities performed, including the timing, during the port call.

10.7. Port Call

The port call (ship-port operation) describes the interaction between the port/terminal and the ship during a port call. A port call consists of anchoring (entering and leaving), berthing, performing some port/terminal services including loading and discharging of cargo, then leaving the berth, and possibly movements to other terminals, berths, or anchorages (in-port movement). The information exchange related to port calls include timestamps needed to support Just-in-Time arrival/Port Call Optimisation, authority reporting, cargo loading/unloading, and information about port and terminal services that are needed during the port call. In the execution phase, also information needed to actually execute the operations must be identified, for instance related to autonomous cranes and mooring systems, electricity services and safety and navigational issues.

10.8. Cargo Handling

Cargo handling includes unloading and loading of cargo and interaction between the terminal and the ship. Crane operations are main services in these activities. During this activity, the transport user needs to get tracking information about the cargo. This information is often given to the user by agents based on information from the ship and terminal. For cargo handling involving autonomous functionalities, this process will lead to an extended information exchange between the ship, the terminal system and the crane. Also, the positioning of the ship, the cargo and the cranes will be essential, in addition to the status of the operation. Also, the autonomous crane will need information about the weight and volume of the cargo, the position it is stored, and where it is to be placed.

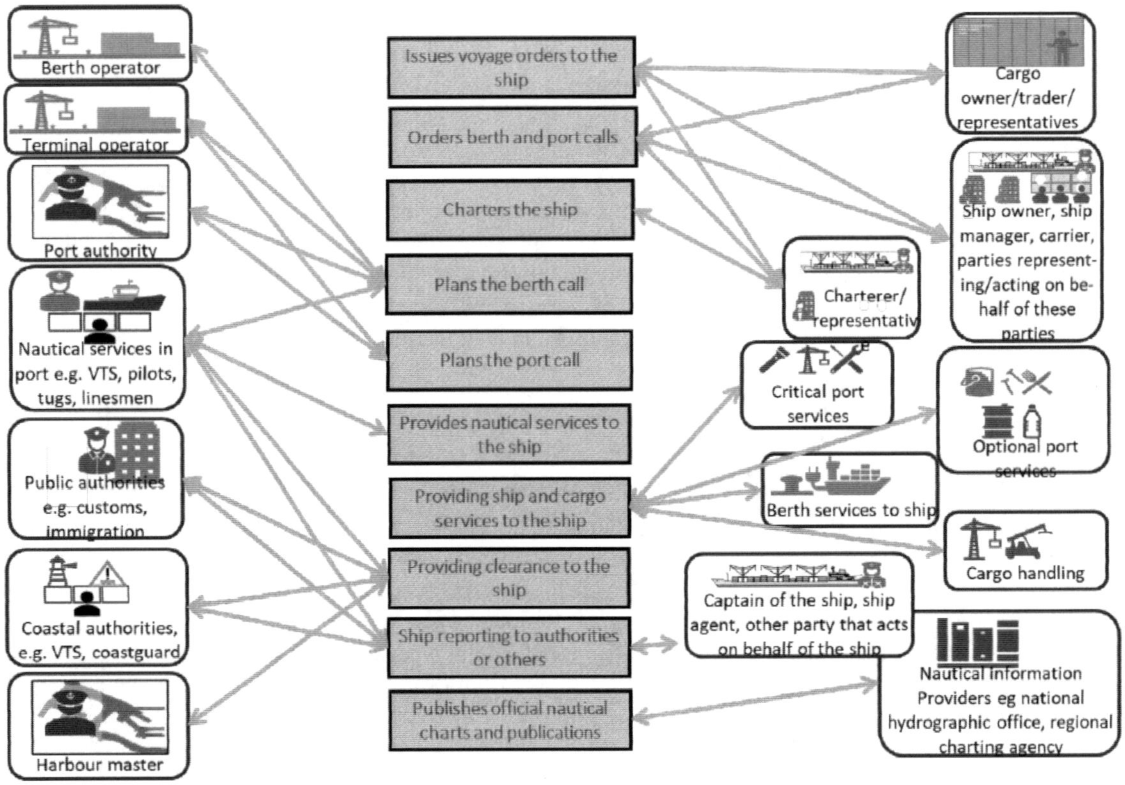

Figure 8. Overview of Actors and Roles

10.9. Cargo Transport Monitoring

During the execution of the cargo transport, the shipping company must ensure monitoring of the cargo handling and the cargo status, monitor the delivery progress, handle deviations during the cargo movement, keep track of commercial operations relevant for the cargo, for instance that the owner has changed, and also to follow up on authorities for customs and cargo clearance (reporting to authorities in accordance with law and regulations about changes and deviations).

When a container is registered for import clearance, the following happens: For customs clearance, each container is assigned a number identifying the container and also the terminal/customs storage and the date. Based on this, a cargo number is generated. This is registered in the customs system by the agent. Then, customs can start their cargo clearance, and after that possibly do inspections. When the terminal gets the generated cargo number from customs, they can release the goods.

11. Actors

Figure 8 gives an overview of the actors related to a port call. The blue boxes in the middle show different roles that the actors must fulfil. Each role can be fulfilled by several actors, and each actor can fulfil several roles.

12. Conclusions

This paper proposes a maritime reference framework containing the different layers that are needed for a "model in the middle" to be able to describe efficient interoperability between systems and stakeholders before, during and after a port call. Especially, the several sub-models are used to describe the commonalities between the different actors and systems to avoid having one large model that could be difficult to define and maintain.

The main benefit of the IMO Reference Model is to ensure a unique definition of concepts related to ship-shore data exchange. This ensures the following:

- Agreement across many stakeholders: IMO as the owner of the IMO Reference Model is supported by several maritime organizations that are involved in the development and maintenance of this model. This ensures its uptake in maritime transport. Also, standardization bodies as ISO, IEC and IHO are involved in the work in addition to organizations representing other domains than the maritime (UNECE and WCO).
- Agreement across many domains: The data elements defined in the IMO Reference Model can be used to ensure interoperability between systems covering different domains, as reporting to port states, terminal and port service systems covering operational data related to a port call, and systems for charts and port assets covering nautical data.
- A common reference model will pave the way for reuse of data in several contexts by different stakeholders, given that the data is clearly defined. This will support the report once – principle of maritime single window systems.
- Clear definitions of a set of data elements that have been agreed on by several stakeholders will ensure efficient interoperability between systems and the possibility to support digitalization of earlier manual processes related to ship reporting and ship operation. However, this will also require a clear description of the system interactions.

Acknowledgments

This work was performed as part of the AEGIS project, that has received funding from the European Union's Horizon 2020 research and innovation program under Grant Agreement 859992.

References

[1] IMO, Equasis [Online] Visited January 2022. **http://www.equasis.org/Fichiers/Statistique/MOA/Documents%20availables%20on%20statistics%20of%20Equasis/Equasis%20Statistics%20-%20The%20world%20fleet%202019.pdf**
[2] IMO Compendium, 2021 [Online] Visited Jan. 2022 **https://www.imo.org/en/OurWork/Facilitation/Pages/IMOCompendium.aspx**
[3] ISO 28005-2, 2021 [Online] Visited Feb. 2022 **https://www.iso.org/standard/74059.html**
[4] Lloyd's Maritime Atlas of World Ports and Shipping Places 2020-2021, 2019. ISBN 9780367427108.
[5] ITPCO Port Information Manual for ship-port interface data, version 3.03, visited Jan. 2022 **https://portcalloptimization.org/images/Port%20Information%20Manual%203.02.pdf**
[6] S-100 Universal Hydrographic Data Model [Online], visited Jan 2022, **https://iho.int/en/s-100-universal-hydrographic-data-model**
[7] Rødseth ØJ. 2016 Integrating IEC and ISO information models into the S-100 Common Maritime Data Structure, E-navigation Underway International 2016; Copenhagen, Denmark.
[8] Rødseth ØJ. 2011 A maritime ITS architecture for e-navigation and e-maritime: Supporting environment friendly ship transport. In 2011 14th International IEEE Conference on Intelligent Transportation Systems (ITSC).

Experimental study on smoke flow behaviour under ships' deck structures in early stages of fire

Arata KIMURA[1], **Yasushi OKA**[2], **Hideyuki OKA**[1] **and Susumu Ota**[1]

[1] National Maritime Research Institute, 6-38-1 Shinkawa, Mitaka, Tokyo 181-0004, Japan
[2] Yokohama National University, 85-7 Tokiwadai, Hodogaya-ku, Yokohama, 240-8501, Japan

E-mail: kimura-a@m.mpat.go.jp

Abstract. In ships' cargo spaces carrying vehicles, so called "vehicle spaces", fire has sometimes taken place and safety measures against such fire is under discussion at the International Maritime Organization (IMO). Early detection of fire is one of the essential elements of such safety measures. Location of fire detectors affects swiftness of detection of fire, and in onshore buildings, fire detectors are generally installed directly on ceilings because there are no obstacles under ceilings and smoke spreads along the ceilings. On the other hand, structures below ceilings, i.e., decks, of vehicle spaces are quite different from those of onshore buildings due to strength members below a deck, such as stiffeners and web beams. The appropriate positions of smoke detectors in a vehicle space, including vertical positions of detectors from a ceiling, has not been well-understood, while the maximum coverage of a detector is determined by an international regulation. In this study, the effect of structures below the ceilings of vehicle spaces on detection of fire has been investigated and the appropriate positions of smoke detectors including vertical positions have been considered, in particular, for smouldering fire sources. For this purpose, a one-sixth scaled deck model was prepared to simulate a part of a vehicle space and fire experiments were conducted using a small-scaled fire source. The effect of the structure on a smoke flow beneath the ceiling was experimentally investigated under the condition that the flow had a weak driving force in a smouldering fire or an early stage of fire. In the experiment, the flow was visualized using a laser and the smoke density expressed with the extinction coefficient was also measured. In addition, a small flaming fire source was used during the experiment, to further investigate a possibility of detection in an early stage of fire. The vertical profiles of smoke density were measured using the laser transmissometer. Furthermore, a smoke flow was visualized using continuous wave laser. As the results, it is concluded that it is appropriate to install smoke detectors at the lower end of the longitudinal stiffener, instead of installing on the ceilings in vehicle spaces.

1. Introduction

In ships' cargo spaces carrying vehicles, so called "vehicle spaces", fire has sometimes taken place and safety measures against such fire is under discussion at the International Maritime Organization (IMO). Early detection of fire is one of the essential elements of such safety measures. In general, location of fire detectors affects swiftness of fire detection, and smoke detectors are used in vehicle spaces. In order to determine an appropriate location of smoke detectors, it is desired to take into account the relevant flow characteristics of smoke under ceiling.

In onshore buildings, fire detectors are generally installed directly on ceilings because there are no obstacles under ceilings and smoke spreads along ceilings. For such structure and for flame combustion, relevant studies focusing on the fire induced smoke flow under ceilings have been reported. In one of the pioneering studies, Alpert performed experiments on the horizontal distribution of temperature and velocity of ceiling flow formed under a flat ceiling, varying three parameters, i.e., ceiling height, horizontal distance from the fire source, and heat release rate. He derived a versatile prediction equation from the results of experiment (Alpert, 1972). Alpert further conducted a theoretical study of ceiling flows and proposed a practical model (Alpert, 1975). Around the same time, Heskestad (1977) theoretically considered the velocity, temperature, and concentration of combustion products in a ceiling flow produced by a fire source whose heat release rate changes as a power of time, and showed that similarity laws exist for respective parameters. By the analysis based on the similarity laws, it was theoretically shown that the time taken for the tip of the ceiling flow to reach the fire detector is hardly affected by the heat release rate of the fire source. Later, Heskestad and Delichatsios (1977) conducted an experimental study to investigate the response characteristics of heat and smoke detectors, and validated the theory proposed by Heskestad (1977). Furthermore, Heskestad and Delichatsios (1978) conducted various experiments in a large space under various conditions of fire source and ceiling height, and derived formulae for predicting the time variation of temperature and velocity in the ceiling flow caused by a fire source having heat release rate proportional to the square of time. Since then, relevant studies have been conducted by Copper (1982), Heskestad and Hamada (1993), Yao and Marshall (2006), and others to investigate the characteristics of ceiling flow under a smooth, horizontal ceiling. From the attenuation properties of temperature and velocity with respect to the distance from the main axis of a fire source to the radial direction, a practical model has been developed and used for the prediction of a response time of heat detector and other applications. In addition, the time for fire detection can be estimated by simplified analytic theory proposed by Tanaka et al. (1985) in relation to heat release rate and horizontal section area of a space, etc. Regarding a past study about fire detection using weak fire source, i.e. smouldering fire, Hotta et al. (1987) experimentally investigated vertical thermal condition and reported that the updraft with smoke originated from the smouldering fire source was prevented from entering into the layer under the ceiling through a thermocline when the excess temperature of the smoke was 2 to 3 K. On the other hand, structures below ceilings, i.e., decks, of vehicle spaces are quite different from those of onshore buildings due to strength members below a deck, such as stiffeners and web beams. The appropriate positions of smoke detectors in a vehicle space, including vertical positions of detectors from a ceiling, has not been well-understood, while the maximum coverage of a detector is determined by an international regulation irrespective of the structures below the ceilings.

In this study, the effect of structures below the ceilings of vehicle spaces on detection of fire has been investigated and the appropriate position of smoke detectors including vertical position has been considered, in particular, for smouldering fire sources. In order to understand the flow characteristics of smoke under ceilings of vehicle spaces, a one-sixth scaled deck model was prepared to simulate a part of a vehicle space and fire experiments were conducted using a small-scaled fire source. The effect of the structure on a smoke flow beneath the ceiling was experimentally investigated under the condition that the flow had a weak driving force in a smouldering fire or an early stage of fire. In the experiment, flow was visualized using a laser, and the smoke density expressed with the extinction coefficient was also measured. In addition, a small flaming fire source was used during the experiment, to further investigate a possibility of detection in an early stage of fire. The vertical profiles of smoke density were measured using the laser transmissometer. Furthermore, a smoke flow was visualized using continuous wave laser. Based on the quantitative and qualitative measurements, the flow characteristics in a smoke layer and suitable vertical position of smoke detector is discussed.

2. Experiments

2.1. Scaled deck model

The $1/6^{th}$ scale deck model was prepared for the experiment. The model simulates a part of internal structure of a pure car carrier including strength members below a deck . The deck model has three zones partitioned by deeper transverse structures, so called "transverse webs". Two of the zones (Zone A and Zone B) are fitted with shallower longitudinal structure, so called "longitudinal stiffeners", and one zone (Zone C) is not fitted with longitudinal stiffeners. Figure 1 outlines the deck model and Zone C is expressed with green in figure 1 (a). The full scale deck structure is assumed to have 3 m in hight, 18 m from the deeper longitudinal structures, so called "centre girder", to the side walls, 3.5 m intervals between the transverse webs, 0.9 m intervals between the longitudinal stiffeners. The depths of the longitudinal stiffeners and the centre girder and transverse webs are 0.3 m and 0.8 m, respectively. Therefore, in the 1/6 scale model, the deck height, distance from the centre girder to the side wall, interval of transverse webs, interval of longitudinal stiffeners, depth of longitudinal stiffeners and depth of the centre girder and transverse webs are 0.5 m, 3 m, 0.583 m, 0.15 m, 0.05 m and 0.133 m, respectively. The deck consists of three layers of 12 mm thick calcium silicate plates to insulate the ceiling. In order to visualize the smoke flow, the centre girder and transverse webs are made of PMMA with a thickness of 2 mm, while heat-resistant glass having 2 mm thick was used for these structures in the area close to the fire source, taking into account the weakness of PMMA to heat. The 1/6 scale model is prepared in accordance with the Froude modelling Scaling Law (Quintiere, J.G. (1989)). As shown in Table 1, physical quantities in model scale can be interpreted into those in full scale. Since the scaling ratio of the model is 1/6 of a full scale deck, the heat release rate (HRR) in the model scale can be converted to the full scale value by multiplying by 88.2 i.e., 6 to the power of 5/2. In the same way, the velocity and time in full-scale are obtained by multiplying values in the model scale by 2.45, i.e., 6 to the power of 1/2, and the temperature and smoke density in the model scale equal to those of full-scale deck.

Table 1. Physical quantities interpretation based on the Froude modelling Scaling Law

Physical quantity	Scaling formula
Length	$l_o : l_m = 1 : a$
Heat release rate	$\dot{Q}_o : \dot{Q}_m = 1 : a^{5/2}$
Velocity	$V_o : V_m = 1 : a^{1/2}$
Time	$t_o : t_m = 1 : a^{1/2}$
Temperature	$T_o : T_m = 1 : 1$
Pressure	$P_o : P_m = 1 : 1$
Characteristic length of fire	$l = \left(\frac{\dot{Q}}{\rho_0 C_p T_0 g^{1/2}}\right)^{2/5}$
(Subscriptions)	O : Full-scale, m : Model scale, a : scaling ratio

2.2. Fire source

In the smouldering combustion experiment, cotton wicks were used varying number and length as smouldering fire source and a weak hot smoke was generated toward the ceiling of the model. The cotton wick used for the experiment was six cotton braided cord manufactured by Nagaya Nensyoku Co., Ltd., taking into account "test fire TF3" in ISO/TS 7240-9:2012. ISO/TS 7240-9:2012 "Fire detection and alarm systems Part 9: Test fires for fire detectors" is the international

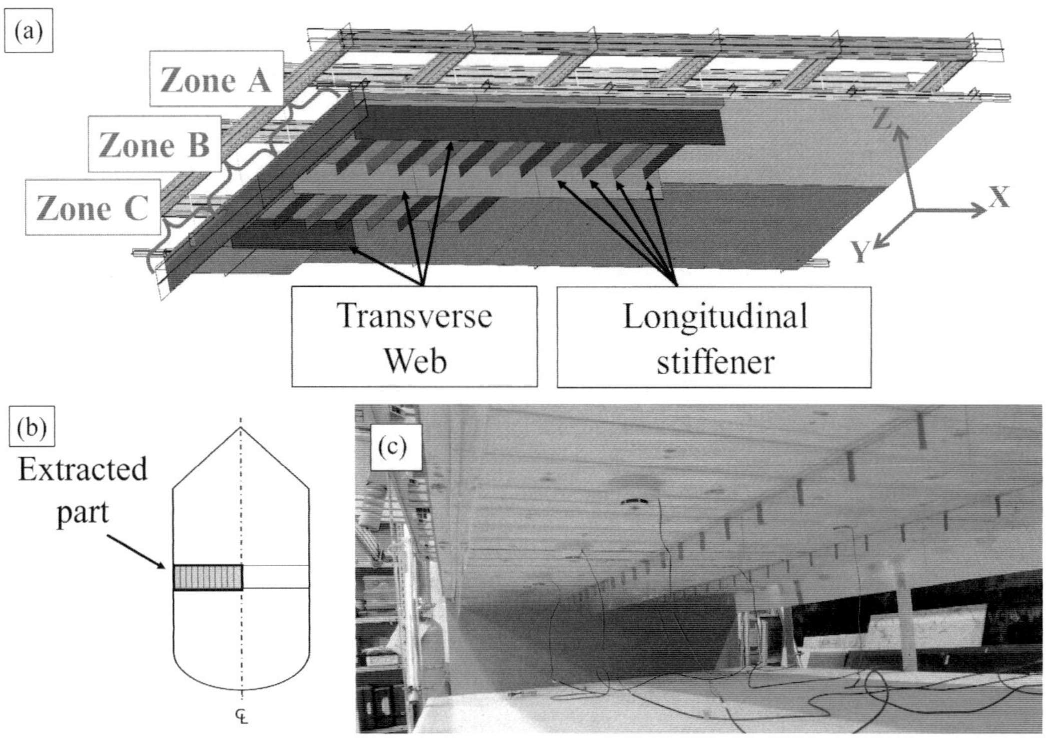

Figure 1. Outline of the $1/6^{th}$ scale model. (a) shows an image of a deck model with the (b) part of the hull extracted. (c) shows a picture of the actual deck model.

standard on the Fire detection and alarm systems, and Test Fire 3 (TF3) is used for testing smoke detection system in case of smoldering fire. A very weak updraft with smoke was sent to the ceiling space through the 15 cm diameter opening on the floor from a combustion chamber installed under the floor of the deck model. Smoke generation rate and heat release rate were varied by changing the number of cotton wicks in four steps from five to twenty pieces, which were suspended in rig shape in the combustion chamber. In the flaming combustion experiment, the smoke generated by the smouldering combustion of cotton wicks was stored in a chamber installed just below the floor of the model, and a diffusion propane burner having inner diameter of 39 mm was installed in the box. The height of the burner was adjusted for each gas flow rate so that the height of the flame formed on the diffusion burner was almost equal to the height of the 15 cm diameter opening on the floor of the model. The heat of combustion per unit mass of the cotton wick was assumed to be 6.4 kJ/g in accordance with the Watanabe et al. (2005), and the heat release rate was calculated from the averaged weight loss rate and duration of combustion during the experiment measured by an electronic balance (A&D, GX-6002A). The calorific value of propane is 46.0 kJ/g from the literature (Tewarson, A. (2008)), and the heat release rate was calculated based on the flow rate of propane supplied by a mass flow controller (Kofloc, 8500), assuming ideal gas and complete combustion. Based on the Froude modelling scaling law, the heat release rate in the model tests can be converted to the full scale as shown in Table 2.

Table 2. Heat release rate of scale model experiment and converted value of full scale fire.

Fuel		Type of Combustion	HRR in model scale [kW]	HRR in full scale [kW]
Cotton wicks, 50 cm in length	5 pieces	Smouldering	0.0299	2.64
	10 pieces		0.0753	6.64
	15 pieces		0.150	13.2
	20 pieces		0.388	34.2
LPG	0.3 L/min	Flaming	0.45	40
	0.5 L/min		0.75	66
	1.0 L/min		1.5	132

2.3. Smoke density measurement and visualization

The ultra-compact digital laser sensor (Keyence Corporation, LX2-V10) was used for smoke density (extinction coefficient) measurement. A measurement equipment consists of a pair of very small light source (emitter) and a digital optical receiver and the light emitter and the receiver were set on a vertical line from the deck (ceiling) to the floor at 11 points (5, 10, 15, 25, 40, 55, 75, 95, 120, 150 and 180 mm from the ceiling face) to measure the optical smoke density. The smoke density was obtained by the following equation:

$$C_S = \frac{1}{L}\log\left(\frac{I_0}{I}\right). \tag{1}$$

Where Cs, is the smoke density expressed by the extinction coefficient. I0, I and L are the intensity of source, intensity of light at the receiver and optical path length between the light source and the receiver, respectively. Figures 2 and 3 show the overview of the compact digital laser sensor and the setting. As shown in Figure 4, the light sources and receivers were installed so that the light path was between the two transverse webs, which divides Zones A and B, and Zones B and C, respectively. The entire system of light emitters and receivers was protected from the intrusion of smoke flowing through Zones A and C. For this reason, the optical path length was 0.583 m, which was the distance between the transverse webs. The vertical distributions of extinction coefficient were obtained at five positions, i.e., 0.3, 0.6, 0.9, 1.2 and 1.65 m from the centre of the fire source as shown in Figure 5. To observe smoke flow around the deck structure, a continuous wave laser having a 2 Watt output of 532 nm wavelength is utilized.

3. Results and Discussions

3.1. Measurement of smoke generation rate of cotton wicks

Firstly, the smoke generation rate of cotton wicks (6 cotton braids) was measured by the following procedures:

(i) 90 pieces of cotton wicks of 50 cm length were hung on a ring of 0.1 m diameter directly under the 1.5 m square smoke collection hood to collect all the smoke generated by smouldering combustion;

(ii) The collected smoke was exhausted through a duct of 0.4 m diameter;

(iii) The exhausted smoke flow rate per unit time was calculated by multiplying the average velocity of the airflow through the duct for smoke vent by the cross-sectional area of the duct. The average velocity of the airflow was estimated based on the output from a bidirectional flow probe and a thermocouple installed in the centre of the duct for smoke vent at a distance of approximately 6 m from the centre of the smoke collection hood; and

Figure 2. Appearance of the Ultra-compact optical receiver used to measure the extinction coefficient.

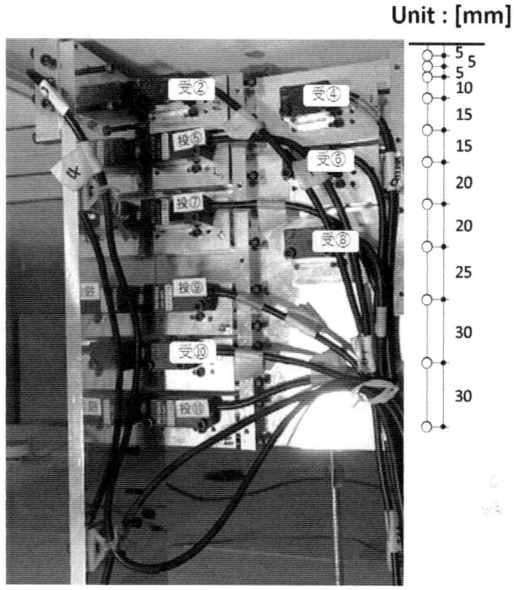

Figure 3. Overview of the apparatus of the extinction coefficient measurement.

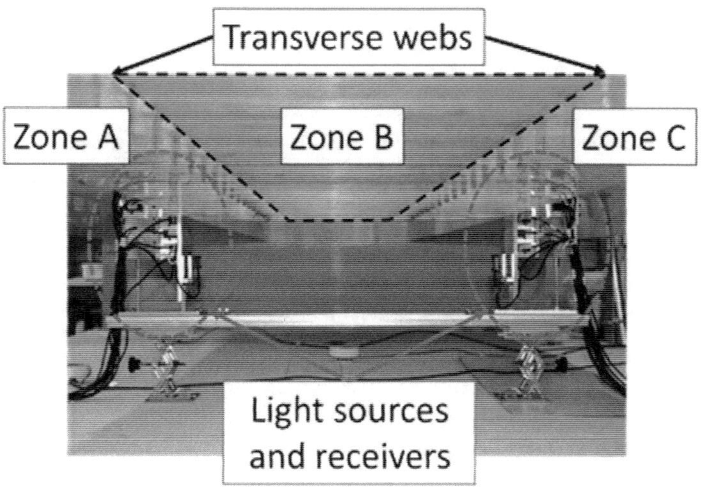

Figure 4. Overview of the measurement of the extinction coefficient passing between two transverse webs in the Zone B.

Figure 5. Scene of the measurement of the extinction coefficient.

(iv) The extinction coefficient of the smoke layer passing through the duct was measured every unit of time with a photoelectric smoke meter, where the optical path length from the light emitter to the receiver of the photoelectric smoke meter was 0.43 m.

The results of smoke emission rate measurements are given in Table 3. In comparison with the previous results of Watanabe et al. (2005), the extinction coefficient and weight loss are almost the same, but the amount of smoke emission is about half of the value provided by Watanabe et al. (2005). It is considered that the reason of the difference between the present measurements and Watanabe et al. is the difference in ambient temperature and humidity during the measurements.

Table 3. Results of smoke emission rate measurements on cotton wicks used in the experiment.

	Stable combustion section		*Extinction coefficient*	*Total smoke emission in the stable section*	*Weight loss in the stability zone*	*Within the stability interval Total smoke/ weight loss*	*Weight Decrease Rate*
	Start [*s*]	*End* [*s*]	*Cs* [1/*m*]	*CsV* [*m*3/*m*]	*W* [*g*]	*CsV*/*W* [*m*3/(*m* − *g*)]	[*g*/*s*]
Test 1	300	900	0.2331	33.98	73.60	0.4617	0.1103
Test 2	300	1400	0.2049	36.02	92.90	0.3878	0.0760
Test 3	300	1600	0.2720	32.03	103.90	0.3083	0.0719
Test 4	300	1200	0.5448	37.16	105.10	0.3535	0.1051
Test 5	300	1000	0.3115	34.53	85.50	0.4038	0.1100
Average			0.313	34.74	92.2	0.383	0.0947
Watanabe et al.		*1st*	0.274	-	111	0.691	-
		2nd	0.372	-	143	0.728	-
		Ave.	0.323	-	127	0.710	-

3.2. Vertical profile of extinction coefficient

Figure 6 shows the vertical profile of extinction coefficients in the smoke layer at 0.3, 0.6, 0.9, 1.2 and 1.65 m horizontally apart from the smouldering fire source (15 pieces of cotton wicks). The profile of extinction coefficient was almost uniform from the ceiling to the bottom of the transverse web, with a decrease in extinction coefficient below that point. Regardless of the number of cotton wicks, at respective measurement points, the extinction coefficients in the region from the ceiling to the lower end of the longitudinal stiffener were within 80 % to 100 % of the value near the ceiling.

Figure 7 shows the profile of the extinction coefficient at x = 1.2 m when the smoke flow was provided from the flaming combustion and smoke was provided by 15 cotton wicks burning. The extinction coefficient was distributed in a uniform manner from the ceiling to the bottom of the transverse web, and even in the lowest case, the extinction coefficient was only reduced to around 80 % of the maximum value. In the case of flaming combustion, similar to the case of smouldering combustion, it was found that a uniform layer of extinction coefficient has been formed down to the lower end of the transverse web.

3.3. Visualization of smoke flow

Figure 8 shows the results of the visualisation of the smoke layer in a B-B' cross-section through the centre of the fire source and across the Zone B. The smoke originated from a fire source has propagated in the floor direction over the centre girder (as a left-to-right flow in the figure) maintaining a height slightly lower than the lower edge of the transverse web during the visualization experiment. The results of the measurement of the vertical profile of extinction

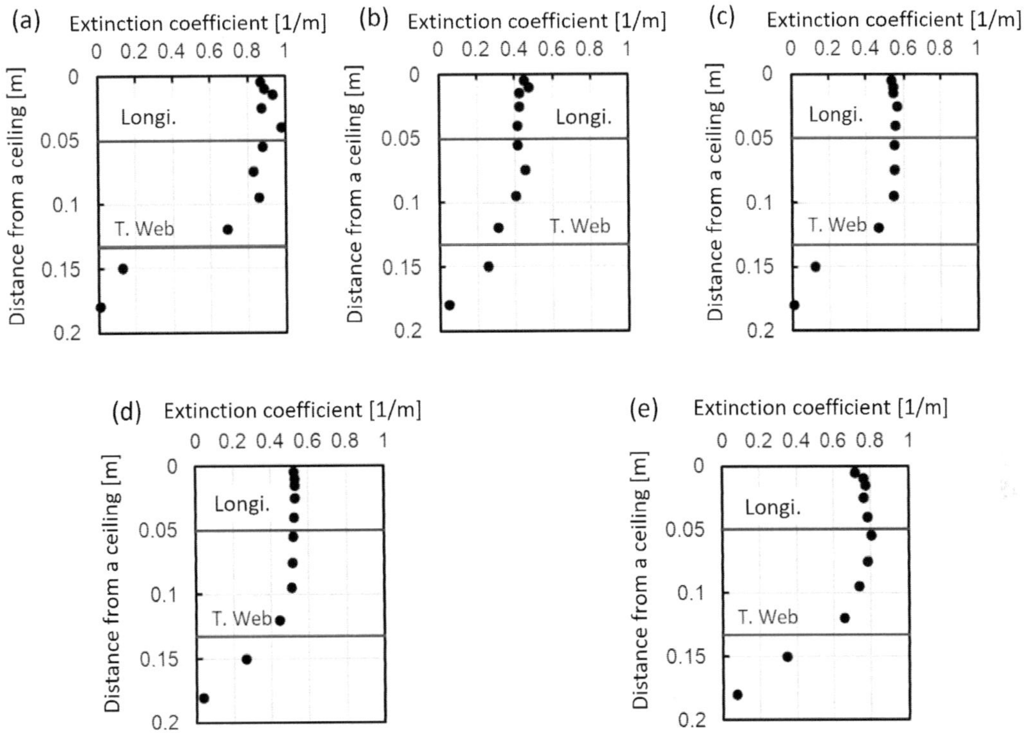

Figure 6. Vertical profile of extinction coefficient in the smouldering fire experiments (15 pieces of cotton wicks). (a), (b), (c), (d) and (e) are data at 0.3 m, 0.6 m, 0.9 m, 1.2 m and 1.65 m from a fire source.

coefficient were consistent with the results of the visualisation. Therefore, it can be said that a vertical position of smoke detectors does not significantly affect on a detection time, taking into account that detection of fire will not be delayed, even if the smoke detector is not installed just on ceiling, as long as the detector exists at any position in the smoke flow.

4. Conclusions

The following findings are obtained as a result of scaled model experiments using a test fire source including the smouldering condition:

(i) In the case of smouldering fire, the vertical profile of extinction coefficient is almost uniform in the smoke layer in the space from the ceiling to the lower end of the longitudinal stiffener. It is observed that the smoke originated from the smouldering fire source is slightly flowed under the longitudinal stiffener and the smoke flow is slowed down by the obstruction of the longitudinal stiffener.

(ii) In case of a flaming fire source, a circulating flow is formed in the afore mentioned space, and a locally dense smoke flow is observed in the space. The vertical distribution of extinction coefficient showed a dogleg shape with the apex of the vertical distribution appearing near the centre of the lower edges of the longitudinal stiffener and the transverse web, in contrast to the case of the smouldering fire source.

It can therefore be said that it is appropriate to install smoke detectors at the lower end of the longitudinal stiffener instead of installing on the ceilings in vehicle spaces, for the reason

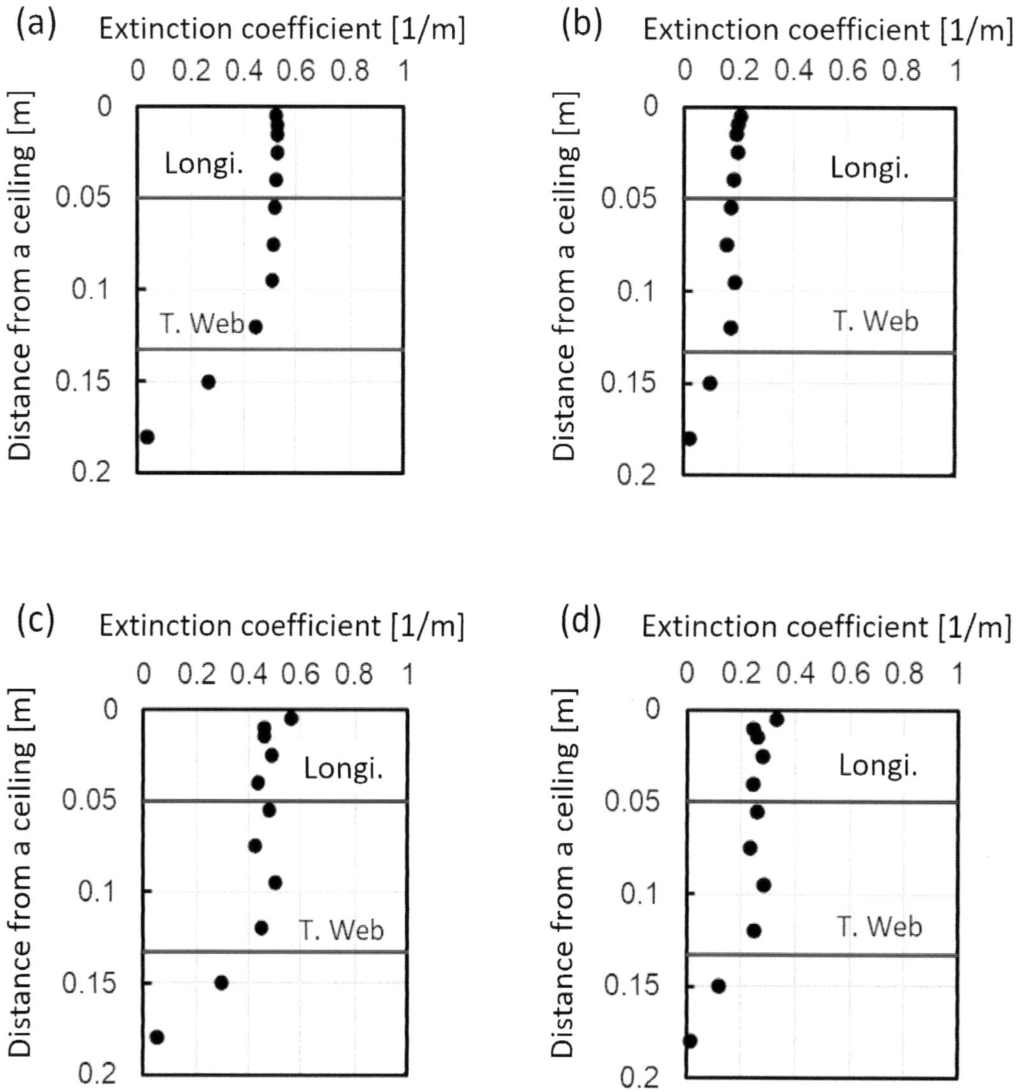

Figure 7. Vertical profile of light extinction coefficient in the smouldering and flaming fire experiments measured at 1.2 m from a fire source. (a) and (b) to (d) are smouldering and flaming (50 cm of cotton wicks, 15 pieces with propane: 0.3 L/min (b), propane: 0.5 L/min (c) and propane: 1.0 L/min (d)) combustions.

that such vertical position of detectors is deemed not to cause significant delay of detection.

Acknowledgments

We would like to thank Mr. Kazuma Baba, undergraduate student at Yokohama National University, for his elaborate experimental works. We would like to thank Mr. Hiroshi Ueno of Nohmi Bosai ltd. for his permission to use the experimental facility and for his cooperation in the measurement of the basic information on smouldering of the cotton wick. This research was financially supported by Japan Ship Technology Research Association (JSTRA).

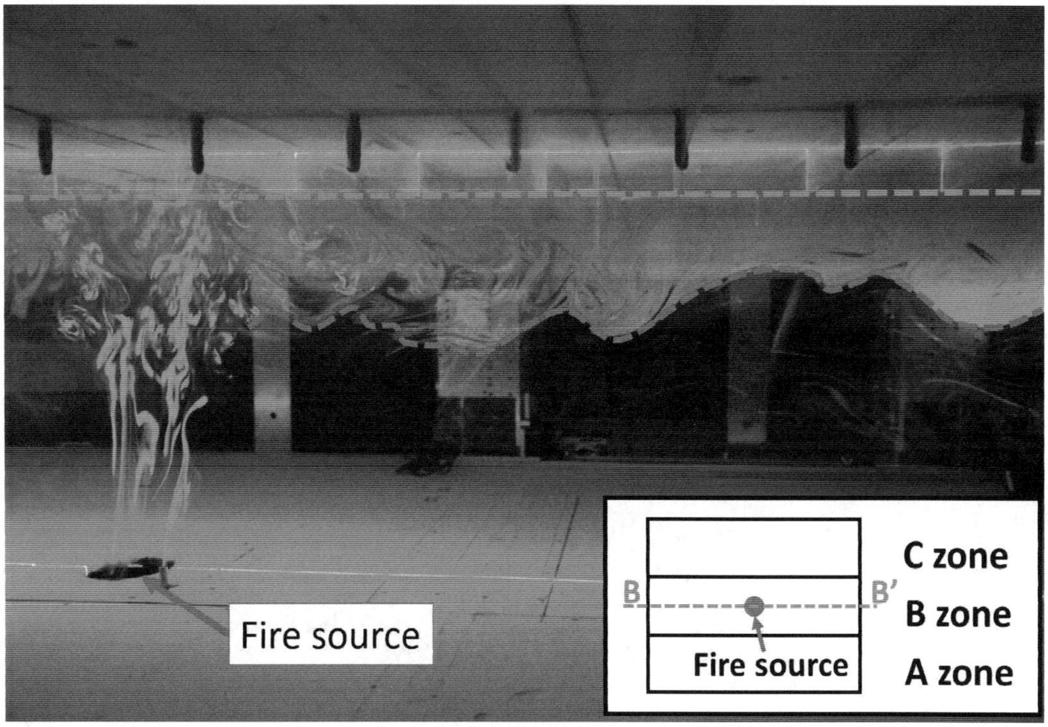

Figure 8. The snapshot of visualization of smoke flow in a B-B' cross- section at the Zone B. The orange dashed line shows lower end of longitudinal stiffeners. The purple dashed line indicates rough boundary between flowing smoke and ambient air.

References

[1] Alpert R L 1972 Calculation of Response Time of Ceiling-Mounted Fire Detectors *Fire Technology* **8** pp 181–195

[2] Alpert R L 1975 Turbulent Ceiling-Jet Induced by Large-Scale Fires *Combustion Science and Technology* **11** pp 197–213

[3] Heskestad G and Delichatsios M A 1977 Environments of Fire Detectors—Phase I: Effect of Fire Size, Ceiling Height and Material (Volume 2. Analysis) (Maryland: National Bureau of Standards) NBS-GCR-77-95

[4] Heskestad G and Delichatsios M A 1978 The Initial Convective Flow in Fire *17th Symposium (International) on Combustion* (Pennsylvania: The Combustion Institute) pp 1113–1123

[5] Cooper L Y 1982 A mathematical model for estimating available safe egress time in fires *Fire and Materials* **6** pp 135–144. Heskestad G and Hamada T 1993 Ceiling jets of strong fire plumes *Fire Safety Journal* **21** pp 69–82

[6] Xiaobo Yao, André W. Marshall 2006 Quantitative salt-water modeling of fire-induced flow *Fire Safety Journal* **41** pp 497–508

[7] Tanaka T and Yamana T 1985 Smoke Control in Large Scale Space; (Part 1 Analytical theories for smoke control in large scale spaces) *Fire Science and Technology* **5** pp 31–40

[8] Hotta H Oka Y and Sugawa O 1987 Interaction Between Hot Layer and Updraft from a Smoldering Fire Source - Part I An Experimental Approach - *Fire Science and Technology* **7** pp 17–25

[9] Quintiere JG 1989 Scaling Applications in Fire Research *Fire Safety Journal* **15** pp 3–29

[10] Watanabe L and Tanaka T 2005 Prediction of Ceiling Jet Smoke Concentration under Two Layer Environment *Fire Science and Technology* **24** pp 151–164

[11] Tewarson A 2008 Generation of heat and gaseous, liquid, and solid products in fires *The SFPE Handbook of Fire Protection Engineering 4th ed* (New York: Springer Science) section 3 chapter 4 p 3-165

Concept design of a digital twin architecture for ship structural health management

Zhiping Cheng*, Eric Tan, Minbo Cai, Allan Ross Magee
Technology Centre for Offshore and Marine, Singapore (TCOMS), 12 Prince George's Park, Singapore 118411

E-mail: chengzp@tcoms.sg

Abstract. Ocean-going vessels frequently experience severe weather conditions. Occasionally, the actual environmental loads are much larger than design values, resulting in deviations from design conditions that may cause serious structural damage. A structural digital twin can enhance predictability and reliability of marine structures, and thus provide vessel operators, owners, and stakeholders with a more accurate overview of a vessel's condition. It can also help reduce lifecycle costs through proposing rational and optimal inspection, maintenance, repair programs, and assist on maximizing service life and mission availability. This paper uses a benchmark vessel to investigate structural behaviors of the vessel through dynamic loading approach. The correlation between structure members and critical responses are established, and high stress concentration areas have been identified. Subsequently, this paper develops a conceptual digital twin architecture. The proposed digital twin predicts responses of the entire structure to measured or predicted environmental conditions with a hybrid model combining numerical simulations with measurements. While the traditional structural monitoring system focuses on the health condition of typical hot spots, the proposed digital twin may improve structural health awareness of the entire hull structure based on a limited number of sensors.
Keywords. ship structural health management, digital twin, dynamic loading approach, critical responses

1. Introduction

Ocean-going vessels frequently experience severe weather conditions. Occasionally, the ship hull will suffer much larger environmental loads than design values, resulting in serious overloading of ship structures and potential structural damage. Heavy casualties and property losses may happen if such failures are not detected in time ([1]). In order to operate ships safer by reducing fatigue damage accumulation and minimizing the possibility of structural damage, it is of great significance to implement the ship structural digital twin. A digital twin is a digital replica of physical asset and its associated environment that is updated through the exchange of information between the physical and virtual systems ([2]). A structural digital twin can monitor ship structural responses to seaway loads and enable to continuously process the measured data to provide warnings and recommendations to vessel operators, owners, and stakeholders concerning structural integrity, maintenance plan, as well as optimized operations that may extend the lifecycle of ship vessels. Meanwhile, a structural digital twin with intelligent diagnosis will enable owners to transit maintenance and inspection strategies from a planned or calendar-based approach to a more condition-based and predictive approach. This may reduce the number and frequency of inspections and allow operators to anticipate and replace damaged parts with minimal resources and downtime, so that maintenance costs will be greatly reduced.

Traditional ship structural monitoring systems are to warn vessel operators that monitored responses, such as ship motions, hull girder stresses, local stresses of typical hot spots, are approaching a level at which corrective action is advisable ([3]). Unfortunately, it is impractical to cover all structural details through this direct sensor-based monitoring at a vessel level due to its scale and complexity. This paper uses a benchmark vessel to investigate ship structural behaviors under environmental loads. The correlation between structure members and critical responses are established, and high stress concentration areas have been identified. Subsequently, this paper develops a conceptual digital twin architecture. The proposed digital twin predicts responses of the entire structure to measured or predicted environmental conditions with a hybrid model combining numerical simulations with measurements. This concept design may improve structural health awareness of the entire hull structure based on a limited number of sensors.

2. Benchmark vessel for structural analysis

The development of structural digital twin entails a ship structural model. Unfortunately, ship structural models are generally confidential and not available in publications. Targeting for open sharing as a benchmark vessel for ship structural research and development, TCOMS developed a ship structural model based on the KVLCC2 hull form. KVLCC2 is a 300K DWT VLCC bare hull designed by KRISO to provide data for hydrodynamic related research and development. The hull form is publicly available on KRISO website. TCOMS conducted the design of KVLCC2 general arrangement and structural scantlings with considerations given to various naval architectural criteria such as stability, maximization of cargo hold capacity, and Common Structural Rules ([4]). The main parameters of KVLCC2 are listed in Table 1. Figure 1(a) demonstrates an isometric view of the KVLCC2 structural model developed with BV Mars 2000 V2.9n and SESAM Genie V7.14. Figure 1(b) is the corresponding global FE model. The FE model consists of a total 209593 nodes and 438769 2-D shell and 1-D beam elements. The mesh size is longitudinal stiffener space, 850mm*850mm. Figure 1(c) illustrates internal structural arrangement details. This paper uses this model to conduct structural analysis and to develop a conceptual digital twin architecture.

Table 1. Main parameters of KVLCC2

Main Dimension	Lpp(m)	320.00	B(m)	58.00	D(m)	30.00
Fully Loaded Departure Condition	Disp. (ton)	3.20E+05	LCG(m)	171.19	R_{xx} (m)	14.45
	Draft (m)	20.80	TCG (m)	0.00	R_{yy} (m)	73.17
	GMT (m)	8.03	VCG (m)	16.29	R_{zz} (m)	74.30
Ballast Arrival Condition	Disp. (ton)	1.44E+05	LCG (m)	176.02	R_{xx} (m)	19.49
	Draft (m)	10.00	TCG (m)	0.00	R_{yy} (m)	85.17
	GMT (m)	20.09	VCG (m)	11.64	R_{zz} (m)	86.83

A right-handed Cartesian coordinate system is utilized in this paper. The wave heading angle is measured counterclockwise from the positive x-axis of the coordinate system. Following sea is defined as 0^0, beam sea as 90^0, and heading sea as 180^0. The motions in six degrees-of-freedom are presented at Center of Gravity (COG) of the vessel. The definitions of the six motion modes, surge, sway, heave, roll, pitch, yaw, are illustrated in figure 1(a).

Uncertainty in the determination of the load components and combinations thereof, along with the idealization of dynamically acting loads as static or quasi-static loads, have been long recognized as the cause of discrepancy of structural analysis results. This paper conducts structural analysis for KVLCC2 using Dynamic Loading Approach ([5]). This approach can reduce the uncertainty of these load factors. The initial step is to perform 3D diffraction/radiation analysis with SESAM HydroD V4.10 to calculate motion RAOs, wave pressure RAOs for a pertinent range of wave heading (0^0

~180⁰ with an increment of 15 degrees) and wave frequency combination (0.10 rad/s~2.00 rad/s with an increment of 0.05 rad/s). The RAO, short for Response Amplitude Operator, is the first order vessel response to unit amplitude, regular, sinusoidal wave. Subsequently, the real part and imagery part of wave pressure RAOs and acceleration RAOs to each unit regular wave are imposed on the 3-D global FE model, respectively. Finally, the stress RAOs of the vessel structure are solved with SESAM Sestra V10.10. As an example, Figure 2 demonstrates global FE analysis results of the fully loaded departure condition corresponding to a unit regular wave with wave heading 180^0 and wave frequency 0.50 rad/s.

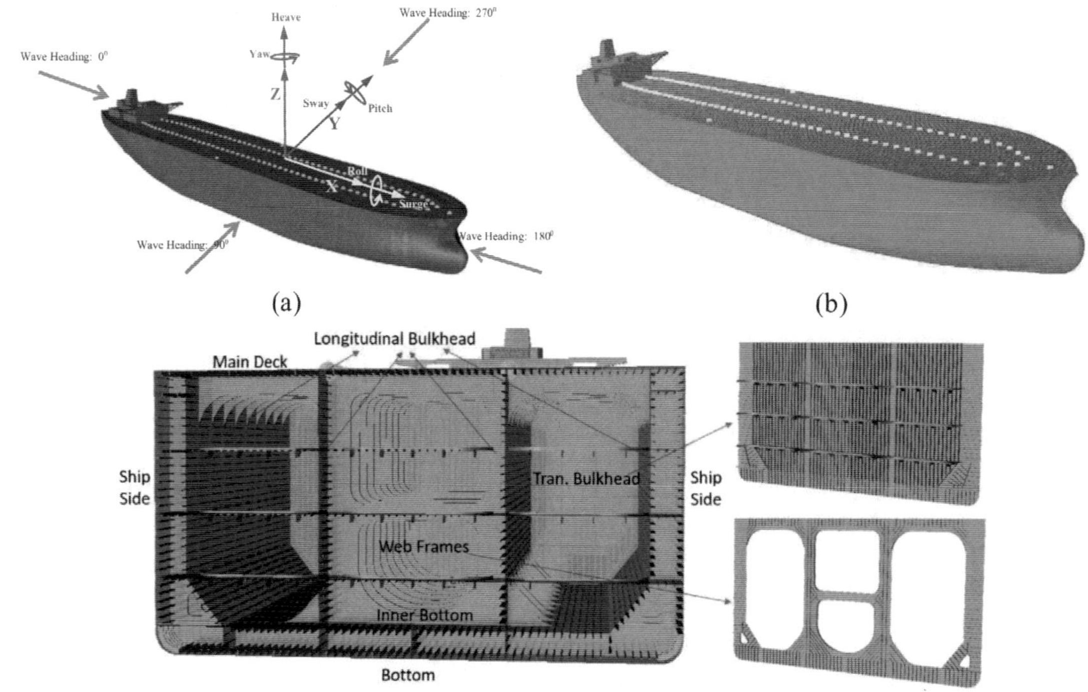

(c)

Figure 1. (a) KVLCC2 structural model, coordinate system and sign conventions; (b) KVLCC2 global FE model; (c) KVLCC2 internal structural arrangement details.

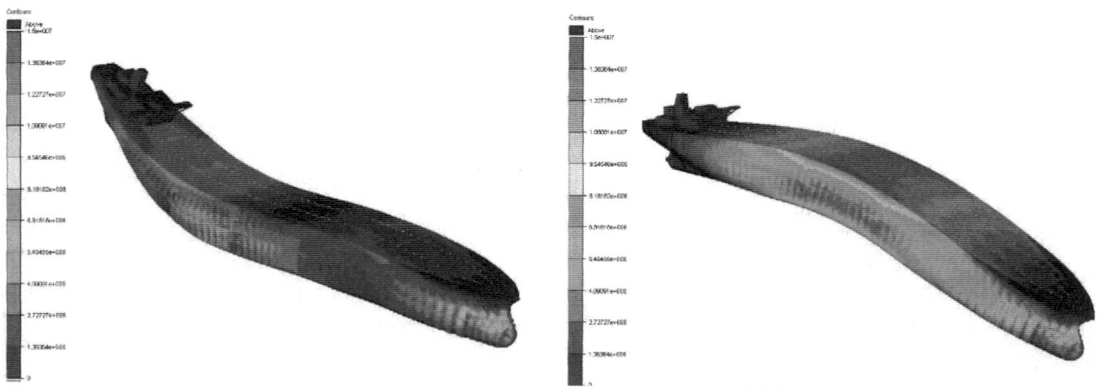

(a) Real part (b) Imagery part

Figure 2. VonMises Stress RAO corresponding to a unit wave with wave heading 180^0 and wave frequency 0.50 rad/s (unit: $N/m^2/m$).

3. KVLCC2 structural response characteristics

Considering the scale and complexity of a vessel, it is impractical and uneconomical to monitor all structural details. A thorough understanding of ship structural response characteristics is essential to correlate individual structural members with critical global responses, and to identify hot spots, therefrom, to achieve cost-effective sensor deployment. Meanwhile, an investigation on ship structural response characteristics will be beneficial to develop data assimilation algorithms to combine monitoring data with numerical simulation results. This paper investigates KVLCC2 structural responses to wave loads using global FE analysis results of the fully loaded departure condition. Typical stress distributions and deformation shapes of individual structural members are screened out from all wave heading ($0^0 \sim 180^0$) and wave frequency (0.20 rad/s ~ 1.40 rad/s) combinations.

As an example, figure 3 illustrates typical VonMises stress contour plots and deformation shapes of longitudinal bulkheads. The deformation in heading sea (wave heading 180^0) is clearly a hogging and sagging deflection, and critical stress responses appear at wave frequency 0.40 rad/s. Table 2 indicates that a heading wave with frequency 0.40 rad/s yields maximum Vertical Bending Moment (VBMT) amplitude at midship, and by inference, critical stress responses of longitudinal bulkheads to heading wave are induced by VBMT. Hence, VBMT is a dominant global response for longitudinal bulkheads in heading sea. One of the structural functions of longitudinal bulkheads is to resist internal liquid pressure from cargo and ballast. At roll natural frequency 0.50 rad/s in beam sea (wave heading 90^0), the maximum roll motion (see table 2) leads to large liquid cargo pressures, and hence, results in high stress responses of longitudinal bulkheads (see figure 3). Therefore, roll motion is a governing global response for longitudinal bulkheads in beam sea.

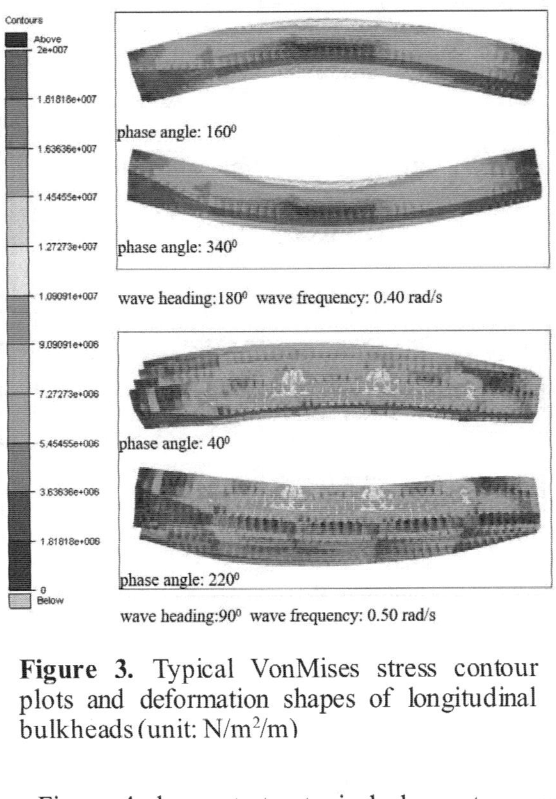

Figure 3. Typical VonMises stress contour plots and deformation shapes of longitudinal bulkheads (unit: $N/m^2/m$)

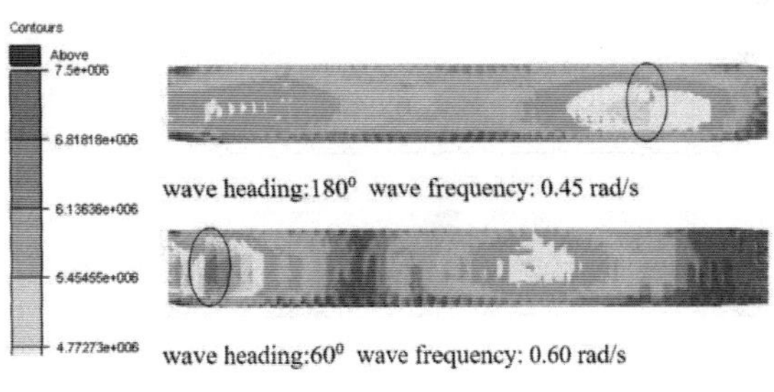

Figure 4. Typical shear stress amplitude contour plots of Inner longitudinal bulkheads (unit: $N/m^2/m$)

Figure 4 demonstrates typical shear stress amplitude contour plots of longitudinal bulkheads. Critical shear stress amplitudes appear around ¾ Lpp in heading sea with wave frequency 0.45 rad/s and around ¼ Lpp in oblique sea (wave heading 60^0) with wave frequency 0.60 rad/s. Longitudinal bulkheads together with ship shell plates, as the web plate of the hull girder beam, take up global shear forces from wave induced loads and weight/buoyancy distribution along the ship length. Table 2 indicates that maximum vertical shear forces happen at 60m to aft perpendicular in oblique sea with frequency 0.60 rad/s and at 250m to aft perpendicular in heading sea with frequency 0.45 rad/s. These critical stress responses result from the maximum vertical shear forces. Therefore, vertical shear forces around ¼ Lpp and ¾ Lpp are governing global responses for longitudinal bulkheads, too.

Table 2. Maximum critical response amplitudes and corresponding wave conditions

CRITICAL RESPONSES	AMPLITUDE	HEADING	FREQUENCY (RAD/S)
ROLL (DEG)	**4.61**	**90^0**	**0.50**
VBMT@160M (N*M/M)	**1.14E+09**	**180^0**	**0.40**
VBMT@160M (N*M/M)	1.25E+09	300^0	0.60
HBMT@170M(N*M/M)	1.13E+09	120^0	0.65
TORS@110M(N*M/M)	1.73E+08	75^0	0.50
VSHF@60M (N/M)	**1.60E+07**	**60^0**	**0.60**
VSHF@250M (N/M)	**1.23E+07**	**180^0**	**0.45**

Note: VBMT – Vertical Bending Moment; HBMT – Horizontal Bending Moment; TORS – Torsional Moment; VSHF – Vertical Shear Force.

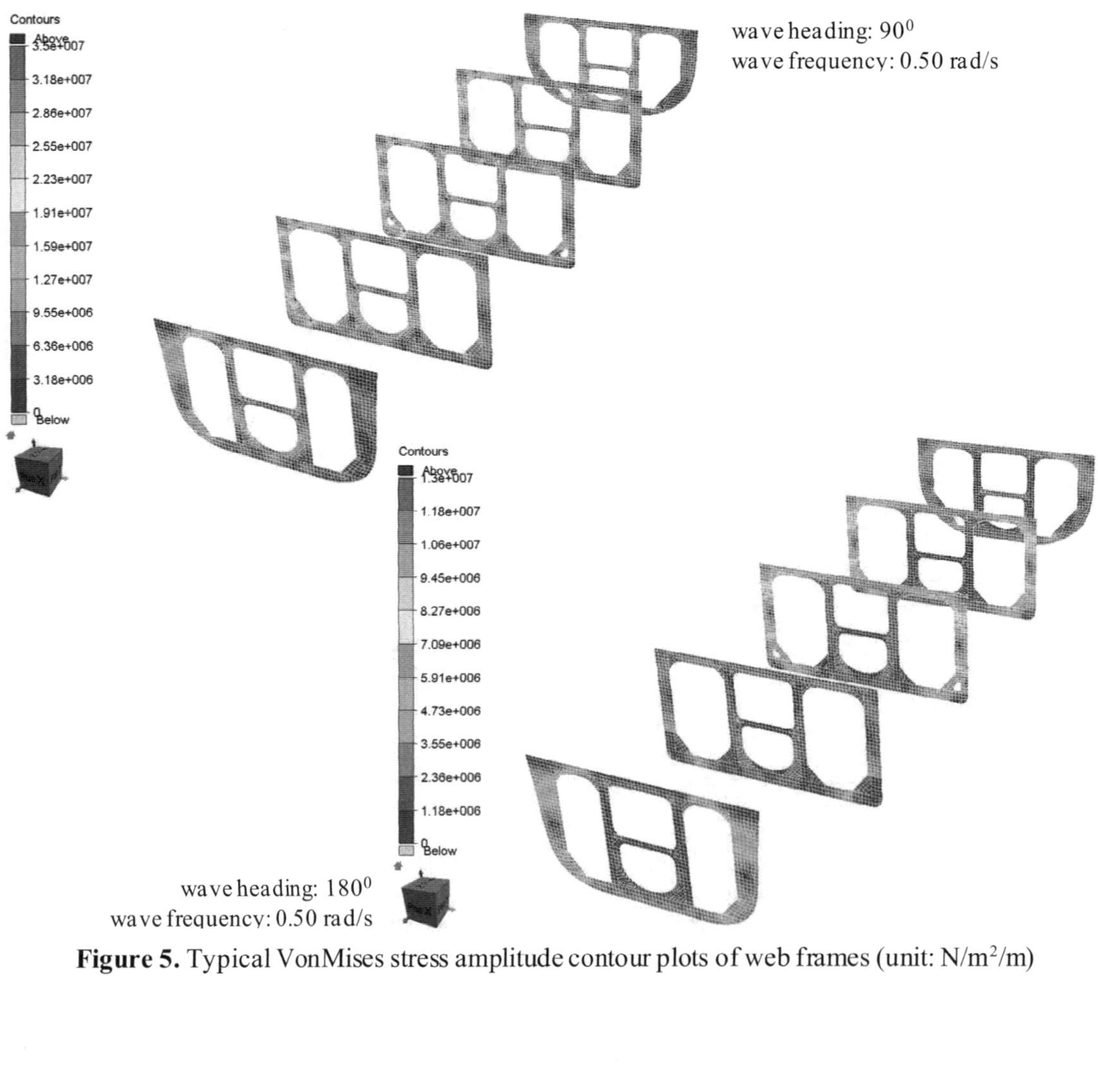

Figure 5. Typical VonMises stress amplitude contour plots of web frames (unit: $N/m^2/m$)

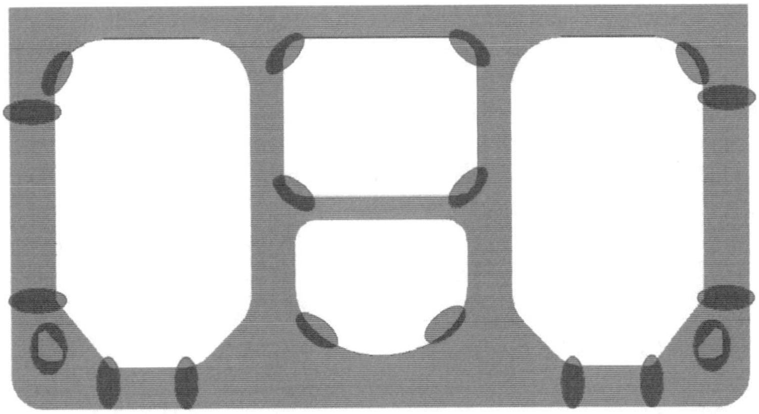

Figure 6. High stress concentration areas on web frames

Governing global responses for longitudinal bulkheads are determined through above analysis. Similar analysis can be conducted to identify governing global responses for other longitudinal structural members, such as main deck, bottom, inner bottom, and ship side. These governing global responses will guide the structural sensor package selection. Stress responses of theses members can be derived using FE analysis together with measured global responses, so that, no need to install plenty of sensors to monitor local stress responses.

Unlike longitudinal structural members withstanding global loads, transverse members, such as transverse bulkheads and web frames, mainly take up local loads from longitudinal stiffeners and transfer them further into the hull girder. It is not applicable to directly correlate transverse structural members with global responses. Sensors should be deployed on hot spots vulnerable to damage to monitor local stresses. These hot spots can be identified by screening high stress concentration areas from global FE analysis results. Figure 5 demonstrates critical stress amplitude contour plots of five web frames locating at the middle of five cargo oil tanks. The high stress concentration areas on these web frames are summarized in Figure 6. A few governing locations, such as bracket toes, hopper knuckles, will be equipped with sensors for local stress monitoring. The stress responses of other stress concentration areas will be evaluated through FE analysis with a fine local FE model. The measured dominant global responses and local stress responses at governing locations will be used to calibrate the FE analysis to ensure the accuracy of evaluations.

4. Concept design of a digital twin architecture

The world is becoming more data-driven nowadays, where physical systems are increasingly connected and mirrored into a virtual space. As more and more industries adopt digital twin to improve performance, marine and offshore industry is compelled to do the same. Digital twin of ship hull structures has attracted interests of researchers in recent years ([6], [7], [8], [9]), and some concepts/products have been implemented to real hull structures, such as DNV's 'Nerves of Steel" and MRAIN's AHMS system.

This paper develops a conceptual digital twin architecture, as shown in figure 7. The proposed digital twin is an on-line system and digital replica of a physical ship unit and its associated environments. It can acquire, manage, and analyze the on-board measured responses to continuously update the knowledge regarding the current health status, *e.g.,* fatigue damage accumulation, of the entire hull structure, and provide decision-making support on safer and more optimal vessel operation, inspection and repairs, and asset integrity management.

At the design stage, a numerical database needs to be established based on ship design data, including a digital model with structural details, the loading manual, as well as hydrodynamic RAOs and stress RAOs of all loading conditions in the loading manual. Hydrodynamic RAOs and stress RAOs can be calculated with dynamic loading approach described in Section 3. The numerical database should be continuously/periodically updated with the inspection data (such as, structural modifications, thickness change of hull structures, identified damaged area and repair history) during the operating stage, so that the digital twin can reflect the real condition of the hull structure. Meanwhile, sensor packages should be deployed onboard to real-time measure following physical parameters. Relevant physical parameters may be determined or further added in accordance with practical conditions of ship and safety need. The methodology developed in Section 3 can be utilized to achieve cost-effective sensor deployment.

- Environmental parameters
  - Wave height, wave period, wave direction;
  - Wind speed, wind direction;
  - Current profile;
- Navigation parameters:
  - Loading condition;
  - Draft, trim and heel;
  - Speed, vessel direction;
- Vessel motion responses:
  - Six degree-of-freedom motion responses;
  - Three translational accelerations at critical locations;
- Structural strength responses:
  - Dominant global responses;
  - Local stresses at critical locations;

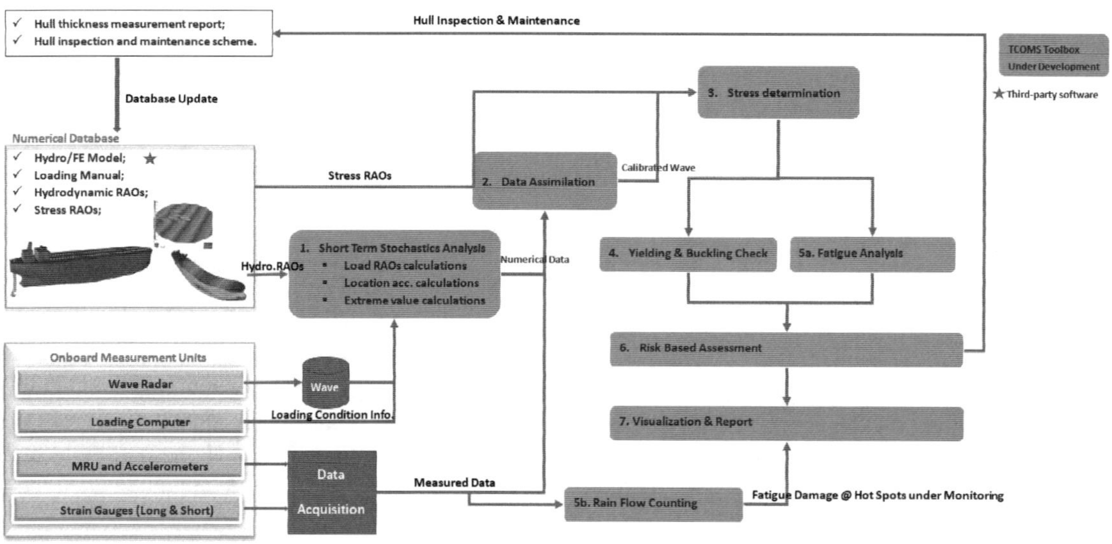

Figure 7. Architecture of the proposed ship structural digital twin

During the operating stage, the conceptual digital twin utilizes a hybrid model, integrating sensor-based monitoring with numerical simulation data, to provide a holistic picture of the entire structural health condition. The key modules of the hybrid model are illustrated in figure 7. For the measured or predicted environment conditions, the proposed digital twin will calculate extreme values of governing global responses through short-term stochastic analysis. The governing global responses of individual structural members are identified using the methodology described in Section 3. Considering limitations of the numerical model and the measuring devices, it is not feasible to make all calculated responses always match well with measured data. The data assimilation module will calibrate a design wave for each individual member by adjusting the corresponding governing global responses to measured values. Thereafter, stress responses of each individual members to measured or predicted environment conditions are determined using the calibrated design wave amplitude and stress RAOs in the numerical database. The measured local stresses of governing locations will be used to verify the accuracy of the numerical results. The digital twin will conduct buckling, yielding and fatigue assessment of the entire structure with the determined stress responses. Meanwhile, the digital twin introduces a risk-based framework to quantify uncertainties and risk factors relevant to data collection, numerical simulation, and operation, so that more reasonable hull inspection and maintenance plan can be provided.

The numerical model adopted in this paper relies on a number of assumptions. For example, the hydrodynamic responses are computed with linear potential flow model under the assumption of small perturbation. This model fails to accurately predict responses of intermittent wetting areas, particularly, under steep waves. For those locations, where the numerical model cannot provide reasonable estimations, sensors will be installed to collect the time history of local stresses. The digital twin will assess the fatigue damage of these locations through time domain fatigue analysis with the measured data.

5. Conclusions

This paper develops a conceptual digital twin architecture, which enables to assess the entire structure health condition with a hybrid model combining numerical simulations with measurements. As a digital replica of physical asset, the digital twin should secure the best possible representation of the real structural behavior to environmental loads. To achieve this target, this paper conducts structural analysis of a benchmark vessel using dynamic loading approach. The ship structural response characteristics are investigated based on structural analysis results. The correlation between structure members and critical global responses are established, and high stress concentration areas have been identified. Subsequently, a guideline is developed on how to achieve a cost-effective sensor deployment. Finally, the paper introduces the key modules and technical procedures of the proposed conceptual digital twin architecture.

Acknowledgements

This research is supported by the National Research Foundation, Singapore and the Singapore Maritime Institute under its Maritime Transformation Programme White Space Funding Support for the Establishment and Operation of the Centre of Excellence for Autonomous and Remotely Operated Vessels (CEAOPS), SMI-2019-MTP-01.

The authors would like to express the appreciation to Dr Chua Kie Hian for his valuable and constructive suggestions during the planning and development of this research work.

References

[1] European Maritime Safety Agency 2020 *Annual overview of marine casualties and incident*
[2] E VamDerHorn, S Mahadevan 2021 Digital twin: generalization, characterization and implementation, *Decision Support Systems*. Vol. **145**
[3] ABS 2015 *Guide for Hull Condition Monitoring Systems*
[4] IACS 2020 *Common Structural Rules for Bulk Carriers and Oil Tankers*
[5] ABS 2018 *Guide for 'Safehull-Dynamic Loading Approach' for vessels*
[6] Konstantinos N Anyfantis 2021 An abstract approach towards the structural digital twin for ship hulls: a numerical study applied to a box girder geometry. *Journal of Engineering for the Maritime Environment*. Vol. **235** p718-736
[7] Mengmeng Wang, Chengye Wang, Anna Hnydiuk-Stefan, Shizhe Feng, Incecik Atilla and Zhixiong Li 2021 Recent progress on reliability analysis of offshore wend turbine support structures considering digital twin solutions. *Ocen Engineering*. Vol. **25**
[8] Xin Fang, Honghui Wang, Wenjing Li, Guijie Liu and Baoping Cai 2022 Fatigue crack growth prediction method for offshore platform based on digital twin. *Ocean Engineering*. Vol. **244**
[9] Eric VanDerHorn, Zhenghua Wang and Sankaran Mahadevan 2022 Towards a digital twin appraoch for vessel-specific fatigue damage monitoring and prognosis. *Reliability Engineering & System Safety*. Vol. **219**

AI-Based Situation Awareness Assessment

Kenneth KWOK[1]*, SS VIRDI[2]

[1] Institute of High Performance Computing (IHPC), 1 Fusionopolis Way, #16-16 Connexis North Tower, Singapore 138632

[2] Singapore Polytechnic (SP), 500 Dover Road, Singapore 139651

E-mail: kenkwok@ihpc.a-star.edu.sg

Abstract. Navigational Safety in maritime operations is a high priority issue. In a recent collaboration between IHPC and SP, an AI-based Behaviour Competency Assessment Tool prototype was developed to aid instructors in assessing soft-skill competencies of watchkeepers conducting navigation, and a proof-of-concept demonstration was performed for Situational Awareness at SP's Singapore Maritime Academy (SMA) training simulator. Use of our tool can help remove human bias by providing detailed behaviour-specific analysis to aid instructors in making their assessments.

1. Problem Statement

It is estimated that 80% of marine accidents can be attributed to human error [1]. Maritime safety and safety of navigation can therefore be enhanced by strengthening the focus on the human element. Specifically, assessment of Behaviour Competency[1], comprising non-technical soft skills, is under-developed because training has traditionally focused on technical competencies and there is room for improvement.

Currently, assessment of Behaviour Competency during simulator training at the Singapore Maritime Academy (SMA) is conducted through human observation of a trainee's performance. This assessment methodology involves a trained instructor's *subjective* evaluation of trainees' non-technical soft skills while they are conducting navigation training. Variations in assessment can occur between assessors (e.g., because of difference in levels of experience) and even for a given assessor from one training session to another. SMA instructors monitor up to 5 concurrent bridge teams in the simulator and manually note down their performance on log sheets. CCTV video cameras in each simulator cubicle capture videos of training sessions, and instructors can use them during after-action-reviews for playback to facilitate discussion and learning. However, the video interface is manual, and events of interest are not automatically flagged out in recordings, making access to specific events troublesome.

2. AI-based Behaviour Competency Assessment Tool

An AI-based Behaviour Competency Assessment Tool prototype was developed that uses video feeds from the simulator to automatically generate assessments of students' behaviour competency with *quantitative* measures for *objective* assessment. A report is generated which provides a summary of the

[1] Behaviour competencies include: 1. Situational Awareness, 2. Communication and influence, 3. Teamwork, 4. Decision Making, 5. Results focus, 6. Leadership, and 7. Management skills

training session with events indexed by time that the instructor can use for debriefing the student and allows the instructor to easily zoom in on specific incidents for deeper review and more effective instruction.

The AI-based Behaviour Competency Assessment Tool prototype (shown in Figure 1) is a software comprising:

- A **behaviour learning and understanding module** which takes in video from cameras in the training simulator and uses computer vision and machine learning to detect and recognise key behavioural indicators from the actions of the students performing their navigation tasks. Behaviour learning is performed on the recognised behavioural indicators to automatically capture behaviours relevant to each behaviour competency and encode them in behaviour scripts.
- A **competency assessment module** which performs performance analysis by recognising behaviours exhibited by students using the learnt behaviour scripts and generates a report of the students' performance quantifying the key behavioural indicators observed and the time-course of observed behaviours. A numerical performance score is also generated for each student from analysis based on a set of behavioural metrics derived from guidelines for Behaviour Competency Assessment and Verification (BCAV) promulgated in IMO and industry publications [2] [3].

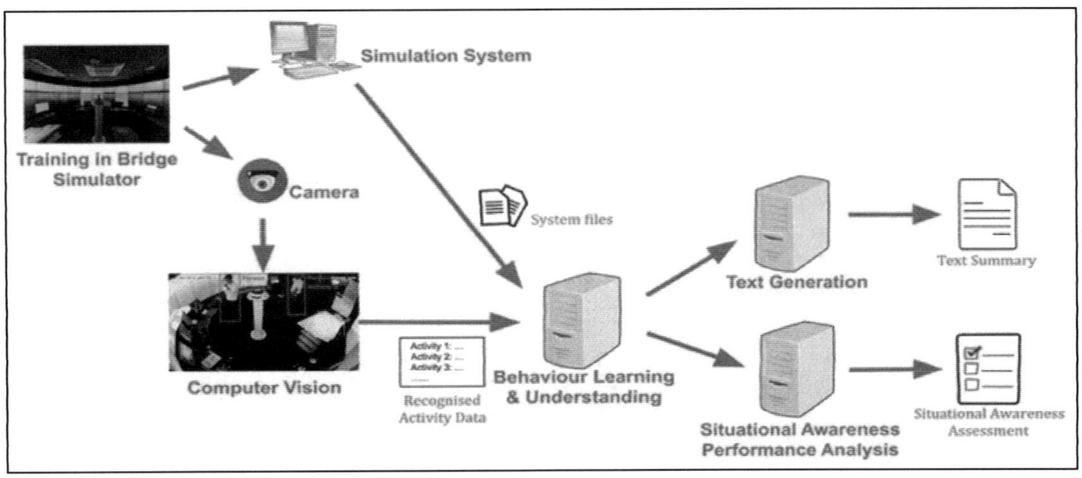

Figure 1. Schematic of the Competency Assessment Tool prototype for Situation Awareness

3. Situation Awareness

For the Proof-of-Concept (POC), the Competency Assessment Tool was used to perform assessments on *situation awareness* while students performed watchkeeping on a coastal navigation task.

According to the American Psychological Association, situation awareness is conscious knowledge of the immediate environment and the events that are occurring in it. Situation awareness involves perception of the elements in the environment, comprehension of what they mean and how they relate to one another, and projection of their future states[2].

In the maritime context, components of situational awareness include [3]:

- Awareness of vessel systems crew
- Awareness of external environment
- Awareness of time

[2] https://dictionary.apa.org/situation-awareness

4. Current work on AI-Based Soft Competency Assessment and Situation Awareness Enhancement

AI-based solutions have begun to emerge in recent years for soft competency assessment. For example, AI systems are used to screen potential candidates for important soft skills (such as, adaptiveness, teamwork and resilience) prior to hiring [4], and to perform assessments of soft skills through chatbot interviews [5] using Natural Language Processing (NLP) technologies.

AI tools for augmenting humans in situation assessment are also being investigated and trialled in both academia [6] [7] and industry [8] [9]. These systems process and fuse sensor data of the maritime environment from several sources (sensors and onboard systems) and present them to crew members for improved situation awareness.

However, to the best of our knowledge, no AI-based system has so far been developed for assessment of soft competencies using video data, and our prototype is the first to perform situation awareness assessment from observation of crew behaviours during training.

5. Method

Working with the domain experts in reviewing video data of navigation training, we identified the following *key behavioural indicators* which are related to situation awareness:

1. Use of RADAR for collision avoidance and during poor visibility condition (Uses Radar)
2. Use of Heading Control (Course Change)
3. Use of Engine Telegraph for controlling ship speed (Speed Change)
4. Conduct Visual Observation of surrounding vessels and navigation landmarks (Visual Observation)
5. Use of ECDIS / Paper Chart for monitoring and plotting position (Plots Position)
6. Use of internal communication (Internal Comms)
7. Use of VHF for external communications (External Comms)

5.1. Computer Vision Module

We extracted 5-second-long examples of these behaviour indicators from videos of simulator training sessions and had domain experts label the clips. We then trained a computer vision model using the labelled clips to recognise the behaviour indicators for 5 of the 7 classes (less indicators 2 and 7[3]) with an accuracy of 92.7% (see Appendix 1 for details).

Specifically, we used a skeleton-based action recognition approach because the dynamics of human skeletons convey significant amount of information for human action recognition which allowed us to achieve much higher accuracy compared to traditional end-to-end two streams-based action recognition models [10]. An open-source pose-estimation model *openpose* [11] was first used to extract a 17-joint human skeleton from each image which is robust to occlusion, and a Spatio-Temporal Graph Convolutional Network (STGCN) [12] was then trained to perform action recognition.

5.2. Behaviour Learning and Understanding Module

Next, we developed the Behaviour Learning and Understanding module for situation awareness analysis from analysing 19 videos of training sessions and extracting behaviour patterns based on recognised behaviour indicators; these sequences of behaviour indicators were encoded into *behaviour scripts.*

[3] Course change (#2) involved a verbal command which is hard to detect using vision, while there were too few examples of external comms (#7) in the training data.

Behaviour scripts enumerate causal sequences of events (which in this case are the recognised behaviour indicators), see Figure 2 for an example.

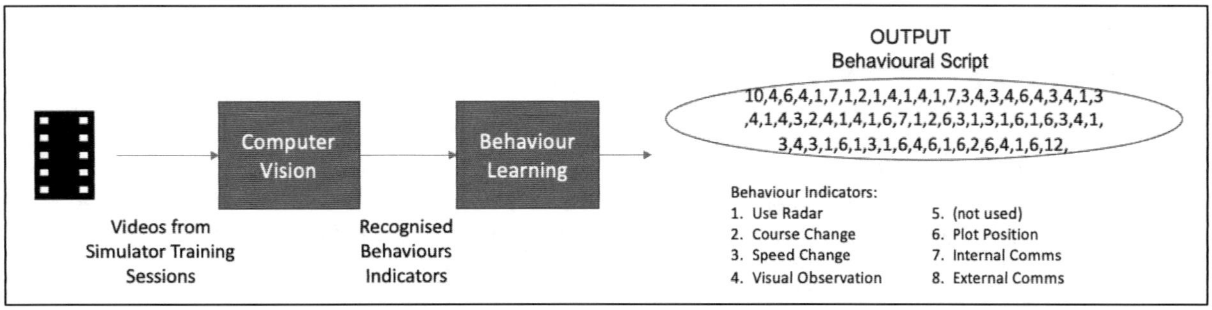

Figure 2. Behaviour Script Learning

Behaviour learning is based on previous work in IHPC on rapid unsupervised causal learning as reported in [13-17]. Essentially, causal learning uncovers the causal relations between events across time. Distinct from deep learning and deep reinforcement learning which rely on big data, causal learning operates at the symbolic level and uses much less data.

To aid in interpretation of the extracted behaviour scripts, they were further segmented into complex events which are shorter sequences of behavioural patterns of the form:

[observation event] … [action event]

where:

- [observation event] initiates the sequence and can be one of "Use Radar", "Visual Observation", or "Plots Position"
- … are other observation events in the sequence
- [action event] terminates the sequence and can be one of "Course Change", "Speed Change", "Internal Comms", or "External Comms"

This definition of complex events (CEs) derives from the intuition that meaningful sequences of events relate observations to actions. Domain experts then reviewed the learnt complex events to give them descriptive labels (see Appendix 2 for examples).

5.3. Competency Assessment Module

To develop our situation awareness competency assessment algorithm, we first related the CEs learnt from the training dataset to the assessment rubric used by the SMA instructors shown in Appendix 3. We then proceeded to group the complex events extracted into 5 Situation Awareness Behaviours (SAB) classes to make our analysis more tractable, as follows:

- Class 1: Monitor own ship and other ships/targets
- Class 2: Assess Closest Point of Approach (CPA)
- Class 3: Events leading to Internal Comms
- Class 4: Collision Avoidance
- Class 5: Events leading to External Comms

Using these 5 classes and an assessment grade for the student's performance (defined for 3 performance bands, 1="Below Expectation", 2="Meet Expectation", and 3="Exceed Expectation" provided by an instructor, we performed a regression of Performance Score against CE counts of the SAB Classes. The resulting regression algorithm would be used for the POC demonstrations on unseen data.

The output of the development phase was therefore:

1. A **knowledge database**: comprising extracted rules, scripts, and complex events representation with grouped into SAB classes.

2. An **assessment algorithm**: tuned after doing correlation analysis using the training dataset to generate a performance score that correlates with the instructor's assessment.

For performance assessment of new unseen videos, the videos need to be processed first by the behaviour learning and understanding module to extracted complex events from the input videos. The extracted CEs are compared to existing complex events in the knowledge database using a similarity metric to match them with the appropriate SAB classes.

A performance score and a report comprising charts showing the distribution of detected Key Behaviour Indicators and Complex Events, and the time-course of observed behaviours using natural language descriptors of the observed CEs at each time point, is generated for each video by our trained system during assessment.

6. Results

We report here on one of the demonstrations performed on unseen data for our POC. The regression model built from the training data was:

[w_1: -0.03800679, w_2: 0.16125468, w_3: 0.03349698, w_4: -0.00528674 w_5: 0.34819296]

For this demo, 5 unseen videos of students' performance in the simulator were obtained from the SMA. However, due to COVID-19 restrictions, not many lessons were being run at the time, and the only data available involved experienced students who largely fell into the "Meet Expectation" and "Exceed Expectation" performance bands. To provide additional resolution on the relative performance of the students, the instructor was asked to provide a ranking of the students as well.

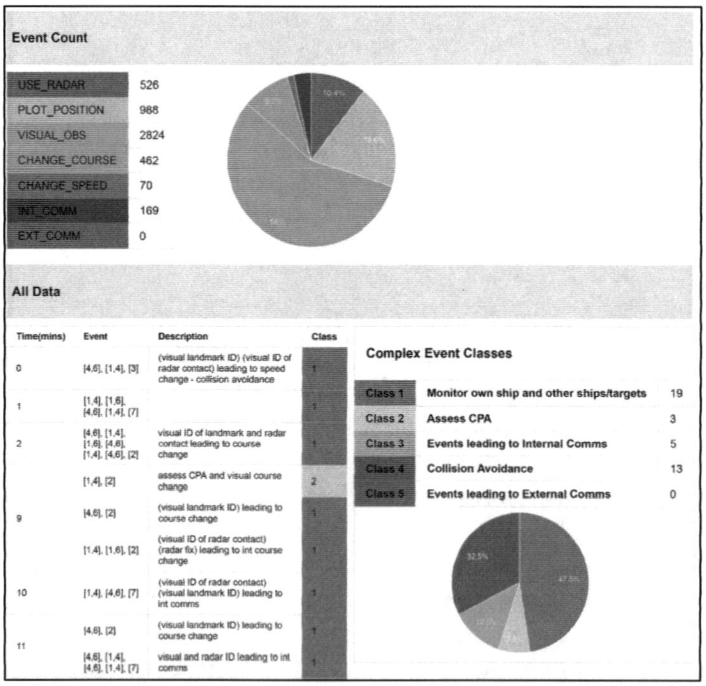

Figure 3. Performance Report

A performance report was generated for each student as shown in Figure 3, comprising counts of observed behaviour indicators[4], and a chronology (time course) of observed behaviours consisting of recognised complex events.

[4] "Course Change" was manually labelled as it was deemed to be a critical behavioral indicator.

An overall performance score was also generated in the range 1-3 [1=Does not meet expectation, 2=Meets expectation, 3=Exceeds expectation] using the regression formula derived previously. In Figure 4, we see the results for 5 students (A to E). The counts for each factor (1 to 5) are the number of times a CE from each SAB class was observed during the assessment for each student, and these were fed into the regression formula to produce a predicted score.

	1	**2**	**3**	**4**	**5**		
	Monitor own ship and other ships/targets	Assess CPA	Events leading to Internal Comms	Collision Avoidance	Events leading to External Comms	**Predicted Score**	**Instructor Ranking**
A	12	1	3	2	0	2.0513	2
B	12	2	4	2	0	2.2461	1
C	15	1	6	2	0	2.0378	3
D	14	1	4	1	0	2.0141	4
E	21	2	6	4	0	1.9605	5

Figure 4. Results from Proof-of-Concept demo

The students' performance ranking from our system was compared with the ranking of the instructor, and we can see that the system's ranking matched the instructor's perfectly in this case. With other sets of unseen videos, we observed that the relative rankings were generally preserved compared to the instructors.

7. Discussion

The POC demonstration showed that an AI-based behaviour competency assessment was able to generate quantifiable performance measures for behavioural competency assessment to aid instructors in making their evaluations. The measures provide a means for instructors to base their evaluations on objective criteria as opposed to current practice which is mostly subjective. The summary of key events in the report was also useful for helping instructors to make more efficient use video recordings from simulator sessions during debriefing.

The AI technology used to develop the Behaviour Competency Assessment Tool included:

- Computer Vision: to recognise key behaviour indicators
- Causal Learning: to extract causally related behaviour patterns that were encoded as behavioural scripts
- Linear Regression: to build the algorithm for generating performance scores from observed behaviours.

Different from purely end-to-end neural machine learning approaches which require massive amounts of training data to train prediction models, we are able to perform behaviour learning to extract behaviour scripts using only 19 videos. This makes our approach a lot more practical for use in real-world training situations where training data can be limited.

For future work, we intend to increase the number of behavioural indicators to increase the richness of the behaviours on which the assessment is based. This includes adding other modalities, such as auditory processing to detect events which are hard to detect using vision alone. For the purpose of the reported demonstration, we had to manually detect and encode "Course Change" because of the importance of this action for situation awareness assessment. For future implementation, this should be done using audio processing, or through direct data feed from the simulator system.

Finally, the assessment of training videos used for behavioural learning should ideally be done by at least 3 domain experts for inter-rater reliability to be established. Assessments for which there is low inter-rater reliability can then be excluded to increase the validity of the assessment.

8. Conclusion

Current assessment of competencies in Navigational Simulators is prone to human bias. Use of AI-based assessment tools to assess competencies can remove such bias by providing detailed behaviour-specific analysis to aid instructors in making their assessments.

Specific contributions of the project include:

- Automatic generation of objective assessment of human behaviour competency in a report with detailed quantitative task analysis and a numerical performance score leading to enhancement of navigational safety training.
- The automation of routine observation and analysis of navigation training complements manual observations, potentially leading to increased efficiency and effectiveness of human assessors in conducting behaviour competency assessments for multiple parallel sessions.
- Knowledge transfer from the maritime domain experts to AI application in terms of identified key behavioural indicators, assessment scenarios and metrics relevant to behaviour competency assessment for the development of the AI-based Behaviour Competency Assessment Tool.

Acknowledgements

This work was funded by Singapore Polytechnic.

References

[1] Berg, N, Storgård, J, Lappalainen, J (2013). "The impact of ship crews on maritime safety", *Centre for Maritime Studies, University of Turku.*

[2] IMO (2017). "IMO Model Course 1.30 Onboard Assessment", *https://www.libramar.net/news/on_board_assessment_imo_model_course_1_30/2021-07-22-4472*

[3] Intertanko and OCIMF (2018). "Behaviour Competency Assessment and Verification (BCAV) for Vessel Operators", *https://www.ocimf.org/document-libary/behavioural-competency-assessment-and-verification*

[4] Adams, M (2020). "3 reasons for using AI to screen for soft skills & how 2degrees, CSIRO & Viventis are doing it", Retrieved from *https://atcevent.com/artificial-intelligence/3-reasons-for-using-ai-to-screen-for-soft-skills-how-2degrees-csiro-viventis-are-doing-it* on 23 Mar 2022.

[5] Safebridge (2019). "SafeMetrix is developing AI-based assessment for seafarer soft skills", retrieved from *https://www.maritime-executive.com/editorials/safemetrix-develops-ai-based-assessment-for-seafarers-soft-skills* on 23 Mar 2022.

[6] Castellini, A, Beltrame, G A, Bicego, M, Bloisi, D D, Blum, J J, Denitto, M & Farinelli, A (2017). "Activity recognition for autonomous water drones based on unsupervised learning methods" *Proceedings of the 4th Italian Workshop on Artificial Intelligence and Robotics, a workshop of the XVI International Conference of the Italian Association for Artificial Intelligence (AIRO@AI*IA)*, pp 16-21.

[7] Tang, R, Zhuo, Z, Zhang, C & Li L (2019). "The Applications of Artificial Intelligence in Situation Assessment and Game Counter measure during Unmanned Air Combat", *Proceedings of IEEE International Conference on Unmanned Systems (ICUS)*, pp. 909-913.

[8] Haun, E (2018). "Maersk to Trial AI-powered Situational Awareness Technology", Retrieved from *https://www.marinelink.com/news/situational-technology436795* on 23 Mar 2022.

[9] Kongsberg (2022). "Situational Awareness: Enhancing safety through intelligent data fusion", retrieved from *https://www.kongsberg.com/no/maritime/about-us/news-and-media/our-stories/intelligent-awareness* on 23 Mar 2022.

[10] El-Nouby, A & Taylor, G W (2018). "Real-time end-to-end action detection with two-stream networks", *15th Conference on Computer and Robot Vision (CRV)*, pp 31-38.

[11] Cao, Z, Hidalgo, G, Simon, T, Wei, S E, & Sheikh, Y (2019). "OpenPose: realtime multi-person 2D pose estimation using Part Affinity Fields", *IEEE transactions on pattern analysis and machine intelligence*, **43(1)**, pp. 172-186.

[12] Yu, B, Yin, H, & Zhu, Z (2017). "Spatio-temporal graph convolutional networks: A deep learning framework for traffic forecasting", *Proceedings of the 27th International Joint Conference on Artificial Intelligence*, pp 3634-3640.
[13] Ho, S B (2017). "The role of synchronic causal conditions in visual knowledge learning", *Proceedings of the IEEE Conference on Computer Vision and Pattern Recognition (CVPR) Workshops 2017,* pp. 9-16.
[14] Ho, S B (2016a). "Principles of Noology: Toward a Theory and Science of Intelligence". Switzerland: Springer International.
[15] Ho, S B (2016b). "Cognitively realistic problem solving through causal learning", *Proceedings of the 2016 International Conference on Artificial Intelligence*, pp. 115-121.
[16] Ho, S B (2014). "On effective causal learning", *Proceedings of the 7th International Conference on Artificial General Intelligence*, pp. 43-52.
[17] Ho, S B, & Liausvia, F (2014). "Rapid learning and problem solving", *Proceedings of the IEEE Symposium Series on Computational Intelligence for Human-like Intelligence*, pp. 110-117.

Appendix 1

Performance of Computer Vision Module

	visual-observation-data	uses-radar-data	uses-internal-communication-data	uses-engine-telegraph-data	plots-position-data
visual-observation-data	125	0	0	13	0
uses-radar-data	4	73	0	1	1
uses-internal-communication-data	0	0	14	0	0
uses-engine-telegraph-data	1	1	0	47	0
plots-position-data	2	1	0	0	49
Acc: 0.927	**0.947**	**0.973**	**1**	**0.770**	**0.980**

Appendix 2

Examples of extracted Complex Event (CE) sequences

CE	**Descriptive Label**
[4], [2]	visual course change
[1], [2]	assess CPA , course change
[4], [3]	visual speed change
[1,4], [2]	assess CPA and visual course change
[1,4], [3]	radar and visual speed change
[4], [2,3,2]	visual couse and speed change - collision avoidance

CE	Descriptive Label
[4], [8]	preparing to embark pilot
[6], [2]	chart course change - regain track

Appendix 3

Relating Complex Events to Assessment Rubric used by Singapore Maritime Academy for Situation Awareness

Monitor own ship and other ships/targets	**Class 1**
2.1 Uses RADAR	[1]
2.2 Uses ECDIS	[6]
2.3 Visual observation of the surrounding traffic	[4]
Assess CPA	**Class 2**
4.1 Visual observation of the surrounding traffic	[4]
4.2 Uses Radar	[1]
	[2], [3]
Inclement weather	**Class 3 – Int Comms**
7.1 Uses Engine Telegraph control, if required (reduce speed)	[3], [2]
7.2 Uses internal communication system – Intercom (post extra watchmen & Lookouts, close hatches, Increased Bridge manning)	[7]
Collision Avoidance	**Class 4**
8.1 Uses Engine Telegraph control, if required	[3]
8.2 Visual observation of the surrounding traffic	[4]
8.3 Order course change	[2]
8.4 Sound horn	
8.5 Uses RADAR	[1]
8.6 Uses ECDIS	[6]
Embarking pilot	**Class 5 – Ext Comms**
11.1 Uses external communication system – VHF	[8], [6]
11.2 Uses Engine Telegraph control, if required (slow down)	[3]
11.3 Changes Course as required	[2]

Development of the Comprehensive Simulation System for Autonomous Ships

M Minami[1], M Kobayashi[1], K Hikida[1] and K Kokubun[1]

[1]National Maritime Research Institute, National Institute of Maritime, Port, and Aviation Technology, 6-38-1, Shinkawa, Mitaka-shi, Tokyo, Japan

e-mail: mminami@m.mpat.go.jp

Abstract. In recent years, there has been a growing interest in autonomous ships, and efforts to realize autonomous ships are underway in Japan. Automated and unmanned ships are expected to improve safety by reducing accidents caused by human with the supports by the automated system. On the other hand, it is necessary to demonstrate that the autonomous ships are safely operated to implement the autonomous ships in society. The National Maritime Research Institute (NMRI) has been studying the evaluation method of safety of the autonomous ships and developing a simulation system as a tool for the evaluation. This paper introduces a comprehensive simulation system including a ship handling simulator, and the evaluation method using the simulation system.

1. Introduction

In recent years, there has been a growing interest in automated and unmanned ships, and the development of autonomous ships and other vessels is underway in Japan. One of the requirements for autonomous ships is to improve safety. About 80% of maritime accidents are caused by human error, such as misjudgment and abnormal behavior in navigation, and autonomous operation is expected to reduce accidents caused by these factors. On the other hand, for autonomous ships to be put to practical use, it is necessary not only to develop the technology, but also to have it accepted by society. Therefore, it is necessary to demonstrate that autonomous ships are safe. Safety in this context means that all assumed risks are within acceptable limits.

To assess risk tolerance and navigation safety of autonomous ships, we are investigating the use of a simulation system consisting of a high-speed simulation system and a full-mission ship handling simulator. In the field of automobiles, which is leading the way in automated operations, developers and surveyors are working together to develop test methods for newly developed technologies, and the performance standards needed to evaluate them are being studied at the same time. We believe that the same process can be used to study the certification method required for realizing the practical application of autonomous ships.

In this paper, the outline of the simulation system, which is scheduled to be completed in March 2023, the study on evaluation methods using the simulation system, and how to make test scenarios are introduced.

2. Safety evaluation of automated navigation system

In evaluating autonomous ships, we decided to adopt the scenario-based evaluation method used to evaluate the safety of Automated Driving Systems (ADS) [1]. Although emergency responses such as parts failure are also subject to evaluation, we focused on external disturbances and considered creating scenarios based on physical principles. Fig.1 shows the flow of identifying hazardous Scenarios of the ADS. Task 1 in Fig. 1 is the target of this research. In scenario-based evaluation, it is necessary to create a large number of scenarios in order to evaluate the validity of required functions within the operational design domain (ODD), the response when the ship deviating from the ODD and the ship response to emergency events.

For the execution of the created scenario, it is necessary to choose which test platform is suitable for each scenario, simulation/virtual or real sea. The simulation/virtual test platform includes a high-

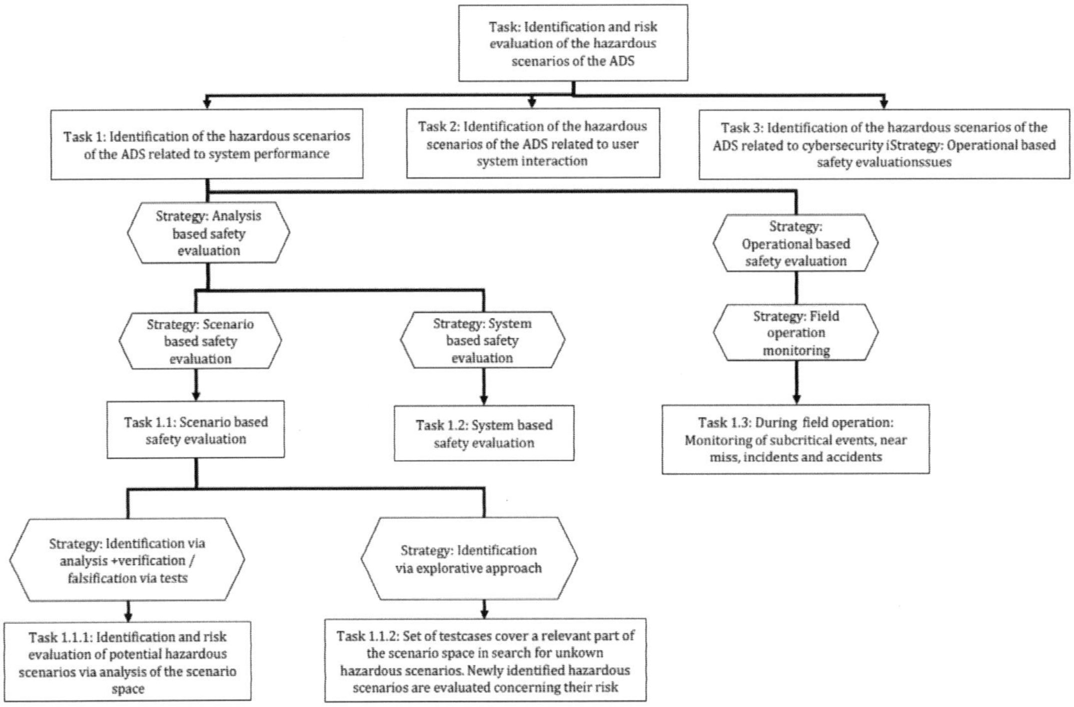

Figure 1 Identifying Hazardous Scenarios [1]

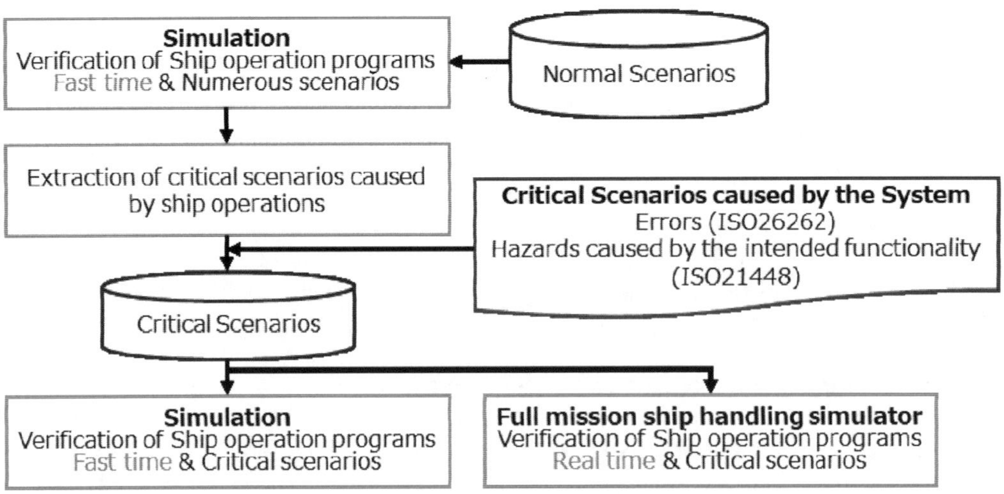

Figure 2 Image of Safety Evaluations (Collision avoidance program)

speed simulation system and a real-time ship handling simulator. The high-speed simulation system is suitable for executing a large number of scenarios in a short time. The use of a real-time ship handling simulator is effective for evaluating scenarios involving humans and scenarios with complicated and danger phenomena. Real sea test platform is suitable for executing scenarios that require a high degree of relevance to the performance of the real system.

Fig.2 shows the image of safety evaluations by simulation for collision avoidance functions. The normal scenarios are created assuming that the automated navigation systems are normal. Due to the large number of these scenarios, it is necessary to run them high-speed. Based on the results of this simulation, critical scenarios, which includes hazardous and difficult important scenario caused by ship operation. In addition, scenarios that take into account system errors according to ISO 26262 and lack of intended functions according to ISO 21448 will be added to the critical scenarios. Since some critical scenarios need to be subjectively evaluated based on the experience of target operations, a full-mission ship handling simulator will be necessary.

3. Study on evaluation method using Simulation

3.1. Scenarios for evaluation

For the creation of disturbance scenarios based on physical principles, the navigation task is decomposed into perceptual, judgement, and control subtasks, and scenarios are created based on the disturbances corresponding to each subtask, i.e., perceptual disturbances, traffic disturbances, and ship motion disturbances. The perception disturbance is caused by the sensor that grasps the position and speed of the other ship and the ship itself, the traffic disturbance is caused by the geographical conditions, traffic flow, and behavior of the other ship, and the ship motion disturbance is caused by the mechanical disturbance acting on the ship such as weather, sea condition, and loading condition.

ISO/DIS34502 for ADS is for limited access highways, and only vehicles, including motorcycles, are considered as moving objects. For MASS, on the other hand, the target of practical application is an area where a wide variety of ships are mixed. In addition, since the time constant of a ship is greater than that of a car, the object must be detected at a greater distance, and prediction of the other ship's behavior is important in decision making. And rules such as CORLEG use subjective expressions to take into account the judgment of the ship operator. Thus, it is necessary to consider the unique characteristics of ships.

Among the functions of the automated navigation system, the safety evaluation of automatic collision avoidance functions begins with a study of normal scenarios, focusing on traffic disturbances. Two scenarios are considered: a series of scenarios of one-on-one encounters to verify basic functions, and a series of scenarios of multiple encounters with other vessels assuming a congested sea area. For one-on-one encounters, we will create a scenario that covers the other ship’s position, course, and speed by limiting the detection range of the sensor. For multiple encounters, since an infinite number of scenarios can be created, when the subject is the collision avoidance function, scenarios should be created from the following perspectives.

1) Extraction of encounter situations that are frequently encountered and those that are difficult to avoid from such as actual track data.
2) Collection of scenarios used in the evaluation of collision avoidance functions from papers related to collision avoidance [2].
3) Extracting scenarios from marine accident cases.

Perception disturbances and ship motion disturbances are also considered separately, and combined with the traffic disturbance scenario in consideration of their frequency of occurrence and degree of influence to form a scenario for evaluation.

3.2. Scenario creation

In this section, we present a traffic disturbance scenario creation method based on 1) in above Section actual track data etc. AIS (Automatic Identification System) data contains information such as ship’s position, speed, heading, MMSI, IMO number, and destination. The data is sorted by MMSI and other ship-specific information and then sorted by time to obtain the ship's track [3].

In addition, for domestic vessels of less than 500 GT, which are not required to be equipped with AIS, the data is estimated from the data of AIS-equipped vessels, and for fishing boats and pleasure boats, the navigation situation in the target area is surveyed by using radar data and interviewing the vessel operators. This procedure is as follows. (Fig.3)

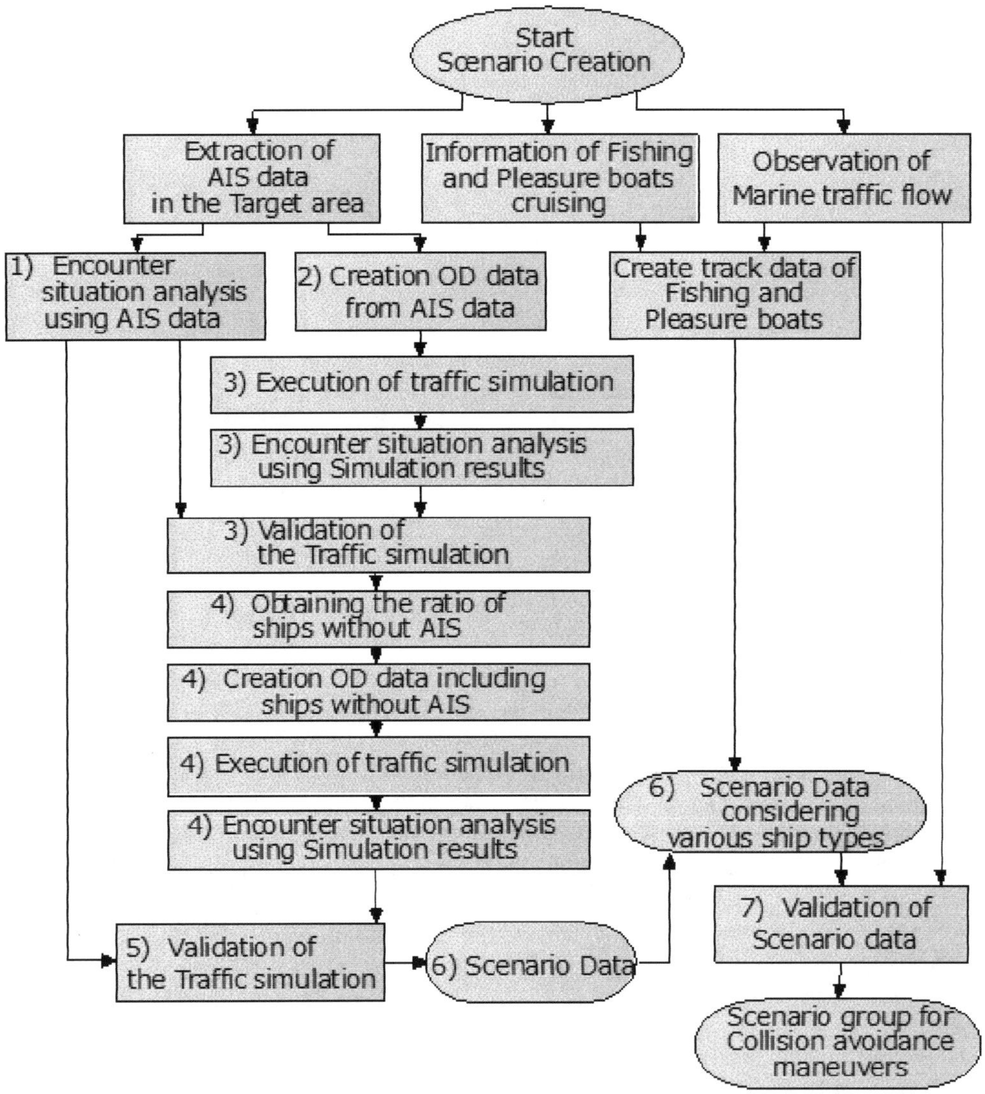

Figure 3 Procedure of Scenario Creation

1) Analysis of encounters between AIS-equipped ships from AIS information
2) Extraction of data for traffic flow simulation in the target area from AIS information
   - Create OD data for major routes for each ship type from track data
3) Execute traffic flow simulation for AIS-equipped vessels only
   - Analyse the encounter situation from the simulation results and compare it to the result obtained from AIS information to validate the simulation.
4) Traffic Flow Simulation Considering Vessels without AIS

- Based on the ratio of AIS-equipped vessels to non-AIS-equipped vessels, add the number of non-AIS-equipped vessels to the OD data obtained in 2) to simulate traffic flow.

5) Check the created scenario
   - Analyse the encounter situation from the simulation results of 4). Verify the results by comparing them with 1).
6) Add fishing boat and pleasure boat data to the scenario
   - Extract scenarios from simulation results considering conditions such as the target sea area and time of day, and create scenarios by adding fishing vessel and pleasure boat tracks estimated from interviews with ship operators.
7) Verification of the validity of the scenario in 6) using radar data, etc.

3.3. Study on Evaluation Indexes

As an evaluation index for collision avoidance functions, it is considered effective to combine quantitative evaluations such as DCPA (Distance to closest point of approach), TCPA (Time to closest point of approach) and conventional collision risk using the track and voyage records obtained during the test. And the method of evaluating compliance with laws and regulations, which has been proposed in recent years [4][5], is also effective. In addition, several evaluation methods have been proposed, such as subjective evaluation and subjective risk by the ship operator [6], and evaluation using OZT (Obstacle Zone by Target) [7].

It is important for autonomous ships to comply with the rules of navigation in order to prevent maritime accidents, but the current rules are ambiguous and difficult to evaluate quantitatively because they are based on the assumption of human operation [5]. In the method for evaluating compliance with laws and regulations described in References [4][5], the rules are expressed in mathematical form using multiple parameters. The parameters are derived from past accident cases and papers describing collision avoidance manoeuvres, and although the determination method is a new issue, it is interesting as a quantitative evaluation method.

In addition, when multiple vessels are involved in an encounter situation, it is necessary to navigate the automated ship in accordance with good seamanship. In the case of a one-on-one encounter, where the relationship is clear, the evaluation should be done according to the rules, and in the case of a complex one, the evaluation should be done from a subjective point of view.

4. Overview of comprehensive simulation system

In the previous chapter, traffic disturbances scenario was described for the evaluation of the collision avoidance manoeuvring system. In the safety evaluation, it is necessary to execute a large number of scenarios. In addition, scenarios that include perception disturbances, ship motion disturbances, and emergency response need to be evaluated. Therefore, evaluation by a simulation/virtual test platform is appropriate, and the test platform must be able to calculate at high speed and must be able to evaluate human involvement, etc. For this reason, we are building a comprehensive simulation system with the following two systems as its core.

1) Ship Handling Simulator (SHS)

   This is a full-mission ship handling simulator. It can evaluate human involvement such as the human-machine interface (HMI), time to transfer navigation authority to the crew in an emergency, and manoeuvring behaviour in case of coexistence with other ships.

2) Fast Time Ship Simulator (FTSS)

   The system performs simulations in sufficiently less time than the actual time to allow exhaustive verification under certain conditions. Note that FTSS is a proper noun.

The SHS and FTSS cover the following functions related to ship operation.

1) Automatic ship operation (Berthing and de-berthing, collision avoidance, rough weather manoeuvring)
2) Remote monitoring and ship operation

3) Emergency manoeuvring (transfer of operation from the machine to the crew)

In order to use SHS and FTSS for certification and development, it is necessary to connect the target automation system (algorithm or system incorporated algorithm that are for automatic collision avoidance etc.). For this purpose, we are considering designing the system to take into account the connection of arbitrary systems.

4.1. Ship Handling Simulator (SHS)

The level of an autonomous ship or automated system can be considered to progress in several stages, from the current level where the crew makes decisions and controls the ship, and the system supports them, to the level of a fully autonomous ship that navigates without human involvement. In other words, in the development stage of a fully autonomous ship, it is necessary to evaluate the safety of the bridge tasks by considering the involvement of the crew. For example, it is necessary to verify the availability of information and time for handing over ship operations from the system to the crew in an emergency. In addition, regardless of the stage of development of automations, it is necessary to consider the coexistence of the ship with other ships operated by operators, and the ship should be operated in a way that does not cause discomfort to those ships. The ship handling simulator is characterized by its ability to evaluate the manoeuvre in such a way that human involvement is taken into account. It is also necessary to verify various conditions, including environments and troubles that are difficult to reproduce in actual sea trials.

Fig.4 shows an image, and Table 1 shows the planned basic specifications of it. In addition, the required functions of the ship handling simulator are as follows.

1) Automatic ship operation
 · Connection of arbitrary automatic ship operation program
 · Incorporation of ship motion model corresponding to evaluation target
2) Remote monitoring and ship operation
 · Communication system
 · Reproduction of information display function
 · Reproduction of communication speed and lack of data, etc.
3) Handing over of ship operation to crew, evaluation of HMI
 · Reproduction of information provision function
 · Reproduction of ship operation switching device and functions
 · Arbitrary arrangement of navigation equipment, etc., and connection to external equipment
4) Incorporation of various types of information
 · Incorporation of a collision avoidance function on other ships
 · Creation of various sensor information
 · Reproduction of information accuracy (noise, missing data, update interval, etc.)
 · Incorporation of engines, thrusters, steering gear
 · Emergency event expression function
 · Reproduce failures of sensors, engines, power supplies, etc.
5) Testing environment
 · Test case creation function
 · Display of ship operation results and results of analysis of various types of indexes

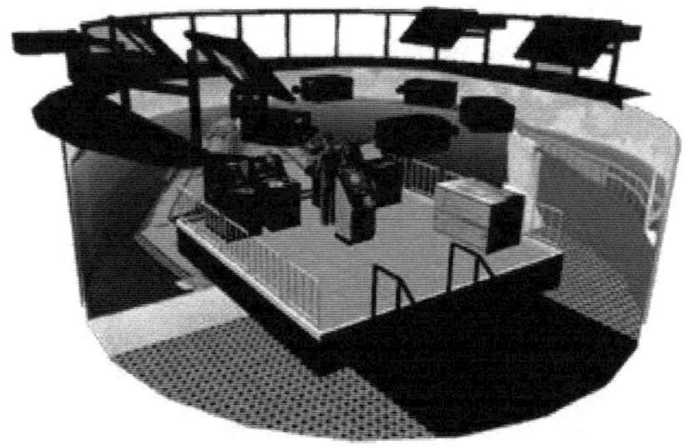

Figure 4 Image of the SHS (Prepared in March 2021)

It includes new functions that are not available in the current ship handling simulator, and we are currently studying the implementation of these functions.

Table 1. Specifications of the SHS (plan)

Visual system	Cylindrical screen r=6.5m, Field of view (H x V) 360deg. x 30deg. Screen for downward view
Bridge	Overhead Panel (Main engine RPM, Speed indicator, Rudder angle indicator, Wind direction/anemometer, clock, inclinometer, etc.) Radar x 2, ECDIS, Steering stand, Repeater compass, Navigation console, Chart table, Binoculars, etc.
Reproduction	Time changeover without steps (day, dusk, night) Ship's Lights and Shapes Characteristics of light of AtoN Weather (including restricted visibility due to fog) etc.

4.2. Fast Time Ship Simulator (FTSS)

In the safety evaluation of automated navigation systems, it is necessary to simulate under various conditions. When human involvement is not required, it is effective to use fast time simulation such as the FTSS, which performs calculations at high speed without real time output.

An overview of the FTSS is shown in Fig.5. Each simulation module that runs on the FTSS, such as the automated navigation system, natural environment, traffic environment, sensors, and ship motion calculation, is connected to the simulation management module through FMI as a Functional Mock-up Unit (FMU), or a Dynamic Link Library (DLL). The entire system functions as the FTSS to test the operation of an autonomous ship. The outline of each module is as follows.

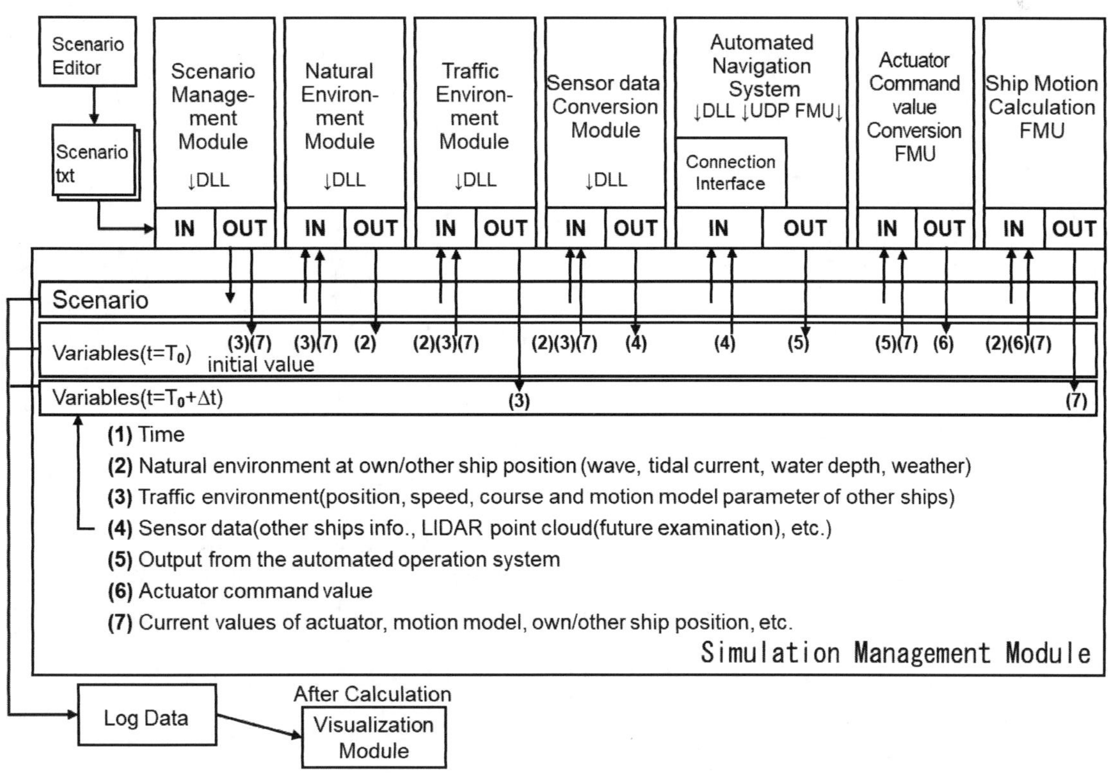

Figure 5 Overview of Operation of Fast Time Ship Simulator

1) Simulation management module
It manages the operation of the FTSS by performing a series of operations such as starting modules, initialization based on scenarios, execution of modules, control of data input and output from modules, log output, output for visualization, judgment of termination based on scenarios, and time update.

2) Scenario management module
The scenario management module reads the configuration items necessary to run the simulation from the scenario file and manages the scenario information so that the simulation management module can use it.

3) Natural environment calculation module
It outputs wind, wave, current, water depth, weather, day and night information from the time and position (XY) of own ship and other ships. In order to improve the calculation speed, it has a function to create a data set by calculating grid data for 24 hours in advance.

4) Traffic environment module
Generate other ships based on the values set in the scenario. Then, each other ship is made to navigate by waypoints or automatically based on the settings.

5) Sensor data conversion module
Based on the true values obtained by other modules for the ship's own/other state (speed, position, motion, etc.) and the natural environment (wind, wave, current, etc.), this module creates and provides sensor data that is consistent with the input of the automated navigation system by superimposing noise that simulates the measurement errors of the measuring instruments.

6) Ship navigation system connection interface
This is an interface for connecting an automated navigation system built with an interface other than FMI to a simulation system using FMI. Since the automated navigation system may have a variety of execution formats and input formats, this interface must be easily modifiable. In addition to the FMI connection, the connection through a network is also assumed at present.

7) Actuator command value conversion FMU
The ship manoeuvring instructions received from the automatic navigation system are converted into a format that the actuator can accept as input, and the mechanical response of the actuator is output to the ship motion

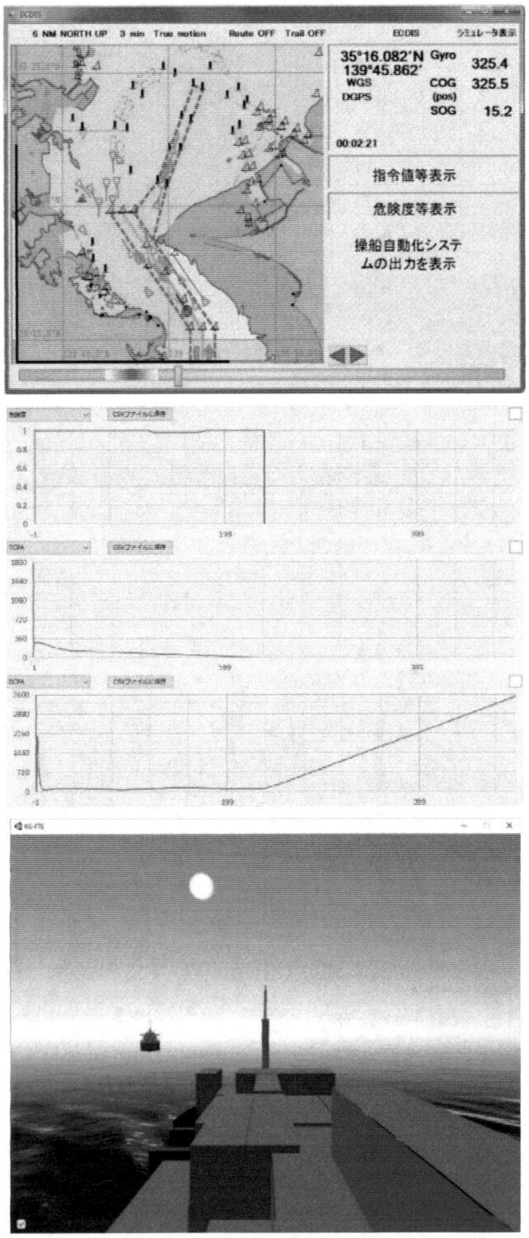

Figure 6 Visualization of Results
(Upper: Display of electronic chart
Middle: Display time-series indexes
Lower: 3D display)

calculation FMU as the current value of the actuator (rotational speed, rudder angle, etc.).

8) Ship motion calculation FMU

The ship motion calculation FMU calculates the time update of the ship's own state quantity from the manoeuvring and natural environment state quantities calculated by the input Actuator command value conversion FMU and Natural environment calculation module, and outputs them to the simulation management module. In addition, we are developing a manoeuvring motion simulation tool that outputs parameters necessary for setting up a mathematical manoeuvring model from the principal particulars and actuator configuration and are considering using the generated parameters for evaluating collision avoidance manoeuvre in this system.

9) Visualization module

The visualization log output from the simulation is visualized by displaying electronic charts, 3D displays, and evaluation indices to provide a function to analyse the execution results in more detail (Fig.6). The evaluation indices are described in Chapter 3.

4.3. Simulation procedure

The procedure for conducting the simulation is shown in Fig.7 using FTSS as an example; in SHS, the basic flow is the same, although the creation of data such as landscape images and display parts of navigational equipment is added during the execution of the calculation.

1) Scenario creation

In scenario creation, the initial conditions of the simulation are set. Here, the state and position of the ship, the planned route (information on the way points), traffic flow such as the layout and speed of other ships, natural environmental conditions such as waves, wind, day and night, and geographical conditions such as water depth and navigational obstacles are set. The scenario file is created as a text file using a scenario editor. 2) Scenario management module in Section 4.2 is responsible for loading the scenario file.

2) Calculation execution

As shown in Fig.7, the simulation calculation updates the data every cycle and continues until the end condition is met. The end condition is determined by the scenario, such as when the last way point is reached. This process is handled by 3) to 8) in Section 4.2.

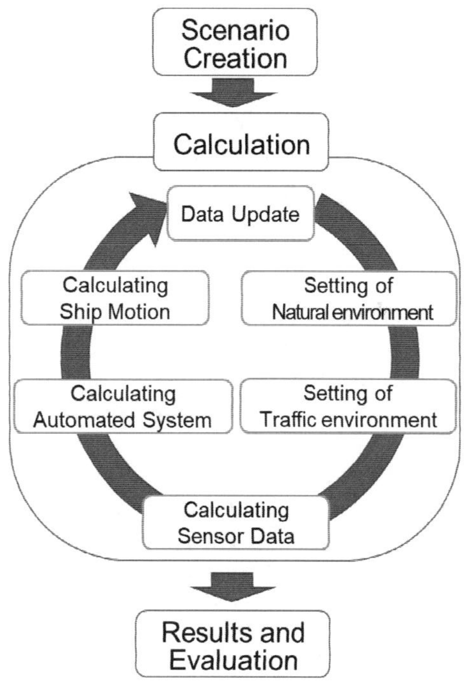

Figure 7 Simulation Execution Procedure

3) Result analysis and evaluation

In addition to the own/other ships' tracks, course, and speed, the output of the automated navigation system can be recorded to evaluate the results of ship operations. We are also studying the construction of an efficient debugging environment that allows easy feedback to the target system during development, such as the ability to reproduce the results from any timing.

5. Conclusions

In the safety evaluation of automated ships, it is important to ensure the sufficiency of the test scenarios, and we decided to use a scenario-based approach, similar to Automated Driving Systems. When creating scenarios for evaluation, it is necessary to consider not only when the system is normal, but also emergency scenarios that include factor such as parts failure and cyber security. In this paper, a method for making disturbance scenarios considering the physical principles of scenario elements. In this method, the number of scenarios reduced to finite number by considering the physical principles, but it is still necessary to run a large number of scenarios, so we developed a FTSS, which is capable of high-speed simulation. We are also developing a full-mission ship handling simulator to evaluate human involved actions, such as transfer navigation authority to the crew in emergencies and coexistence with a ship operated by humans. Safety evaluation is indispensable for the practical use of autonomous ships, and these standards will serve as a guide for future development of them. In cooperation with other companies involved in the development of autonomous ships, we will establish safety evaluation methods and develop the comprehensive simulation system, to support a development of autonomous ships.

References

[1] ISO/DIS34502 Road vehicles-Scenario-based safety evaluation framework for Automated Driving System, 2021.
[2] (For example) Kinzo INOUE, Kenji MASUDA and Wataru SERA. 1997 Guidelines to Assess the Safety of Marine Traffic -1, The Journal of Japan Institute of Navigation, 98, pp. 225-234.
[3] Kenji Masuda, Daichi Hara and Keisuke Nishizawa: Overview and Data Composition of AIS, Navigation, No. 188, 2014.
[4] K. Woerner et al.: Quantifying protocol evaluation for autonomous collision avoidance, Autonomous Robots, Vol. 43, April 2019.
[5] P.K.E. Minne: Automatic testing of maritime collision avoidance algorithms, 2017.
[6] Shinya NAKAMURA, Naoki OKADA, Satoru KUWAHARA, Koji KUTSUNA, Takuya NAKASHIMA, and Hideyuki ANDO: Study on Automatic Collision Avoidance System and Method for Evaluating Collision Avoidance Manoeuvring Results, MTEC/ICMASS2019,2019
[7] Chihiro Nishizaki, Jun Kayano, Ruri Shoji and Hayama Imazu: Encounter Features of Collision Risk in the West Side Sea Area of Izu Oshima Using OZT, Journal of the Japan Institute of Navigation, Vol. 139, 2018.

Acknowledgments

This research is being carried out as commissioned research in the “Safety Evaluation of MEGURI 2040 (Unmanned Ship) Project” being conducted by the Japan Ship Technology Research Association, with the support of the Nippon Foundation. We wish to express our deep gratitude to all those concerned. The authors are grateful ClassNK to support and useful discussions. We would like to thank Dr. FUKUTO for assisting in the studies.

Journal of Physics: Conference Series **2311** (2022) 012013 doi:10.1088/1742-6596/2311/1/012013

Application of Dynamic-Task-Based Hazard Identification Method to Remote Operation of Experimental Ship Shinpo

R Miyake[1], J Kudo, E Ishimura, H Itoh, T Yuzui, M Shiokari, S Kawashima, K Hirata, Y Niki, M Kobayashi, R Sawada and S Inaba

National Maritime Research Institute, National Institute of Maritime, Port and Aviation Technology, 6-38-1 Shinkawa, Mitaka, Tokyo 181-0004, Japan

[1] r_miyake@m.mpat.go.jp

Abstract. Recently, the development of autonomous ships has been progressing worldwide; thus, it is essential to ensure the safety of these ships for their successful realization. In previous studies, task-based hazard identification (HAZID) methods were developed as safety assessment methods for autonomous ships to identify hazards related to the tasks of ship operators. In this paper, a dynamic-task-based HAZID (DTB-HAZID) method, which is an improvement on the previously developed task-based HAZID method, is proposed, and the results of the risk analysis obtained from applying this method to a small experimental ship, Shinpo, are reported. The DTB-HAZID method was designed to analyze time-related hazards such as task delays and the simultaneous occurrences of tasks. To achieve this, the activity diagram, which is a type of unified modeling language (UML) model, was introduced to create task flow diagrams. In the risk analysis, first, the tasks of the operator who remotely navigated Shinpo were analyzed, and the corresponding task flow diagrams were created. Then, the time-related hazards were identified with reference to these diagrams. Finally, the obtained hazards were reviewed, and the applicability of the proposed method in hazard analysis from the perspective of time was discussed.

1. Introduction

The development of autonomous ships is expected to reduce the number of human errors that result in accidents, as well as the burden of seafarers. Ships are generally developed in several stages, starting with the design stage, which involves conceptual design, initial design, and detailed design, followed by the practical application stage; risk analysis is required at all these stages. To date, as a part of risk analysis, hazard identification has been conducted in every stage of design, mainly focusing on the physical aspects of ship systems.

In contrast, in an autonomous ship, tasks performed by humans such as cognition, decision-making, and actions such as course change are being automated by a system composed of hardware and software [1]. Some of these tasks are also executed in combination with humans and other automated systems. For example, a human checks and executes a collision avoidance (CA) plan proposed by an automated system. One of the hazard analysis techniques is Systems-Theoretic Process Analysis (STPA) based on Systems-Theoretic Accident Model and Processes (STAMP). STAMP/STPA is an analysis method that focuses on the interaction between various elements including hardware, software, and humans, and is considered suitable for such autonomous ships. Wróbel et al. [2] applied STPA to a simple structure of an autonomous ship and identified hazards. The simple structure consisted of the autonomous ship and its surrounding environment such as shore facility etc. Valdez Banda et al. [3] applied STPA to two autonomous ships concepts for urban in Finland and described the identified hazards and their safety controls. The application to two ships resulted in 15 hazards and proposed 73 safety controls for them. However, STAMP/STPA does not explicitly deal with the functions of the components, or the information required to perform the functions. Besides, this also causes difficulties in the dynamic aspects of the analysis, which will be discussed below.

As a method to resolve the issue, therefore, Shiokari et al. [4] proposed a modeling method for a system including software, which can be used with existing hazard identification methods, and Shiokari et al. [5] applied this method to a hypothetical autonomous ship created with reference to a real domestic small ferry and identified the hazards in the system. In addition, in the model diagram of the hypothetical autonomous ship, Ishimura et al. [6] focused on the tasks of each component of the model and proposed a task-based hazard identification (HAZID) method that supported the identification of hazards related to the tasks and interactions among the components. These methods helped to identify hazards in terms of incomplete task execution, equipment failure, and deviation from the operational design domain (ODD).

However, these methods only focused on the static aspects of the tasks. Practically, task execution has conspicuous dynamic characteristics, hence the sequence or timing of execution, as well as the time required for an individual to perform the task are important aspects for consideration. Particularly, in ship operations, the order, timing, and duration of tasks are important factors in task execution, and problems in these areas can be significant hazards.

In system development, it is assumed that the concept of ship operation by an automated system, including cooperation with humans, will be considered in the initial design stage. However, it is difficult to use conventional analysis techniques to identify hazards and develop countermeasures by considering the timing, sequence, and order of the tasks in an actual operation at that stage. Therefore, to identify the hazards arising from the dynamic characteristics of task execution and develop countermeasures, the system and human operations need to be represented in terms of time. Miyake et al. [7] proposed a dynamic-task-based hazard identification (DTB-HAZID) method that introduced a task flow diagram having the notations of the activity diagram, a unified modeling language (UML) model, into the method of Ishimura et al. [6].

In this paper, we reported the results of applying the DTB-HAZID method as a safety analysis method in the remote operation of a small experimental ship, Shinpo. First, a task flow diagram obtained from the task analysis of the remote operation is presented. Then, the results of hazard identification using this diagram are presented, and the characteristics of the obtained hazards are reviewed.

2. Shinpo and remote operation experiment

2.1. Overview of Shinpo

Shinpo is an experimental ship managed by the National Maritime Research Institute (NMRI). Table 1 shows the principal particulars of Shinpo, and Figure 1 provides an overview of them. Shinpo has one rudder, a fixed-pitch propeller (FPP), and a bow thruster. Figure 2 shows the configuration of the control system of Shinpo. As illustrated in Figure 2, Shinpo has sensors such as a Global Navigation Satellite System (GNSS) and an anemometer, onboard network cameras, a long term evolution (LTE) communication system, a programmable logic controller (PLC) for controlling the actuators of the ship, a touch panel display to input control values to the PLC and display response values from the actuators, and a control PC connected to the PLC.

Table 2 shows an outline of the Shinpo control systems. The ship control system (SCS) mainly controls the actuators using the PLC. It also has functions to switch between automated control using the PLC and manual control, to issue an alarm when any malfunction occurs and to prevent erroneous operations. The waypoint navigation system (WNS) automatically steers the ship to a target course along predefined waypoints by controlling the rudder. The remote control system (RCS) connects the ship and remote operation center via LTE communication to enable remote control.

Table 1. Principal particulars of Shinpo.

Particulars, Notation [unit]	Value
Length overall, Loa [m]	16.50
Length between perpendiculars, Lpp [m]	14.90
Breadth (1.0 WL), B [m]	4.60
Depth, D [m]	1.50
Draft, d [m]	0.75
Gross tonnage	17

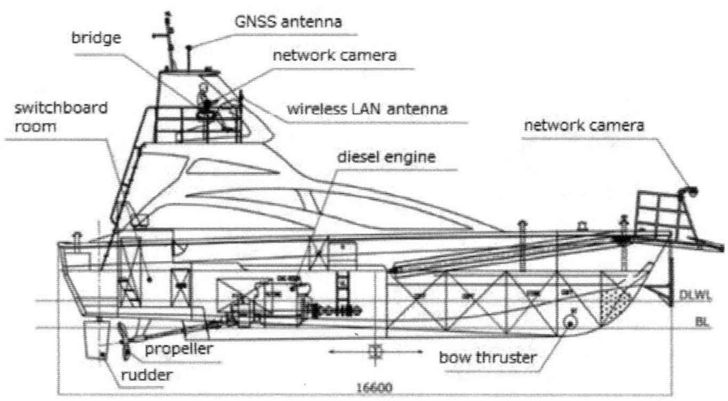

Figure 1. Overview of Shinpo.

Table 2. Outline of the control systems of Shinpo.

System name	Outline
Ship control system (SCS)	Basic system used by the PLC to control the actuators of the ship
Waypoint navigation system (WNS)	Automatic system that steers the ship to a target course along predefined waypoints by controlling the rudder
Remote control system (RCS)	System that connects the ship and the remote operation center via LTE communication for remote operation.

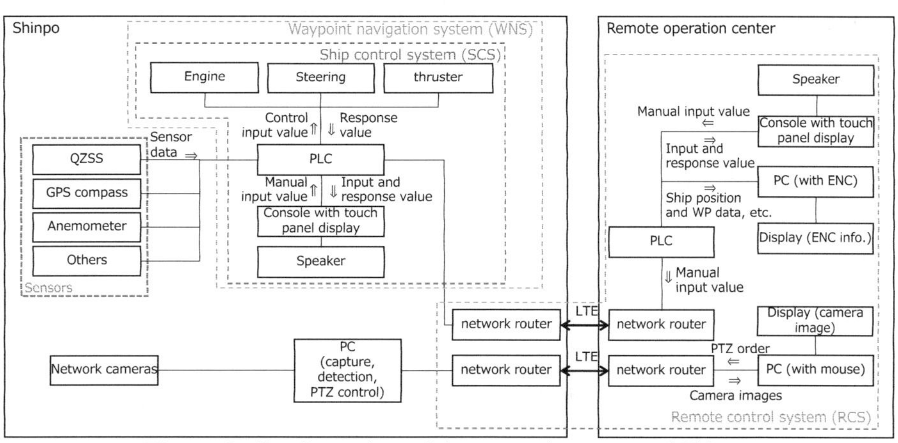

Figure 2. Configuration of the control systems of Shinpo.

2.2. Overview of the remote operation center

Figure 3 shows the remote operation center on the shore, with the configuration of the control systems illustrated in Figure 2. The center has a PLC, a console with a touch panel display for operating the SCS and WNS, an electronic navigational chart (ENC) display to show waypoint positions and ship positions, a camera image display, a speaker that announces information from the WNS, and PCs. The camera image display has four partial images. The lower two partial images are omnidirectional (ODR) camera images and display the landscape around Shinpo with the markers of the ships detected from camera images to support the monitoring of the navigational environment around Shinpo. The upper row images are panoramic-tilt-zoom (PTZ) camera images, and the landscape of the location which the remote operator points on the PTZ camera images or the ODR camera images is displayed. The left of the three displays in front of the remote operator is a spare display, which is beyond the scope of this study. Table 3 shows the information available to the remote operator during a remote waypoint operation (WPO). The system time in seconds is also shown on each display to confirm the real-time validity of the information displayed.

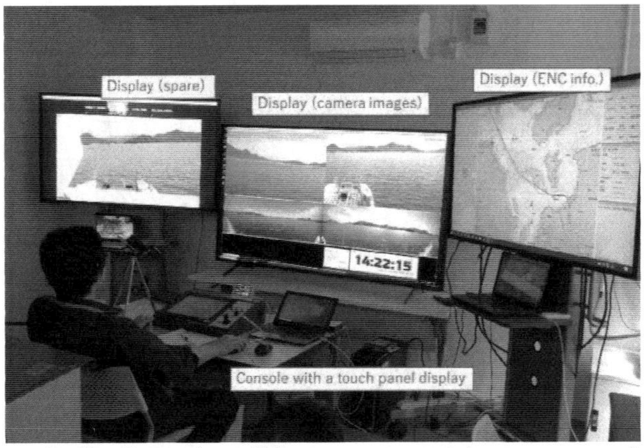

Figure 3. Configuration of the remote operation center on shore.

Table 3. Information available to the remote operator.

Device	Contents	Detailed information
Touch panel display*	SCS information	· Shinpo's status information. · Control command value and response value of equipment, such as those of a rudder. · System error warnings.
	WNS information	· Waypoint setting information. · Shinpo's status information on the waypoint operation. · System error warnings.
Camera image display	ODR images	· Landscape around Shinpo. · Ships' markers detected from camera images.
	PTZ camera images **	· Landscape specified on the panoramic camera image.
ENC display	Position information **	· ENC. · Positions of waypoints. · Position and heading of Shinpo. · System error warnings.
Speaker	SCS voice information	· Warning in emergencies.
	WNS voice information	· Distance and bearing to the targeted waypoint. · Changing the course angle toward the targeted waypoint. · System error warnings.
	Response voice information from the person on board	· Information on other ships. · Prediction of other ships' behaviors. · System error information and countermeasures.

* System screens need to be switched by operators.
** Moving and zooming in and out of the displayed screen requires the operation of the remote operator.

2.3. Overview of remote operation experiment

Figure 4 shows the experimental sites and the planned route. The experimental site is in the Inland Sea of Japan, whereas the remote operation center is in Tokyo. The planned route was approximately 20 minutes one way connecting 13 predefined waypoints.

The ship operation for the experiment consisted of the following phases: start of the operation, manual departure from a pier by an operator on board, WPO by the remote operator (hereinafter "remote WPO") from waypoints 1 to 6, automatic berthing operation at a virtual pier, manual departure from the virtual pier by the remote operator, remote WPO from waypoints 7 to 13, automatic berthing at the pier in the marina, and the end of the operation. In this study, we focused on the remote WPO phase.

According to an experimental protocol, the main procedures in the remote WPO phase are as follows:

(1) At a sufficiently distant point after leaving the marina, the remote operator starts the WNS via the RCS.
(2) The remote operator controls the joystick on the console to increase the ship speed.
(3) The WNS controls the rudder to maintain the ship on the planned course.
(4) When the ship approaches the target waypoint, the WNS announces the changing of the course angle toward the next waypoint and requests the remote operator's approval.
(5) The remote operator approves the course change.
(6) The WNS changes the course and the target waypoint to the next waypoint.
(7) When the ship approaches the final waypoint (WP6 and WP13 on the going and returning trips, respectively), the WNS requests the remote operator's approval to stop each trip.
(8) The remote operator completes the remote WPO phase and shifts to the next phase.

During the above operations, the remote operator maintains a constant lookout for surrounding ships and performs CA maneuvers when necessary. The remote operator also continues to monitor the system to ensure that everything is working properly. Although the captain and the system engineers are on board, they will not control Shinpo, only intervening in case of an emergency according to their assigned roles, as mentioned in Section 3.1. Instead, they take on the role of experimenters, which includes securing the safety of the ship during the remote WPO.

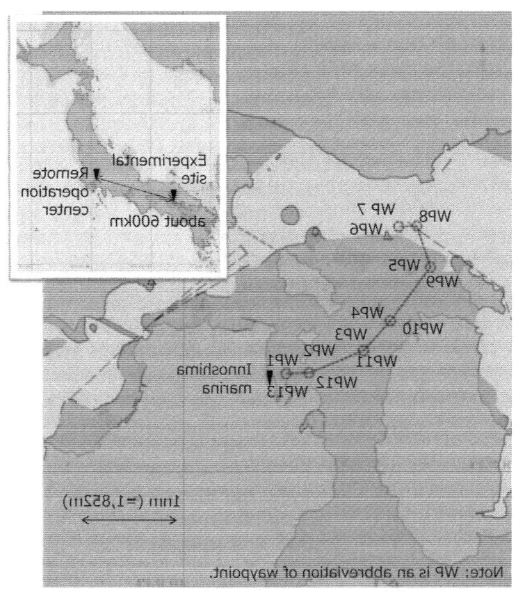

Figure 4. Experimental sites and planned route for remote operation.

3. Task analysis of remote operator

3.1. *Roles assigned to the systems and remote operator*

Table 4 shows the roles assigned to the systems and humans in the remote WPO. The remote operator in the remote operation center maneuvered Shinpo based on the information obtained from the displays and the speaker, as mentioned in Table 3. The experimenters on board provided supplementary information to the remote operator, such as information on other ships' movements and emergency responses, upon request from the remote operator.

Table 4. Roles assigned to the systems and humans.

	On shore	On board
Humans	Remote operator ・System monitoring. ・Determining when shifting to the next phase is required and executing it. ・Determining when switching to manual operation is required and executing it. ・Determining when changing the target course to the next waypoint is required and executing it. ・Determining when CA is required and executing it.	Experimenters ・System monitoring support. ・Monitoring support and providing information to the remote operator. ・Determining when suspending the remote operation is required and executing it.
Systems	・Provision of information acquired on board. ・Sending commands to the ship through the remote operator. ・Issuance of system error warnings.	・Execution of ship operation following the planned route. ・Acquisition and transmission of camera and sensor information to shore. ・Issuance of system error warnings.

3.2. *Task analysis of remote operator during the remote WPO phase*

We observed the experiment, organized the tasks of the remote operator, and specified the sequence of operations and other relevant information. We identified the following main tasks of the remote operator: maneuvering, watchkeeping, positioning, and monitoring the systems, as shown in Table 5. The main tasks "maneuvering," "watchkeeping," and "positioning" were tasks observed not only in remote operations but also in ordinary ship operations.

These main tasks were divided into multiple subtasks, as shown in Tables 6–9. The subtasks were classified according to cognition, behavior, and system operation procedures. Furthermore, the subtasks were composed of detailed activities performed using cognition and human behaviors such as gathering information from camera images and sensors, decision making, and system operation.

Table 5. Main tasks during the remote WPO.

Maneuvering	Maneuvering Shinpo via the systems.
Watchkeeping	Watching the movements of other ships around Shinpo.
Positioning	Confirming the positions of Shinpo and the target waypoint.
Monitoring the systems	Monitoring the system properly.

Table 6. Overview of the task "maneuvering."

Subtask	Activity
Start operation	・Determining when to start the remote WPO. ・Setting the initial target waypoint. ・Order for starting WNS. ・Setting the engine revolution.
Collision avoidance	・Determining when CA is necessary. ・Determining means of CA.

	・Adjusting course angle or engine revolution. ・Determining when CA is completed. ・Order for stopping WNS and executing CA manually.
Shifting to the next waypoint.	・Confirmation of the distance to the targeted waypoint and the course change angle. ・Determining when to shift to the next waypoint. ・Approval of shifting to the next waypoint (execution of course change).
Emergency stop	・Determining when the remote WPO is completed. ・Order for stopping WNS.

Table 7. Overview of the task "watchkeeping."

Subtask	Activity
Detecting obstacles	・Confirmation of the ODR camera images (landscape) ・Detecting obstacles. ・Request for supplemental information on obstacles to the experimenters and confirming the same.
Identifying ships	・Confirmation of the ODR camera images (landscape and ships' markers). ・Confirmation of the ships' markers detected from the camera images. ・Confirmation of the PTZ camera images. ・Moving and zooming operation of PTZ camera. ・Identifying the obstacle as either a ship or an object. ・Request for supplemental information on the presence of other ships to the experimenters and confirming the same.
Gathering ships' information	・Confirmation of PTZ camera images. ・Moving and zooming operation of PTZ camera. ・Gathering information on other ships. ・Request for supplemental information on the presence of other ships to the experimenters and confirming the same.
Predicting ships' behaviors	・Determining the passing distance to other ships. ・Request for supplemental information on the behavior of other ships to the experimenters and confirming the same.

Table 8. Overview of the task "positioning."

Subtask	Activity
Confirming the Shinpo's position	・Confirmation of the ODR camera images (landscape). ・Moving and zooming operation of ENC. ・Confirmation of environment around Shinpo. ・Confirmation of the position of Shinpo. ・Confirmation of mapped position information on ENC display.
Confirming the waypoint's position	・Operation of touch panel display for displaying the WNS. ・Moving and zooming operation of ENC. ・Gathering the targeted waypoint information. ・Confirmation of the targeted waypoint position. ・Confirmation of mapped position information on ENC display.
Confirming the relative position of Shinpo and waypoint	・Confirmation of the relative position of Shinpo and waypoint. ・Request for supplemental navigation information to the experimenters and confirming the same.

Table 9. Overview of the task "monitoring the systems."

Subtask	Activity
Monitoring of remote communication delay	· Checking the live monitoring of network communications. · Checking the communication delay.
System error monitoring	· Confirmation of warning issuance.
Securing safetyof the ship	· Sharing information on any errors with people on board. · Emergency stops of the systems.

3.3. Task flow diagram for the remote WPO phase

Based on the system specifications and the observation results of the experiment using videos, a task flow diagram of the system and the remote operator during the remote WPO was generated. A notation of the simplest flow diagram might be a flowchart and could represent the simple process of the system and the operator's activities. However, we thought that the notations of the flowchart were not enough to represent the complicated operator's activities during the remote WPO because the operator performed the task "maneuvering" concurrently with the task "watchkeeping," "positioning" and "monitoring the systems." Therefore, the notations of the activity diagram, a type of behavioral diagram in the UML [10], were adopted for the task flow diagram.

Activity diagrams model the behaviors of systems and their related objects and indicate the flow of an action. They have the following elements, as shown in Figure 5. The "send signal action" element expresses the action of sending a signal to the system, the "receive signal action" element denotes that the system executes some action upon receiving the signal, and the "fork/join" elements denote the parallel execution of the actions.

Figures 6 to 9 show the activity diagrams of the main tasks of the remote operator. The final state of each diagram indicates the completion of the main task and transition to another phase, such as the manual departure by the remote operator, which is not within the scope of this paper.

The sub-activity element, "CA," in Figure 6, represents a decomposition of the subtask "CA" in Table 6 and provides a detailed procedure for performing it. The activity element "gathering ships' information" in Figure 7 represents one of the several subtasks of the remote operator, namely gathering ships' information, in Table 7. The expansion region elements with indefinite numbers in Figure 7 represent multiple iterations of the framed activity, which means that "watchkeeping" is continuously performed to monitor and deal with all the surrounding ships during the voyage.

Practically, the entire remote control system is composed of multiple systems on board and on shore, with functions configured according to their respective roles, and with many detailed information flows, such as those of sensor data and the interactions between the systems. However, for simplicity, in the task flow diagram, the activity diagrams are composed of two partitions: systems and the remote operator. In addition, the activity elements in the system partition are restricted only to the essential elements that interact directly with the remote operator. Considering the configuration of the systems in Figure 2, for example, though the activity element "display of ODR camera images" has the following activities: "network cameras capture images," "ships are detected from the camera images," and "images are transmitted to the remote operation center via LTE," prior to "display to the remote operator," these are not represented in Figure 7. Furthermore, to describe parallel action of the remote operator, the pathways between the "fork/join" elements are divided into each group which has clearly successive procedures, for example, the process of gathering information from the camera images after operating the camera by the mouse, and of cognitive behaviors including detection, identification, and prediction. It means that unless performances of all groups are complete, the next action will never start.

By using the task flow diagram, it is possible to understand the overall task flow in the entire system and the major interactions between the systems and the remote operator. In summary, using task flow diagrams to represent the timing, sequence, and order of the tasks supports hazard analysis for the time aspects of the operation.

The main tasks have close interactions with each other. However, it should be noted that the main task diagrams represent the interior activities of the main tasks and do not represent the interactions among them. Specifically, in the task "maneuvering," the remote operator performs the subtask "CA" based on the consequences of the cognition or decision-making of the other main tasks, such as "watchkeeping" and "confirmation of navigational position," but the relationships between these main tasks are not explicitly shown in the figures.

Notation Description		Description
	Initial	Start of a flow
	Activity	A task to be performed.
Activity	Sub-Activity	A task organized and specified subordinate behaviors.
	Send Signal Action	An action of sending a signal.
	Receive signal action	An action is executed after acceptance of a send event.
	Decision Node/ Merge Node	Branching of conditions/ Merging of branched conditions.
	Fork Node/ Join Node	Parallel execution of actions.
	Activity Final Node	Completion of an activity.
	Interruptible Activity Region	All tokens in the region are terminated when certain interrupts (i.e. cancel request) are raised.
Expansion Node	Expansion Region	A region executed several times on the incoming data.

Figure 5. Legends for activity diagrams in Figs. 6–9.

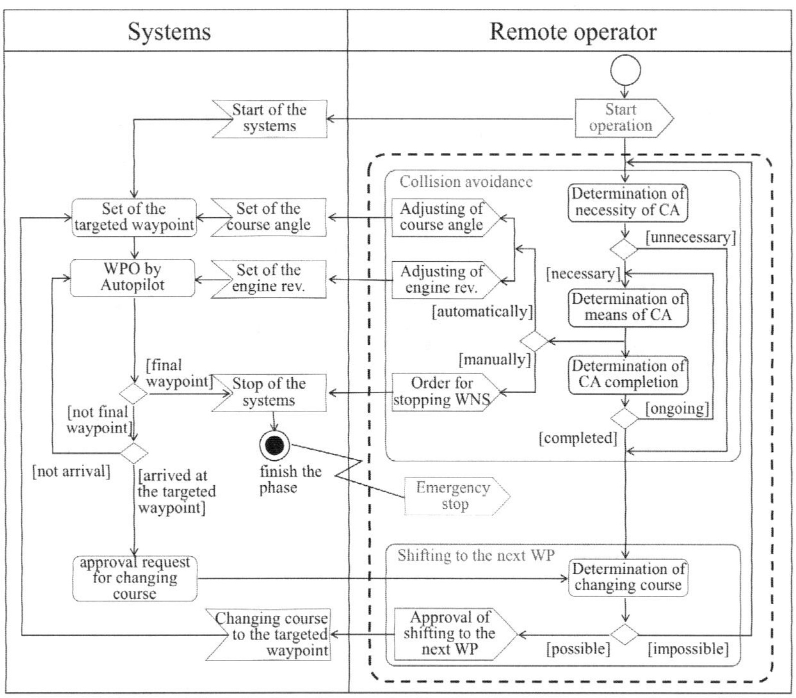

Figure 6. Activity diagram of the task "maneuvering."

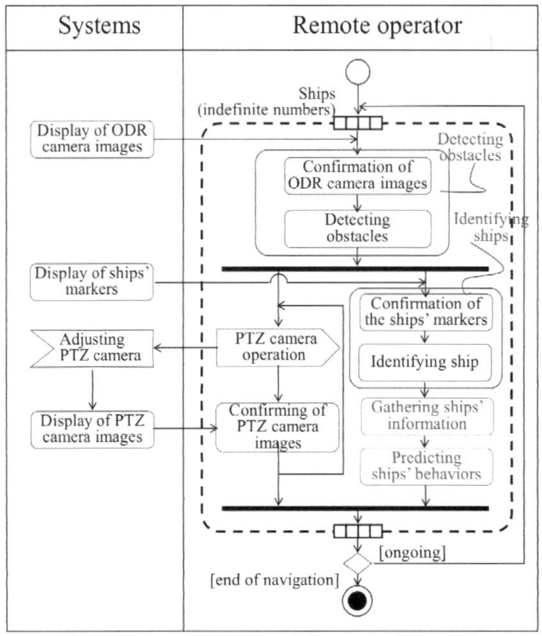

Figure 7. Activity diagram of the task "watchkeeping."

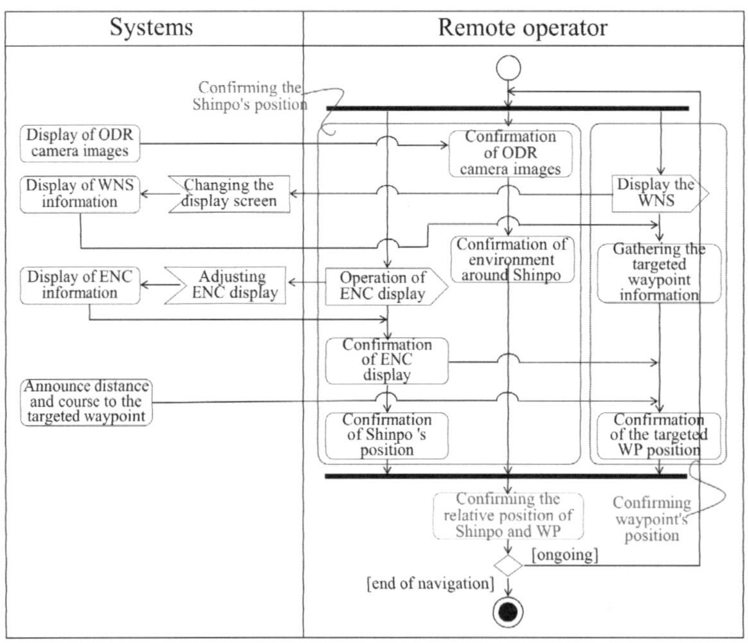

Figure 8. Activity diagram of the task "positioning."

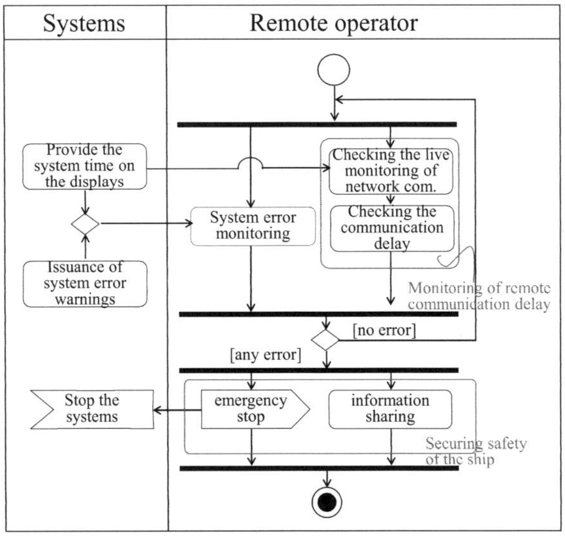

Figure 9. Activity diagram of the task "monitoring the systems."

4. Hazard identification (HAZID)

4.1. Overview of DTB-HAZID

DTB-HAZID was conducted with reference to the task flow diagram defined in the previous section from the viewpoint of deviations from the normal procedure and status, such as incomplete task execution, failed task execution, swapping the order of the tasks, and deficiency in information due to equipment failure. The Structured What If Technique (SWIFT) method was used in a HAZID online conference meeting on the DTB-HAZID. The SWIFT method uses brainstorming.

In HAZID, the causes, consequences, and indices regarding the consequences and probability of hazards are referred to. These indices are useful for clarifying the order of priority of countermeasures to be implemented to deal with the identified hazard. However, it is not always practical or possible to assess the severity and frequency of the consequences of hazards for systems that are envisioned in the early stages of design. Therefore, in this study, instead of ranking and screening all the identified hazards, we tried to improve the safety of Shinpo by taking all the possible countermeasures against them.

Incidentally, as mentioned in Section 3.3, the main task diagrams are not enough to represent the interactions among them. We attempted to express those interactions by the notations of the activity diagram and the use case diagram, a type of behavioral diagram in the UML. However, it was difficult to represent them because the multiple main tasks were executed in parallel or their sequence was complicated and irregular, for example, in the relationship between the main task "maneuvering" and "watchkeeping," the whole task of "watchkeeping," was successively performed during collision avoidance and the manner for avoiding was changed according to the change in the circumstances. Therefore, DTB-HAZID was conducted after all analysts understood that the main tasks were closely related to each other, for instance, by watching the remote ship operation in advance.

From the results of DTB-HAZID, 30 hazards were identified, including some duplicates. The number of hazards identified for each main task is presented in Table 10. Some hazards, such as the lack of camera images, were identified as common hazards in the main tasks. Tables 11 to 13 show some of the identified hazards for the three main tasks, their causes, consequences, and the corresponding safety measures to deal with them. Table 14 lists some of the common hazards found for multiple main tasks.

4.2. Discussion

Most of the hazards were caused by mistakes in handling previous tasks, incompletion of the previous tasks, or a lack of the necessary information needed to perform the task. In addition, they led to consequences such as failures in starting subsequent tasks or collisions with other ships. Safety measures against such hazards included system redundancy, support to facilitate the perception of remote operators, and training to improve the skills and understanding of remote operators in handling the system. It was suggested that for maintaining the safety of remote ship maneuvering, each task needed to be performed smoothly, and information from the system should be correctly presented to the remote operator.

By referring to the task flow diagram, the task flow of the remote operator and the triggers for task execution (e.g., presentation of camera images) became clear. The hazards related to the sequence of operations performed by the system and the human operators, as well as the timing and duration of each task, were identified. In particular, the introduction of the activity diagram notations into the task flow diagram made it possible to represent the parallel processing of tasks, and facilitated the identification of hazards produced by the simultaneous execution of tasks.

For these reasons, the proposed DTB-HAZID method, with a task flow diagram, was useful for identifying hazards and developing safety measures. In addition, although a diagram that can represent the interrelationships among the main tasks is the subject of future work, the diagram would facilitate the identification of hazards caused by dependencies between tasks.

Table 10. Number of hazards identified for each main task in the HAZID meeting

Main task	Number of hazards
Maneuvering	12
Watchkeeping	10
Positioning	0
System monitoring	3
Common hazards	5
Total	30

Table 11. Examples of identified hazards for the task “maneuvering.”

Hazards	**Necessity for CA against other ships, determined by remote operator, is mistaken/ cannot be determined/ late.**
Causes	· Remote operator cannot detect obstacles, identify ships, or predict the ships’ behaviors properly.
Consequences	· Collision with other ships.
Safeguards	· Utilization of radar and automatic tracking function on Automatic Radar Plotting Aid (ARPA).
Hazards	**Means of CA against other ships, determined by remote operator, is mistaken/ cannot be determined/ late.**
Causes	· Remote operator cannot predict the ships’ behaviors. · Remote operator cannot determine the necessity for CA promptly. · Among multiple ships, the wrong ship or means are determined for CA.
Consequences	· Collision with other ships. · Deviation from the planned route caused by inputting a large course angle.
Safeguards	· Understanding the performance of Shinpo. · Compliance with the above in the system manual. · Utilization of radar and automatic tracking function on Automatic Radar Plotting Aid (ARPA).
Hazards	**Remote operator incorrectly inputs the course angle for CA.**
Causes	· Remote operator has no knowledge on how to set the course angle. · Input error by remote operator. · Input values are not sent to the actual ship due to communication errors.
Consequences	· Collision with other ships.

Safeguards	· Training the remote operator in the use of systems. · Compliance with the above in the system manual. · Redundant communication channels. · Presentation of actual measured values of the ship and confirming their validity. · Announcing the input value from the system.

Table 12. Examples of identified hazards on the task "watchkeeping."

Hazards	**Remote operator cannot detect obstacles from the ODR camera images.**
Causes	· No display of ODR camera images. · Remote operator is not aware of obstacles from the camera images. · Other ships are not in the camera images.
Consequences	· Remote operator cannot identity the obstacle as a ship or an object.
Safeguards	· Eliminate blind spots in the camera images.
Hazards	**Remote operator cannot recognize a ship behind another.**
Causes	· Remote operator incorrectly confirms ODR camera images. · A ship is completely hidden behind another and is not captured on the ODR images.
Consequences	· Delay in CA. · Remote operator incorrectly approves shift to the next waypoint.
Safeguards	· Utilization of radar and automatic tracking function on Automatic Radar Plotting Aid (ARPA). · Utilization of high-resolution ODR cameras.
Hazards	**Remote operator needs a lot of time to operate PTZ cameras.**
Causes	· Unfamiliarity of remote operator with the systems. · Difficulty in operating camera systems.
Consequences	· Remote operator cannot detect obstacles, identify ships, gather information on ships, or predict the ships' behaviors. · Aforementioned tasks are performed late.
Safeguards	· Training the remote operator in the use of camera systems. · Automatic calculation of direction through which ships proceed in from camera images and presentation of the result.
Hazards	**Remote operator needs to handle multiple jobs at the same time.**
Causes	· Simultaneously handling operations such as inputting commands for CA and camera operations for detecting or tracking other ships.
Consequences	· Remote operator cannot execute CA. · Delay in watchkeeping for other ships.
Safeguards	· Utilizing other means of operation such as VR headsets and operating pedals.

Table 13. Examples of identified hazards on the task "system monitoring."

Hazards	**Remote operator cannot check live monitoring and delay in network communication.**
Causes	· A clock at hand is out of order. · Remote operator is busy with other tasks and has no time to check the clock.
Consequences	· Remote operator cannot handle live monitoring and delay in network communication. · Remote operator makes decisions under conditions different from those the ship is under.
Safeguards	· Automating live monitoring and checking delays in network communication by the system. · Issuance of alarms in case of network error.
Hazards	**Remote operator cannot check live monitoring and delay in network communication.**
Causes	· Multiple alarms go off at the same time. · Due to similar alarm sounds, the remote operator cannot understand which alarm was raised.
Consequences	· Delay in error handling and securing the safety of the ship. · Ship's safety cannot be ensured due to operating out of the ODD.

	· Lack of understanding of the system on the part of the remote operator
Safeguards	· Description of system errors and error handling in the system manual.
Hazards	**Remote operator cannot check live monitoring and delay in network communication.**
Causes	· Insufficient system checking before starting operation. · Time deviation during voyage.
Consequences	· Delay in error handling and securing safety of the ship. · Remote operator cannot handle live monitoring and delay in network communication. · Remote operator makes decisions under conditions different from those the ship is under.
Safeguards	· Synchronization with GNSS. · Synchronization with time management server via the web.

Table 14. Examples of identified hazards related to multiple main tasks.

Hazards	**Remote operator cannot operate SPS and WNS on touch panel display or there is a delay in operation.**
Causes	· Unfamiliarity of remote operator with the systems. · Difficulty in understanding the SCS and WNS operations.
Consequences	· Remote operator cannot switch on the display of the SCS and WNS. · Remote operator cannot obtain necessary information when needed or provide the required inputs. · In the worst case, collision with other ships.
Safeguards	· Using a redundant input device and input method such as one that inputs numerical values using the buttons on the touch panel display and inputs commands via a throttle lever or steering wheel.
Hazards	**No production of camera images or delay in them**
Causes	· Disconnection or delay in network communication · Failure or malfunction of cameras on board.
Consequences	· Remote operator cannot detect other ships. · In the worst case, collision with other ships.
Safeguards	· Automating live monitoring and checking delays in network communication between the ship and shore. · Automating issuance of alarms in case of network errors. · Redundant communication channel. · Experimenters on board stop the operation by force in case of any communication errors.

5. Conclusions

In this study, we analyzed the tasks and information flow of the remote WPO of a small experimental ship, Shinpo. Based on the results, task flow diagrams that used the notations of the activity diagram, a type of UML behavioral diagram, are presented. By using these diagrams, it was possible to identify time-related hazards, such as task delays and the simultaneous occurrence of tasks.

The proposed DTB-HAZID method may help analysts understand the task flow of systems, including the human-activities related to the system procedures and facilitate hazard identification for risk analysis. The proposed method will contribute to enhancing the safety of autonomous ship systems, including remote operation systems.

References

[1] Itoh H et al. 2021 Risk Assessment of Autonomous Ship Systems *ClassNK Technical Journal* no 4 pp 11–23

[2] Wróbel K et al. 2018 System-theoretic approach to safety of remotely-controlled merchant vessel *Ocean Engineering* vol 152 pp 334-345

[3] Valdez Banda O.A. et al. 2019 A systemic hazard analysis and management process for the concept design phase of an autonomous vessel *Reliability Engineering and System Safety* vol 191
[4] Shiokari M et al. 2020 Towards the Development of Risk Analysis Method for Autonomous Ships *Proc. Conf. on the Japan Society of Naval Architects and Ocean Engineers (Tokyo)* vol 30 pp 393–396 [Japanese]
[5] Shiokari M et al. 2021 Application of Risk Analysis Method with System Modeling to Conceptual Design of Autonomous Ships *Proc. Conf. on the Japan Society of Naval Architects and Ocean Engineers* vol 32 pp 355–366 [Japanese]
[6] Ishimura E et al. 2021 Task-Based Hazard Identification for a Hypothetical Autonomous Ship *Proc. the 91st Conf. on the Japan Institute of Marine Engineering (JIME)* pp 169–170 [Japanese]
[7] Miyake R et al. 2021 Task Analysis Based on Remote Control Trial of Small Experimental Ship "Shimpo" *Proc. the 91st Conf. on the Japan Institute of Marine Engineering (JIME)* pp 171–172 [Japanese]
[8] R. Rosness et al. 2016 WHAT DO YOU DO WHEN YOU BUILD SAFETY? ISBN: 978-82 - 14-06185-7 aveilable at https://www.sintef.no/globalassets/sintef-teknologi-og-samfunn/rapporter-sintef-ts/successful-operations-guide.pdf.(accessed on 2022.3.22)
[9] Herrera I.A et al. 2009 Comparing a multi-linear (STEP) and systemic (FRAM) method for accident analysis *Reliability Engineering and System Safety* ISBN 978-0-415-48513-5 (London: Taylor & Francis Group)
[10] Inoue T 2018 *Thorough Use of UML by Diagram* (Japan: Shoeisha)

Objective and intelligent training assessment package for maritime training in simulator

O Sam[1], D Le[2], G A Ang[2], Y Liu[2], D Zhang[2]

[1] SMA Singapore Polytechnic, 500 Dover Rd, W1, Singapore
[2] CEMS Singapore Polytechnic, 500 Dover Rd, Innovillage INV C35/36, Singapore

Email: Osman_SAM@sp.edu.sg

Abstract. Training in simulator has been used in maritime domain for years due to its cost-effective nature. Different scenarios with varied visibility, traffic density, weather conditions, etc. could be easily set up in the simulators. However, the existing build-in training and assessment package in the simulators is rigid as it does not take into consideration the change in traffic and navigational environment. Human experts are always involved to observe and assess the performance of trainees. An objective and intelligent training and assessment package is proposed to assist the experts in training to ensure objectivity and consistency. To identify the gaps and determine the relevance and the adequacy of the existing build-in assessment in the simulator, online survey has been distributed to the shipping companies and associations. The survey has gathered inputs on the actual shipboard practices when the seafarers are navigating in Singapore waters. This includes different types of vessels, vessel sizes, and external environment conditions such as visibility. The outcome of this analysis is used to form the baseline of a common practice at sea and used in the training and assessment package, which provides consistency in setting the critical training parameters. The training and assessment package reads the data from an advanced navigation research simulator and is able to sense changes and update itself in accordance with user-defined environment. Several assessment modules such as CPA under different traffic density and visibility, minimum distance from other vessels, rate of turn at different speed and loading conditions, will be covered by this training and assessment package.

1. Introduction

Singapore being one of the busiest port, volume of sea traffic is heavy; and being flanked on all sides by Malaysia and Indonesia, there is limited availability of navigable waters [1]. Areas of traffic congestion and the high risk of near misses and incidents in congested waters can be expected especially near pilot boarding grounds and traffic separation schemes. Navigating in and around Singapore waters and straits have always been challenging. Thus, there is a need to address the concern and mitigate the risks to the mariners.

One of the primary means in enhancing navigation and ship-handling competency which could eventually lead to a safer navigation is the use of a full mission ship-handling simulator (FMSS) for training and assessment. Simulation training allows the trainers to replicate scenarios with different settings such as varied traffic density and weather conditions, which raises the competency training to another level and training using simulators has allowed STCW sea-time requirement for sea cadets to be drastically reduced.

To guarantee effectiveness and consistency in training, it is essential that the training and assessment are guided by certain standards for safe navigation so that delivery and rigour of training and assessment would be of similar intensity regardless of the occasion and the personnel who conducts the training or the assessment. The system should also be able to auto-update the training and assessment rules appropriate to the prevailing conditions and circumstances and finally auto-generate the training and assessment reports with minimal intervention from the trainer or the assessor. However, the existing build-in training and assessment package in the simulators does not consider the change in traffic and navigational environment and it is still highly dependent on human assessors' observation of the trainees. To covers this shortage, we developed an objective and intelligent training and assessment package and present it in this paper.

The study has been done in two phases. The target of phase 1 is to set a baseline of a common practice at sea. Online surveys have been distributed to the shipping companies and associations. The survey has gathered about 179 inputs on the actual shipboard practices when the seafarers are navigating in Singapore waters. Based on the analysis of the input from the survey, in phase 2, we developed a software which provide objective assessments. Several assessment modules have been included such as CPA under different traffic density and visibility, minimum distance from other vessels, rate of turn at different speed and loading conditions. The training and assessment package is able to read the data from an advanced navigation research simulator (ANRS) and sense changes and update itself in accordance with user-defined environment.

The paper is constructed as follows. In Section 2, existing work on training and assessment in maritime simulators has been reviewed. The methodology used is presented in Section 3. Section 4 shows the results and Section 5 concludes the paper.

2. Related work

Simulators have been used in maritime domain for years to train and assess the trainees [2]. As a supplement to onsite training, the trainees can acquire certain level of competencies from the simulator-based training for their onboard roles [3]. The simulator-based training has its unique advantages such as providing an immersive yet safe environment to experience different scenarios, offering flexibility to playback for debriefing. However, [4] also points out the assessment of simulator-based training could be challenging. Among off-the-shelf simulators, only some comes with automated assessment packages and the assessment packages do not allow trainers to modify the assessment criteria. In addition, the criteria may not be the best practice for Singapore Strait. The target of the study presented in this paper is to bridge this gap. The developed training and assessment package can be deployed to different simulators and the assessment criteria are set based on the actual shipboard practices.

There are several safety-critical parameters such as speed, Closest Point of Approach (CPA), Under Keel Clearance (UKC), rate of turn, safe passing distance from navigational aids, and danger areas, that are usually monitored in the assessment. For example, the vessel type and speed impact the planning of collision avoidance manoeuvres of vessels and the safe route [5]. CPA, which gives the estimated minimum distance (DCPA) between two vessels the estimated time (TCPA) when they reach this closest point, is an essential factor for navigational safety in collision avoidance [6]. Another factor highly related to the ship's manoeuvrability is UKC, which is defined as the vertical distance between the lowest part of the ship's hull and the seabed [7]. The shipmaster must make sure the UKC always remains sufficient [8]. The recommended UKC also varies in different conditions, e.g., deep water, shallow water, and at anchor. Rate of turn indicates the rate a vessel is turning. Constant rate-of-turn are required as the minimum standard of competence for masters and chief mates on ships of 500 gross tonnage or more [9]. Last but not least, the distance from a conspicuous object such as buoys and beacons is also required for navigators [10]. In this work, all the above-mentioned parameters have been taken into consideration to provide a comprehensive assessment to assist the trainers/assessors.

3. Methodology

As mentioned, the study can be categorized into two distinct phases. The first phase is to obtain data so as to achieve a standard that can be used as a reference to assess if the current navigation is safe. The second phase is to develop a software which interacts with the ANRS to get the real-time manoeuvring information of the ship and compares with the achieved standard in phase 1 for assessment.

In phase 1, an online survey which consists of 38 questions has been distributed to ship officers who have worked or are working on different types and sizes of ships navigating through the Strait of Singapore. This online survey had garnered 179 responses of which about 50% were from practicing shipmasters. The survey parameters include 1) speed of ship in different traffic density, different conditions of visibility, and different sea areas; 2) CPA in different traffic density, different conditions of visibility, danger areas, and during collision avoidance; 3) UKC in deep water, shallow water, and at anchor; 4) means or methods of positioning during transit, at anchor; 5) rate of turn at different speed, and different condition of loading; 6) near misses/dangerous passing (in subsequent survey) ships, navigational aids, and danger areas.

4. Results and discussion

4.1. Phase 1 survey results

One of the questions being investigated is “in a risk of collision situation during the transit, what CPA would you deem to be safe and appropriate when passing ahead of the other vessel?”. In this section, the result from this question is presented as an example. Three broad factors are analyzed from the data collected for this question, namely the main factor – CPA, the ship’s type, in this case, container ship, and the ship’s length which is investigated by a scale of 50 meters increment. The scale being set to 50 meters makes it more flexible to combine and achieve different categories of length.

To ensure the robustness and veracity of the survey data collected and, hence on the standard arrived at, the survey data were compared with the AIS data. The comparison for the investigated question has been summarized in Table 1 regarding minimum, maximum, mean values, and standard deviation, from which we can see that the mean CPA values derived from AIS is smaller than the values collated from the survey responses. For example, for container ships of all length, the mean AIS value is 1.04 miles and from the survey is 1.16 miles.

We also applied three weightings schemes to different respondents as shown in Table 2. This is to investigate if there is any difference in the output when different weightings are assumed for respective designations. In the first weighting scheme, the four designations carry equal weightings. In the second scheme, a weighting of 1.5 was allocated to the master, 1.3 to chief officer and trainer. The third weighting scheme assigns a weighting of 2 for the master, 1.6 for the chief officer and 1.4 for the trainer. In all schemes, the weighting for others is 1.

The results for the same question but with different weighting schemes are shown in Table 3. Besides the minimum, maximum values, median values and the average values for different weighting schemes are included in Table 3. It was noted that the difference in weighted average values for different categories of ship in terms of length is insignificant.

The collected survey data has been further checked about the distribution of the responses (Figure 1). Figure 1 (a) presents the counts of responses for different CPA values. For example, 31 participants had submitted response on CPA of 1.0 nm. Figure 1 (b) shows the mean and standard deviation for each ship length category. E.g., the first bar is for ship length 101-150(m), the mean of this category is 1.3 and SD is 0.53 (the black line). Figure 1(c) plots the distribution curve of the bar chart in Figure 1(a) while Figure 1(d) gives the distribution curves by different designations. From Figure 1(d), it shows that despite of designation, the participants tend to select CPA 1.0 nm.

Based on the above analysis and findings, a CPA of 1.0 mile is decided to be used as the standard in the simulation training and assessment when passing ahead of a ship in a collision avoidance situation. The same methodology has been used to determine the reference safe standard for the other navigational parameters such safe passing distance and UKC.

The next part of this phase is to define to what constitute "near misses" or "dangerous situations". An internal survey amongst the trainers within the academy was conducted and an example of the obtained matrix for the training and evaluation criteria are given in Table 4.

Table 1. Comparison between survey data and AIS data

Ship Type	Navigation	Item		Length (m)					
				101 - 150		251 - 300		All Length	
				Survey	AIS	Survey	AIS	Survey	AIS
Container	Passing ahead of the other vessel	CPA	Total Vessels	9	158	25	443	84	2,077
			Total Messages		4,735		13,104		60,981
			Max	2.00	2.00	2.00	2.00	2.00	2.00
			Min	0.70	0.10	0.50	0.10	0.30	0.10
			Mean	1.30	0.99	1.19	1.04	1.16	1.04
			SD	0.53	0.50	0.45	0.48	0.48	0.49

Table 2. Weighting schemes

Weightings	Master	Chief Officer	Trainer	Others
WA (1)	1	1	1	1
WA (2)	1.5	1.3	1.3	1
WA (3)	2	1.6	1.4	1

Table 3. CPA responses under different weighting schemes.

Ship Type	Survey Question	Item	Length (m)								
			101 – 150 (9)			251 – 300 (25)			All Length (84)		
			Value		SD	Value		SD	Value		SD
Container	In a risk of collision situation during the transit, what CPA would you deem to be safe and appropriate when passing ahead of the other vessel	CPA	Max	2.0		Max	2.0		Max	2.0	
			Min	0.7		Min	0.5		Min	0.3	
			Med	1.0		Med	1.0		Med	1.0	
			WA1	1.30	0.53	WA1	1.19	0.45	WA1	1.16	0.48
			WA2	1.32	0.5	WA2	1.18	0.44	WA2	1.15	0.48
			WA3	1.34	0.5	WA3	1.18	0.44	WA3	1.15	0.38

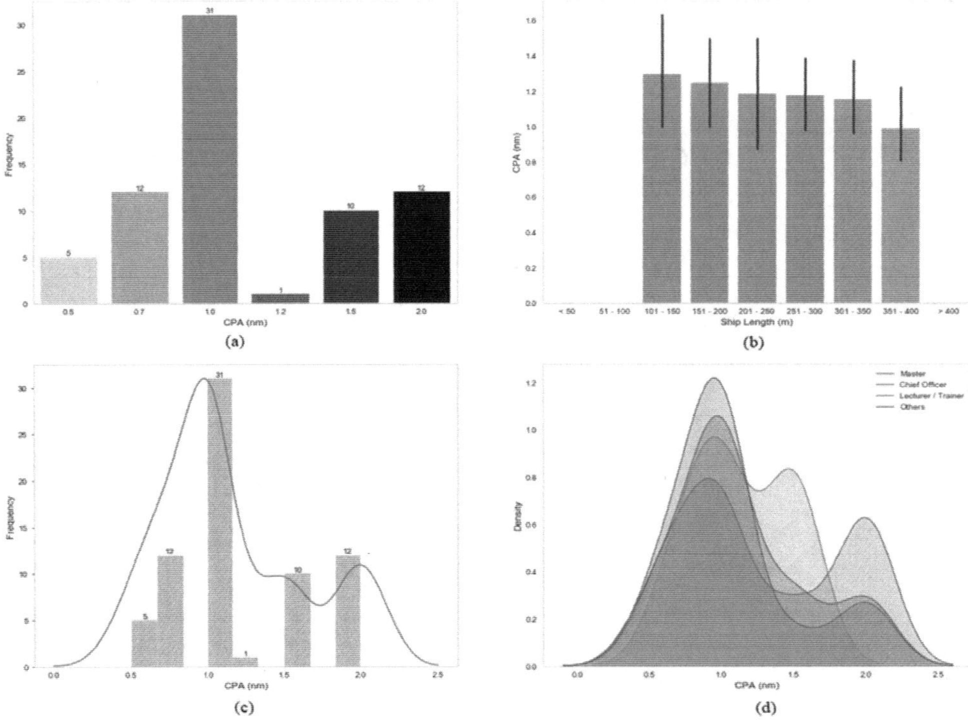

Figure 1. Distribution of the responses for CPA.

Table 4. An example of training and evaluation criteria for near misses/ dangerous situations.

Survey Questions	Survey Response	
	Normal Situations	Near Misses Dangerous Situation
In a risk of collision situation during the transit, what CPA would you deem to be safe and appropriate when passing ahead of the other vessel?	1.0 nm	0.3 nm
In a risk of collision situation during the transit, what CPA would you deem to be safe and appropriate when passing astern of the other vessel?	0.5 nm	0.1 nm
In a risk of collision situation during the transit, what CPA would you deem to be safe and appropriate when passing clear of the other vessel in an overtaking or head-on situation?	0.5 nm	0.2 nm

4.2. Phase 2 software development

Based on the safe standards obtained from phase 1, in phase 2 the development of a software program has been done. The program could generate a report on the conduct of the navigation exercise. As in this study, the ANRS simulator is used, this software is plugged into the simulator when running the exercise. The developed software is able to describe in detail the actions taken by the OOWs and compare them with safe standards. Compliance or non-compliance with the safe standards is clearly shown in the report and points/marks will be updated accordingly. The developed software is versatile and easily ported over and plugged into any simulation system. A screenshot of the developed GUI of this software is as shown in Figure 2.

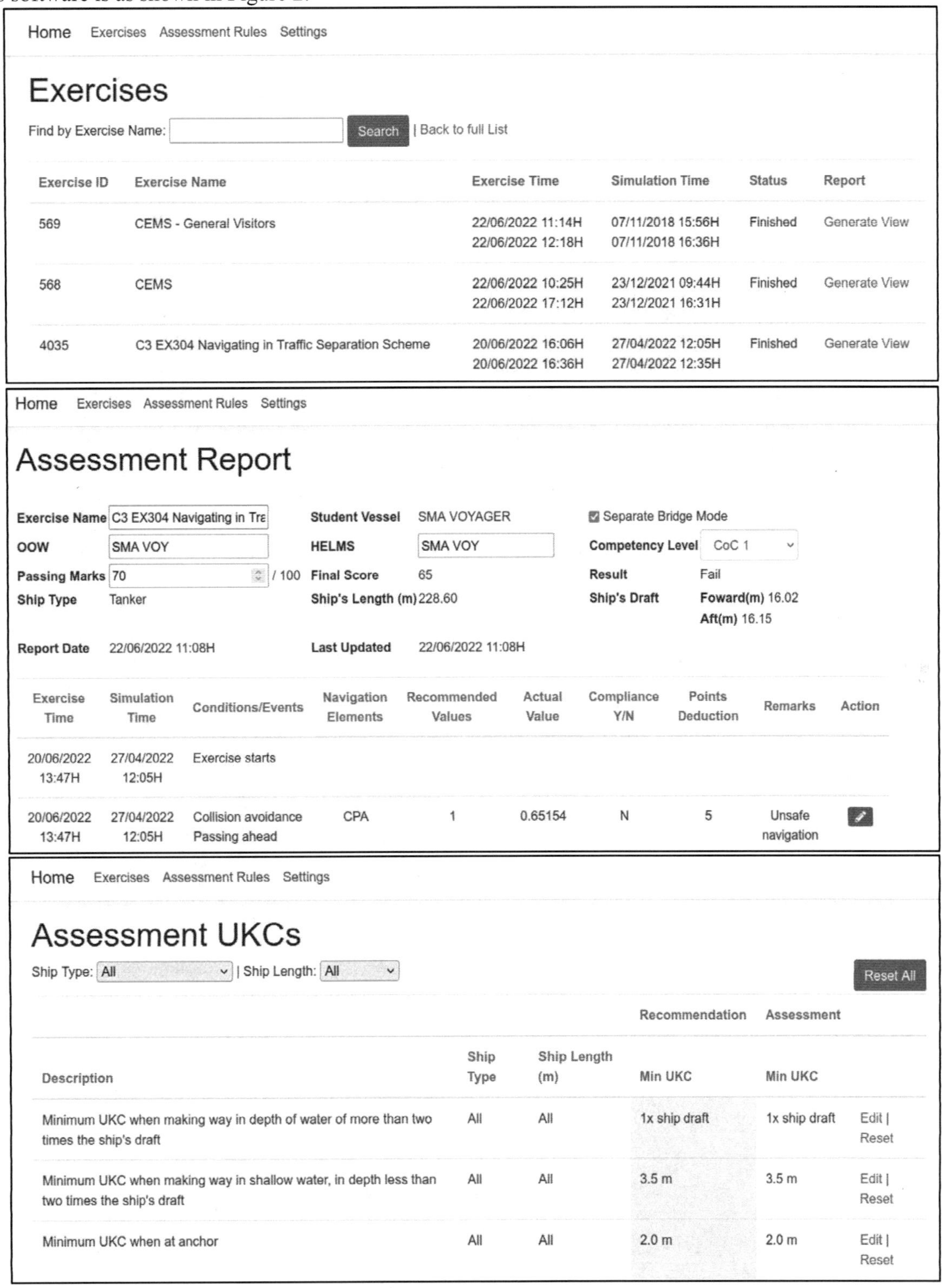

Figure 2. GUI of the assessment software.

5. Conclusions

In this paper, a package to provide an objective and intelligent ship-handling simulation training and assessment has been developed, where the assessment criteria are obtained based on online surveys from shipping companies and associations. As a result, the evaluation of participants requires minimum intervention from the trainers and assessors and, hence, lead to a very high degree of consistency. The minimal intervention required from the trainers and assessors would allow them more time and better evaluate other aspect of training that may include situational awareness and measure of human-machine interface.

References

[1] MPA Singapore. Singaporean Nautical Charts. Available from: https://www.mpa.gov.sg/web/portal/home/port-of-singapore/services/charts-tidal-info-atons-and-hydrography/singaporean-nautical-charts.

[2] Maung, C.T., Simulation training and assessment in maritime education and training. 2019.

[3] Kim, T.-e., et al., The continuum of simulator-based maritime training and education. WMU Journal of Maritime Affairs, 2021: p. 1-16.

[4] Øvergård, K.I., et al., Maritime bridge crew training, in Simulators for transportation human factors. 2017, CRC Press. p. 281-310.

[5] Pietrzykowski, Z. and M. Wielgosz, Effective ship domain–Impact of ship size and speed. Ocean Engineering, 2021. 219: p. 108423.

[6] Sang, L.-z., et al., CPA calculation method based on AIS position prediction. The journal of navigation, 2016. 69(6): p. 1409-1426.

[7] Northern Territory Government, Regional Harbourmaster's Direction for Under Keel Clearance requirements for Port of Darwin. 2015.

[8] Business, S. Maintaining Safe Under Keel Clearance -Shallow Water Navigation Safety Guideline. Available from: http://shipsbusiness.com/shallow-water-navigation.html.

[9] International Maritime Organization, The Manila Amendments to the Seafarers' Training, Certification and Watchkeeping (STCW) Code. 2010.

[10] MPA Singapore, Singaporean notices to mariners 2021. Monthly Edition No. 2

Journal of Physics: Conference Series **2311** (2022) 012015 doi:10.1088/1742-6596/2311/1/012015

Standardization of Risk Assessment in Navigation

Jens Froese

Technical University Hamburg

Jens.Froese@tuhh.de

Abstract. Marine traffic safety assessed by quantitative instead of qualitative methods is increasingly required. There are mature methods available to conduct a quantitative risk assessment such as IWRAP [1], resulting in collision and grounding probabilities. There is, however, the difficulty to determine the so-called causation factor having an impact on the results but there is currently no reliable method to fix it. To globally collect relevant empirical data to allow a statical determination could make risk analysis results less questionable but comparable.

1. Introduction

Formal requirements and public sensitivity increasingly focus on marine traffic safety in coastal areas. Times are over when ships simply could navigate where water depth al-lows. Nowadays a wide range of interest groups stick out their hands to exploit coastal zones or to prevent from exploitation including shipping to preserve nature. In busy sea areas traffic lanes become increasingly squeezed and an increased exploitation level may result in an increased risk of ships to collide with fixed or floating objects or to ground. To ensure safe vessel traffic, navigational risks must be known. One can only manage what can be measured.

Methods to assess navigational risk mean-while escaped the nursery. There are qualitative methods identifying possible hazards and mitigation options jointly with a rough estimation of consequences and quantitative methods to more accurately calculate risks based on specified risk scenarios (risk models). Authorities increasingly ask for a quantitative approach.

The definition of risk is

Risk = probability x consequences
[annual frequency x average costs]

It is cumbersome but achievable to capture consequence costs, especially when insurance providers support. For comparison purposes collision and grounding probabilities often are sufficient as long as ship types and cargoes do not differ too much.

2. Quantitative Assessment of Navigational Risk

The International Maritime Organization (IMO) published "Guidelines for formal Safety Assessment (FSA) [2], a structured and systematic methodology to be used as a frame-work to support in the evaluation of new regulations for maritime safety and protection of the marine environment or in comparing be-tween existing and projected regulations. For quantitative risk assessment IMO recommends applying the IALA Risk Management Tool for Ports and Restricted Waterways, the

IWRAP Mk2 tool [3] which allows to quantify the frequencies of collisions and groundings in any traffic area. As explained in the introduction, the probability of collisions and groundings is only one fraction of risk. In case of continuous risk assessment conducted to monitor the safety of a sea area or waterway it is sufficient to only once calculate the risk and then to restrict to probabilities for consecutive assessments as long as nature and composition of traffic does not significantly change.

The IWRAP-label "IALA Risk Management Tool…." is misleading as the tool allows only to determine collision and grounding frequencies but not consequence costs, which needs to become assessed separately based on input from IWRAP-results. This shall not deteriorate the enormous benefits of the tool. Whoever sat weeks to investigate collision and grounding frequencies manually, will certainly appreciate this extremely helpful tool, very efficiently to be used.

IWRAP is a probabilistic approach on the basis of geometric (spacial) distribution of traffic and navigational obstacles. The IWARP theory is described in [4]. The traffic distribution can become captured by AIS-counts of ships. To allow for mathematical processing a statistical distribution is selected which best fits the histogram-data of traffic (see fig. 1). Frequently a normal distribution can be ap-plied, however it depends on how the distribution matches the histogram-data and on the scenario to become investigated.

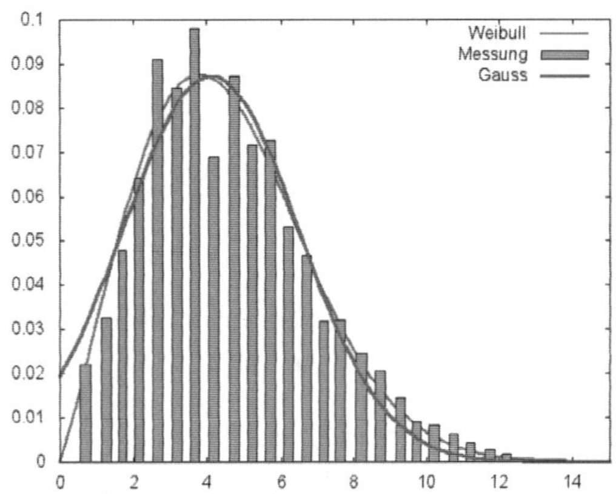

Figure 1: Traffic histogram and statistical distributions

3. Probability of Collisions

Risk models and formulae depend on the scenario describing encountering, overtaking, crossing or lane merging traffic, collisions between vessels and fixed obstacles or groundings. To elucidate the approach encountering traffic is used as an example below. Figure 2 shows the Fujii-model for encountering traffic and figure 3 the associated formula to calculate the collision frequency [5] [6].

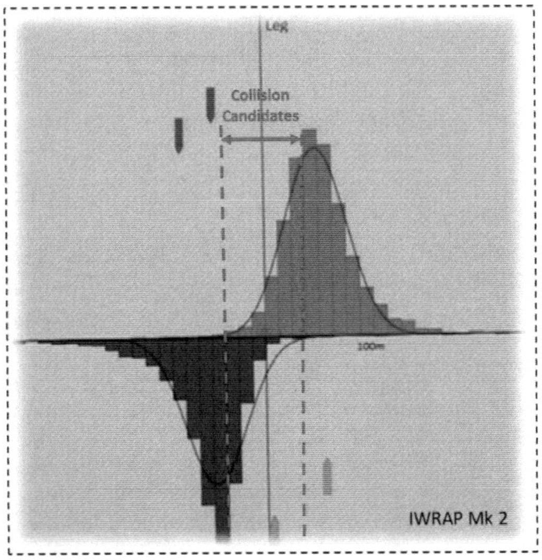

Figure 2: Fujii-Macduff-Model for encountering traffic

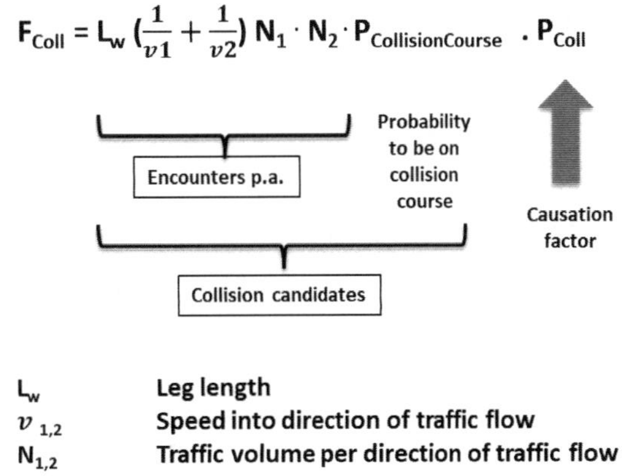

Figure 3: Collision frequency of encountering traffic (Fujii-Macduff) [7]

The elements of the formula are the encounters per time span, usually 1 year, and the probability to be on collision course, calculated on the basis of the AIS-counts providing the traffic distribution. The result shows the collision candidates, however, not all ships having the opportunity to collide, will do so. In the vast majority of cases on one or both ships an evasive action will be taken. Thus, the causation probability or causation factor is added to achieve the fraction of the collision candidates eventually colliding.

4. Causation Factor

The causation factor mirrors all possible human and technical collision causes, from human failure to steering failure. Recommended values can be found in literature [7]:

$P_{human} = 2 \cdot 10^{-4}$

$P_{steering} = 6.3 \cdot 10^{-5}$

$P_{propulsion} = 1.5 \cdot 10^{-4}$

IWRAP also provides default values for the causation probability. It seems, however, problematic to generalize the default factor. The conditions governing e.g., the human causation factor may be found in insufficient training, negligence, fatigue, attitude, illness or insobriety. It also depends on the bridge team management and organization of ships. It can be assumed that human failures on cruise vessels occur more seldom than on container feeder vessels with minimum manning and always running behind terminal arrival schedules. It can also be assumed that there is a dependence between weather conditions and human failure rates when e.g., in dense fog an otherwise competent and engaged navigator lacks experience in radar-based collision avoidance.

The most reliable way to determine the causation factor is from empirical data. Also, an analytical approach appears possible. There are suitable methods available, such as

Fault tree analysis

Failure mode and effect analysis

Bayesian believe network.

An analytical approach requires to consider all possible causes and conditions to individually estimate the particular likelihood of occurrence. As long as there are no statistical data available, this is extremely difficult and experience shows that usually the cumulated probability is considerably too high.

There is a better availability of empirical data from some of the technical failures, such as loss of propulsion or steering than from failure of navigational equipment such as e.g., RADAR.

It is obvious that the difficulty to determine the causation factor is the weakness of the quantitative assessment of navigational risk. The only way to improve the accuracy of the causation probability is to collect suitable empirical data to allow for statistical analysis.

5. Causation Factor Database

The causation probability depends on a wide variety of conditions. To reliably base causation factors on representative statistical data, traffic and incident data covering all relevant navigational conditions must be collected over some years. This is only possible by global cooperation. It is required to depict e.g.

- Waterway/sea area
  - Geometry (routes, exclusion zones, traffic separation schemes, regulations)
  - Underwater morphology
  - Current
  - Aids to navigation
- Traffic
  - Ship types
  - Ship dimensions
  - Ship speeds
  - Cargoes carried
  - Traffic distribution per ship class
- Environmental conditions
  - Wind
  - Seastate
  - Visibility
  - Ice
- Incidents
  - Types
  - Consequences

The format and content of the databank and the methods to retrieve results must follow an agreed standard.

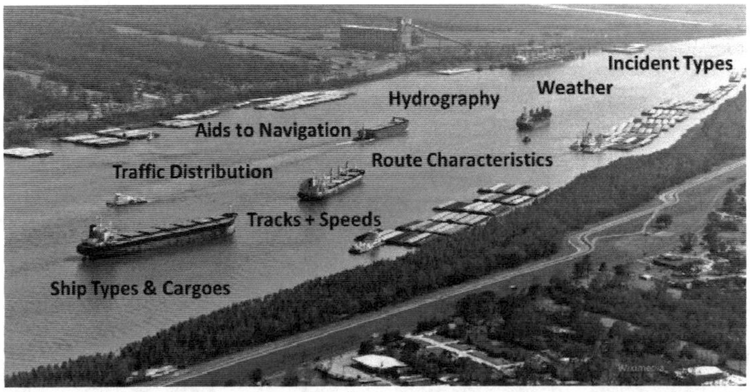

Figure 4: Conditions contributing to the causation probability

By collecting appropriate data from all parts of the world such a databank would soon become sufficiently mature to deliver reliable causation factors on demand based on an equally standardized retrieval format describing the relevant conditions of the area. The capability and the reliability of causation factors will continuously improve over time.

There are already some marine traffic databases providing at least a fraction of the a.m. information, which could contribute. In Europe this is the European Marine Casualty Information Platform (ECIMP), operated by the European Maritime Safety Agency (EMSA) in Lisbon. Amended by a consequences database to easily retrieve damage costs of injuries/fatalities, material and environment by collisions and groundings a very powerful "Navigational Risk Database" could become generated making risk assessment easy and results reliable and comparable.

6. Conclusion

The reliability of quantitative assessment of the likelihood of collisions and groundings as part of the assessment of navigational risk depends on statistical data allowing determination of the causation factor mirroring the local navigational and traffic conditions. Reliable deductions from statistical data requires a sufficiently large statistical population, which usually is not available for a specific navigational area of interest. By collecting traffic data globally according to an agreed taxonomy of relevant parameters data shortcomings could become mastered allowing to significantly increase the reliability of quantitative risk assessment in navigation.

References

[1] IALA, Recommendation O-134 on IALA Risk Management Tool for Ports and Restricted Waterways, Edition 2, May 2009
[2] IMO, Revised Guidelines for Formal Safety Assessment (FSA). MSC-MEPC.2/Circ.12/Rev.2, 9 April 2018
[3] IMO, Degree of Risk Evaluations. SN.1/Circ.296 7 December 2010
[4] Friis-Hansen, Peter, Basic Modelling Principles for Prediction of Collision and Grounding Frequencies, Technical University of Denmark, 1 August 2007
[5] Fujii, Shiobara, The Analysis of Traffic Accidents, Journal of Navigation, Volume 24, October 1971
[6] Macduff, T. The Probability of Vessel Collisions, Ocean Industry, September 1974
[7] Rasmussen, Glibbery, Melchild, Hansen, Jensen, Lehn-Schiøler, Randrup-Thomsen, Quantitative assessment of risk to ship traffic in the Fehmarnbelt Fixed Link project, Journal of Polish Safety and Reliability Association, Summer Safety and Reliability Seminars, Volume 3, Number 1-2, 2012

Evolution of Safety in Marine Systems: From System-Theoretic Process Analysis to Automated Test Scenario Generation

Tom Arne PEDERSEN[1*], Åse NEVERLIEN[1], Jon Arne GLOMSRUD[1], Imran IBRAHIM[2], Sigrid Marie MO[3], Martin RINDARØY[4], Tobias TORBEN[5], Børge ROKSETH[5]

[1]Det Norske Veritas (DNV), Group Research and Development, [2]Det Norske Veritas (DNV), MI Advisory, [3]Brunvoll, Department Engineer Marine Cybernetics, [4]SINTEF Ocean, Ships and Ocean Structures, [5]Norwegian University of Science and Technology (NTNU), Department of Marine Technology

Corresponding author: Tom Arne Pedersen; e-mail: tom.arne.pedersen@dnv.com; Tel.: +47 952 80 695

Abstract. The technological development ongoing in the maritime industry is making the ground for remotely and even autonomously operated vessels in the future. This is a result of increased data collection, processing and inter-connectivity capabilities. The industry is working towards increased safety, improved efficiency of the ship's operation, improved environmental performance and a more cost-effective shipping. New technologies are developed in order to reach these goals, and DNV as a Class society is developing frameworks for assurance of such systems. The certification of ships and vessels with a high degree of automation or autonomy needs an increased focus on software, an understanding of the human-to-machine interaction and the resulting ability to solve complex operations in a secure way. In this paper, a method for high-level risk analysis of the safety aspects of autonomous vessels combined with automatic simulation-based testing of a control system, is proposed.

1. Introduction

The Fourth IMO Greenhouse Gas study [1] estimated that the greenhouse gas (GHG) emissions from shipping has increased from 977 million tons in 2012 to 1,076 million tons in 2018. In 2018, IMO adopted an internal strategy [2] with a vision to reduce the GHG emissions from international shipping by at least 50% by 2050 and to reduce the carbon intensity of international shipping by at least 40% in 2030, towards 70% in 2050, compared to 2008. Even though there were an increasing demand for shipping from 2008 to 2014, the GHG emissions from international shipping decreased due to a period of rapid carbon intensity reduction [1]. The demand for shipping has continued to increase, but with a moderate improvement in carbon intensity, the emissions have declined since 2008, but in the period 2014-2018 the trend shows an emission growth, see Figure 1. To reach IMO's 2050 GHG reduction ambition, a large share of GHG reduction will have to come from alternative low-carbon fuels, energy-saving technology, and speed reductions of ships [1]. This is where autonomy can play a role as automated systems may gradually improve the environmental performance and enable more cost-effective shipping [3]. One way to lower the energy consumption of shipping is to reduce the energy consumption. For a ship, energy consumers can be divided into three groups: energy required for ship

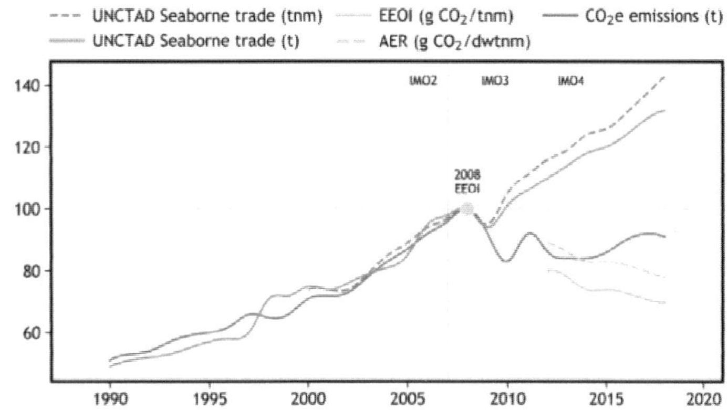

Figure 1: International shipping emissions and trade metrics, indexed in 2008, [1].

propulsion, hotel loads and for on-board operations [3]. In the first group, ship resistance is the main contributor and reduction of ship speed will change the carbon intensity. According to [3], two ships operation at 6 knots would in sum consume 30 to 50% less energy than one ship operating at 12 knots, depending on design and environmental conditions.

Ship navigation may be broken down in four sub-tasks: condition detection, external/internal situation awareness, action planning and action control. All ships must detect the actual situation, meaning own ship location, nearby objects that may pose a threat, status and conditions of on-board equipment that may potentially affect the vessel's ability to maneuver etc. The gathered information is then analyzed to give a complete situation awareness, before making a plan to best handle specific situations. Finally, the plan is effectuated. An autonomous navigation system (ANS) consists of several different controllers and algorithms, each controlling different aspects of the vessel. When putting these smaller control systems together, the resulting system is a complex system with emergent properties [4]. Emergent properties are properties that are not there initially, but may only be observed when the different controllers and subsystems are put together. In other words, they are not observable by looking at the individual pieces, but instead they emerge at a system level. In addition to being a complex system, the ANS also interact in a complex environment with an infinite number of different traffic situations.

In the perspectives of assurance and testing, ensuring that ANS algorithms are safe and do not cause accidents is crucial. Testing may be done in real life using the actual ship, in the virtual world using simulators, or in combination. Real-life testing only would be too time consuming, especially considering the of situations needed to be tested. In addition, critical situations may be too difficult or dangerous for real life testing, thus a combination of simulation-based and real-life testing would be the preferred solution. If enabled in the control system, simulation-based testing may also be performed by running the control system in closed loop with the simulator faster than real time, empowering accelerated testing.

[5] introduced System-Theoretic Process Analysis (STPA) as part of the verification and testing process of maritime automation systems such as dynamic positioning (DP) systems. STPA was used to identify safety constraints on system interactions and the constraints were translated into verification objectives.

[6] explored the feasibility of applying an STPA safety and security co-analysis on novel cyber-physical systems (CPS). They studied the STPA safety and security co-analysis to assess the risks of CPS collectively by adopting widely accepted security analysis methods into the STPA analysis process.

In [7], STPA was applied to an SAE (Society of Automotive Engineers) level four vehicle with the aim to uncover potential hazards that have not been discovered by traditional methods. The STPA analysis resulted in 242 requirements, where 11 of these requirements were not handled in the current implementation.

A method for automatic simulation-based testing of control systems for autonomous vessels was proposed in [8]. The method formulated the requirements using formal Signal Temporal Logic (STL) and used a Gaussian Process (GP) model to estimate the robustness score and the uncertainty level for the untested cases in the simulation. The GP model was then used to guide the case selection towards cases with low robustness or high uncertainty. The methodology incrementally run new simulations until the parameter space covered to a desired confidence level, or until a case falsified the requirements.

In this paper, the results from [5], [6] and [8] are combined. A high-level risk analysis is performed using STPA. The outcome of the STPA is a set of loss scenarios with corresponding safety constraints for preventing losses. Signal Temporal Logic (STL) is used to express the STPA constraints as formal requirements.

2. Background Theory

2.1. System-Theoretic Process Analysis

Nancy Leveson introduced Systems-Theoretic Accident Model and Processes (STAMP) [9] by looking at safety as a control problem. She stated that safety can be controlled by applying sufficient constraints to a system. She also stated that reliable system components do not necessarily result in a safe system. In [10], Thomas argued that most major accidents today are not caused by component failure, but instead caused by complex and often unexpected interactions among components operating as designed. Flawed system design, human error or missing software requirements are also important factors, and to handle this, Leveson developed System-Theoretic Process Analysis (STPA) designed to analyze socio-technical systems characterized by complex human-machine interaction. Leveson argued that accidents would not occur if the system avoids hazards [11]. In the following, loss [11], accident and hazards [6] are defined:

Definition 1. **Loss**: A loss involves something of value to stakeholders. Losses may include a loss of human life or human injury, property damage, environmental pollution, loss of mission, loss of reputation, loss or leak of sensitive information, or any other loss that is unacceptable to the stakeholders.

Definition 2. **Accident**: An accident is an undesired or unplanned event that results in a loss.

Definition 3. **Hazard**: A hazard is a system state or set of conditions that, together with a particular set of worst-case environmental conditions, will lead to an accident or a loss.

Losses can also relate to negated positive goals, e.g. an efficiency goal can be expressed as a loss of efficiency.

According to [11], it is prescribed to split the STPA analysis into 4 steps, as seen in Figure 2:

1. Define purpose of the analysis.
2. Model the control structure.
3. Identify unsafe control actions.
4. Identify loss scenarios.

The first step of the STPA analysis is to define what losses should be prevented. Not only safety goals, such as preventing loss of human life/injury, property damage and so on, but also efficiency requirements will come into play when performing the STPA analysis.

According to [6], it is not straightforward to transition into the second step of modelling the control structure. To mediate this challenge, they proposed to identify functional requirements as an intermediate step to convert the constrains into requirements to facilitate the development or assessment of the control structure. [6] states that STPA defines a control structure consisting of two important control system properties - i.e., controllability (control input and control actions) and observability (feedback and process model). Functional requirements include requirements for both

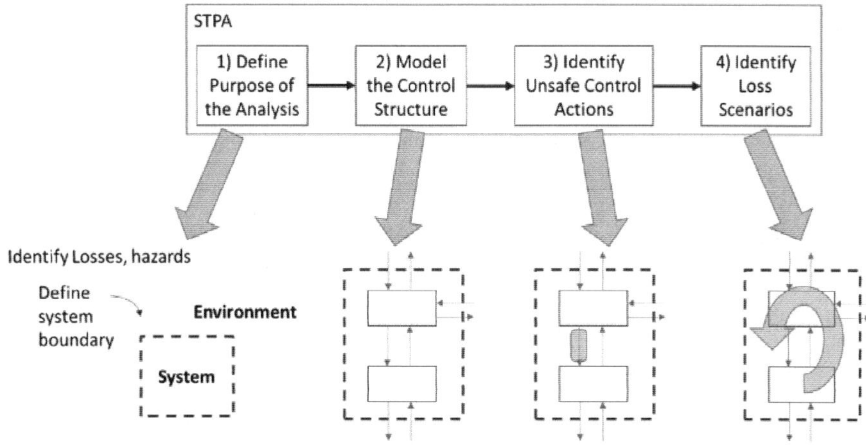

Figure 2: Overview of STPA analysis, adopted from [11].

properties. To control or mitigate the hazards, functional requirements address both the hazard and relevant control process.

The second step is to model the control structure of the system, given the system properties and system boundaries. The control structure consists of feedback control loops representing the functional relationships and interactions in the system. A set of interface arrows may be seen in lower part of Figure 2. The downwards pointing arrows generally contain control actions while the upwards pointing arrows generally contain feedback data. Controller responsibility may be assigned when the control structure is defined.

The third step examines how control actions may be inadequate and potentially lead to the hazards and losses defined in the first step. In STPA, these control actions are defined as unsafe control actions (UCA). In this paper, inadequate control action (ICA) is used to pinpoint that not only safety is the concern, and thus, in step 3, this is explored.

The fourth and final step identifies causal factors to how ICAs might emerge in the system. According to [11], loss or causal scenarios are identified to explain:

1. How incorrect feedback, inadequate requirements, design errors, component failures, and other factors could cause inadequate control actions and ultimately lead to losses.
2. How adequate control actions might be provided, but not followed or executed properly, leading to a loss.

By identifying the loss scenarios, additional requirements may be identified as well as mitigative actions that may additionally lead to design recommendations and changes. The scenarios may also be used for defining test cases and creating test plans to bring in evidence that the scenarios are managed properly by the control system.

2.2. Simulation-based testing

A simulator may be used to predict some aspect or behavior of a system or process to be used for different purposes, e.g. understanding a process, documentation or for collecting evidence demonstrating some goals or needs. Testing is an important contribution in gaining assurance and finally trust in cyber-physical systems. The following aspects of testing are relevant for simulation-based testing (inspired by [12] and [13]):

1. Testing of functional behavior
2. Testing of performance
3. Testing of robustness

2.2.1. Testing of functional behaviour Functional testing may be described as verifying the existence of the functions related to the planned operation and the needed functional interaction between the systems and sub-systems. It can be viewed as verifying the functional suitability of the system. To perform functional testing, a clear and concrete specification is needed for the intended functions and interactions with the operating environment or other parts of the system. To trust the testing it is important that the simulation adequately simulates the needed functions, operational scenarios and interactions.

2.2.2. Testing of performance According to *[12]*, performance testing is defined as "type of testing conducted to evaluate the degree to which a test item accomplishes it's designated function within given constraints of time and other resources". Functional and performance testing are not totally separable, but they are rather distinctly different aspects. The performance aspect is not always relevant when considering a function, but often functions cannot fully be tested without building an opinion of the performance and vice versa, testing performance without including the functional behaviour is not possible.

2.2.3. Testing of robustness The third aspect of testing is testing of robustness. Robustness is defined as "the degree to which a component or system can function correctly in the presence of invalid inputs or stressful environmental conditions" *[13]*. Invalid inputs may come from defects in a component or system such as a sensor, preventing it from working as intended. Environmental disturbance could be caused by wind, current, waves and weather conditions, but also lighting conditions. It may also be more discrete and random events, e.g. happening in the operating environment, by the system operators. Robustness may be related to system inadequacies in general, that the system is not being capable of handling the operative environment, either due to the operative envelope exceeding the system's capability or that the system's capability is reduced. When testing robustness, exploratory testing where the tester actively controls the design of new tests based on already performed tests *[13]*, will be important as it is difficult on beforehand to know where defects or other inadequacies in the system may hide.

When simulator-based testing is used for collecting evidence that a specific control system is performing as intended in different situations, the simulator is connected to the control system, e.g. using a classical Hardware-In-the-Loop (HIL) [14] or Software-In-the-Loop (SIL) set-up. In both set-ups, the control system should not be able to detect any difference between being connected to the real asset or the simulator. HIL and SIL are discussed next.

2.2.4. Hardware-In-the-Loop testing A control system interacts with its surroundings through input/output (I/O) communication channels. These inputs are from operator stations, other control systems, and sensors measuring parameters and dynamical states. The inputs are used, together with the internal models in the control system, to calculate control signals which are sent to the actuators. In a HIL set-up, the actual I/O's are replaced with simulated I/O's from a HIL simulator running in real-time. The surroundings of the control system, such as the dynamical systems, sensors and actuators, are imitated by the HIL simulator. The HIL simulator responds to control signals and provide consistent measurements, see Figure 3.

2.2.5. Software-In-the-Loop testing SIL testing allows for testing of the behavior of the control system without physically connecting the hardware to the simulator. The control software is either running on emulated hardware or as a separate process on the computer running the simulation. As the control software is not running on real-time hardware, it may be possible to control time, such that the control system may run in sync with the simulation possibly faster than real time. When using simulation models to generate trust, is it important that the simulation models can be trusted, being the next topic.

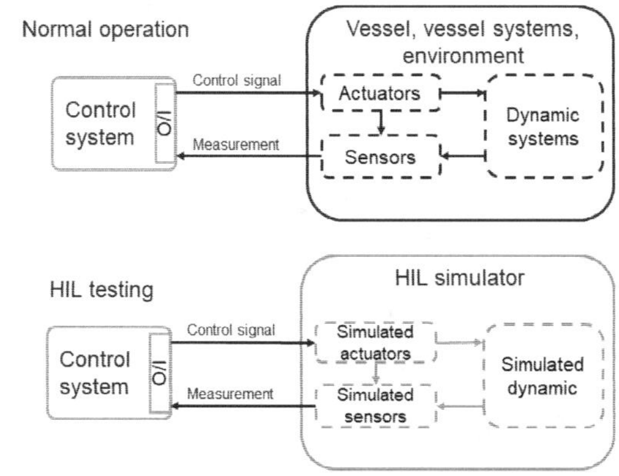

Figure 3: HIL testing conceptional setup, excluding other control systems and operators, courtesy of Marine Cybernetics AS.

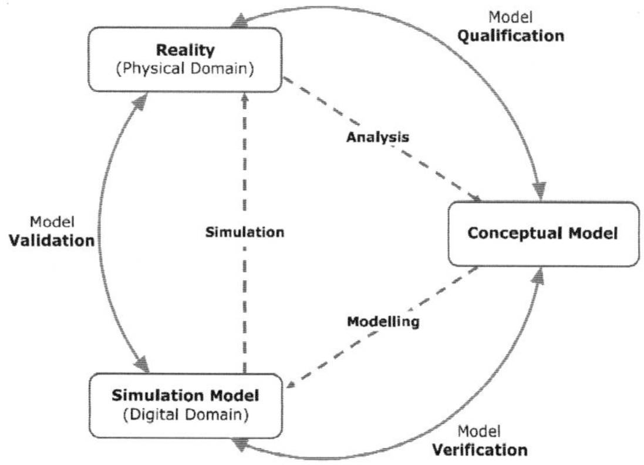

Figure 4: Relationship between reality, conceptual model and simulation model, together with verification and validation activities, [15].

2.2.6. Assurance of Simulation models In June 2021, DNV released a recommended practice (RP) providing a framework for assuring simulation models *[15]*. The RP defines what is necessary in terms of competence and assurance activities to minimize the risk of using a simulation model in a given use case to an acceptable level.

Before developing a simulation model, it is vital to first establish a conceptual model of the reality. The conceptual model gives a common understanding of the features to include, how to model these features and the high-level architecture of the model [15]. All three elements, the reality, the conceptual model and the simulation model, are important and essential to be included in the assurance of a simulation model, providing trust in the model and the predictions it makes about the reality in a specific use case (Figure 4).

More information about assurance of Simulation models is found in [15].

2.2.7. Automatic testing using Gaussian Processes and System Temporal Logic An automatic simulation-based testing approach was proposed in *[8]* which is briefly presented. For a deeper understanding of the proposed method, the reader is recommended reading the paper and the references therein.

An overview of the main components of the proposed testing methodology and how they interact can be seen in Figure 5. A simulator with all its simulator models, is described by a set of parameters, e.g. initial conditions, input signals and configurations. The results obtained from running a simulation is dependent on the value of these parameters. Let P be a parameter set defined by the Cartesian product $P = P_1 \times P_2 \times \dots \times P_n$. The parameter set contains all combinations for a simulator with n parameters and each parameter p_i has an associated P_i, where $i \in (1,2,\dots,n)$. Testing all the simulations in P may not be relevant nor possible. Instead, a case Σ is defined as a collection of k parameters $p_1, p_2, \dots, p_k$ with corresponding parameter sets $P_1, P_2, \dots, P_k$. Normally, one would keep most of the parameters in a simulation constant, and only let a few of the parameters vary. For each test case, the fixed parameters are described, as well as the parameters that may vary by the range values they may take. Each sub-case will then hold a unique set of varying and fixed parameters.

Figure 5 shows an overview of the main components of the proposed testing methodology. The test case describes the complete simulation in addition to the Signal Temporal Logic (STL) metric used for the evaluation. If applicable, also other evaluation criteria's will be given as part of the test case. Prior to simulation an initial set of sub-cases are generated using Latin Hypercube Sampling (LHS) [16] by splitting the predefined parameter space into n_{seed} slots, LHS selects samples such that no two samples occupy the same hypercube. The initial sub-cases are simulated, evaluated and the used to

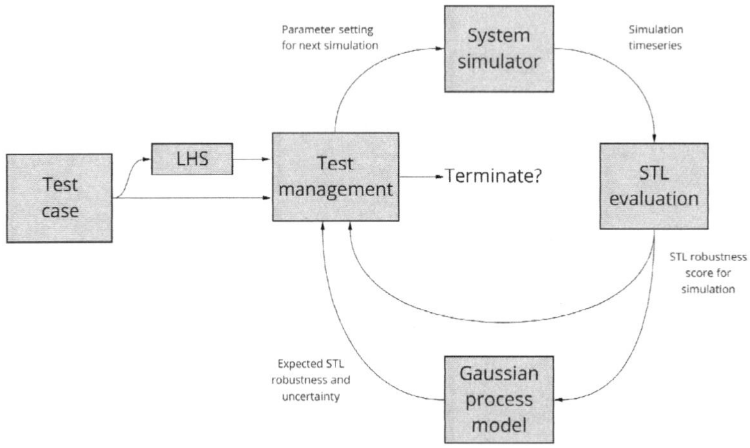

Figure 5: An overview of the main components of the proposed testing methodology and how they interact, adopted from [8].

establish an initial prior of the Gaussian process (GP) model. New parameter sampling points are then selected based on an updated prior of the GP model using a combination of

- points with low expected robustness, and
- points with large uncertainty.

By letting κ be an *exploration vs. exploitation* trade-off parameter, the next parameter set to simulate may be given by

$$\boldsymbol{p}_{next} = argmin_{\boldsymbol{p} \in \boldsymbol{P}}\ \bar{\rho}(\boldsymbol{p}) - \kappa\sqrt{var(\rho(\boldsymbol{p}))}, \tag{1}$$

where $\bar{\rho}(\boldsymbol{p})$ is the expected robustness and $var(\rho(\boldsymbol{p}))$ is the associated covariance. The choice of κ may affect the number of simulations needed to obtain a sufficient confidence level. For small values of κ, the search is guided towards areas with low robustness which is good for fast falsification. With large values of κ, the search is guided towards unexplored areas with high uncertainty which is good for fast verification. Any test cases producing a low STL robustness metric indicates a falsification. If the expected STL robustness metric and uncertainty is above a certain limit, verification is concluded as a success. This will continue until any sub-case is falsified or all sub-cases are verified.

A brief description of STL and GP are given next.

2.2.8. Signal Temporal Logic STL is used to evaluate a real-valued signal *[17]* against a temporal logic formula to see if it satisfies a specific behavior by using predicates, is defined as:

$$\pi ::= f(y) \leq c, \tag{2}$$

where $f(y)$is a scalar, real-valued function which maps the input signal y (an output from the simulation) to a real-valued scalar, and c is a real-valued scalar. The logical predicates can either be true or false.

To explain formulating STLs, the following example is given. Assuming the speed of a vessel is controlled by a speed controller, using speed feedback from three different sources. If one source fails, the controller shall still be able to control the vessel at a constant speed. This may be expressed using STL as

$$\varphi_{speed} = \square\left(\left|U_{vessel} - U_{ref}\right| \leq C_{\Delta U_{vessel_{max}}}\right), \tag{3}$$

where $\square$ is the *always* operator such that φ is true if φ is true for all times $t \in T$, U_{vessel} is the vessel speed over ground and U_{ref} is the vessel speed reference.

One advantage with STL is the STL robustness metric which gives a quantitative indication of how robustly a signal satisfies an STL formula. Using predicates, STL provides a language to specify behaviors of the system, and with the robustness metric, STL measures how well a behavior satisfies this formula.

A metric and signed distance are key concepts in STL defined next.

Definition 4. **Metric** [18]: A metric on a set S is a positive function $d: S \times S \rightarrow \mathbb{R}_{\geq 0}$ such that

$$\forall s, s' \in S, d(s, s') = 0 \Leftrightarrow s = s', \tag{4}$$

$$\forall s, s' \in S, d(s, s') = d(s', s), \tag{5}$$

$$\forall s, s', s'' \in S, d(s, s'') \leq d(s, s') + d(s', s''), \tag{6}$$

In this paper the Euclidean metric $d(s, s') = \|s - s'\|$ has been used.

Definition 5. **Signed Distance** [18]: With the metric d, the distance of a point $s \in S$ from a set $A \subseteq S$, the Signed Distance is defined as:

$$Dist_d(s, A) := \begin{cases} -\inf\{d(s, s') | s' \in A\} \; if \; s \notin A \\ \inf\{d(s, s') | s' \notin A\} \quad if \; s \in A \end{cases}. \tag{7}$$

The $Dist_d(s, A)$ is the shortest distance from s to any point in A. The signed distance states the robustness of $s \in A$, meaning how robustly a point belongs to a set. The signed distance is positive or negative in case a point is within or outside the set, respectively. The magnitude represents how far away s is from the boundary of A.

2.2.9. Gaussian Process model A Gaussian Process model is used to estimate the STL robustness metric the parameter combinations not already tested. According to *[19]*, a GP is a collection of random variables, any finite number of which have (consistent) joint Gaussian distributions defined by its mean function $m(x)$ and covariance function $k(x, x')$ as

$$f \sim \mathcal{GP}(m, k), \tag{8}$$

where f is a stochastic function, and $f(x)$ is the value of the function f at the location of the input argument x. There exists several covariance functions, also known as *kernel functions*, where the squared exponential covariance function probably is the most commonly used, given as

$$k(\boldsymbol{x}_i, \boldsymbol{x}_j) = \sigma^2 \exp\left(-\frac{1}{2l^2}\|\boldsymbol{x}_i - \boldsymbol{x}_j\|^2\right). \tag{9}$$

It has two parameters, the variance σ^2 and the length scale l, which represents how much $\|x\|$ has to change to significantly change $f(x)$.

Assume an input $\boldsymbol{x_i} \in \mathbb{R}^{n_i}$, an output $y_i \in \mathbb{R}$ and a noisy observation set $D = \{(\boldsymbol{x_i}, y_i) | i = 1, \ldots, n\}$, then consider the unknown function $y = f(\boldsymbol{x_i})$ which can be learned from the observations $(\boldsymbol{x_i}, y_i)$. Assuming the noise in the observations is independent and Gaussian distributed, then

$$y(\boldsymbol{x_i}) = f(\boldsymbol{x_i}) + \epsilon_i, \; \epsilon_i \sim \mathcal{N}(0, \sigma_n^2), \tag{10}$$

where $\epsilon_i \in \mathbb{R}$ is a zero-mean white Gaussian noise with variance σ_n^2.

With the defined observation set, the value of $f(\boldsymbol{x})$ at n_* test points are to be predicted, given by $f_* \in \mathbb{R}^{n_*}$. A joint Gaussian probability distribution can then be defined, where $\boldsymbol{y}$ is the known function values of the training set and $\boldsymbol{f}_*$ is a set of function values corresponding to the test set inputs X_*:

$$\begin{bmatrix} \boldsymbol{y} \\ \boldsymbol{f}_* \end{bmatrix} \sim \mathcal{N}\left(\boldsymbol{0}, \begin{bmatrix} K(X, X) + \sigma_n^2 & K(X, X_*) \\ K(X_*, X) & K(X_*, X_*) \end{bmatrix}\right), \tag{11}$$

where $K(X, X)$ is the $n \times n$ matrix training set covariances, $K(X, X_*)$ and $K(X_*, X)$ are the training-test set covariances and $K(X_*, X_*)$ is the test set covariance.

For prediction, a Bayesian inference approach is used to calculate the conditional probability distribution for $\boldsymbol{f}_*$ given $\boldsymbol{y}$. This results in the distribution

$$\boldsymbol{f}_* | X, \boldsymbol{y}, X_* \sim \mathcal{N}\left(\bar{\boldsymbol{f}}_x, cov(\boldsymbol{f}_*)\right), \tag{12}$$

where

$$\bar{\boldsymbol{f}}_{\boldsymbol{x}} = \mathbb{E}[\boldsymbol{f}_* | X, \boldsymbol{y}, X_*] = K(X_*, X)[K(X, X) + \sigma_n^2 I]^{-1} \boldsymbol{y} \tag{13}$$

and the covariance matrix is given as

$$cov(\boldsymbol{f}_*) = K(X_*, X_*) - K(X_*, X)[K(X,X) + \sigma_n^2 I]^{-1} K(X, X_*). \tag{14}$$

The next step is to update the posterior predictions of the robustness values at each point in $\boldsymbol{P}$ based on the observed robustness values. Given that n_{obs} observations have been made at points $\boldsymbol{P_{obs}} \in \mathbb{R}^{k \times n_{obs}}$, the observed robustness values are collected in the vector $\rho_{obs} \in \mathbb{R}^{n_{obs}}$. The robustness values at each point in $\boldsymbol{P}$ and the observed points at $\boldsymbol{P_{obs}}$ are jointly Gaussian with distribution

$$\begin{bmatrix} \rho_{obs} \\ \rho \end{bmatrix} \sim \mathcal{N}\left(\boldsymbol{0}, \begin{bmatrix} K_{obs} + \sigma_\epsilon^2 I & K_{cross} \\ K_{cross} & K \end{bmatrix}\right), \tag{15}$$

where $K_{obs} \in \mathbb{R}^{n_{obs} \times n_{obs}}$ is the covariance matrix for the observation points and $K_{cross} \in \mathbb{R}^{n \times n_{obs}}$ is the cross covariance between points $\boldsymbol{P}$ and $\boldsymbol{P_{obs}}$. The conditional probability distribution of ρ given ρ_{obs} is given as

$$\rho | \rho_{obs} \sim \mathcal{N}(\boldsymbol{K_{cross}}[\boldsymbol{K_{obs}} + \sigma_\epsilon^2 \boldsymbol{I}]^{-1},\ \boldsymbol{K} - \boldsymbol{K}_{cross}[\boldsymbol{K_{obs}} + \sigma_\epsilon^2 \boldsymbol{I}]^{-1} \boldsymbol{K}_{cross}^T), \tag{16}$$

where each point in $\boldsymbol{P}$ have an associated expected value and uncertainty. In the remainder of this paper, let $\bar{\rho}(\boldsymbol{p})$ and $var(\rho(\boldsymbol{p}))$ refer to the expected value and variance of the STL robustness at point $\boldsymbol{p}$.

In [8], it is proposed to let the simulations run iteratively and update the GP until a desired confidence level is achieved, or until the requirements are falsified. For a sufficient confidence level, the criterion is that the minimum of the p_{conf} probability confidence interval of f_ρ is greater than zero. The search will be terminated in a *Verified* state when

$$p_{conf} = min_{p \in P} \bar{\rho}(\boldsymbol{p}) - n_{conf} \sqrt{var(\rho(\boldsymbol{p}))} \geq 0 \tag{17}$$

where n_{conf} is defined as the number of standard deviations associated with the confidence level p_{conf}. Similarly, if the search process finds a case which falsifies the requirements, a case with robustness lower than zero, the search will terminate in a *Falsified* state.

3. Analysis

A line fishing vessel is used as a case for exploring the proposed method. A complete analysis of the control structure for the fishing vessel, is not the scope of this paper and instead a brief presentation is given.

The analysis starts with a description of the operation, then a summary of the results from the STPA analysis are presented. At the end, an example of how the result may be formulated as formal signal temporal logic requirement is given.

3.1. Operation

The long-line fishing vessel MS Geir is used as a use case in the SEAOPS project [20]. The vessel is today navigated by a captain, either manually or using a conventional autopilot. The autopilot is controlling the course of the vessel while the speed of the vessel is controlled using pitch and rpm levers. In the SEAOPS project, a controller is developed that shall automatically follow a path or control the course/speed for optimal setting of the long-line, and at the same time minimize wear and tear or damages to fishing equipment or vessel machinery and minimize fuel consumption.

The analysis focuses on the line setting phase. The vessels speed through water (STW) should be between 8.9 and 9.2 knots for optimal bait setting conditions. If the speed of the vessel is too high, some fishing-hooks might be set without bait., which can result in reduced catch. If the speed. is too low, fishing line setting time increases, resulting in lower efficiency

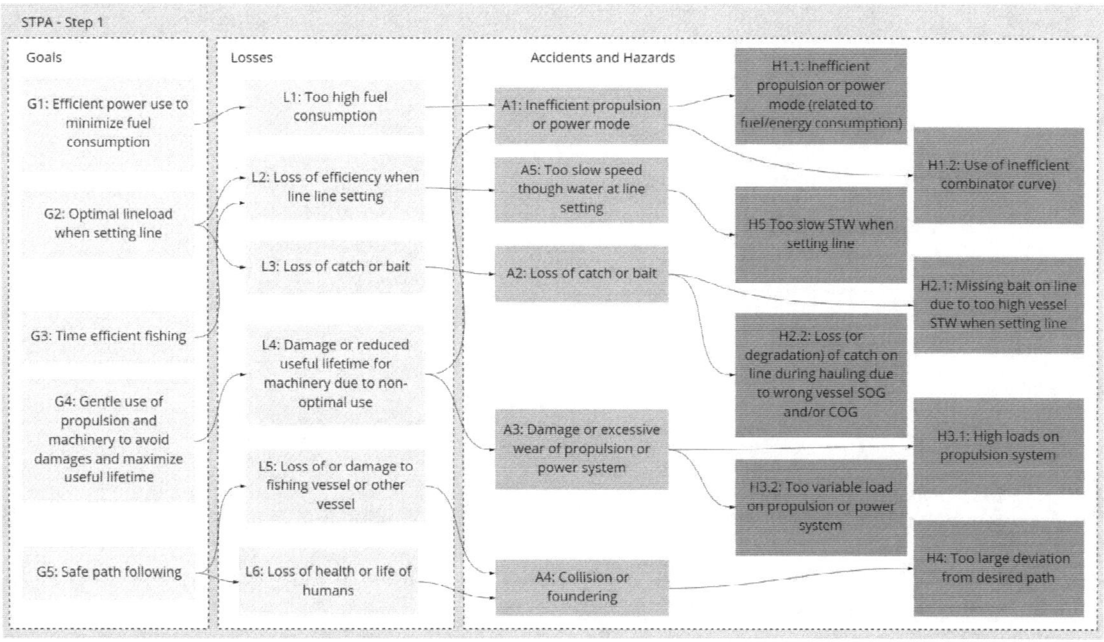

Figure 6: Goals - Losses - Accidents and Hazards

3.2. System Theoretic Process Analysis

3.2.1. Step 1: Goals - Losses - Accidents and Hazards During step 1 of the STPA, five operative goals were identified (see also Figure 6): G1 - Efficient power use to minimize fuel consumption, G2 - Optimal line load when setting line, G3 - Time efficient fishing, G4 - Gentle use of propulsion and machinery to avoid damages and maximize useful lifetime, and G5 - Safe path following. The corresponding losses were: L1 - Too high fuel consumption (G1), L2 - Loss of efficiency when line setting (G2). L3 - Loss of catch or bait (G2/G3), L4 - Damage or reduced useful lifetime for machinery due to non-optimal use (G4), L5 - Loss of or damages to fishing vessel or other vessels (G5) and L6 - Loss of health or life of humans (G5).

The detailed operation was analysed and accidents were identified and connected to the relevant losses. For each accident, hazards were identified:

- A1 Inefficient propulsion or power mode (L1, L4)
  - H1.1 Inefficient propulsion or power mode (related to fuel/energy consumption)
  - H1.2 Use of inefficient combinator curve
- A2 Loss of catch or bait (L3)
  - H2.1 Missing bait on line due to too high vessel STW when setting line
  - H2.2 Loss (or degradation) of catch on line during hauling due to wrong vessel SOG and/or COG
- A3 Damage or excessive wear of propulsion or power system (L4)
  - H3.1 High loads on propulsion system
  - H3.2 Too variable load on propulsion or power system
- A4 Collision or foundering (L5, L6)
  - H4 Too large deviation from desired path
- A5 Too slow STW line setting (L2)
  - H5 Too slow STW when setting line

In the following, H4: Too large deviation from desired path is selected presenting the analysis and for covering the most severe loss, see Figure 6.

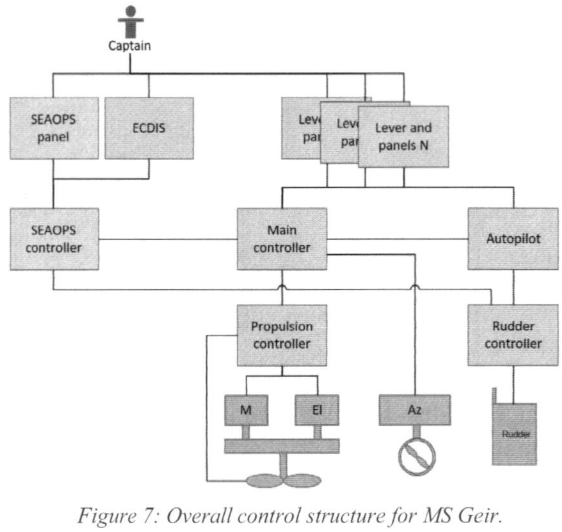

Figure 7: Overall control structure for MS Geir.

Treating the hazards as control problems, functional requirements for monitoring and control are identified:

- H4 Too large deviation from desired path
  - FR4.1 Desired path needs to be known.
  - FR4.2 Actual vessel position/heading/ course/speed need to be known.
  - FR4.3 Vessel position/heading/course/ speed need to be controlled.

The next step in the analysis is to identify the control structure.

3.2.2. Step 2: Control Structure The functional requirements identified in step 1 is mapped towards functions in the control system. The functional requirements connected to H4 are used as examples. FR4.1: Desired path needs to be known is mapped to be the responsibility of the captain, as he/she shall give way-point inputs to the SEAOPS controller ECDIS or similar. FR4.2: Actual vessel position/heading/course/speed need to be known may be estimated by e.g. a vessel state estimator in the SEAOPS controller based on feedback from the global positioning system (GPS). The last functional requirement is FR4.3 Vessel position/heading/course/speed need to be controlled, which is the control objective of the SEAOPS controller.

When all identified functional requirements are mapped as responsibilities of specific functions in the control system, the control structure can be developed by identifying the controllers, their control inputs, control actions and feedback information on which they depend on. The overall control structure for the fishing vessel MS Geir is shown in Figure 7 with a short description. The green boxes, the SEAOPS panel and the SEAOPS controller, will be developed during the project, and is target for the analysis and later testing. The blue boxes are existing panels, levers and control systems on board MS Geir, while the grey boxes are the components that are controlled by the on board control systems, e.g. main motor (M), electric motor (El), azimuth (Az) and rudder. The captain controls the SEAOPS controller using the SEAOPS panel, while route plans are sent to the SEAOPS controller using ECDIS. The SEAOPS controller is connected to the main controller and the rudder controller. Other levers and panels and Autopilot are disengaged when SEAOPS controller is in control. The captain will always have the possibility to disable the SEAOPS controller and take direct control of the vessel. The SEAOPS controller will be responsible for deciding the optimal propulsion mode together with optimal combinator curve for (pitch/rpm) and controlling course and speed using the main propeller, rudder and azimuth.

3.2.3. Step 3/4: Inadequate Control Actions and Design-specific causes Inadequate control actions (ICAs) are identified in step 3. An ICA is defined as a control action that in a particular situation and

Journal of Physics: Conference Series **2311** (2022) 012016 doi:10.1088/1742-6596/2311/1/012016

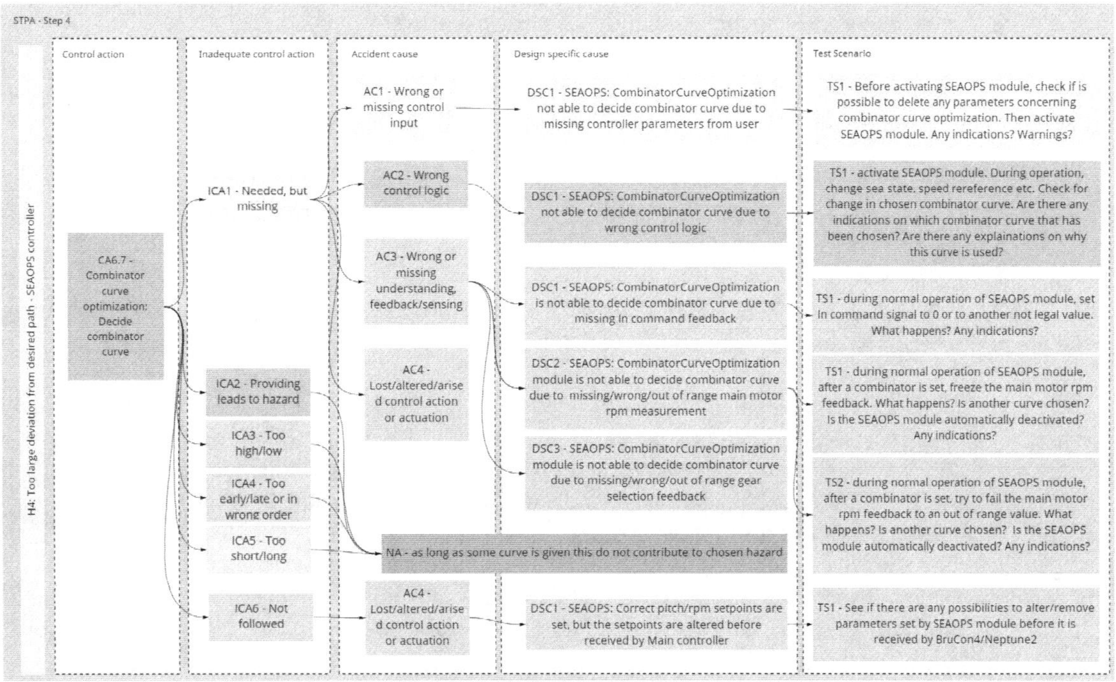

Figure 8: Step 4 - Design-Specific cause

in worst-case scenario leads to a hazard. *[11]* lists four ways a control action may be inadequate or unsafe:

- Not providing the control action leads to a hazards Providing the control action leads to a hazard.
- Providing a potentially adequate control action too early, late or in a wrong order leads to a hazard.
- The control action lasts too long or is stopped too soon leads to a hazard.

To further help the analysis, the following inadequate control actions were additionally added:

- Providing too high or too low control action leads to a hazard.
- Not following the control action leads to a hazard.

The last ICA may seem similar to the first ICA but covers the case if a proper control action is implemented but for some reason not followed. This is done for consistency in the analysis. If the control action is handled by a lower-level controller included in the analysis, the ICA can cover potential corruption of the information over the communication, but this overlaps with the accident cause 1 (AC1) explained later in this paper. If the lower-level controller is not included, the ICA can cover any case of inadequate implementation of a control action (see specifically accident cause 4 (AC4) also explained later). With identified ICAs, the controller constraints may be derived directly, which is the last step (4) in the analysis, that is to define causal scenarios or loss scenario which describes the causal factors that can lead to an inadequate control action and to hazards. According to [11], two types of loss scenarios should be considered:

- How would inadequate control action occur?
- How would control actions be improperly executed or not executed, leading to hazards?

The SEAOPS controller includes several control functions and the focus in the following is the control function *Combinator curve optimization: Decide combinator curve*. Parts of the analysis is shown in Figure 8. The list of ICAs are used together with the following defined accidents causes based on [11]:

- AC1 - Wrong or missing control input.
- AC2 - Wrong control logic.

- AC3 - Wrong or missing understanding/ feedback/sensing
- AC4 - Lost/altered/arised control action or actuation.

A design specific cause (DSC) answering ICA1 - Needed, but missing and AC1 - wrong or missing control input, may be written as

DSC1 - SEAOPS: CombinatorCurve-Optimization not able to decide combinator curve due to missing controller parameters from user.

The DSC may be tested using the following Test Scenario:

TS1 - Before activating SEAOPS module, check if is possible to delete any parameters concerning combinator curve optimization. Then activate SEAOPS module. Any indications? Warnings?

By combining ICAs with accident causes, it is possible to identify a set of DSCs which again may be used to define Test Scenarios. In this example, ICA2 to ICA5 have been defined as Not Applicable (NA), using the argument that if any curve has been chosen, then this will not lead to the chosen hazard which is H4 - Too large deviation from desired path. If the hazard had been something else, e.g. low efficiency, then ICA2 to ICA5 would have been used.

The identified Test Scenarios may then be used for testing the control system. Some of the test may be performed manually on the actual vessel, while others may form the basis for simulator- based testing. Some of the scenarios will also be suitable for automatic testing using the STL robustness metric, which is the next topic.

3.2.4. Development of STL robustness metric The hazard *H4 - Too large deviation from desired path* may be chosen for exploring the use of STL robustness metric together with the use of GP to guide the test sub-case selection. Looking into CA6.7, ICA1 and AC3, see Figure 8, the following DSC can be transformed to STL robustness metric:

DSC2 - SEAOPS: CombinatorCurve-Optimization module is not able to decide combinator curve due to missing/wrong/out of range main motor rpm measurement.

The steps needed for transforming the test scenario to test requirement and then to STL formulation are given in Table 1.

4. Conclusion

In this paper, the results from [5], [6] and [8] have been combined to develop evaluation requirements in addition to develop a method for automating exploratory testing to be used for assurance of complex system.

The next step will be to implement these requirements as evaluation criteria into a test system to demonstrate a procedure for simulation-based testing of the SEAOPS controller using a test setup with the controller connected in closed loop with a simulator. The use of GP to guide the selection of new sub-cases to test together with STL for evaluation is believed by the authors to be an efficient way to automate testing of complex systems such as autonomous navigation.

Acknowledgement

This work was supported by the Research Council of Norway through the project "SEAOPS", project number 10203745.

Table 1: From STPA to STL, CA6.7ICA1.AC3.DSC2.TS1

Test scenario	Set desired speed reference for line setting and activate SEAOPS module. After a combinator curve is set, freeze the main motor rpm feedback. What happens? Is another curve chosen? Is the SEAOPS module automatically deactivated? Any indications?
Test requirement	For all tests, the SEAOPS module shall be activated. Set a speed reference within line setting range. Freeze main motor rpm feedback. Check several different speed references, all within line setting range. For all rpm speed feedback, freeze the feedback's one by one. - Vessel speed should remain approximately constant after the failure has been initiated. - Propeller pitch should not be changed after the failure has been initiated. - Propeller rpm should not be changed after the failure has been initiated. - Check for warnings.
STL formulation	Parameterization: $U_{vessel_{ref}} \in [8.0, 9.5]$[knots], where $U_{vessel_{ref}}$ is desired speed reference for the vessel. $C_{\Delta U_{Vessel_{max}}}$ is maximum allowed speed variations, $C_{\Delta P_{max}}$ and $C_{\Delta \omega_{max}}$ are maximum allowed pitch and propeller speed variation, respectively. Parameter to be adjusted: $U_{vessel_{ref}}$ $\varphi_{speed} = \square\left(\lvert U_{vessel} - U_0\rvert \leq C_{\Delta U_{Vessel_{max}}}\right)$ $\varphi_P = \square(\lvert P - P_0\rvert \leq C_{\Delta P_{max}})$ $\varphi_\omega = \square(\lvert \omega - \omega_0\rvert \leq C_{\Delta \omega_{max}})$ U_{vessel}, P and ω are vessel speed, propeller pitch and propeller speed feedback, respectively, while U_0, P_0 and ω_0 are vessel speed, propeller pitch and propeller speed at the moment when the main motor rpm feedback to the controller was frozen.

References

[1] IMO, "Fourth IMO Greenhouse Gas Study," *International Maritime Organization,* 2020.

[2] IMO, "The Initial IMO Strategy on Reduction of GHG Emissions from Ships," *International Maritime Organization,* 2018.

[3] B. J. Vartdal, R. Skjong and A. L. St. Clair, "Remote-Controlled and Autonomous Ships," DNV, 2018.

[4] O. I. Haugen, "Properties of evidence and evidence generation," *Marine Technology Society, Dynamic Positioning Conference,* 2018.

[5] B. Rokseth, I. B. Utne and J. E. Vinnem, "Deriving Verification Objectives and Scenarios for Maritime Systems Using the Systems-Theoretic Process Analysis," *Reliability Engineering and System Safety,* 2018.

[6] J. A. Glomsrud and J. Xie, "A Structured STPA Safety and Security Co-analysis Framework for Autonomous Ships," in *Proceedings of the 29th European Safety and Reliability Conference,* 2019.

[7] G. Koelln, M. Klicker, T. Schmid and S. Schmidt, "System Theoretic Process Analysis for a Vehicle SAE Level Four," in *Proceedings of the 30th European Safety and Reliability Conference*, 2020.

[8] T. Torben, J. A. Glomsrud, T. A. Pedersen, I. Utne and A. Sørensen, "Automatic Simulation-based Testing of Autonomous Ships using Gaussian Processes and Temporal Logic," *Journal of Risk and Reliability,* 2022.

[9] N. Leveson, Engineering a Safer World: Systems Thinking Applied to Safety, The MIT Press, 2012.

[10] J. Thomas, Extending and Automating a Systems-Theoretic Hazard Analysis for Requirements Generation and Analysis, PhD thesis, Massachusetts Institute of, 2013.

[11] N. Leveson and J. Thomas, STPA Handbook, The MIT Press, 2018.

[12] ISO, "ISO/IEC/IEEE 29119-1:2013 Software and systems engineering — Software testing — Part 1: Concepts and definitions," *ISO/IEC/IEEE,* 2013.

[13] ISTQB, "Standard glossary of terms used in Software Testing," *International Software Testing Qualifications Board,* 2010.

[14] R. Sjetne and O. Egeland, "Hardware-in-the-loop testing of marine control systems," in *46th Conference on Simulation and Modeling (SIMS 2005)*, 2005.

[15] DNV, "Recommended Practice: Assurance of Simulation Models," *DNV-RP-0513,* 2021.

[16] M. D. McKay, R. Beckman and W. Conover, "A comparison of three methods for selecting values of input variables in the analysis of output from a computer code," *Technometrics,* pp. 239-245, 1979.

[17] O. Maler and D. Nickovic, "Monitoring temporal properties of continuous signals," *In Lecture Notes in Computer Science,* p. 152–166, 2004.

[18] G. Fainekos and G. Pappas, "Robustness of temporal logic specifications for continuous-time signals," *Theoretical Computer Science,* pp. 4262-4291, 2009.

[19] C. E. Rasmussen and K. I. Williams, Gaussian Processes in Machine Learning, the MIT Press, 2006.

[20] "SEAOPS," January 2022. [Online]. Available: https://www.dnv.com/research/review-2020/featured-projects/seaops-ship-operations-simulation.html.

[21] H. Rashmi, S. Yako, K. Post and S. Nuesch, "Systems Theoretic Process Analysis for Layers of System Safety," *INCOSE International Symposium,* vol. 29, pp. 895-909, 2019.

Exploring navigator roles and tasks in transitioning towards supervisory control of autonomous collision avoidance systems

G K van de Merwe[1,2], S C Mallam[2], Ø Engelhardtsen[1] and S Nazir[2]

[1]Group Research and Development, DNV, Veritasveien 1, 1363, Høvik, Norway
[2]Faculty of Technology, Natural Sciences and Maritime Sciences, Department for Maritime Operations, University of South-Eastern Norway, Raveien 215, 3184, Borre, Norway

koen.van.de.merwe@dnv.com

Abstract. This study aims to systematically map and assess performance requirements for collision avoidance manoeuvring for two cases. A case where the navigator performs collision avoidance, and a case where collision avoidance is performed by collision avoidance system where the navigator acts as its supervisor. An appraisal of collision avoidance manoeuvring was performed based on three data sources: the collision avoidance regulations, a ferry operator's procedures, and interviews with navigators including in situ observations. A framework was established in which the gathered data was structured and analysed using a cognitive task analysis approach. Based on the results, performance requirements and information needs were established. Further work will focus on detailing the navigator's information needs and the corresponding system's transparency requirements to support effective human performance.

1. Introduction

Recent trends in the maritime domain point toward the application of advanced automation that assume responsibility for functions currently performed by humans. The reasons for pursuing autonomy in maritime shipping are diverse, but the prospect of reduced manning for the sake more efficient operations has sparked interest in autonomous systems within the industry. Having fewer humans onboard a vessel reduces the need for human support functions. As such, vessels can potentially be made lighter, transport more cargo, and perform voyages more efficiently resulting in reduced operating costs and a reduced environmental footprint [1]. In addition, autonomy is often introduced as a means to increase safety of the vessel by reducing the potential for human error. Data from the European Maritime Safety Authority suggests that most of the causes of maritime accidents between 2014 and 2019 were attributed to human actions [2]. Although the attribution of accidents to human actions can be disputed [3], some argue that by removing human operators from the sharp end of ship operations, a reduction in the likelihood of marine incidents should occur [4]. Whilst this may be true in terms of reducing individual risk (i.e., humans are removed from hazardous locations and can therefore not be harmed), the relocation of control from vessel to land can introduce new and unknown risks [5,6].

Autonomous navigation is an area of autonomous shipping receiving a significant amount of attention. Within autonomous navigation, collision avoidance is seen as an important steppingstone towards enabling vessels to navigate without human intervention. As collision avoidance is a complex and multi-faceted operation, solving the collision avoidance problem is critical to realize successful autonomous shipping. Ideally, a vessel under autonomous navigation can safely navigate from port to port, resolving potential conflicting situations it may encounter throughout its journey without human intervention.

Collision avoidance is internationally regulated through the collision avoidance rules [7]. The collision regulations "COLREGs" were developed to provide the "rules of the road" for maritime vessels, much like the rules for driving a car. The COLREGs were developed to increase safety for sea-going vessels by providing navigators with a ruleset that allows for predictable vessel behaviour. However, the rules were also developed to provide a level of flexibility and allow for interpretation, such that they are applicable to a wide range of situations. For example, Rule 8 "Action to avoid collision" states: *"any action to avoid collision shall be taken in accordance with the Rules of this Part and shall, if the circumstances of the case admit, be positive, made in ample time and with due regard to the observance of good seamanship"* [7]. Furthermore, Rule 2 "Responsibility" states: *"In construing and complying with these Rules due regard shall be had to all dangers of navigation and collision and to any special circumstances, including the limitations of the vessels involved, which may make a departure from these Rules necessary to avoid immediate danger"* [7]. The former uses the term "good seamanship" as a way of stating that a minimum level of competency is expected to seamen in handling conflicting situations. The latter even states that the rules can and should be deviated from, when necessary, to avoid any immediate danger. Although these statements in the rules may make sense to humans, and especially experienced seamen, they can be challenging to program into a system responsible for collision avoidance. Porathe [8] notes that the rules were not developed with autonomous vessels in mind and ensuring sufficient coherency in their interpretations by autonomous and conventional vessels can pose a challenge.

The typical aim of automation is to replace human manual control with automated control through the application of computing devices [9]. However, there are several issues related to the assumption that a user tasked with the role of a supervisor of automation will be able to effectively execute this role. As has been argued by Strauch [10], even sophisticated systems require human supervision for cases that are outside its operational design domain (e.g., weather limits) or when failures occur that need to be handled. In such cases, the human may take on the role as a supervisor, ready to intervene when needed, i.e., a "fallback ready user" [11]. As such, using advanced autonomy for ship navigation functions, e.g., in autonomous conflict avoidance, implies changes to the role of the human navigator from an active task performer to a more passive supervisor. However, the new role depends on the chosen operational concept. One example is a conventional vessel equipped with a function that provides the onboard navigator with decision support in traffic conflict situations. Another example is an autonomous and unmanned vessel remotely supervised by an operator in a remote-control centre. Regardless of the chosen concept, an understanding of the implications of the changes it brings is essential for understanding how to support humans in their new role.

2. Methodology

2.1. Framework

To understand how collision avoidance is accomplished, a framework was established to guide the analysis (see figure 1). The framework was based on an approach that was developed to manage major accident risk through safety barrier management [12]. Understanding risks and managing the barriers that prevent a hazard from developing form an essential part of major accident prevention. Therefore, it is important to understand the elements that play a role in the prevention and mitigation of hazards developing into hazardous events. In our case, the prevention of major maritime accidents, i.e., ship collisions, rests almost entirely on the tasks performed by the navigator. It is the navigator that perceives the environment, estimates collision risk, and takes measures to prevent collisions from occurring (albeit supported by technology in all steps). Therefore, mapping and assessing this role and its tasks in relation to collision risk prevention and mitigation is an important step in understanding how this safety barrier is realized. Furthermore, understanding how the realization of this barrier will be affected with the introduction of technologies that assume (parts of) this role will be crucial. As such, the following task elements were captured through task analysis [13] to be able to derive performance requirements for collision avoidance:

- *Key task steps:* What are the key tasks to be performed in collision avoidance?
- *Critical information:* What information is essential to support and perform the tasks?
- *Task responsible:* Who is performing the task?
- *Time requirements:* When or for how long shall the task be performed?

A similar approach has previously been applied to identify performance requirements for safety barriers in major accident risk in the oil and gas industry [14].

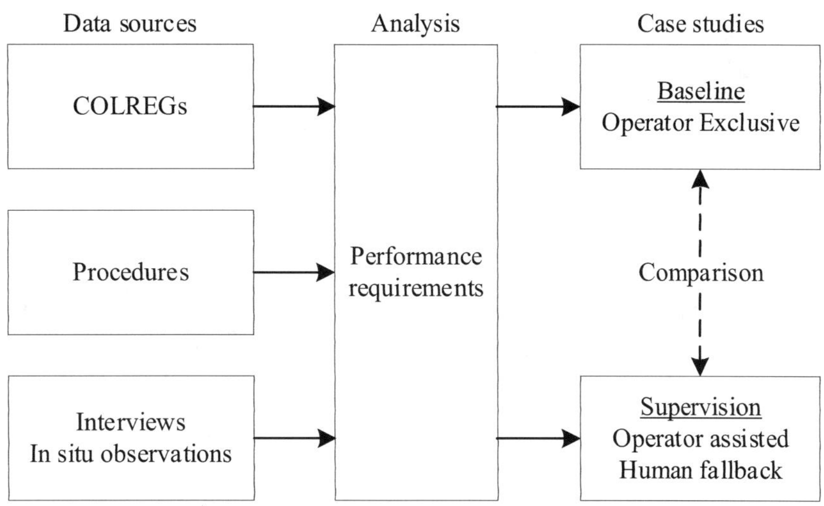

Figure 1. The framework used for the analysis.

2.2. Case studies

To analyse these changes, two case studies were established (see table 1). First, a baseline case was developed where a human navigator performed all tasks related to collision avoidance. In this case, the human navigator is present on the bridge and collision avoidance tasks are exclusively performed by the navigator as per present-day best practice, i.e., also called "operator-exclusive" [15]. Second, a hypothetical supervision case was developed where an autonomous collision avoidance system (CAS) was envisioned to be able to perform all collision avoidance scenarios. However, a human supervisor is still present on the bridge to oversee the performance of the system, i.e., also called "operator-assisted" [15]. If needed, the supervisor can take over manually, i.e., thereby assuming a "human fallback" role [11].

Table 1. The case studies

Baseline	**Supervision**
Operator-exclusive: The navigator performs collision avoidance tasks, supported by technology. There is no collision avoidance system.	Operator-assisted: The autonomous collision avoidance system performs collision avoidance tasks and is periodically supervised by a navigator present on the bridge.
	Fallback: The navigator performs the fallback task if needed, i.e., takes over manually.
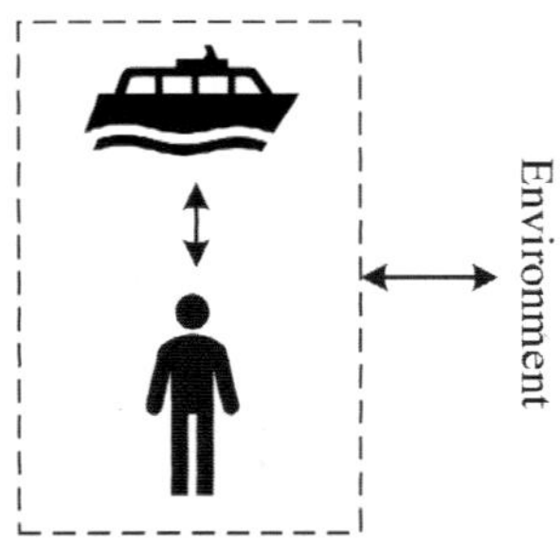	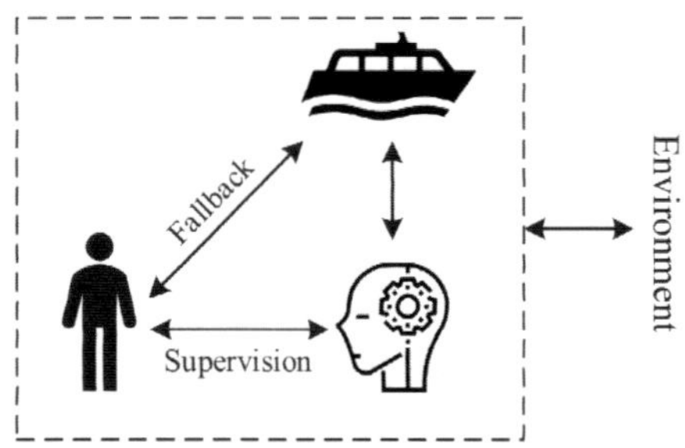

2.3. Scenario development

Scenarios were developed to probe the different aspects and tasks related to conflict avoidance manoeuvring. As the COLREGs describe collision manoeuvring in both general terms (e.g., generic collision avoidance actions) as well as specific cases (e.g., crossing) the individual rules were used as scenarios for the analysis. A subset of the rules was chosen for this study, these comprised of:

- Rule 2 – Responsibilities
- Rule 5 – Look-out
- Rule 6 – Safe speed
- Rule 7 – Risk of collision
- Rule 8 – Action to avoid collision
- Rule 13 – Overtaking
- Rule 14 – Head-on situation
- Rule 15 – Crossing situation
- Rule 16 – Action by give-way vessel
- Rule 17 – Action by stand-on vessel
- Rule 19 – Conduct of vessels in restricted visibility

Rules describing non-operational aspects of collision avoidance, i.e., applicability (rule 1), general definitions and responsibilities (rules 3 and 18), statements for information (rules 4 and 11), rules for specific situations such as narrow channels and traffic separation zones (rules 9 and 10), rules for sailing vessels (rule 12), and rules about lights, shapes, signals, exemptions, and verification (rules 20-41) were out of scope. However, as the rules are not independent, the content of some of the rules were considered in the analysis for the other rules. For example, rule 18 states the responsibilities between various types of vessels, vessel operations and vessel limitations and defines who has right of way given these factors. This information was considered in the tasks analysed in rule 8f – action to avoid collision, where the priority between vessels is determined by understanding which vessel is not to impede the safe passage of another vessel. As such, the content of rule 18 was applied in those rules in which this was considered by the navigator, i.e., when determining which is the stand-on vessel and which is the give-way vessel in taking action to avoid collision.

2.4. Data gathering

The data used to establish the performance requirements were based on three information sources. Combined, the following three data sources aimed to provide a comprehensive as possible picture of the performance requirements related to conflict avoidance actions:

1. An appraisal of the COLREGs,
2. A review of a ferry operator's operational procedures for collision avoidance, and
3. Interviews including in situ observations with ferry navigators on duty.

The interviews also collected the views of the navigators around supervising an autonomous collision avoidance system.

First, the COLREGs were analysed to extract the abovementioned performance requirements. The purpose of the COLREGs is to provide guidance to navigators on how to prevent collisions at sea, and to serve as a basis to apportion blame when collisions do occur [16]. Although much has been written about the rules and the language used to describe them (e.g., [8,16–18]), for all purposes, they consist of a set of explicit and implicit statements about what tasks a navigator should (not) perform, how to perform them, when to perform them and what to (not) do in special cases. For this study, these descriptions provided a basis to identify the performance requirements as described above: key task steps, critical information, task responsible and time requirements.

Second, the operational procedures of a ferry operator were reviewed and analysed to support and supplement the COLREGs analysis. Specific focus was on written documentation on the operator's interpretation of the rules and the roles, responsibilities, and actions relevant in avoiding collisions.

Third, navigators on duty were interviewed to better understand how collision situations are handled in practice. Interview questions were developed for the selected rules. In addition, in situ observations were performed on the bridge of a ro-ro ferry to supplement the interview data. The information gathered in this activity was used to supplement identified key tasks, equipment used, and any time requirements associated with collision avoidance manoeuvring.

As part of the interviews, the navigators' potential interaction with an autonomous collision avoidance system was addressed (i.e., the supervision case). A series of questions, based on the MITRE Human-Machine Teaming Systems Engineering Guide [19], was developed focusing on identifying information needs when humans need to team up with advanced automation to perform a task.

3. Results

3.1. Establishing performance requirements

The established framework provided a systematic approach to mapping requirements to human and system performance for existing and hypothetical future navigation cases. The obtained data was captured and structured in a spreadsheet and an excerpt is provided in table 2 for rule 7.

The example in table 2 describes the performance requirements for the navigator and the CAS for the two cases for rule 7. The performance requirement is the culmination of a set of individual performance requirements, described in section 2.1, and combined into a sentence. This sentence forms a statement describing the expectations to the performance of the navigator and the combined CAS/Navigator in a supervisory role. Note that the performance requirement presented in table 2 is not intended to be a proposal for an alternative interpretation of COLREGs, or in any way intended to replace the rules. For our study, it served as a means to be able to compare the baseline- and the supervision performance requirements. For this, a reformulation of the COLREGs to support the analysis of this study was beneficial.

Table 2. Example of performance requirements established for COLREGs rule 7.

Original description of COLREGs Rule 7. Risk of Collision [7]	
Description	*"a. Every vessel shall use all available means appropriate to the prevailing circumstances and conditions to determine if risk of collision exists. If there is any doubt such risk shall be deemed to exist.* *b. Proper use shall be made of radar equipment if fitted and operational, including long-range scanning to obtain early warning of risk of collision and radar plotting or equivalent systematic observation of detected objects.* *c. Assumptions shall not be made on the basis of scanty information, especially scanty radar information.* *d. In determining if risk of collision exists the following considerations shall be among those taken into account:* *(i) Such risk shall be deemed to exist if the compass bearing of an approaching vessel does nor appreciably change;* *(ii) such risk may sometimes exist even when an appreciable bearing change is evident, particularly when approaching a very large vessel or a tow or when approaching a vessel at close range."*
The reformulation of the rule for the purpose of this study	
Baseline Navigator	When underway, the navigator shall continuously determine the presence of collision risk using all available means and as early as possible by using long-range radar scanning and systematic observations of detected objects. The navigator shall avoid relying on insufficient data but continuously obtain an adequate level of information to estimate risk of collision by obtaining information from multiple independent and reliable sources. When target vessel is detected, the navigator shall ascertain changes to (compass) bearings over time to determine collision risk. In addition, the vessel's size, presence of a tow, and distance to target vessel shall be taken into account.
Supervision CAS	When underway, CAS shall continuously determine the presence of collision risk using all available means and as early as possible by using long-range radar scanning and systematic observations of detected objects. CAS shall avoid relying on insufficient data but continuously obtain an adequate level of information to estimate risk of collision by obtaining information from multiple independent and reliable sources. When target vessel is detected, CAS shall ascertain changes to (compass) bearings over time to determine collision risk. In addition, the vessel's size, presence of a tow, and distance to target vessel shall be taken into account.
Navigator	When underway, the navigator shall continuously be able to verify whether CAS has determined the presence of collision risk. This means that the navigator shall be able to verify if and how CAS is using relevant sensors, if it is detecting relevant objects and performing systematic observations of these, and if collision risk for these objects is estimated as early as possible. The navigator shall also be able to verify the reliability of CAS' sensor information, collision risk prediction and target identification.
Navigator/ fallback	Upon identifying insufficient collision risk capabilities by CAS, a request to intervene or a critical system alarm, the navigator shall immediately take over manually. This includes CAS' availability and reliability of sensor information, ability to detect and plot targets, reliability of collision risk estimation, and inaccuracies in type and size of target estimation.

The baseline performance requirement was formulated first and was based on the three combined data sources from the COLREGs appraisal. Care was taken to ensure that the performance requirement reflected the intentions of the COLREGs rule, but also that it contained the individual performance requirements captured through the procedural input, the interviews, and the observations.

Once the baseline performance requirements were established, three sets of performance requirements were developed for the supervision case: one for the CAS, one for the navigator supervising the CAS and, one for the navigator in a fallback position should the CAS' performance be insufficient (see table 2).

First, the performance requirement for the CAS was obtained by replacing the "who" in the baseline performance requirement, i.e., the word "navigator" was replaced by "CAS". This was done under the assumption that the COLREGs apply to all vessels, and therefore also for vessels equipped with a CAS. Although specific rules may be developed for autonomous vessels sometime in the future, this was not assumed here. As such, this means that as the CAS is taking over the collision avoidance task from the navigator, the performance requirement for the navigator then applies to the CAS. Therefore, the performance requirement for the CAS is the same as for the navigator except for the entity executing the task.

Second, the performance requirement for the navigator in a supervising position was determined by appraising the tasks of the navigator in this position. In the supervision case, the navigator had a supervisory role, and it was assumed that the CAS was able to perform the collision avoidance task, but that periodic verification of its performance was necessary. Therefore, the performance requirement for the supervising navigator consisted of the words "shall be able to" indicating that it shall be possible for the navigator to assess the performance of the CAS and to decide whether to intervene or not.

Third, the performance requirement of the fallback task is there to provide the requirements to the navigator in case the CAS leaves its operational design domain. The fallback performance requirement therefore consists of a set of criteria for the navigator for when to intervene and if so, what to do. For example, due to weather conditions, the CAS may exit is operational design domain, i.e., it enters conditions it was not designed for, thereby resulting in unreliable collision risk estimations and an inability to meet its required performance. As such, a navigator may use this situation as a trigger to initiate the fallback task, e.g., by taking control manually.

3.2. Identifying information needs

The mapping and assessment exercise performed in this study provided a generic information basis, systematically performed, that has many application areas, e.g., identifying training needs, establishing HMI requirements, function and task distribution between human and system, criteria for initiating fallback, and establishing procedures. For our purposes, we focused on establishing the performance requirements to highlight the expected behaviour of the navigator and/or CAS in future autonomous collision avoidance situations. The advantage of this approach is that expectations to performance can be established without considering how the performance can be achieved; we have simply described what we expect the CAS and navigator to perform, using an approach based on task analysis. The next step, therefore, is to use the information from the task analysis to define how effective performance can be supported. This can be achieved by analysing the task steps from the task analysis and define the information needed to perform the tasks.

As described in table 2, the performance requirement for rule 7 for the navigator in the supervision role was *"When underway, the navigator shall continuously be able to verify whether CAS has determined the presence of collision risk"*. The question is, how is the navigator able to perform this task? What information is needed to support the navigator in effectively verifying the CAS' performance? A starting point is the performance requirement of the CAS. Here, the requirements to the CAS are described such that it can meet the description for rule 7. As can be seen in table 2, the performance requirement includes statements such as *"by all available means"*, *"as early as possible"*, *"systematic observation of detected objects"*, *"independent and reliable sources"*, etcetera.

These statements, in addition to the information captured in the interviews, were used to identify task steps, and subsequently information needs or the navigator. For example, to be able to verify that CAS is performing a systematic observation of detected objects, the CAS should display the objects it has detected. Furthermore, as the rule states that collision risk shall not be based on scanty information, it follows that some kind of reliability information should be provided about CAS' sensor information. In all, the information needs identified through this analysis, should allow an navigator to have sufficient information to supervise the CAS and assess whether its performance is sufficient or whether a fallback should be performed.

The information needs identified for the supervision case can be extended to specific criteria for when CAS' performance is not deemed sufficient. Table 3 provides the example of rule 7 where criteria (task cues) are described based on which the supervising navigator can decide to initiate fallback, and what the fallback task is. In this case, when targets are not detected, or collision estimations are not performed, or collision risk is not estimated sufficiently, the navigator may decide to take over manually. Likewise, the system may provide alarms, or provide the supervising navigator with a take-over request upon which fallback can be initiated.

Table 3. Tasks and information needs for rule 7 "Determining risk of collision".

	Operator-assisted		**Operator-fallback**	
Rule	**Task steps**	**Information needs**	**Task cue**	**Fallback task**
7a	Be able to: - Verify detected objects by CAS - Verify if CAS has observed a risk of collision	- Which sensors are available to CAS - Which sensors are used by CAS - Which vessels/objects form a collision risk	- Insufficient sensors available to CAS - Incorrect sensors used by CAS - Collision risk not determined - Request to intervene from CAS - CAS critical alarm	Determine risks manually
7b	Be able to: - Verify that CAS is using relevant sensors - Verify that systematic observations of detected objects occur - Verify that relevant targets are detected	- Detected objects in the short to long range - Plotting of targets - Estimated risk of collision	- Targets detected too late - Targets not plotted - Collision risk not determined	
7c	Be able to: - Verify the reliability of CAS' sensor information	- The reliability of sensor information	- Sensor information unreliable	
7d	Be able to: - Verify CAS' collision risk prediction methods, e.g., the use of compass bearing	- The reliability of collision risk estimates - Changes to compass bearing of target vessel over time - Target vessel type and size	- Collision risk not determined - Collision risk unreliable - Target vessel type and size incorrect	

4. Discussion and further work

This study aimed to systematically map and assess the performance requirements for collision avoidance two cases: a manual case and an autonomous collision avoidance case with human supervision. This study demonstrated that by performing a systematic analysis of tasks, with input from COLREGs, procedures, navigators, and observations of collision avoidance manoeuvres, performance requirements could be established. A further analysis, for a hypothetical autonomous collision avoidance system including human supervision and fallback, established performance requirements and information needs for effective supervision and fallback decisions.

In the baseline case, the navigator interacted directly with the vessel and its environment. As there is no autonomous collision avoidance system, there are no verifications and fallback decisions to be made by the navigator. This means that the decisions regarding collision avoidance manoeuvres were the result of the navigator's information processing capacities, decision-making strategies and task execution only [20–22]. However, for the supervised autonomy case, control is executed by a collision avoidance system with the navigator as supervisor. Although direct control is possible, e.g., through manual intervention, this is only performed when the system is malfunctioning, when it is outside its operational design domain or at the navigator's discretion.

As the navigator is physically located on the bridge, the navigator can directly perceive the environment and manoeuvre the vessel, as per the baseline case. For the supervision case, the tasks to be performed in verifying CAS' performance, are a comparison between the system's proposed conflict avoidance solution, and the navigator's solution. That is, the navigator's task is about deciding which of the two options, automated or manual control, is the best option, e.g., is the safest, fastest, or least disruptive. The amount of effort to perform this verification depends, amongst others, on the level of expertise of the navigator and the level of experience with the collision avoidance system. For example, a navigator experienced working with the system may recognize patterns in the information provided by the automation and quickly be able to assess its performance, thereby calibrating his/her reliance on the system [23].

Although not discussed in the example for rule 7 above, an interesting finding from the interviews was that navigators, based on their experience with other highly automated systems, state they trust a system as long as it behaves like they would have done. In other words, navigator expect to be able to evaluate the performance of the system by assessing the vessel's physical behaviour. If the vessel does not move as they expect this would be considered a cue for fallback. This seems to imply that a "compatibility" between human and system behaviour could be an important element in building trust.

This may have important consequences for completely unmanned autonomous vessel supervised from a remote location. Although not part of this study, for these cases the operator is likely to be completely dependent on the information transferred from the vessel and will likely not be able to perform independent verification of the vessel's input parameters, information processing, decisions, and actions. As such, this may have consequences for how effective supervision can be performed and how trust in the system can be attained. Possibly, a form of "cognitive compatibility" can assist where, as long as the system processes its information and comes up with decision outcomes in a way that is compatible with human information processing, human performance may be maintained [24].

Nevertheless, overreliance on such a system may result in failures of monitoring or decision biases. That is, misuse of automation is using automation when the task could better be performed by a less automated alternative [25]. Conversely, disuse is the neglect or underutilisation of automation. Not using automation in cases where it would improve the likelihood of successful performance may have severe consequences, e.g., in situations with reduced visibility where the system may perform better than humans (rule 19). Avoiding automation misuse and disuse therefore become important elements for future use of automation in autonomous shipping. Proper information provision and scenario training have been proposed to mitigate these effects [26]. Also, transparency of the system may be important to allow for understanding of the automation's information processing, decision making and future actions [26–28]. As such, further work will focus on detailing the navigator's information needs and the corresponding system's transparency requirements necessary to enable proper supervision and human fallback.

Acknowledgements
The authors would like to thank the ferry operators and their navigators for the opportunity to perform interviews and observations.

References

[1] Kretschmann L, Burmeister H C and Jahn C 2017 Analyzing the economic benefit of unmanned autonomous ships: An exploratory cost-comparison between an autonomous and a conventional bulk carrier *Research in Transportation Business and Management* **25** 76–86
[2] EMSA 2020 *Annual Overview of Marine Casualties and Incidents* (Lisbon: European Maritime Safety Agency)
[3] Wróbel K 2021 Searching for the origins of the myth: 80% human error impact on maritime safety *Reliab Eng Syst Saf* **216** 107942
[4] Wróbel K, Montewka J and Kujala P 2017 Towards the assessment of potential impact of unmanned vessels on maritime transportation safety *Reliab Eng Syst Saf* **165** 155–69
[5] Fan C, Wróbel K, Montewka J, Gil M, Wan C and Zhang D 2020 A framework to identify factors influencing navigational risk for Maritime Autonomous Surface Ships *Ocean Engineering* **202**
[6] Ramos M A, Utne I B and Mosleh A 2019 Collision avoidance on maritime autonomous surface ships: Operators' tasks and human failure events *Saf Sci* **116** 33–44
[7] IMO 1977 *Convention of the international regulations for preventing collisions at sea (COLREGS)* (Montreal, Canada: International Maritime Organisation)
[8] Porathe T 2019 Safety of Autonomous Shipping: COLREGS and Interaction between Manned and Unmanned Ships *Proceedings of the 29th European Safety and Reliability Conference (ESREL)* Proceedings of the 29th European Safety and Reliability Conference (ESREL) pp 4146–53
[9] Bainbridge L 1983 Ironies of automation *Automatica* **19** 775–9
[10] Strauch B 2018 Ironies of Automation: Still Unresolved After All These Years *IEEE Trans Hum Mach Syst* **48** 419–33
[11] Society of Automotive Engineers 2021 *Taxonomy and Definitions for Terms Related to Driving Automation Systems for On-Road Motor Vehicles*
[12] DNV 2014 *Barrier management in operation for the rig industry - Good Practices*
[13] Kirwan B and Ainsworth L K 1992 *A Guide to task analysis* (London; Washington, DC: Taylor & Francis)
[14] Ludvigsen J T, van de Merwe K, Klemsdal le-Borgne E and Teigen T 2018 Applying an operational safety barrier framework in a major oil and gas field development project *Safety and Reliability – Safe Societies in a Changing World* 28th European Safety and Reliability Conference, ESREL 2018 (Trondheim, Norway: CRC Press)
[15] Rødseth Ø J, Lien Wennersberg L A and Nordahl H 2021 Towards approval of autonomous ship systems by their operational envelope *J Mar Sci Technol*
[16] Stitt I P A 2002 The COLREGS – Time for a Rewrite? *The Journal of Navigation* **55** 419–30
[17] Belcher P 2002 A sociological interpretation of the COLREGS *The Journal of Navigation* **55** 213–24
[18] Weber H 1995 Clarification of the Steering and Sailing Rules of the COLREGS *The Journal of Navigation* **48** 289–92
[19] MITRE 2018 *Human-Machine Teaming Systems Engineering Guide* (MITRE Corporation)
[20] Klein G A and Calderwood R 1991 Decision models: some lessons from the field *IEEE Trans. Syst., Man, Cybern.* **21** 1018–26
[21] Lintern G 2010 A Comparison of the Decision Ladder and the Recognition-Primed Decision

Model *J Cogn Eng Decis Mak* **4** 304–27
[22] Rasmussen J 1986 *Information processing and human-machine interaction: an approach to cognitive engineering* (New York: North-Holland)
[23] Kraus J, Scholz D, Stiegemeier D and Baumann M 2020 The More You Know: Trust Dynamics and Calibration in Highly Automated Driving and the Effects of Take-Overs, System Malfunction, and System Transparency *Hum Factors* **62** 718–36
[24] Hilburn B, Westin C and Borst C 2014 Will Controllers Accept a Machine That Thinks Like They Think? The Role of Strategic Conformance in Decision Aiding Automation *Air Traffic Control Quarterly* **22** 115–36
[25] Parasuraman R and Riley V 1997 Humans and Automation: Use, Misuse, Disuse, Abuse *Hum Factors* **39** 230–53
[26] Beck H P, Dzindolet M T and Pierce L G 2007 Automation Usage Decisions: Controlling Intent and Appraisal Errors in a Target Detection Task *Hum Factors* **49** 429–37
[27] Chen J Y C, Lakhmani S G, Stowers K, Selkowitz A R, Wright J L and Barnes M J 2018 Situation awareness-based agent transparency and human-autonomy teaming effectiveness *Theor Issues Ergon Sci* **19** 259–82
[28] van de Merwe K, Mallam S and Nazir S 2022 Agent Transparency, Situation Awareness, Mental Workload, and Operator Performance: A Systematic Literature Review *Hum Factors*

Levels of autonomy for ships

Ørnulf Jan Rødseth[1,*], Lars Andreas Lien Wennersberg[1], Håvard Nordahl[1]

[1]SINTEF Ocean, Trondheim, Norway

[*] Corresponding author, e-mail: OrnulfJan.Rodseth@sintef.no

Abstract. This paper gives a summary of previously published papers on the definition of autonomy for ships, how this relates to different crewing regimes, and the terminology to be used. A conclusion is that autonomy should be retained as a descriptive term, but that we should distinguish between "full autonomy" and "constrained autonomy", where the latter is the more relevant term for ships today. The proposed classification of autonomy is related to both degree of automation and degree of human control and will be presented as a matrix with generic classes of autonomy. This matrix is also transformed to a set of more practically useful levels of autonomy based on likely organization of crew on land or on the ship. The paper has mainly been written based on our work with maritime autonomous surface ships (MASS) but is also applicable to other types of surface vessels, e.g. inland waterway vessels.

1. Introduction

Self-guiding mobile robots have a long history and one of the first examples was developed as early as in the 1950's [1], although the term "autonomous" was not used in that paper. The term "autonomous robot" seems to appear around the 1970's [2] and one can assume that the discussions on the meaning of autonomy in the context of robotics started soon after. This discussion is ongoing, and the current proposal in the car industry is to depreciate the term "autonomous" and use "driving automation systems" instead [3]. Section 2 will give some background to this discussion and provide a rationale for continuing to use the term "autonomous ship". This includes a more detailed definition of both automation and autonomy in the context of ships.

The next point that often cause confusion is the relationship between ship autonomy and crew presence, either on the ship itself or in a remote control center (RCC). Section 3 explains why human oversight will be necessary for many years to come. Different combinations of crew attendance on the ship or in the RCC, and the corresponding required capabilities in the automation system will be classified. The term "operational envelope" will also be explained as an alternative to the corresponding term "Operational Design Domain" in the car industry.

The final subject that will be discussed is the levels or degrees of autonomy. Section 4 will propose a classification scheme, which includes the concept of "constrained autonomy". This refers to a combination of automation and human intervention that overcomes some of the problems related to a safe human-automation interface. This classification scheme will also be developed into a list of more application-oriented levels of autonomy.

Section 5 gives a brief example of assignment of autonomy levels and section 6 provides a summary of conclusions and a brief outlook on further research.

2. Autonomous or automated?

Autonomy versus automation has been discussed extensively in the literature and many different proposals have been made for definitions of both terms [4].

The word "autonomy" is derived from the Greek word "autonomia" which means independent, or more literally, living by one's own laws [5]. In engineering, the term has been used to describe the ability of an engineering system to make its own decisions, without the need for the involvement of an exogenous system or operator [6]. However, this is in practical terms indistinguishable from the definition of automation: "pertaining to a process or equipment that, under specified conditions functions without human intervention" [7]. Several authors have suggested additional criteria to distinguish automation from autonomy, but there are weaknesses with many of these approaches [4].

The ambiguity of the terms has led the car industry, represented by the Society of Automotive Engineers [3] and the British Centre for Connected & Autonomous Vehicles [8], to suggest that the term "driving automation" should be used instead of autonomy. On the other hand, the International Maritime Organization (IMO) uses the term autonomous, as e.g. in "Maritime Autonomous Surface Ships" (MASS). The inland waterway community, represented by the Central Commission for the Navigation of the Rhine (CCNR), follows the SAE example, and refers to "levels of automation" [9].

In [4] it is concluded that autonomy as a term is indeed relevant for ships and that it is possible to provide definitions that clarify the distinction between automation and autonomy. An important argument in favor of this is that "autonomy" is already being used extensively, e.g. by IMO and the maritime industry. Furthermore, the term autonomy can be used to describe ship automation functions that requires a new approach to approval because it:

1. Implements automated control and monitoring responsibilities that are today allocated solely to humans and where specialist training is required as, e.g. specified in the STCW code [10].
2. Is designed to operate safely without human supervision for certain periods or under certain circumstances, which is not allowed today.

As automation already is common on ships and inland vessels, e.g. as autopilots or engine automation, it is necessary to define the basic difference between today's automation and autonomy. The main points to consider are:

1. Ships needs many different processes to function, e.g. energy production, stability, fire safety and navigation. Not all need to have the same level of autonomy.
2. Ship processes can have different levels of autonomy during different voyage phases, e.g. open sea versus port navigation or lock passing.
3. As will be discussed in section 3, it is highly likely that autonomous ships will be supervised by humans. These humans may be onboard or in the RCC, so autonomy does not imply that the ship needs to be uncrewed.

In the context of autonomous ships, the following definition of automatic has been developed by ISO [11]: "*process or equipment that, under specified conditions, can function without human control.*" This says that automation <u>can</u> control the relevant process, but it may not be allowed or approved to do so, without continuous supervision. This can be exemplified by an autopilot that safely can steer the ship for hours or days, as long as there is a human present, that can disable the autopilot to take evasive manoeuvres, when necessary.

Arguably, the most significant change from today's automation to autonomous operation is that autonomous processes must be approved to be operated without human supervision or intervention. This has led to the following definition for autonomy: "*In the context of ships, autonomy means that one or more of a ship system's processes or equipment, under certain conditions, is designed and verified to be controlled by automation, without human assistance*"[11]. They key phrase here is <u>designed and verified</u> to emphasize the need for formal approval of the automation system to operate without human supervision or control.

3. Uncrewed, remote controlled, or automated ship?

In popular literature, autonomous ships evoke idea of ships sailing completely without human control and with no crew, neither onboard nor in the RCC. In reality, the picture is much more complex. None of the known MASS projects aim for full autonomy today. Although many are aiming for uncrewed operation of the ships, all will use an RCC for supervision.

The reasons for this are usually the result of a cost-benefit assessment [12]:

1. Ships are few, large, and expensive, and the relative cost of using an RCC is small compared, e.g. to that of an autonomous car.
2. The high value of the asset as well as potentially severe consequences from any accidents, will make owners very reluctant to leave them unsupervised.
3. Sensors and object classification can in principle be developed to an arbitrary high reliability, but at a cost. Once the RCC is in place, it will be more cost-effective to use the operators to handle the rare and difficult cases.

However, there are also reasons why automation may not be able to handle all problems [13]:

4. In encounters with crewed ships, it will be very challenging, if not impossible, to develop automation that sufficiently well can predict what the other ship will do in all situations. Until new regulation and technology is developed, a human is in most cases needed to handle situations that are outside the automation's capabilities.

Furthermore, ships move relatively slowly, and often in open and uncongested waters. In such cases, difficult situations can be detected in time for a human operator to be alerted to the need for intervention, and to allow the operator to get sufficient situational awareness to take a safe action to avoid or correct the situation. The general operating conditions also mean that the need for intervention is a relatively rare occurrence. This means that a team of RCC operators can operate more than one ship each. In the MUNIN project, experiments indicated that one operator can supervise six ships [14], but this requires a specialized expert team that can take over control in situations that are too complex or time-consuming for the first-line operators.

Some important benefits of uncrewed ships are increased cargo capacity, capital cost savings, and energy saving from removing the accommodation section [15]. This benefit is obviously relatively larger the smaller the ship is. For large ships, the benefit will be small, so autonomous ship projects seen today are all related to small ships on shorter voyages. Also, on today's larger ships with long voyage durations, crew sizes are probably near the minimum needed to operate and maintain technical systems onboard. Thus, autonomy on larger ships is more likely to be used for decision support, to increase safety, or to allow crew to go to sleep during night-time passages in open sea. Also in these cases, one may want to have an RCC as backup.

This means that there are several combinations of onboard automation, onboard crew, and supervision from RCC that are likely. This is shown by the four rows in Table 1, where respectively automation and/or RCC are used to implement some of the responsibilities that today are handled by the crew. The columns represent different variants of the crew onboard the ship: **Full** is a full normal crew. **Periodically** is crew that can be away from control positions for some time, e.g. when sleeping during night. **Uncrewed** is no crew on the ship at all. Note that the latter may also include the case where there are safety personnel and/or passengers onboard that do not control the normal ship processes, but can be involved, e.g. in passenger safety functions.

Table 1. Onboard crew cooperating with automation and RCC

	Full	Periodically	Uncrewed
Automation	AO	CA	FA
Automation & RCC	AO & RS	CA	CA
RCC	RS	RC	RC
Crew only	As today	-	-

The codes in the table cells show the most likely operating modes:

- **RS:** Remote supervision from RCC, crew in charge.
- **AO**: Automatic operation of some processes, but crew in charge at all times.
- **RC**: Remote control from RCC, RCC in charge.
- **CA**: Constrained autonomous, crew, RCC, and automation share responsibilities. RCC is needed when the crew is not at the control position.
- **FA**: Fully autonomous. No crew neither on the ship nor in RCC.

These codes are further explained in section 4.

Note that there are other, less likely, operating modes for most combinations that have been omitted from the table, e.g. that RCC do remote control also for a fully crewed ship.

As was discussed above, it can be assumed that very few ships will be fully autonomous. Thus, control of autonomous ships will be a shared responsibility between automation and humans, and the design of the human-automation interface will be an important factor. This is common to a range of other types of "Industrial Autonomous Mobile Robots" (IAMR) that operate in controlled or semi-controlled environments, such as autonomous mining trucks and automatic guided vehicles [16].

The car industry has other challenges. For cost reasons, they cannot easily rely on an RCC, and the high speed of cars and the complexity of the environment makes it difficult to safely hand control over from automation to human if something unexpected happens. This segment has defined the concept of the operational design domain (ODD) as the "*operating conditions under which a given driving automation system or feature thereof is specifically designed to function*"[3],[8]. For the higher automation degrees (3 to 5) defined by SAE, the automation is expected to handle the full ODD. For levels 3 and 4 the ODD is limited, and the driver must handle situations outside the ODD. The driver must also be able to act as a fallback on level 3 if the automation system fails. There have been some discussions on whether this type of interaction between automation and human is safe [17], and this is one reason why it has been proposed that ships may need an alternative to the ODD, which has been called the "operational envelope" [12].

Another reason to use an operational envelope instead of ODD is that autonomous ships almost always will use a combination of automation and human control. Thus, the operational envelope is defined as "*conditions and related operator control modes under which an autonomous ship system is designed to operate, including all tolerable events*", where "autonomous ship system" is defined as "*all elements that interact to ensure effective functioning of the autonomous and non-autonomous processes and equipment that are necessary to perform the ship's operation or voyage*" [11].

The definition of the operational envelope also requires a definition of a fallback state which is a "*designed state that can be entered through a fallback function when it is not possible for the autonomous ship system to stay within the operational envelope*".

This allows an integrated description of automation and crew responsibilities in the different situations the ship is designed to encounter [12]. Developments are also underway to find methods where these concepts can be integrated in a more formal description of autonomous ship system capabilities and how this can be used in approval [18].

4. How to define levels of autonomy?

There are at least as many proposals for *levels* of autonomy as there are definitions of autonomy [19]. This section will use the definition of autonomy as presented in section 2, and the interaction between human and automation as discussed in section 3 to describe a general classification scheme for ship autonomy. More details, and the background for the description, can be found in [12].

The classification scheme is based on four degrees of respectively automation and human control, as shown in Table 2 on the next page.

The degree of human control describes the level of attention that the operator is giving to the operation at hand and is specified as the maximum time the operator needs to reach the control position, gain situational awareness and be ready to perform actions to maintain safety (T_{MR} – maximum response time).

The degree of automation describes the automation systems ability to be in control of a given operation, and to maintain safety for a specified future time (T_{DL} - response deadline), without any human assistance.

Table 2. Degrees of automation and human control

Degree	Automation	Human control
0	Low, $T_{DL} \approx 0$	None, $T_{MR} = \infty$
1	Partial, $T_{DL} > 0$	Available, $T_{MR} >\approx 20$min
2	Constrained, $T_{DL} > t$	Discontinuous, $T_{MR} >\approx 1$min
3	Full, $T_{DL} = \infty$	Continuous, $T_{MR} \approx 0$

The main idea behind **constrained** automation is that even if human intervention is sporadically needed, this can be made much safer and operator friendly if the automation system is able to issue an alert before an action by the operator may be needed. This requires that the automation system can determine T_{DL}, which in turn requires that the capabilities of the automation system is well defined, and possibly constrained to functions where T_{DL} can be reliably measured. When this is combined in a system where the automation system also knows the operators' maximum response time T_{MR}, the automation can issue an alert when $T_{DL} \leq T_{MR}$. This type of cooperation between human and automation is called *constrained autonomy* [18]. An example of how this could work is that the automation system has a limited set of pre-defined operator modes, each associated with a known T_{MR}. Some examples with the corresponding human control mode from Table 2 in parenthesis could be "on bridge, in control" (Continuous), "on bridge, not supervising" (Discontinuous), or "not on bridge, sleeping" (Available).

Partial automation also has a non-zero T_{DL} but differs from constrained automation in that T_{DL} cannot be determined by the automation system. This makes it impossible to issue an alert to the operators. Operators can still use their own judgement as to if, and for how long, they can leave the control position, but still need to return often enough to ensure that the ship is still safe. This puts additional strain on the operators and makes it impossible, e.g. to allow operators to go to sleep during nigh.

The actual value of T_{MR} will depend on the organization of the RCC and the crew. There are many cases where it is useful to use more or fewer categories for T_{MR} than shown in Table 2.

The categorization used in Table 2 allows the definition of four useful classes of human-automation interfaces as illustrated in Figure 1. Here columns represent human control degrees and rows automation degrees. The unlabeled cells represent combinations that cannot be sustained.

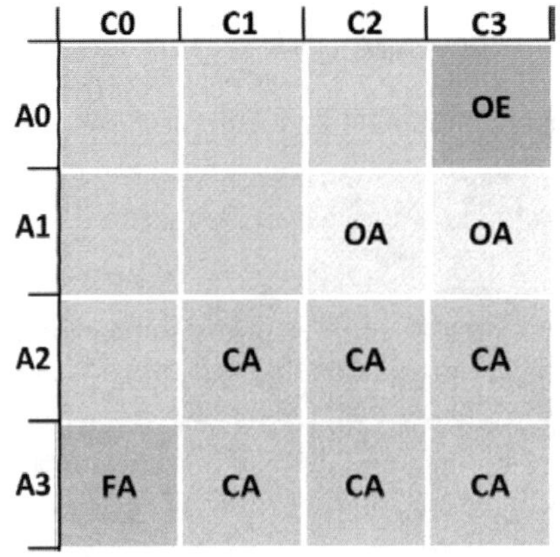

Figure 1. General classification of operations

The labeled cells represent four different classes of viable human-automation interfaces:

- **OE** (Operator Exclusive): Limited automation, operator must be always present.

- **OA** (Operator and Automation): Automation can control the system, but with continuous attention from the crew. The crew must use their own judgement as to how far away from the control position they can be.
- **CA** (Constrained Autonomy): Automation can control the process for a known period without human attention. The human will be alerted in time to get safely back to control, when needed.
- **FA** (Full Autonomy): Automation can control the process for as long as needed without human attention.

This diagram is independent of if crew are in an RCC or on the ship, but one would assume that C1 is the most useful for onboard crew, while C2 may be most appropriate for RCC. Note that FA could have been used in the whole bottom row. However, as humans are in a monitoring or control function for C1 to C3, these are classified as CA.

This classification scheme provides a systematic and unambiguous definition of how humans and automation interact, e.g. for use in system documentation and approval. It may, however, be less useful in describing the actual implementation of automation in a specific autonomous ship system. This will require more details on the definition of time constants T_{MR} and T_{DL}, and how the RCC and onboard crew are organized. For this purpose, one may want to use a more descriptive scale. Several different methods have been proposed as the basis for such a scale [19] but here we will briefly describe a scale based on the one proposed by the Norwegian Forum for Autonomous Ships [20]. This tries to catch the most relevant combinations of automation and human control, which could be called levels of autonomy. Each level has got an abbreviated form as well as a full name and the code in parenthesis corresponds to the autonomy class codes used in Fig. 1:

- **DC: Direct control** (OE): This is the situation on ships today where the crew has full control of the ship and uses relatively simple automation and decision support functions.
- **AO: Automatic operation** (OA): Examples of this could be dynamic positioning, automatic berthing, or automatic crossings where automation performs operations under continuous supervision.
- **RS: Remote supervision** (OE/OA): A ship with conventional crew, and with DC or AO autonomy level onboard, is continuously supervised from shore, e.g. for increased safety.
- **RC: Remote control** (OA): In this case the ship would be remotely controlled from shore, either all the time or, e.g. during night time. This normally uses the AO autonomy class, as OE is more work intensive for the RCC operators, and one will want to avoid this, if possible.
- **PU: Periodically unattended** (CA): The ship can steer itself automatically for extended periods, e.g. in open waters and calm weather. Crew is available onboard to handle more complex situations, but can be away from the controls and, possibly at sleep during nighttime.
- **CA: Constrained autonomous** (CA): Uncrewed operation with constrained autonomy onboard but with operators in RCC that can handle more complex situations. This corresponds to PU onboard.
- **FA: Fully autonomous** (FA): The ship handles all foreseeable situations by itself and there is no crew neither on ship nor in RCC. This is not very realistic today, except in very simple cases.

There are many more combinations, but the above examples are the most relevant today, as one will normally optimize the ship and RCC crew responsibilities to the capabilities of the automation system.

5. An example

To illustrate the concepts described in the previous section, we will look at a hypothetical example: A container feeder vessel trading between the cities Stavanger and Bergen on the west coast of Norway, and Rotterdam in the Netherlands. The voyage is illustrated in Figure 2.

The labels show a simplification to six general operations or voyage phases:

1. Port operations in Rotterdam, including movement between terminals, cargo operations etc.
2. Departure from or arrival to port area Rotterdam, possibly including e.g. waiting or anchorage.

3. Open sea passage.
4. Arrival to or departure from Stavanger or Bergen, relatively small ports with limited traffic.
5. Port operations in Stavanger or Bergen.
6. Sheltered water transit between Bergen and Stavanger. Some narrow passages and significant local traffic.

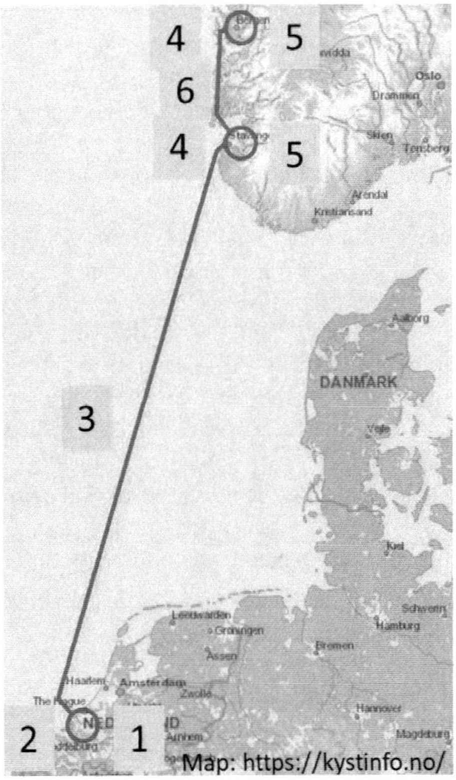

Figure 2. Example route

We can then consider two realizations of this example. One is a crewed vessel where autonomous control is used onboard to assist crew and to allow them to go to sleep during sea passage at night-time. The second is an uncrewed ship with supervision from a continuously manned RCC.

Table 3 shows an example of likely levels of autonomy for three different function groups through the six voyage phases and for the two different crewing options. The leftmost column shows the voyage phases (VP). To the right of that are two groups of three different functions for respectively the crewed ship and the RCC controlled ship. The three function groups are **stability**, e.g. ballasting, water ingress and stability monitoring; **energy**, e.g. main and auxiliary engine operations; and **navigation**, e.g. outlook, position fixing and maneuvering. The cells are labeled with the level of autonomy as suggested at the end of section 5. The shaded cells indicate where constrained autonomy can be used, which corresponds to PU (periodically unattended) for the crewed ship and CA (constrained autonomous) for the RCC operated ship.

On the crewed ship, stability will probably be automated, but supervised during port operations, and using constrained autonomy otherwise. Energy production will likely be using constrained autonomy in all phases. However, dependent on the type of cargo handling equipment, it may be on a lower level during port operations. Navigation will use constrained autonomy during the deep-sea passage. Here, traffic is relatively low and ships far apart and easy to detect on radar. Thus, the automation systems can take care of most situations and have time to alert a sleeping crew if the situation becomes difficult. In ports one probably wants to use direct control and in sheltered transit crew supervised automatic operation.

Table 3: Autonomy levels

	Crew sleeping, $T_{MR} \approx 20$ minutes			RCC operators, $T_{MR} \approx 1$ minute		
VP	**Stability**	**Energy**	**Navigation**	**Stability**	**Energy**	**Navigation**
1	AO	PU	DC	RC	CA	RC
2	PU	PU	AO	CA	CA	CA
3	PU	PU	PU	CA	CA	CA
4	PU	PU	AO	CA	CA	CA
5	AO	PU	DC	RC	CA	RC
6	PU	PU	AO	CA	CA	CA

On the RCC controlled ship, the automation degrees are similar, although the autonomy levels have different names.

An important difference between crewed and RCC operations is that all navigation phases are moved to a higher automation degree when the RCC in charge. The more extensive use of constrained autonomy is possible because the RCC crew will have a lower T_{MR} than a sleeping crew. This means that automation can be relied on in more operations, as human assistance is more easily available. The operations inside the port will also generally need a higher automation level (AO) than for the crewed ship (OE). This is to keep the workload for the RCC operators at a reasonable level.

Note that this example does not consider need for technical maintenance, equipment interventions, or similar issues.

6. Summary and conclusions

Autonomy and autonomy levels are being discussed in many different transport modes, where ships belong to the group of "industrial autonomous mobile robots". Ships are characterized by being large, expensive, and slow moving in environments with relatively little traffic. For this reason, autonomous ships will normally have humans involved in the control of the ship. This means that it is more important to focus on the effective cooperation between humans and automation, rather than on creating automation that can control the ship fully independent of humans.

The car is very different: Cars operate in very complex environments at high speeds, and the main target is fully automatic control. While the car industry uses the ODD as the format for describing the automation system's capabilities, it is proposed to use the "operational envelope" for autonomous ship systems, where also human responsibilities are included in the envelope.

As was discussed in section 2, the term autonomy is a useful concept when applied to ships, but it must be given a very clear definition to distinguish it from automation. Our proposal is to base the differentiation on whether the automation is *designed and verified* to operate without human assistance. However, this raises a new question: How do you design and verify automation to qualify as autonomous? One possibility is to aim for a *constrained* autonomous system. This means that functionality is limited to that which the automation system can reliably assess in terms of complexity related to its own capabilities. By careful design, this will allow the automation system to alert the operator in time to take over control, when complexity is about to exceed capabilities.

Constrained autonomy can be used to construct a cooperative human-automation system, where the human is warned about situations where intervention may be needed. This avoids that the human continuously must monitor the ship and its environment to detect any developing problem. This, in turn, allows us to build complete systems with automation, onboard crew, and RCC crew that effectively and safely can control autonomous ships.

SINTEF Ocean is currently engaged in several research projects that are testing out the concepts of the operational envelope and constrained autonomy on a selection of use cases. These projects are also investigating how UML (Unified Modelling Language) can be used to capture the ideas of the operational envelope and the different classes of autonomy in a more formal way than, e.g. the "concept of operation" (CONOPS) that is being suggested as the basis for approval by several classification societies and administrations [18].

Acknowledgements

The work presented in this text has received funding from the European Union's Horizon 2020 research and innovation programme under grant agreement No 815012 (AUTOSHIP) and grant agreement No 859992 (AEGIS).

References

[1] Walter, W. G. (1951). A machine that learns. Scientific American, 185(2), 60-64.

[2] Fu, K. (1971). Learning control systems and intelligent control systems: An intersection of artificial intelligence and automatic control. IEEE Transactions on Automatic Control, 16(1),

70-72.
[3] SAE J3016 (2021) Taxonomy and definitions for terms related to on-road motor vehicle automated driving systems, Ed. 2. SAE International.
[4] Rødseth, Ø. J., & Vagia, M. (2020, November). A taxonomy for autonomy in industrial autonomous mobile robots including autonomous merchant ships. In IOP conference series: materials science and engineering (Vol. 929, No. 1, p. 012003). IOP Publishing.
[5] Online (2022), Online Etymology Dictionary, Retrieved January 2022 from https://www.etymonline.com.
[6] Albus, J., Antsaklis, P., Meystel, A., Passino, K., Samad, T. (1998). Setting the stage: some autonomous thoughts on autonomy. In: IEEE ISIC/CIRA/ISAS Joint Conference.
[7] IEC (2013) IEC 60050-351, International Electrotechnical Vocabulary - Part 351: Control Technology.
[8] BSI (2020), BSI PAS 1883:2020 – Operational Design Domain (ODD) taxonomy for an automated driving system (ADS) – Specification, The British Standards Institution 2020.
[9] CCNR, Automated Navigation – Definitions of levels of automation in inland navigation, Resolution 2018-II-16.
[10] IMO (1978), International Convention on Standards of Training, Certification and Watchkeeping for Seafarers, 1978 – as amended.
[11] ISO (2022), ISO/PRF TS 23860, Ships and marine technology – Vocabulary related to autonomous ship systems, Final text for proof-reading.
[12] Rødseth, Ø.J., Nordahl, H., Wennersberg, L.A.L, Myhre, B., Petersen, P. (2021) Operational Design Domain for Cars versus Operational Envelope for Ships: Handling Human Capabilities and Fallbacks, Proceedings of the 31st European Safety and Reliability Conference, ISBN 9789811820168.
[13] Rødseth, Ø.J., Wennersberg, L.A.L, Nordahl, H. (2021), Improving safety of interactions between conventional and autonomous ships, Proceedings of the 1st International Conference on the Stability and Safety of Ships and Ocean Vehicles, 7-11 June 2021, Glasgow, Scotland, UK.
[14] MacKinnon, S.N, Man, Y., Baldauf, M. (2015), Final Report: Shore Control Centre, MUNIN Deliverable D8.8. URL: http://www.unmanned-ship.org/munin/news-information/downloads-information -material/munin-deliverables/ (accessed January 2022).
[15] Kretschmann, L., Burmeister, H. C., & Jahn, C. (2017). Analyzing the economic benefit of unmanned autonomous ships: An exploratory cost-comparison between an autonomous and a conventional bulk carrier. Research in transportation business & management, 25.
[16] Vagia, M., & Rødseth, Ø. J. (2019). A taxonomy for autonomous vehicles for different transportation modes. In Journal of Physics: Conference Series (Vol. 1357, No. 1, p. 012022). IOP Publishing.
[17] Inagaki, T., & Sheridan, T. B. (2019). A critique of the SAE conditional driving automation definition, and analyses of options for improvement. Cognition, technology & work, 21(4), 569-578.
[18] Rødseth, Ø. J., Wennersberg, L. A. L., & Nordahl, H. (2021). Towards approval of autonomous ship systems by their operational envelope. Journal of Marine Science and Technology.
[19] Vagia, M., Transeth, A.A. and Fjerdingen, S.A., (2016). A literature review on the levels of automation during the years. What are the different taxonomies that have been proposed? Applied ergonomics, 53, pp.190-202.
[20] Rødseth, Ø. J., & Nordahl, H. Ed. (2017, October). Definitions for autonomous merchant ships. Norwegian Forum for Autonomous Ships, Version 1.

The Importance of documenting autonomous tests

Pauline Røstum Bellingmo[1], Kay Fjørtoft[1], Ulrik Jørgensen[1], Per Erik Kvam[1], Inger Lill Bratbergsengen[2] and Jose Vicente Perello Gisbert[3]

[1]Department of Energy and Transport, SINTEF Ocean AS, Postbox 4762 Torgarden, NO-7465 Trondheim, Norway
[2]Kongsberg Seatex AS, Pirsenteret, Havnegata 9, NO-7010 Trondheim, Norway
[3]TEC-ESN, European Space Agency, Keplerlaan 1, 2200 AG Noordwijk aan Zee, Netherlands

E-mail: pauline.bellingmo@sintef.no

Abstract. This paper presents how autonomous tests can be documented and why this is important. A test area in Norway, more specifically the Trondheimsfjorden Test Area for Autonomous Ships, is used as a pilot to conduct tests with autonomous vessels and for demonstrating the procedure of documenting results. There are typically three stages in such a documentation process; 1) To register and inform about a planned test on the fjord, 2) To inform about ongoing tests and to document test results by collecting data from the vessel and from the sensor infrastructure, 3) To show historical tests and be able to do analytics or conduct learning from previous tests. The Trondheimsfjorden Test Area has been instrumented with communication and navigation infrastructure, a control centre for control and monitor of the install infrastructure and for remote operation of a ship, and a data centre for planning autonomous tests, storing data, and for sharing of test results. By documenting test results in a standardized format, this can be used to verify new technology and solutions, share knowledge and experiences, and for documentation procedures and guidelines used for the purpose. A demonstration held in The Trondheimsfjorden Test Area showed the importance of streamlining the process of conducting autonomous test and documenting them in a standardized format. This work is based on the results from the research project NAVISP-EL3-005 "Trondheimsfjorden Test Area for Autonomous Ships". The Navigation Innovation and Support Programme (NAVISP) is the programme of the European Space Agency to support the competitiveness of the European industry in the wide field of positioning, navigation and timing while supporting member states in enhancing national objectives and capabilities in the sector.

1. Introduction

Even with a slower grow in transport demand than expected before the Covid-19 pandemic, the total transport activity is still expected to more than double from 2015 to 2050 [1]. Maritime transport is today the main sector regarding tonnage transport. Autonomous shipping is one promising element in future transport, as it is counted to be safer and more sustainable than road-based transport. Sailing unmanned, the autonomous vessels will be lighter by reducing hotel loads for humans and can reduce the speed compared with conventional shipping without increasing crew and management cost. This can lower the energy consumption of the vessel and increase the cargo to weight ratio. Additionally, autonomous shipping is expected to have lower operating costs than traditional shipping, e.g., by reducing the number of crew onboard and creating more space for goods to carry. Operators at a control centre can also manage several vessels in parallel, which will be an added value regards cost. The autonomous shipping sector is still at an early stage, but many ongoing projects are developing innovation and technology. The technology needs to be tested and verified before it can be applied on board vessels. A test area can be used for such a purpose, where typically navigation and communication technology can be tested in a small scale to achieve a higher technology readiness level.

Another challenge regarding the implementation of autonomous vessels is to develop the regulations in parallel with the technological innovation. The International Maritime Organization's (IMO) has conducted a Regulatory Scoping Exercise for use of Maritime Autonomous Surface Ships (MASS) identifying gaps in the current regulations [2], For instance, the regulations are developed with the assumption of having a master onboard the vessel. A master is responsible for the vessel, its operations, and the crew and cargo at the vessel. The introduction of autonomy will require new regulations since the master might not be onboard the vessel (unmanned).

The general situation of today's testing regime is that plans and results from autonomous tests are not shared, they are in best cases described in a power-point presentation after the tests have been finalised where the results only are available internally. A reason for this is that the test infrastructure is usually experimental, types of tests between the sites may differ significantly, and the users are developing new technology that is part of the innovation process before a product is ready for commercialization. Moreover, it is hard to compare results from different tests. Another drawback is that procedures for documenting results are not well established.

In this paper we seek to identify the importance of documenting autonomous tests. Documented test results can be used to test and verify the quality of service of technology and solutions for maritime operations, and to give input to the development of the regulations in parallel with the technological development. As different actors are testing new technology and solutions for autonomous vessels, the test results should be standardized and categorized making it easy to compare results. The knowledge and experience regarding autonomous innovation should be shared, since this will accelerate the technological and regulatory development, moving one step closer to a higher level of autonomy.

To demonstrate how tests can be conducted and documented, the Trondheimsfjorden Test Area for Autonomous Ships (TTA) located in Norway has been used as a pilot. This test area is instrumented with sensor infrastructure, control centres, and a platform for documenting and reviewing previous tests and historical data. In this way, TTA will play a key role in the contribution of autonomous shipping development and can be considered an important tool and laboratory for the maritime industry when developing technological innovation.

2. Background

2.1. Autonomous Tests

In this paper, the term "autonomous tests" include tests of both vessel-related systems and infrastructure related to autonomy. For instance, this can involve testing different levels of autonomy for a vessel. The IMO has defined four degrees of autonomy [2]:

(i) Degree One: Ship with automated processes and decision support.

(ii) Degree Two: Remotely controlled ship with seafarers on board.

(iii) Degree Three: Remotely controlled ship without seafarers on board.

(iv) Degree Four: Fully autonomous ship.

Another example of an autonomous test can be testing communication infrastructure. The duration of the tests can range from a few minutes to several weeks or years.

2.2. Test Areas for Autonomous Ships

International Network of Autonomous Ships (INAS) has defined a test area as "a facility for doing in situ tests of MASS and MASS systems which may or may not be open for any user" [3]. A test area's main purpose is to facilitate for tests, to offer a reliable site equipped with technological infrastructure and a data storge platform for documentation, and to assure that a test can be done together with conventional operations and are following legal regulations. A test area can include both land- and space-based sensor technology that can be used for navigation, communication, and monitoring. For instance, radars installed onshore can provide vessels and traffic surveillance with improved situational awareness. Additionally, a test area may be an information centre providing historical data from e.g., traffic, weather, sensors in the area, and test results.

In order to test and verify new technological capabilities and applications, several test areas for autonomous ships have been developed [3]. They exist in several countries, e.g., Norway, the United Kingdom, Finland, Belgium, and the USA to mention some. The responsible organizations of these test areas seek cooperation for the development of autonomous vessel, and are all members of INAS, a network for sharing information on autonomous ships. INAS actively gives inputs to authorities for the development on standards and regulations for MASS.

One of these test areas is the TTA that was established in 2016 as the world's first test area for autonomous ships [4]. The test area has been approved by Norwegian maritime authorities for test purposes. It is an area covering more than 17 000 km2 and covers open sea to inner fjord. The test area is located in the center of Norway, see Figure 1.

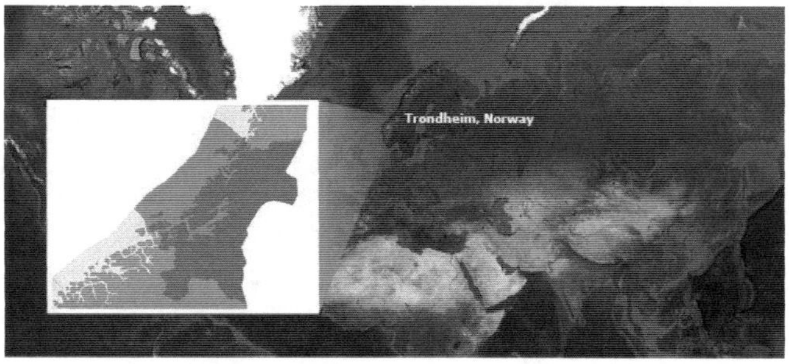

Figure 1: Trondheimsfjorden Test Area for Autonomous Ships in Trondheim, Norway marked as red.

2.3. Guidelines for MASS Tests

There exist several guidelines for tests of MASS. The IMO's Maritime Safety Committee (MSC) has approved interim guidelines for MASS trials (MSC.1/Circ. 1604)[5]. The guidelines are made to assist authorities and relevant stakeholder to ensure that MASS trials are conducted safely, securely, and protects the environment. Taking into consideration and complementing IMO's Interim Guidelines for MASS trials, the EU Operational Guidelines for Trials of Maritime Autonomous Surface Ships (MASS) [6] has been developed. This is a guidance for authorities and applicants that want to perform MASS tests. The guidelines provide an example for developing relevant documentation when applying for MASS tests, including test plan and reporting descriptions.

There exist specific guidelines on reporting of results of e-navigation testbed [7, 8]. The guidelines states that the test results should be harmonized such that the results can be shared and easily compared so that the maritime community can learn from the tests. A template for reporting the results of testbeds is included in the guidelines. In the UK, an industry code of practice has been developed for vessel up to 24 meters that includes practical guidance, standards, and best practice [9]. In Belgium/Flanders it exists a code of practice for automated vessel technologies on the inland waterways in Flanders [10].

2.4. Related work

We have seen that there exist several guidelines describing how MASS tests should be conducted and documented (see Section 2.3). However, to the authors knowledge, there is a lack of platforms or applications that enables documenting autonomous test results in a standardized format and sharing the results available in the market. The Designing the Future of Full Autonomous Ship Project (DFFAS Project) aims to conduct crewless maritime autonomous surface ship demonstration trials in domestic coastal areas [11]. 30 Japanese companies are taking part in the project. The project aims to move towards the standardization of technology and establish systems and infrastructure.

3. Important Tasks When Conducting Tests

When conducting a test with autonomous vessels, the test can be divided in three phases, i.e., before the test, during the test, and after the test. Figure 2 illustrates different tasks that should be done depending on the test phases.

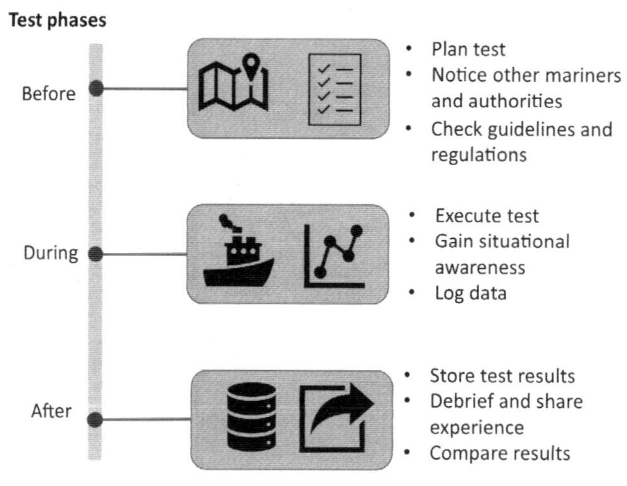

Figure 2: Different tasks to consider for different test phases.

Before a test is executed, it is necessary to plan the test. This could include checking the weather forecast and weather statistics, see available sensor infrastructure and resources in the area, and check traffic statistics and status. For instance, getting an overview of the communication system available in the test area may be essential to decide if the test area is suitable for the planned test purpose. Additionally, other mariners and authorities should be noticed that an autonomous test is going to be performed. Moreover, one should verify that the test plans are in accordance with relevant guidelines and legislations and apply for relevant permissions to conduct the test. A description of what may be included in a test plan is described in [6].

During the execution of a test, it is important to gain situational awareness. This can include accessing live camera feed from the test area, viewing local wind measurements at port, and checking infrastructure status, e.g., the position accuracy or possible communication failures in the area. Additionally, data from the test should be logged.

After a test has been conducted, the test results should be stored such that the results and experience gained can be shared with others, or for documenting purposes. The test results should include the location, time period, test type, equipment and technology used, level of autonomy, traffic picture, weather conditions, the logged data, and lessons learned, as illustrated in Figure 3.

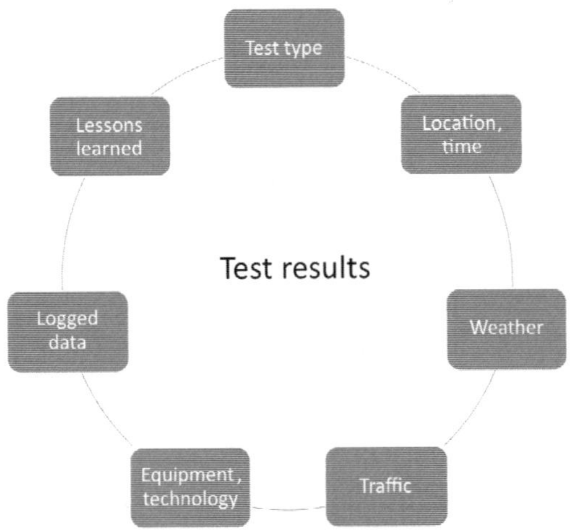

Figure 3: Information that should be included when documenting results from autonomous tests.

When inspecting results from previous tests, the location and time of the test are useful for navigating to relevant tests. The test type can be divided in different categories, e.g., communication or navigation tests. The test type is highly important when inspecting results since it gives a clear indication of what has been done during the test and helps navigate to the relevant test results. Moreover, information on the equipment and technology used during the test specifies in more details what has been done. Let's say a communication test has been conducted. To learn from this test, it is essential to know what type of communication technology was used, e.g., a Maritime Broadband Radio. As mentioned in Section 2.1, a vessel can be operated under different levels of autonomy, where each level will have different requirements to systems and technologies used. The weather and traffic conditions experienced during a test give an indication of the complexity of the test performed. For example, testing an autonomous vessel in a traffic dense area impose a risk and navigational challenge for the vessel. Other information documented that may be relevant for other stakeholders includes the contact information, permits used in test, and the logged test data.

The above-mentioned information included in the test results enables categorizing of test results such that one can easily compare the results with other autonomous test, regardless of where the tests were performed. Furthermore, perhaps the most important information when reviewing previous test results would be the key findings and lessons learned from the tests. This information allows other actors to learn from the previous tests and can contribute to accelerate the development of autonomous vessels. Taking the autonomous vessel Mayflower[1] developed by the non-profit research organization ProMare and IBM as an example. This is a fully autonomous vessel using different on board sensors to obtain situational awareness and

[1] https://mas400.com/

makes operational decisions by itself. The plan is to sail across the Atlantic to gather ocean data. During this project, they have run in to some challenges and issues. Some lessons learned from the Mayflower project includes [12, 13]:

(i) Design to follow COLREGs, but other ships may not.
(ii) Data quality matters.
(iii) Think creatively about how to obtain data.
(iv) There's an advantage in a component-based system.
(v) Sometimes hard coded-rules are necessary.

Looking in to the first lesson learned (i), it is about the how autonomous vessels handle the maritime regulations in their navigation system. At sea there are specific regulations for preventing collisions called COLREGs. The developers of Mayflower have designed the navigation system such that it will follow these rules. However, according to Brett Phaneuf, the director of the Mayflower Autonomous Ship project, many ships do not follow these regulations [13]. This issue has also been highlighted previously, e.g. in [14]. This makes it difficult to design a decision making system for an autonomous vessel. This experience is important for other actors developing an autonomous vessels, knowing that the decision making can not be solely based on the rules, but on how the vessels actually behave. A test area equipped with an application providing traffic information could be used to learn the typical ship behavior. For instance by inspecting historical AIS data one can learn patterns and behavior modes of the traffic. However, ship behavior may change depending on the region of interest. This means that one may need different behavior models for different regions.

4. Instrumenting Trondheimsfjorden Test Area

When conducting autonomous tests, it is important to have a well-established sensor infrastructure in the test area. Several measurements from different sensors and data sources give the users a complete picture of the test situation and make it easier to reproduce a test and compare with other tests. Having sensor data available is essential for documenting autonomous tests. To perform and document test of autonomy, the Trondheimsfjorden Test Area has been instrumented with several technologies and systems. The infrastructure includes systems for communication, navigation, control and monitoring, and information sharing.

Communication infrastructure can be used for communication between vessels or between a vessel and shore (e.g., a control centre or port). In TTA, several communication systems with different coverage areas and characteristics have been installed. The communication infrastructure includes a Mobil Broadband Radio (MBR) and a VHF Data Exchange System (VDES) Satellite terminal in addition to existing 4G and 5G networks.

When operating a vessel, the navigation requirements, e.g., position accuracy, depend on the operation that is being performed. For instance, if a vessel is doing a docking operation, the position accuracy requirements will be higher than when sailing in a coastal area with less traffic. For testing different operational scenarios, a well-established navigation infrastructure is essential. In TTA there has been installed a navigation infrastructure including coastal surveillance radars, Automatic Identification System (AIS) base station, Global Navigation Satellite System (GNSS) monitoring station, and Differential GNSS (DGNSS) reference stations.

When performing tests with the sensor infrastructure, systems for monitoring and controlling the sensor data is needed. TTA has been instrumented with a Test Area Control Centre (TACC) with central monitors that can monitor and control the infrastructure. For instance, a DGNSS central monitor has been installed to test and validate the position accuracy. Moreover, to test different levels of autonomy, a remote control centre is needed. The TACC is installed with applications enabling remote control of vessels.

4.1. Data Centre

As described in Section 2.4, when conducting an autonomous test, it is important to plan the test, check guidelines and regulations, store test results, and share and compare knowledge and experiences from autonomous tests. In TTA, a Data Centre has been developed to support users of the test area with information and functionality to assist maritime operations and to collect test results. The Data Centre can be used when planning, conducting, and post-processing autonomous tests. A screenshot of the Data Centre is shown in Figure 4. At the top of the window is the Area map, which is a graphical representation of the data where sensors are installed and dynamic data is displayed. The time window is a time slider to be used when the data can be represented with a time series, such as weather status last week. At the bottom of the page is a list of the resource available in the area. The menu sections present Data Centre functionalities Some of the data is integration from external WMS services, such as AIS data from the Norwegian Coastal Administration.

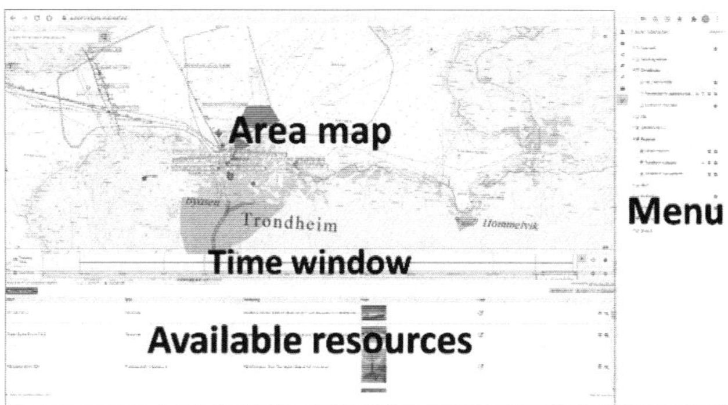

Figure 4: Screenshot of Data Centre.

The following functionalities are included in the Data Centre: Maps and satellite images, area data (ocean farms, protected areas, quays), reference routes, infrastructure coverage area (communication, navigation, observation), and guidelines and permissions overview. Moreover, the Data Centre includes weather (live and historic), traffic (live and historic), infrastructure status (e.g., communication availability), availability of resources (e.g., vessels, drones, service personnel, control room), test results (documenting and statistics), and statistics and analysis. From the available data it will be possible to achieve support in operational planning, for example if a MASS is heading for a quay at a terminal, the MASS operator can receive information (e.g., reference route, ongoing activities) as well as status from available sensors (e.g., local wind sensor at the quay) that can be used for the purpose. Some of the information and functionalities in the Data Centre are imported from other data sources.

5. Demonstration

The TTA has been used as a pilot to demonstrate how an autonomous test can be documented using the installed infrastructure and resources. The goal was to demonstrate remote control of a vessel and demonstrate automatic control where a vessel follows a pre-defined route. The purpose was to find out if this was possible and the performance of a vessel following a route automatically.

Before the test, the Data Centre was used to plan the operation of the test. This included checking the weather forecast to see if the weather was suitable for conducting the test, whom to inform, and status of the infrastructure. Moreover, recommended guidelines, procedures, and regulations were checked using the Data Centre. Based on this information, the route was defined and inserted into the controller for the vessel and at the TACC. Once the details of the test had been decided, the upcoming test was registered using a form in the Data Centre notifying other mariners and authorities that an autonomous test was going to be performed.

The Ocean Space Drone was used to perform the demonstration. The TACC was used to monitor and remotely control the vessel. The vessel and the TACC communicated using MBR. This communication link was used to send video stream and navigation data from the vessel to the TACC, and used to send GNSS corrections and control commands from the TACC to the vessel. The vessel received IALA beacon corrections (RTCM 2.3 format) from the DGNSS reference station installed and GNSS corrections (RTCM 3 format) using MBR from the TACC. Position data (NMEA messages) from the vessel was logged during the demonstration.

Using the logged position data from the vessel, the cross-track error of the vessel was computed. This is illustrated in Figure 5. Cross-track error is the perpendicular distance from actual position to intended track line of the pre-defined route. The pre-defined route is illustrated as the green solid line, while the actual track is the wide line with varying colors. The different colors indicate how large the cross-track error was at that location, where red symbolize a small error and blue a large error (see legend box in plot). During the demonstration, the vessel was run under different control modes between different waypoints (WPs), i.e. remote control from TACC from WP4 to WP5, manual control by on-board skipper from WP1 to WP2, and by autopilot for the rest of the route.

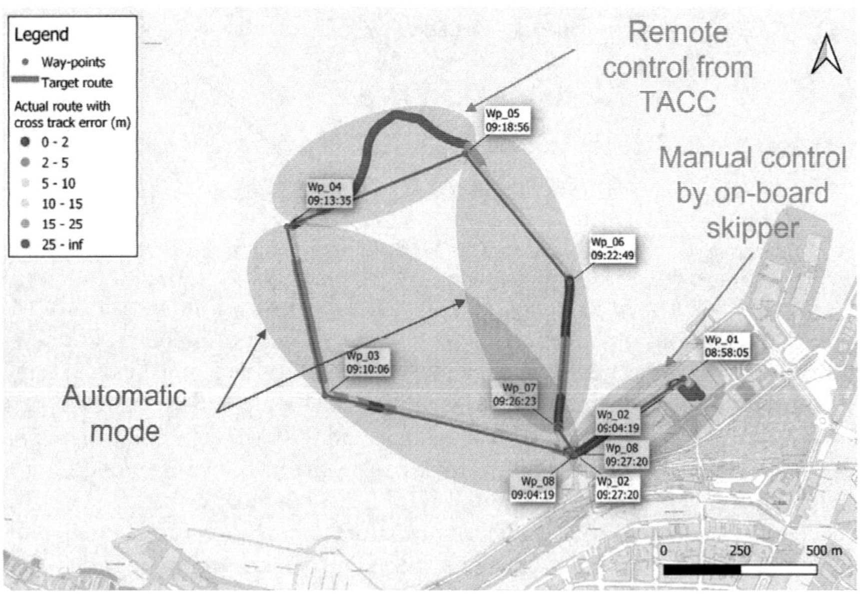

Figure 5: Cross-track error of vessel during the demonstration.

The vessel was successfully remotely controlled from the TACC. This can be seen between WP4 and WP5, where the vessel was steered away from the pre-defined route by intention, hence the large cross-track error on this leg. To see the performance of the autopilot, i.e. how well the vessel can follow the intended route in automatic mode, we inspect the cross-track error for WP2 to WP4 and WP5 to WP8. During automatic mode, the cross-track error is mainly below 10 m. Many factors will impact the cross-track error, for example navigation system accuracy, environmental conditions like wind and current, track complexity, and the vessel's maneuverability. One can see that the cross-track error differs between the different legs in automatic mode. During these different legs, the course of the vessel is different, meaning that the environmental conditions will affect the vessel differently for these legs. Another observation from the figure is that the cross-track error is slightly higher near the waypoints when the vessel is changing its course. This can be due to the fact that initial route does not consider the turning radius of the vessel. Two of the lessons learned from this demonstration is that (1) the vessel can navigate under different autonomy levels and (2) the established infrastructure simplifies in situ testing.

The test results were documented and stored in the Data Centre ensuring that results are shared. Parts of the test results are summarized in Table 1.

Table 1: Summary of test results from demonstration.

Test results	
Location	63.4N, 10.4Ø (Waypoint 1)
Time	06.12.2021, 08:30-10:00
Test type	Automatic navigation and remote control
Technology,equipment	Vessel, Test Area Control Centre, MBR, DGNSS navigation system
Weather	Calm
Traffic	Low
Test report	Project report NAVISP EL3-D5.2.pdf
Logged data	Position data from the vessel
Lessons learned	Can navigate under different autonomy levels. The established infrastructure simplifies in situ testing.

6. Discussion

In this paper, the benefits of documenting test results have been described and we have demonstrated how a test area can be used to conduct and document autonomous tests. So, how do we move on from a pilot and ensure that test results are documented and shared?

In theory, an autonomous test can be conducted at any location with permission from the authorities. However, by conducting an autonomous test in a test area, the test user will have access to a well-established infrastructure and resource in the area, as opposed to having to install and develop these on their own. Utilizing a test area simplifies the process of planning, executing and documenting a test. Users of a test area will save money and time using an established test area. "Being able to get out and test the sensors and systems in real life is the ultimate (test). You can get far with simulations at the laboratory, but out here there are weather and wind, and can verify if the systems behave as expected." - Geir Håøy, CEO in Kongsberg Gruppen, about the TTA[2]. A challenge of a test area is to make sure that the resources available in the test area are in fact known and easily accessible for parties that would benefit from using the test area.

Today, test plans and results from autonomous tests are not shared. To do so, there is need for a common platform for documenting and sharing results and experiences, like the Data Centre developed in this project. A challenge would be to ensure that test users utilize such platforms for documentation and sharing. A solution could be to encourage users of a test area to document the test results in a standardized format, e.g., by using a data centre developed for test areas. Another obstacle for sharing test results, is that the results may be confidential.

[2] https://www.mn24.no/nyheter/i/Jx4Wv8/viste-frem-trondheimsteknologi-til-toppsjefen?

Due to high competition in the maritime industry, many companies fear the consequences of sharing detailed test results. However, test results from the academia and research are typically shared openly. The whole maritime industry will be interested in viewing and learning from the experiences obtained in the test results that are shared, also between different regions and test areas.

The Data Centre developed in this project receives data from the vessel after a test has been performed. Getting data in real-time from the vessel is difficult due to security restrictions and limitations on the communication bandwidth when transmitting data. On the other hand, it can be argued whether there is a need for having real-time data in the Data Centre.

A test area will have different stakeholders with different needs. This contributes to the requirements for a data centre and the documentation process. In TTA, we have different stakeholders, including maritime authorities, infrastructure owners, service providers, classification, clusters, and research and education stakeholders. These stakeholders have been important for defining the requirements and functionality of the Data Centre. The involvement of existing and new stakeholders will be important for further development and use of the test area.

7. Conclusion

Autonomous shipping is important to meet the increasing transport demands of the future in a sustainable manner. To develop autonomous shipping, it is important to test and document autonomous tests. By documenting the test results in a standardized format, it is easy to categorize and compare the test results. Sharing test results is important so that others can learn for the experience gained. Moreover, documentation from autonomous test can help verify technological maturity regarding regulations and requirements. These are important steps for the development of autonomous vessels. Test areas for autonomous ships simplify the process of conducting tests and documenting test results since they have an established test infrastructure and may include a data centre for storing and sharing test results. Thus, utilizing well-established test areas are important to accelerate the development of sustainable transport means.

Acknowledgments

This work is based on the results from the research project NAVISP-EL3-005 "Trondheimsfjorden Test Area for Autonomous Ships" [15], which is founded by the European Space Agency under the Navigation Innovation and Support Program (NAVISP), and the research project IMAT [16], which founded by the Research Council of Norway. NAVISP is a program to support the competitiveness of the European industry in the wide field of positioning, navigation, and timing while supporting member states in enhancing national objectives and capabilities in the sector. These projects are based on a cooperation between Kongsberg Seatex AS, SINTEF Ocean AS, the Norwegian University of Science and Technology, Kongsberg Norcontrol AS, and Massterly. The Data Centre developed in this project is based on Adaptive, an application from Asplan Viak.

References

[1] International Transport Forum 2021 Itf transport outlook 2021 URL `https://doi.org/https://doi.org/10.1787/16826a30-en`
[2] IMO 2021 Outcome of the regulatory scoping exercise for the use of maritime autonomous surface ships (mass) (msc.1/circ.1638)
[3] International Network for Autonomous Ships 2021 Autonomous ships test areas URL `http://autonomous-ship.org/testarea.html`
[4] TESTSITEtrd 2021 About URL `https://testsitetrd.no/about/`
[5] IMO 2019 Interim guidelines for mass trials (msc.1/circ. 1604)

[6] EU, EMSA and European Commission 2020 Eu operational guidelines for trials of maritime autonomous surface ships (mass)
[7] IMO 2014 Guidelines on harmonization of testbed reporting (msc.1/circ.1494)
[8] IALA 2016 *IALA Guideline 1107. Planning and reporting of e-Navigation Testbeds* 2nd ed
[9] Maritime UK 2021 *Maritime Autonomous Surface Ships (MASS) UK Industry Conduct Principles and Code of Practice* 5th ed
[10] De Vlaamse Waterweg nv 2018 Smart shipping. code of practice for testing in flanders URL https://www.vlaamsewaterweg.be/smart-shipping-english-version
[11] SAFETY4SEA 2021 Autonomous ship project ready for trial URL https://safety4sea.com/autonomous-ship-project-ready-for-trial/
[12] Kahn J 2021 The mayflower autonomous ship holds lessons for creating better a.i. — fortune URL https://fortune.com/2021/11/30/a-i-lessons-from-mayflower-autonomous-ship-eye-on-ai/
[13] Briscoe N 2020 'aye, ai, captain.' ibm's autonomous ship seeks to revolutionise modern transport URL \url{https://www.irishtimes.com/news/science/aye-ai-captain-ibm-s-autonomous-ship-seeks-to-revolutionise-modern-transport-1.4397984?localLinksEnabled=false}
[14] Roche P 2020 *Norton Rose Fulbright* URL https://www.nortonrosefulbright.com/en/knowledge/publications/5fedab67/the-collision-regulations-and-autonomous-shipping
[15] European Space Agency (ESA) 2021 005 - trondheimsfjorden test area URL https://navisp.esa.int/project/details/59/show
[16] The Research Council of Norway 2021 Integrerte maritime autonome transportsystemer (imat) - prosjektbanken URL https://prosjektbanken.forskningsradet.no/en/project/FORISS/295647

Autonomous ship concept evaluation – Quantification of competitiveness and societal impact

Håvard Nordahl[1], Dag Atle Nesheim[1], Elizabeth Lindstad[1]

[1]SINTEF Ocean, Postboks 4762 Torgard, N-7465 Trondheim, Norway

E-mail: haavard.nordahl@sintef.no

Abstract. The prospect of large-scale international adoption of autonomous ships has led to expectations of reduced costs and emissions for waterborne transport of goods. This is commonly attributed to the possibility of removing manning from the ship, which enables more efficient ship designs and reduced operational costs. So why have we not seen a multitude of autonomous ship building projects? There are several reasons for this, including immature technology and regulations. However, there is another reason which has received less attention; the lack of quantifiable evidence for the benefits arising from investing in autonomous ships. There are some case studies on the impact of autonomy on transport cost, but there is no established method for evaluating the effects of an investment in autonomous ships. This paper will present Key Performance Indicators (KPIs) developed to enable such quantification. Furthermore, the developed KPIs are chosen not only to enable quantification of benefits but also to be calculable based on data which it is reasonable to assume that are available or obtainable at a concept stage.

1. Introduction

The EU has had a long-standing goal of shifting transportation of cargo from road to rail and water [1]. This is a result of the high societal costs caused by road transportation through accidents, congestion, pollution, health impact and other negative side effects, estimated to 820 billion Euro for the EU28 countries in 2016 [2]. The total external cost of transportation was estimated to be 987 billion Euro in 2016, where 31% of these costs are caused by freight transportation [2]. Unfortunately, to date, no development towards a shift of cargo flows from road to waterborne or rail transport is observed [3], [4].

We believe that a key reason for this failure is that investments needed to set up an alternative to road transport are too high. This when considering that the alternative must meet the customers' key demands to the transportation service (e.g., lead time and shipment frequency), at an equal or lower cost, to be competitive. There is hence a need for increasing the competitiveness of alternative transport modes, such as waterborne transport.

New concepts such as autonomy are expected to increase waterborne transport competitiveness by opening for new ship- and transport system concepts [5], [6], [7]. Still, the impact of such novel concepts on competitiveness needs more studies [8], [9], [10]. Such studies need methods for quantifying the impact of new ship and transport system concepts on competitiveness. Furthermore, some concepts may prove to yield insufficient in terms of competitiveness, but still give significant societal benefits. In such cases it is conceivable that the would-be investors could obtain subsidies from public financing programs, provided that the societal benefits can be quantified. An example is the construction of the Yara Birkeland, which will remove 40,000 truckloads annually. The construction cost for the ship is estimated to 250 million NOK and the Yara Birkeland project was supported by the Norwegian Government agency ENOVA by a 133.6 million NOK grant [11]. The grant was given due to an estimated reduction of 772 ton CO_2 equivalent emissions annually [12]. It should be noted that the reduction in CO_2 equivalent emissions is mainly a result of the ship being fully electric. It should also be noted that the estimated construction cost is for the construction contract with the yard VARD. This means that other investments, such as infrastructure for charging, cargo handling, etc., is not included in the estimated 250 million NOK.

Much is written about definitions of MASS and levels of autonomy; this is not a debate that we will visit in this paper. However, for the purpose of discussions in this paper we consider a MASS to be a ship without crew and which is supported by a Remote Control Center (RCC) and automatic facilities services as shown in Figure 1. Automatic facilities services are the services that are required to support the operations of a MASS and examples are connectivity, auto-mooring and automated cargo handling. A formal definition of automatic facilities services is given in [13].

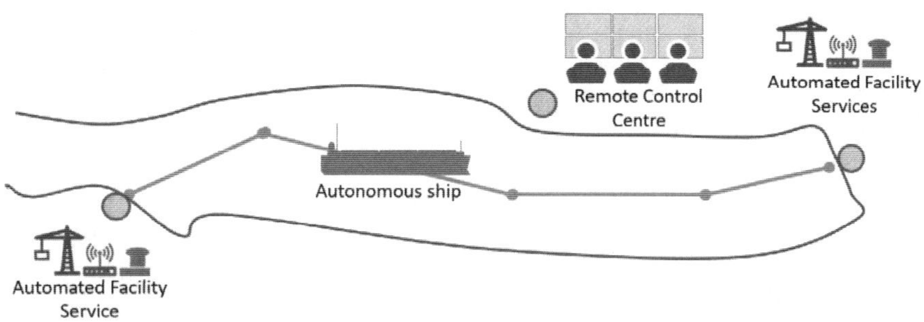

Figure 1. The autonomous ship, RCC, and Automatic Facilities Services. The figure is taken from [14].

The structure of this paper is: First, we present our method in section 2, followed by the results in section 3. The results are organized as KPIs and calculation methods for competitiveness evaluation in subsection 3.1, and for societal impact evaluation in subsection 3.2. An example of how two of the most central KPIs can be calculated is given in section 4. Finally, we discuss the results and conclude the paper in section 5.

2. Method

This article is mainly based on the work conducted within the AUTOSHIP project [15]. The AUTOSHIP project has two demonstration cases, one covering inland water ways shipping and one covering short sea shipping. Both demonstration cases are based on ships that are in operation today, and the project will upgrade these ships and demonstrate autonomous operations with them. The project has mapped the supply chains of both use-cases and identified the involved stakeholders [16]. Moreover, we have performed an analysis of the stakeholders that are involved in investments in autonomous ships, and their main objectives for- and rationale behind such investments [17]. The identified objectives may hold a somewhat different meaning for the different stakeholders. As a result, the different stakeholders may also evaluate whether their objectives are met, differently. We have approached this problem by dividing the main questions "is my objective met/is this a rational decision" into more precise sub-questions capturing what each stakeholder wants to verify when assessing the investment at hand. We then created KPIs to represent quantitative answers to said questions.

As an example of this approach consider the objective *Reliability* and which questions must be answered to demonstrate a concept's achievement of this objective. The following are some examples:

- How does a specific investment or set of investments impact the ability to keep slot times, delivery times and pick-up times?
- How much of the time is the solution operational?

These sub-questions can be answered quantitatively by defining KPIs. The necessary calculation formulas for evaluating the KPIs were also derived. This enabled us to define input parameters for the KPI calculation and evaluate if the KPI could be calculated at a concept stage or not, by applying the following criteria:

- The KPI must be possible to calculate based on data that is likely available at a concept study stage or,
- The KPI must be possible to estimate based on data that is generated by some model, where the input data to the model must be available at a concept stage or,
- The KPI must be possible to estimate based on data that is produced by some simulation tool, and the simulation tool can only depend on input data that is available at a concept stage.

3. Results

The evaluation of an investment in a MASS should first and foremost evaluate the competitiveness of the considered transportation service. In addition, it should be evaluated if the considered investment will give significant societal benefits as this could result in obtaining subsides, potentially balancing the cost-benefit assessment result. In the following we will give KPIs for evaluating competitiveness and societal impacts.

3.1. KPIs for competitiveness

Competitiveness is about meeting the customers' demands to the service, and about cost. If we consider the view of a transportation service customer, it is not that important how the transportation provider implements the service if it meets certain demands and is cost competitive. Therefore, it might not be important for the customer whether the transport is by road, sea, rail, or a combination, if the alternatives perform equally in terms of the customer criteria for the service. It is therefore important to be able to evaluate the transportation service performance for typical customer criteria, and a business case for investing in MASS should quantify the impacts of the investment on the customer demands to the service, including costs.

So, what do we know about customer demands to transportation services? In [17] we studied the customer objectives, and suggested the following customer demands as being most important: Reliability, flexibility, and cost. In the following we will give our definitions of KPIs and show how they can be estimated.

3.1.1. Reliability. Reliability, in short, is about on time deliveries and pickup. We have defined KPIs that can express these factors and a common denominator for these KPIs is that their estimation requires supply chain analysis. In the following we will give the KPI definitions, and briefly discuss how they can be estimated.

The estimation of on time delivery or pick-up can be done by estimating delays. If the number of delays, and duration of delays during a given time are estimated, delays can be measured by two KPIs; the average duration of delays ($\bar{t_{\mathrm{d}}}$), and the percentage of deliveries (or pick-ups) that are delayed P_{d}:

$$\bar{t_{\mathrm{d}}} = \frac{\sum t_{\mathrm{d}}}{n_{\mathrm{d}}} \quad (1)$$

$$P_{\mathrm{d}} = \frac{n_{\mathrm{d}}}{n} 100\% \quad (2)$$

where t_{d} is the duration of each delay, n_{d} is the total number of delays, and n is the total number of either deliveries or pick-ups.

To estimate the duration and number of delays, it is necessary to analyze the supply chain. This can be done through simulations where the transportation need is modelled along with the terminals, cargo handling, transporters, and factors disrupting the transporters. In [18] we presented a simulation method and models intended for this purpose. When such simulations are conducted, it is possible to detect delays, measure their durations and frequency. The simulations will be based on a schedule, and on scenarios for weather or other environmental or stochastic conditions influencing operations and causing delays. It will be necessary to simulate over a considerable time to get good estimates.

3.1.2. Flexibility. Possibly the main competitive edge of trucks is flexibility. They can transport small quantities on one truck, and large quantities by several trucks. They have relatively short lead time, high frequency of delivery, and can reach almost any destination which are connected through land or ferry. To evaluate competitiveness in terms of flexibility, we therefore define the KPIs frequency, lead time, and ship capacity utilization. These KPIs can be estimated through the same type of supply chain simulations as discussed for the reliability KPIs and given in [18]. Lead time can be estimated by measuring the duration from when a cargo unit enters the supply chain until it exits the supply chain, frequency by measuring the number of shipments over a time period, capacity utilization by measuring the ship capacity utilization for each simulated voyage.

3.1.3. Transportation cost. Probably the most important factor for determining the competitiveness of a service is the cost of the service. For transportation, the customer is interested in comparing the total cost of transporting their goods between certain origin and destination points. With the lack of new business models related to the introduction of autonomous shipping [19], we have decided to exclude the profit element from our cost definitions, thus focusing on the actual cost of transporting a certain cargo unit between a start and end point. Such transportation can be done either by one transport mode only or by an intermodal transportation chain. If the customer compares two transport services, where one service provider transports the goods by truck between the two points and the other transports the goods by a combination of MASS and truck between the same two points, both the MASS and truck transportation costs must be included in the latter case. However, as we have identified a need for extending the knowledge on the impact of MASS on competitiveness, transportation by MASS will be the focus of the more detailed discussions of cost estimation in this paper.

The transportation cost KPI is therefore taken to be the cost of transporting one cargo unit between an origin and a destination. This is similar to the *Unit Cost* as given by [20], which is the sum of costs (including repositioning costs) divided by the parcel size. The *Unit Cost* is an estimate of the cost of transporting one unit of a certain cargo type for a certain ship. As we are interested in estimating the cost per transported unit between certain points, including all transportation legs (for example truck or MASS transportation legs), we adapt the *Unit Cost* and define the transportation cost KPI *cost per unit* C_{pu}:

$$C_{\mathrm{pu}} = \sum_{i=1}^{n} \frac{C_{\mathrm{leg},i}}{n_{\mathrm{units},i}} \quad (3)$$

where $n_{\mathrm{units},i}$ is the number of units at leg i, and $C_{\mathrm{leg},i}$ is the sum of costs for transportation leg i, including the costs associated with the transfer of cargo between transportation legs.

To estimate $C_{\mathrm{leg},i}$ it is necessary to include all costs related to owning and operating the transportation service. For ships, we find a well-established cost categorization in [20]: Operating costs, periodic maintenance, voyage costs, capital costs, and cargo handling costs. For MASS, we suggest a slight adaptation of some of the cost categories to account for the differences between MASS and conventional ships. The following definitions are taken from [20] and our changes are highlighted by using *italic* font for changed text and [...] for omitted text which can be found in the source for interested readers:

- Operating costs, which constitute the expenses involved in the day-to-day running of the ship – essentially those costs such as *remote-control center, insurance, and maintenance by boarding crews,* that will be incurred whatever trade the ship is engaged in.
- Periodic maintenance costs are incurred when the ship is dry-docked for major repairs, usually at the time of its special survey. [...]
- Voyage costs are variable costs associated with a specific voyage and include such items as *Automated Facility Services (AFS),* fuel, port charges and canal dues.
- Capital costs depend on the way the ship has been financed. They may take the form of dividends to equity which are discretionary, or interest and capital payments on debt finance, which are not.
- Cargo-handling costs represent the expense of loading, stowing, and discharging cargo. They are particularly important in linear trades.

These cost categories can be sorted into two higher level categories; costs which are incurred because of performing the freight work (voyage costs and cargo-handling costs), and costs which are incurred independently of performing freight work (operating costs, periodic maintenance, and capital costs). While the former can be calculated per transportation leg performed by a ship or a MASS, the latter is more conveniently estimated as annual costs and divided amongst performed transportation legs during a year, based on the relative duration of performing each transportation leg. This gives the following definition of C_{leg} for conventional ships and MASS:

$$C_{\mathrm{leg},i} = C_{\mathrm{VOYEX},i} + C_{\mathrm{CH},i} + \frac{\left(C_{\mathrm{CAPEXyear}} + C_{\mathrm{PM}} + C_{\mathrm{OPEX}}\right)t_{\mathrm{leg},i}}{t} \quad (4)$$

where $C_{\mathrm{VOYEX},i}$ is the voyage cost of leg i, $C_{\mathrm{CH},i}$ is the cargo-handling costs of leg i, $C_{\mathrm{CAPEXyear}}$ is the annual capital costs, C_{PM} is the annual periodic maintenance costs, C_{OPEX} is the annual operational costs, t is the duration of all performed transportation legs in a year, $t_{\mathrm{leg},i}$ is the duration of the transportation leg i.

At a concept stage, some of these costs are not immediately available, and some of the required data must be generated. We will discuss the cost items C_{CAPEX}, C_{OPEX} and $C_{\mathrm{VOYEX},i}$ in some detail, while the estimation of $C_{\mathrm{CH},i}$ and C_{PM} can be found in [20].

Estimating the cost of acquiring a ship is not straightforward. Historically, the newbuild and second-hand market are both highly volatile due to immense variations in supply and demand [20]. It is possible to create regression models for estimating new-building costs of conventional ships within different market segments, however, it should be noted that one must take great care and consider the current and projected market situation and adjust such estimation results accordingly. Estimating the new build cost for a MASS is even more challenging as the only MASS built to date for commercial trade, is the Yara Birkeland [21]. It is built for operating manned initially, and then to gradually transition into autonomous operations. Though the estimated construction cost of the Yara Birkeland is published [11], it is as much a research and development project as a commercial project, and hence, the Yara Birkeland estimate is hardly representative for the cost of future MASS construction.

Some studies have suggested that construction costs will be comparable to that of conventional ships [5], so an option is to assume a certain percentage of the cost of a corresponding conventional ship. Another option is to use what we know and develop a cost model. Both options imply significant uncertainty, and it remains to see which will give the best estimates. We have chosen to develop a cost model based on a regression model of conventional ship newbuild costs. Based on 474 general cargo and container ships, the SATS project [22] have developed the following cost estimation formula (in million Euro) for new builds of conventional ships:

$$C_{\mathrm{CAPEX}} = (0.0053 - 0.0000001DPL)(0.8LDT + 0.2PWR) \tag{5}$$

where DPL is the displacement of the ship in ton, LDT is the ship lightweight in ton, and PWR is the installed power in kW. Equation (5) for estimating C_{CAPEX} will capture some of the cost differences between MASS and conventional ships. Namely reduced steel weight and reduced energy consumption. However, other differences such as technology for MASS, increased redundancy in systems, and costly machinery requiring low maintenance, are not captured. One would therefore need to adjust the estimate by adding these factors:

$$C_{CAPEX} = (0.0053 - 0.0000001DPL)(0.8LDT + 0.2PWR) + C_{\mathrm{tech}} + C_{\mathrm{M}\delta} \tag{6}$$

where C_{tech} is the total additional technology and redundancy cost, and $C_{M\delta}$ is the additional cost for non-conventional machinery. The additional machinery cost can be estimated by:

$$C_{\mathrm{M}\delta} = PWR(k_{\mathrm{M}} - k_{\mathrm{convM}}) \tag{7}$$

where k_{M} and k_{convM} are factors converting installed power to Euro, respectively for the installed machinery and a conventional machinery. One source for these factors is found in table 1 of [23]. Note that this source provides the factors in USD. Also note that C_{CAPEX} in equations (5) and (6) is the estimated investment cost and must be converted to an estimated yearly cost of the investment $C_{\mathrm{CAPEXyear}}$ before $C_{\mathrm{leg},i}$ is estimated in equation (4).

Automated facility services (AFS) will give rise to additional costs for MASS compared to conventional ships. Depending on the situation for each case that is under evaluation, they may be part of the MASS investment, or considered to be available services at a charge, for example per usage or per year. Most AFS are resources that could be used by several autonomous ships, in some cases even conventional ships. Considering that the conventional versions of some AFS (like cargo handling) often is a per usage cost, we include AFS in the estimation of voyage costs. The estimation of voyage costs is given in [20] as:

$$C_{\mathrm{VOYEX}} = C_{\mathrm{fuel}} + C_{\mathrm{PD}} + C_{\mathrm{TP}} + C_{\mathrm{CD}} \tag{8}$$

where C_{fuel} is the fuel cost, C_{PD} are the port costs and light dues, C_{TP} are the costs related to tugs and pilotage, C_{CD} are canal dues. For MASS, we add the cost from the AFS that are utilized during the voyage C_{AFS}. The VOYEX cost estimated in [20] can therefore be updated to:

$$C_{\mathrm{VOYEX}} = C_{\mathrm{fuel}} + C_{\mathrm{PD}} + C_{\mathrm{TP}} + C_{\mathrm{CD}} + C_{\mathrm{AFS}} \tag{9}$$

If we consult the estimation of operational costs in [20], we find that for a conventional ship it can be estimated as:

$$C_{\mathrm{OPEX}} = C_{\mathrm{M}} + C_{\mathrm{ST}} + C_{\mathrm{MN}} + C_{\mathrm{I}} + C_{\mathrm{AD}} \tag{10}$$

where C_{M} is the crew cost, C_{ST} is the cost of stores and consumables such as lubricants, C_{MN} is the cost of maintenance and repairs, C_{I} is the cost of insurance, C_{AD} is the cost of administration and general costs. For MASS we can adjust this to:

$$C_{\mathrm{OPEX}} = C_{\mathrm{RCC}} + C_{\mathrm{MN}} + C_{\mathrm{I}} + C_{\mathrm{AD}} \tag{11}$$

where C_{RCC} is the annual cost of the RCC and replaces the cost of crew (C_{M}), C_{ST} is removed as stores are related to the presence of crew, and as it is expected that MASS will depend on less maintenance intensive machinery and as such consume considerably less consumables such as lubricants. For MASS C_{MN} should also include an estimated cost for required technology upgrades and maintenance. There are some available sources for estimating the cost items of C_{OPEX}, [20] being one such source. However, the cost items C_{RCC} and C_{MN} for autonomous ships are less discussed in existing literature, though one source for estimating these is [5].

3.2. KPIs for societal impact

Some investments in transportation services may have considerable impacts on the society by bringing down external costs. The main categories for external costs of transportation in Europe are accidents (29%), congestion (27%), air pollution (14%), climate (14%), noise (7%), well to tank (5%) and habitat damage (4%), [2]. If we compare road transportation to waterborne transport, we find that accidents, congestion, noise, and habitat damage related costs would be more or less eliminated if a waterborne transportation service option is chosen over road transport (though inland water ways do cause some minor costs related to accidents and habitat damage). We can also see that the main external costs of waterborne transportation are from air pollution, climate (GHG emission), and well-to-tank (the external costs related to supplying the ships with fuel).

Based on this we define the KPI EC_{δ} for evaluating the annual impact on external costs from an investment in MASS that would reduce truck transport:

$$EC_{\delta} = Lk_{\mathrm{ship}} + L_{\delta}k_{\mathrm{truck}} \tag{12}$$

where L is the yearly ship transport in ton-km, L_{δ} is a negative value representing the yearly reduction in truck transport resulting from the introduction of the ship and is given in ton-km for HGV (heavy goods vehicle), and vkm (vehicle kilometer) for LCV (light commercial vehicle), k_{truck} is the factor converting ton-km or vkm to external costs in Euro, k_{ship} is the factor converting ton-km to external costs of ship transport in Euro. Notice that L_{δ} is a negative value because it represents the reduction of truck transport, and that a negative value of EC_{δ} means that the shift of transportation from road to sea gives a reduction in external costs, corresponding to the negative value of EC_{δ}. The factors k_{ship} and k_{truck} are found in [2]. The way of estimating EC_{δ} in equation (12) would give the difference between average external costs of truck transport and waterborne transport. If we instead want to investigate the difference between average external costs of truck transportation and the external costs of a specific MASS transportation service concept, we can estimate EC_{δ} as:

$$EC_{\delta} = L_{\delta}k_{\mathrm{truck}} + CO_{2\mathrm{eq,TTW}}k_{\mathrm{CO_2}} + \sum_{i=1}^{n} P_i k_i + Lk_{\mathrm{w2t}} \tag{13}$$

where $CO_{2\text{eq,TTW}}$ is the tank-to-wake emitted CO_2 equivalents in ton per year for the ship concept, k_{CO_2} is a factor converting ton CO_2 to external costs in Euro, which is set to 100 in [2], P_i is air pollutant i in kg per year for the ship concept, and k_i is a factor converting pollutant i to external costs in Euro. Values for k_i can be found in table 14 of [2]. L is the yearly MASS concept transport in ton-km, and k_{w2t} is a factor converting ton-km to the cost of well-to-tank in Euro. In [2] k_{w2t} is set to 0,001 for average IWW vessels, and 0,0006 for average maritime freight ships.

It should be noted that the estimated average societal cost of well-to-tank emissions for ships in [2] is given as a function of ton-km. Which means that the estimates are independent of the fuel type and amount of consumed fuel. Which makes the estimate rough, and independent of the specific ship concept. It is however stated that GHG emissions contributes to 60-65% of the well—to-tank external costs. As can be found in [23], the well-to-tank GHG emissions varies considerably for various fuel types. In fact, new fuel types that are currently being researched as means to reduce global warming are estimated to produce well-to-tank GHG emissions ranging from 0 to 30 times as much as for more traditional maritime fuels. EC_δ can therefore be adapted to better capture the change in external costs for the specific ship concept as follows:

$$EC_\delta = L_\delta k_{\text{truck}} + (CO_{2\text{eq,TTW}} + CO_{2\text{eq,WTT}})k_{\text{CO}_2} + \sum_{i=1}^{n} P_i k_i + 0.4Lk_{\text{w2t}} \quad (14)$$

where 40% of the estimated well-to-tank cost from [2] is included to account for the well-to-tank costs that are not related to GHG emissions, and the cost from well-to-tank GHG emissions is included by adding $CO_{2\text{eq,WTT}}$.

For IWW, EC_δ should be extended to include costs related to accidents and habitat damage, such that EC_δ becomes:

$$EC_\delta = L_\delta k_{\text{truck}} + (CO_{2\text{eq,TTW}} + CO_{2\text{eq,WTT}})k_{\text{CO}_2} + \textstyle\sum_{i=1}^{n} P_i k_i + L(0.4k_{\text{w2t}} + k_{\text{accident}} + k_{\text{habitat}}) \quad (15)$$

where k_{accident} is a factor converting ton-km to the external cost of accidents in Euro and k_{habitat} is a factor converting ton-km to the external cost of habitat damage in Euro. The factors are found in [2]. Note that some of the factors in [2] are given in Eurocent.

In some cases, the investment might not be in a transport solution that competes with truck transportation. For rail, the expression $L_\delta k_{\text{truck}}$ in equations (14) and (15) can be changed by replacing k_{truck} with k_{rail} and calculating L_δ as the yearly reduction in ton-km by rail, as [2] also provides values for k_{rail}. For comparing maritime shipping alternatives, the expression $L_\delta k_{\text{truck}}$ can be replaced with $-(CO_{2\text{eq,TTW}} + CO_{2\text{eq,WTT}})k_{CO_2} - \sum_{i=1}^{n} P_i k_i - 0.4Lk_{\text{w2t}}$ for the base case ship(s). Furthermore, for IWW, $-0.4Lk_{\text{w2t}}$ can be replaced with $-L(0.4k_{\text{w2t}} + k_{\text{accident}} + k_{\text{habitat}})$.

Some of the data needed for the estimation of EC_δ at a concept level for a MASS, can be produced by simulations that estimates the required energy consumption for propulsion (and other loads). Energy consumption can be converted to emitted CO_2equivalents and pollutants P_i, by using conversion factors. Conversion factors for converting energy to CO_2, to CH_4 in CO_2 equivalents, and to N_2O in CO_2 equivalents, for a range of machinery types, are found in [23]. One source for conversion factors that converts energy consumption to pollutants P_i is found in [24]. Energy consumption estimation by simulations, for propulsion of ships, is discussed in [25] and [18].

4. Example calculation of investment and societal costs

To give some example calculations of C_{CAPEX} and EC_{δ} in a case study of autonomous ship concepts, we will use the study of the impacts from replacing truck transportation with autonomous ship concepts in [26]. The study in [26] was done as part of the SATS project [22] where autonomous ship concepts were developed and compared to a real conventional ship, and to truck transportation, for container transportation between two short sea ports. The purpose was to study the impact on cost and emissions from design changes made possible by autonomy, and to study if autonomy made shipping more competitive to trucks. The autonomous ship concepts were developed by a naval architect and included GA drawings, hydrodynamic models (speed-power curves), and dimensioning of machinery. Thus, the study is an excellent source for realistic weight, resistance (water and waves), installed power, and cargo capacity impact from design changes made possible by autonomy. And this is what is needed to show the impact of autonomy on C_{CAPEX} and EC_{δ} for a specific case.

The study in [26] does not consider differences in machinery type. Since our proposed KPIs C_{CAPEX} and EC_{δ} captures differences in machinery choices, we have made a change to the case studied in [26] before using it in our example. The change is that the autonomous ship concept has hydrogen fuel cells as energy source instead of diesel, while the conventional ship machinery is unchanged. The data required for estimating the impact on C_{CAPEX} and EC_{δ} from replacing the diesel machinery with hydrogen fuel cells are taken from the study of alternative fuels and engine technologies in [23].

Not all required data points for the KPI estimations are given in [26], however, this data is available to the authors of the present paper as they contributed to the study and co-authored [26]. We will highlight the data that this applies to in the discussion throughout the example. This means that the example calculations in this section are an extension of the previous study in [26], and that some previously unpublished data from the study in [26] are made available in this paper.

4.1. CAPEX estimation

The study in [26] compares a conventional ship with 4 autonomous ship concepts, and a truck, for the transportation of cargo between two points. In the example CAPEX estimation, we will use the conventional ship and the *concept 100%* ship from [26]. This means that we are comparing ships with the same main dimensions.

The ship particulars are given in Table 1 of [26], however the ship light weight (LDT) is missing as well as the displacement (DPL). These are available to the authors of the present paper. The additional cost for autonomy related technology is also not given in [26], but is available to the authors of the present paper. The machinery for the conventional ship is kept the same as in [26], a diesel engine running on MGO, and for the *concept 100%* ship we have changed the machinery to fuel cells running on liquid hydrogen. This is different from [26] and is done to include the impact of non-conventional fuel and engine types in the CAPEX estimate.

The impact of choosing liquid hydrogen over MGO, $C_{\mathrm{M}\delta}$, can be estimated by inserting the installed power for the ship and the factors k_{M} and k_{convM} from [23] in equation (7) for $C_{\mathrm{M}\delta}$. For the conventional ship this is not needed, as $C_{\mathrm{M}\delta}$ will be zero. For the concept ship with fuel cells, we estimate $C_{\mathrm{M}\delta}$ as:

$$C_{\mathrm{M}\delta} = PWR(k_{\mathrm{M}} - k_{\mathrm{convM}}) = 1700(2700 - 400)USD = 3\,910\,000\ USD \quad (16)$$

Assuming an exchange rate of 0.9 Euro per USD, this gives a $C_{\mathrm{M}\delta}$ of 3.5 million Euro. The data for estimating C_{CAPEX} is summarized in Table 1.

Table 1. Input data to CAPEX estimation, based on [26] and [23].

Parameter	Conventional	*Concept 100%*
L_{pp} (m)	85.0	85
Breadth (m)	15.8	15.8
Design draft (m)	5.4	5,4
C_b	0.825	0,7
DPL	6133	5220
LDT (ton)	1970	1670
PWR (kW)	2400	1700
C_{tech} (Million Euro)	0	0,925
$C_{M\delta}$ (Million Euro)	0	3.5

The CAPEX for the conventional ship can then be estimated by inserting the respective data from Table 1 into equation (6):

$$C_{\mathrm{CAPEX,conv}} = (0.0053 - 0.0000001 * 6133)(0.8 * 1970 + 0.2 * 2400) = 9.6\mathrm{MEuro} \quad (17)$$

While the CAPEX for the *concept 100%* ship can be estimated by inserting the respective data from Table 1 into equation (6):

$$C_{CAPEX,con} = (0.0053 - 0.0000001 * 5220)(0.8 * 1670 + 0.2 * 1700) + 0{,}925\mathrm{MEuro} + 3.5\mathrm{MEuro} = 12.4\mathrm{MEuro} \quad (18)$$

Which means that in this example the concept ship with hydrogen fuel cells has a 2.8 million Euro higher investment cost than the conventional ship. It is worth noting that the investment cost would be slightly lower for the autonomous ship than the conventional if the machinery was identical.

4.2. External cost reduction estimation

The estimation of the KPI EC_δ for quantifying the societal impact of investing in the ship concept in [26], requires that we know the distance that the concept ship sails, and that the truck drives, the number of transported containers, and the weight of the containers. It also requires the conversion factor for estimating GHG emissions for the energy consumed by the ship (or tank-to-wake) and for well-to-tank for the ship, and the factors for converting the truck travel distance to euro, CO_2 equivalents to euro, and ship travel distance to euro. These values are given in Table 2 and are taken from the case study of the ship *concept 100%* in [26]. The estimated energy is not given in [26], but is available to the authors of the present paper. The conversion factor for energy to well-to-tank (WTT) and tank-to-wake (TTW) emissions is taken from [23]. The conversion factors k_{truck}, $\mathrm{k_{CO_2}}$ and $\mathrm{k_{w2t}}$ are taken from [2].

Table 2. Parameters for EC_δ estimation based on scenario B in [26] where the ship concept 100% carries 50% of its TEU capacity.

Parameter	Truck	*Concept 100%*
Travel distance (km)	236	194
TEU weight (ton)	14	14
Transported TEU	104	104
Energy (MJ)	-	144 121
$k_{e2GHG_{WTT}}$ ton $CO_{2\text{eq}}$/MJ	-	0.0001508
$k_{e2GHG_{TTW}}$ ton $CO_{2\text{eq}}$/MJ	-	0
k_{truck} (Euro/ton)	0.04	-
k_{CO_2} (Euro/ton $CO_{2\text{eq}}$)	-	100
K_{w2t} (Euro/ton km)	-	0.0006

If we insert all these parameters to the equation for EC_δ (equation 14), we find that the ship *concept 100%* would reduce external costs by an estimated 12 190 Euro for each trip that it sails, when this results in moving the transport from road to sea. The calculations are as follows:

First, we calculate the reduction in truck ton km:

$$L_\delta = -236 * 14 * 104 = -343\ 616 \text{ ton km} \tag{19}$$

The additional shipping ton kilometer:

$$L = 194 * 14 * 104 = 282\ 464 \text{ ton km} \tag{20}$$

And the emitted $CO_{2\text{eq}}$ by the ship:

$$CO_{2eq} = energy * (k_{\text{e2GHG}_{\text{WTT}}} + k_{\text{e2GHG}_{\text{TTW}}}) = 144121 * (0 + 0{,}0001508) = 21.73 \text{ ton CO}_{2\text{eq}} \tag{21}$$

Before the reduction in external cost is estimated by inserting all values into equation (14):

$$EC_\delta = -343616 * 0{,}042 + 21.73 * 100 + 0{,}4 * 282464 * 0{,}0006 \ = -12\ 190 \text{ Euro} \tag{22}$$

Note that P_i from equation 14 was set to zero in this example since the concept was assumed to be using hydrogen fuel cells for energy generation, which results in emitting zero pollutants P_i.

5. Discussion and conclusion

Deciding to invest in a transportation service requires that one is confident that the service will be competitive and attract customers. Therefore, we took on the transportation service customer perspective when we developed KPIs for evaluating the competitiveness of autonomous ship-based transportation service concepts. Additionally, there are expectations for autonomous ships having a positive impact on the society and external costs. If autonomous ships could be shown to have such impacts, it might give rise to subsidy programs. Such as ENOVA which subsidies investments with positive environmental impacts, Yara Birkeland being an example [12].

There are methods for comparing alternatives based on multiple criteria, such as grey relational analysis (GRA) where criteria scores are weighed and aggregated into one score which is used for comparison of alternatives. GRA is applied in [27] to investigate "...*beneficial (competitive) impact of the Northern Sea Route...*". Another well-known method is to monetize benefits and costs and perform a cost benefit analysis. It is also possible to compare criteria by criteria, and simply do a qualitative assessment. Whichever approach is chosen, the KPIs proposed in this paper can be applied.

External costs generated from well-to-tank for new fuels, such as hydrogen, are missing in [2]. Some alternative sources exist, such as [23], however, therein, the only well-to-tank impact on society that is considered is the emitted CO_2 equivalents. More research is therefore needed on external costs related to well-to-tank for modern fuels.

Remote control center costs, autonomous technology costs, Automatic Facilities Services costs, and maintenance costs for autonomous ships are also cost items that there are few published sources for estimating. Further studies on these topics are therefore needed.

There are some shortcomings in our proposal for KPIs related to life-cycle evaluations. What is missing is the end-of-life, or decommissioning costs (both economic and environmental). Studies on how these costs are influenced by different novel ship concepts are therefore needed.

Regarding access to reliable, high-quality data in the estimation of KPIs in the design phase, we are relying on simulation tools utilizing available calculation parameters. This is particularly relevant for KPIs related to emissions. Several initiatives are currently underway, such as in the projects AUTOSHIP [15] for ship designs and AEGIS [28] for logistics chains. Such tools will be a vital part of producing reliable KPIs, especially in early phases such as the design phase. One will never be able to make true calculations (such as Annual Energy Ratio or Energy Efficiency Operational Indicator) before the MASS is in actual operations but the closer the estimates derived from such tools; the better decision support can be provided. The KPI formulas will be the same, but the quality of the estimates will increase in correlation with the development of accurate simulation tools.

Positive cost benefit assessments are key to large scale uptake of MASS. In this regard, societal KPIs, expressing the actual positive societal impact of MASS will be ever more important in terms of providing decision support to assessment of public support and subsidies. A clear and internationally accepted set of societal KPIs is an important contributor in that regard.

Acknowledgments
The work presented in this paper is part of the AUTOSHIP project funded by EU Horizon 2020 with grant number 815012. Thanks to Lars Andreas Lien Wennersberg for comments and contributions.

References

[1] Kallas S 2011 White Paper on Transport: Roadmap to a Single European Transport Area—towards a Competitive and Resource-efficient Transport System. *Brussels: Office for Official Publications of the European Communities*.
[2] Essen H, Wijngaarden L, Schroten A, Sutter D, Bieler C, Maffi S, Brambilla M, Fiorello D, Fermi F, Parolin R, and Beyrouty K E 2019 Handbook on the external costs of transport, version 2019 — 1.1. *Delft: European Commission, Directorate-General for Mobility and Transport.*
[3] European Commission 2021 Report from the commission to the European parliament and the council — Seventh monitoring report on the development of the rail market under Article 15(4) of Directive 2012/34/EU of the European Parliament and of the Council, *Brussels: European Commission*.
[4] Pastori E, Brambilla M, Maffii S, Vergnani R, Gualandi E, and Skinner I 2018 Research for TRAN Committee – Modal shift in European transport: a way forward. *Brussels: European*

Parliament, Policy Department for Structural and Cohesion Policies.

[5] Kretschmann L, Burmeister H, and Jahn C 2017 Analyzing the economic benefit of unmanned autonomous ships: An exploratory cost-comparison between an autonomous and a conventional bulk carrier. *Research in Transportation Business and Management*, **25**, 76-86.

[6] Munim Z 2019 Autonomous ships: A review, innovative applications and future maritime business models. *Supply Chain Forum*, **20**, 266-279.

[7] Akbar A, Aasen A K, Msakni M K, Fagerholt K, Lindstad E, and Meisel F 2021 An economic analysis of introducing autonomous ships in a short-sea liner shipping network, *International Transactions in Operational Research*, **28**, 1740-1764.

[8] Bălan C 2018 The disruptive impact of future advanced ICTs on maritime transport: A systematic review, *Supply Chain Management: An International Journal*, **25**, 157-175.

[9] Gu Y, Goez J, Guajardo M, and Wallace S 2021 Autonomous vessels: state of the art and potential opportunities in logistics. *International Transactions in Operational Research*, **28**, 1706-1739.

[10] Ziajka-Poznańska E, and Montewka J 2021 Costs and benefits of autonomous shipping - a literature review. *Apllied Sciences* , **11**, 4553

[11] YARA 2018 YARA selects Norwegian shipbuilder VARD for zero-emission vessel Yara Birkeland. https://www.yara.com/corporate-releases/yara-selects-norwegian-shipbuilder-vard-for-zero-emission-vessel-yara-birkeland/ visited 27.01.2022.

[12] ENOVA 2017 Zero emission logistical solution. https://www.enova.no/om-enova/om-organisasjonen/teknologiportefoljen/zero-emission-logistical-solution/ visited 24.01.2022.

[13] ISO, International Organisation of Standardisation, 2021 ISO/TS 23860:2021 (E): Ships and marine technology -- Vocabulary related to Autonomous Ship Systems, ISO (in print).

[14] Wennersberg L A L, Nordahl H, Rødseth Ø J, Hjørungnes S R, Faivre J, Liavaag E J, Bolbot V, and Leenders E 2020 AUTOSHIP deliverable D3.2: Autonomous ship design methods and test principles. AUTOSHIP H2020.

[15] AUTOSHIP H2020. 2022 The AUTOSHIP project, https://www.autoship-project.eu/the-project/ visited 13.01.2022.

[16] Wennersberg L A L, Nordahl H, Rødseth Ø J, Bolbot V, and Theotokatos G 2020 Analysing supply chain phases for design of effective autonomous ship technology in new transport system solutions. OMAE2020.

[17] Nordahl H, Nesheim A, Colella M, and Wennersberg L A L 2021 AUTOSHIP deliverable D8.1: Generic business case model and KPIs, AUTOSHIP H2020.

[18] Nordahl H, Kisialiou Y, and Wennersberg L A L 2021 AUTOSHIP Deliverable D7.5: Complete supply chain and logistics model, AUTOSHIP H2020.

[19] Fonseca T, Lagdami K, Schröder-Hinrichs J 2021 Assessing innovation in transport: An application of the Technology. *Transport Policy*, **114**, 182-195.

[20] Stopford M 2008 *Maritime economics* (3 ed.), London: Routledge

[21] Yara B 2019 Yara Birkeland press kit. https://www.yara.com/news-and-media/press-kits/yara-birkeland-press-kit/ visited 24.01.2022.

[22] SATS 2020 Shifting cargo to sea with Autonomous Transport Systems. The Norwegian Research Council project number 280626.

[23] Lindstad E, Gamlem G, Rialland A, and Valland A 2021 Assessment of Alternative Fuels and Engine Technologies to Reduce GHG. *In SNAME Maritime Convention. OnePetro*

[24] ICF International 2009 Current methodologies in preparing mobile source port-related emission inventories, U.S. Environmental Protection Agency

[25] Dæhlen J S, Sandvik E, Rialland A I, and Lagemann B 2021 A Method for Evaluating Ship Concepts in Realistic Operational Scenarios using Agent-based Discrete-Event Simulation. *20th International Conference on Computer and IT Applications in the Maritime Industries (COMPI21)*, 141-150.

[26] Gribkovskaia V, Borgen H, Holte E A, Lindstad E, and Nordahl H 2019 Autonomous ships for

Journal of Physics: Conference Series **2311** (2022) 012020 doi:10.1088/1742-6596/2311/1/012020

coastal and short-sea shipping. *In SNAME Maritime Convention.*
[27] Sur J M, and Kim D J 2020 Multi criteria evaluation of beneficial effect of commercializing Northern Sea Route on Europe and Asia countries. *The Asian Journal of Shipping and Logistics*, **36**, 181-189.
[28] AEGIS H2020 2022 Say hello to AEGIS (Advanced, Efficient and Green Intermodal Systems), https://aegis.autonomous-ship.org/ visited 26.01.2022.

Journal of Physics: Conference Series **2311** (2022) 012021 doi:10.1088/1742-6596/2311/1/012021

Evaluation of effectiveness of the STAMP / STPA in risk analysis of autonomous ship systems

Tomoaki YAMADA[1*], Makoto SATO[1], Rikiya KURANOBU[1], Ryo WATANABE[1], Hiroko ITOH[2], Megumi SHIOKARI[2], Tomohiro YUZUI[2].

1 Nippon Kaiji Kyokai, Japan

2 National Maritime Research Institute, National Institute of Maritime, Port and Aviation Technology, Japan

Corresponding author: Tomoaki YAMADA; e-mail: t-yamada@classnk.or.jp; Tel.: 81 3 52262737 Fax: 81 3 52262736

Abstract. To ensure safety of Maritime Autonomous Surface Ships (MASS), the importance of risk assessment is emphasized in classification societies, but the assessment method is not specified, and standardization of the method is expected. In the risk analysis of conventional ship, we focus on equipment failure mainly on hardware, but in MASS, it is necessary to analyze not only hardware but also large-scale and complicated systems including software and human beings. STAMP / STPA is an analysis method for large-scale and complicated systems, and it may be applicable to MASS risk analysis. In this paper, we will verify the effectiveness of STAMP / STPA in the initial stage of MASS design and consider it from a certification point of view.

1. Introduction

Maritime Autonomous Surface Ship (hereinafter MASS) is a large-scale and complicated system consisting of many functional groups. When determining a concept of a MASS, it is important to organize a wide variety of onboard tasks from the perspective of "to what extent" and "which tasks" are autonomous (automated or remotely controlled) in the MASS [1]. For example, system configuration for maneuvering tasks of MASS is constituted conceptually of a group of sensors for collecting information, a group of cognitive functions for analyzing and recognizing information on other ships and obstacles from the input information, and a group of control functions that control the direction and propulsion of the ship based on the voyage plan approved before departure and the action plan for avoiding obstacles on the route. In addition, a communication system between the MASS and the Remote Operation Center (hereinafter ROC) is required for remote monitoring and emergency response.

The necessity of new goal-based MASS instrument is being discussed at International Maritime Organization (IMO). From certification point of view, in the case of MASS, a highly complicated system will be installed, and it is difficult to set uniform requirements for such a large-scale system. Even for systems with the same function, the requirements and performance standards specified for certification might differ depending on the conditions under which it is operated. To ensure goal-based safety, risk management from the early design stages is important.

The interim guidelines issued by IMO [2] and guidelines issued by some flag states [3]-[5] specify the implementation of risk assessment. Guidelines for MASS have already been issued by multiple classification societies [6]-[9], and risk assessment is also emphasized in all of them. For example, in ClassNK guidelines [6] it is necessary to carry out risk assessment depending on the development phase of the autonomous ship system.

On the other hand, there are no clear rules or guidelines regarding the method used for risk assessment, and the choice is up to the system supplier. Therefore, there are some cases where risk assessment is not always performed in the form intended by third-party certifiers such as classification societies, and standardization of risk assessment methods according to the development phase is expected in the industry.

In particular, in large-scale and complicated systems, not only failure of individual components / subsystems, but also hazards hidden in the interaction between components / subsystems may lead to accidents such as collisions, grounding, and sinking. In the conventional risk analysis for ships, hazards have been identified by focusing on equipment failures using a system configuration diagram mainly consists of hardware, but in risk analysis for MASS, it is important to comprehensively extract hazards based on the above characteristics. For that purpose, by aggregating hardware, software and human beings, etc. that compose MASS are regarded as one big system, and it is necessary to clearly define the task that each component of the entire system is in charge of, the processes for the exchange of information between the components / subsystems, and the processes that require human approval from the initial stage of design [10]-[12].

As a safety analysis method for large-scale and complicated systems, system-theoretical approaches such as STAMP / STPA (Systems-Theoretic Accident Model and Processes / System Theoretic Process Analysis) [13] are attracting attention. STAMP / STPA is an analysis method that considers hazard factors in units of interacting functions [13], and is considered to be useful for risk analysis in the early stages of designing large-scale and complicated systems. It is also considered useful to select an appropriate analysis method according to the design stage [14],[15]. The application of STAMP / STPA to the autonomous ship system has already been studied in multiple papers [15]-[21].

In this paper, we evaluate the effectiveness of STAMP / STPA to a risk assessment in the initial stage of MASS design, and also considered how STAMP / STPA works in goal-based safety evaluation process from a certification point of view.

2. Analysis

2.1. Subject for analysis

In this study, the unmanned bridge ship that operate automatically under remote monitoring at ROC is analyzed. STAMP / STPA was performed on "the hypothetical autonomous ship" defined by Shiokari et al. (2021). Figure 1 shows the conceptual structure of "the hypothetical autonomous ship" set in this paper, and Table 1 shows the outline. In this paper, we consider only the mechanism from a technical point of view, and do not deal with legal issues (such as seafarers' allocation and qualification requirements).

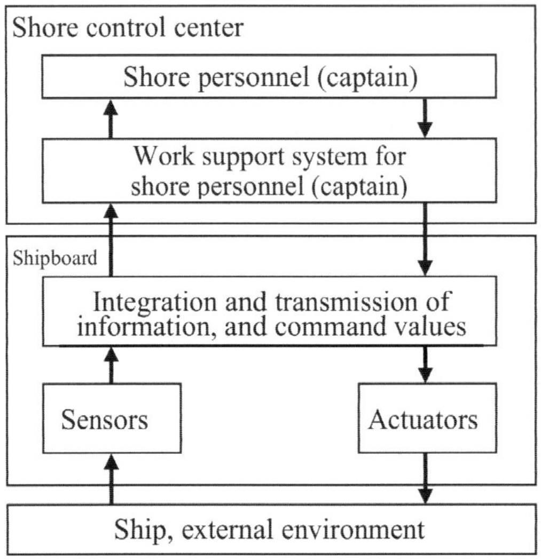

Figure 1. Conceptual structure of the hypothetical autonomous ship [11]

Table 1. Outline of the hypothetical autonomous ship

Degree of autonomy	Unmanned bridge - Autonomous "planning of operation and decision for ship maneuvering" and "condition check for equipment and automation system" - The decision maker is a remote operator (captain) on ROC. - Installed "emergency response system" for emergency response onboard
Target of autonomy	Navigation task on bridge (automation and remote monitoring)
Size of MASS	less than 500 GT class
Assumed route	Short-distance regular route
Others	- No ballast adjustments will be made during the voyage. - Mooring relies on port equipment.

2.2. Method (STAMP / STPA)

Figure 2 shows the basic STAMP / STPA method. In this study, risk analysis for the hypothetical autonomous ship was performed using this procedure. Information-technology promotion agency's STAMP Workbench [22] is used for analysis. This analysis was carried out by multiple people with different specialties, those who are familiar with risk assessment methods including STAMP / STPA, those who have experience in designing automated systems, and those who have experience for certification work in classification societies.

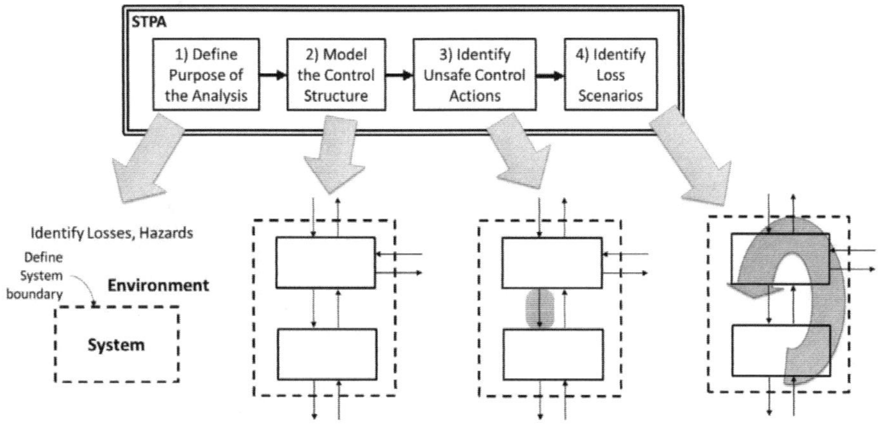

Figure 2. Overview of the basic STAMP / STPA Method [13]

3. Result

3.1. (Step 1) Define Purpose of the Analysis

As the first step of STAMP / STPA, hazards and safety constraints (hereinafter SC) were extracted after identifying losses for the autonomy of ship maneuvering tasks. The results are shown in Table 2. In this analysis, in order to consider the safety constraints that should be achieved as the entire MASS, we started with the highest concept of "ship".

Table 2. Hazard identification and safety constraints

ID	Hazard	ID	Safety constraints
H1	The MASS is too close to other ships and drifting objects.	SC1	The MASS should not approach too close to other ships nor drifting objects.
H2	The MASS is too close to the quay.	SC2	The MASS should not approach too close to the quay.
H3	The MASS is too close to shallow water.	SC3	The MASS should not approach too close to shallow water.
H4	The MASS loses stability.	SC4	The MASS should not lose stability.
H5	The MASS tilts abnormally.	SC5	The MASS should not tilt abnormally.
H6	There is something wrong with the fire prevention / extinguishing equipment of the MASS.	SC6	There should be no abnormality in the fire prevention / extinguishing equipment of the MASS.
H7	The acceleration of the MASS is too large.	SC7	The acceleration of the MASS should not be too large.
H8	The MASS blacks out.	SC8	The MASS should not black out.
H9	An engine abnormality occurs on the MASS.	SC9	There should be no abnormality in the MASS's engine.
H10	An abnormality occurs in the maneuvering and propulsion equipment of the MASS.	SC10	There should be no abnormalities in the maneuvering and propulsion equipment of the MASS
H11	An abnormality occurs in the fuel oil system of the MASS.	SC11	There should be no abnormality in the fuel oil system of the MASS.

3.2. (Step 2) Model the Control Structure

Figure 3 shows the modeled control structure diagram (hereinafter CS diagram). Regarding communication between the MASS and ROC, each onboard system independently transmits information to ROC.

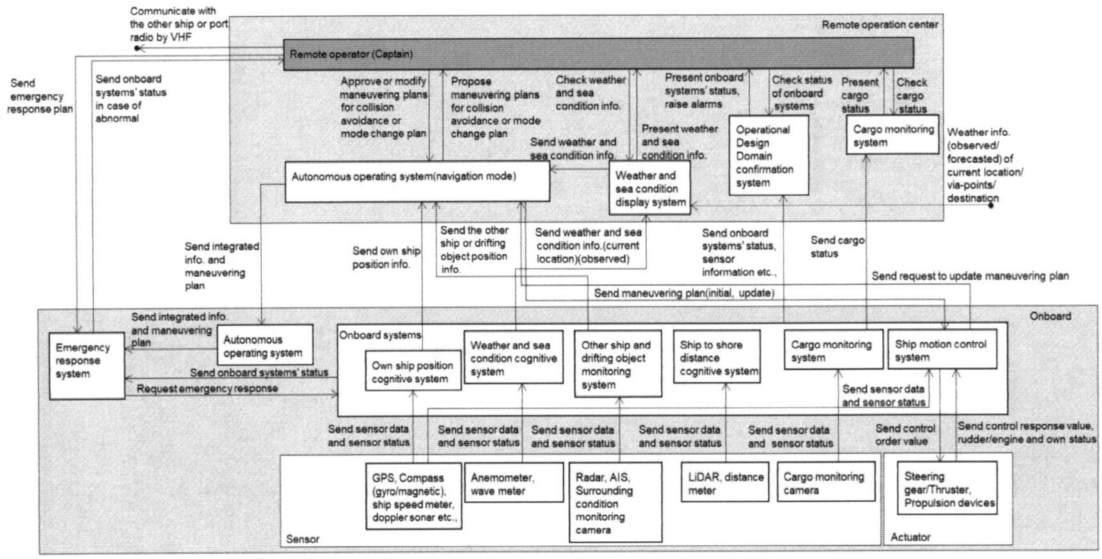

Figure 3. Control structure diagram of the hypothetical autonomous ship

3.3. (Step 3) Identify Unsafe Control Actions (UCA)

Table 3 shows typical UCAs identified in this study. Since the purpose is not to focus on equipment failure but to comprehensively identify hazards for the entire MASS system, we selected UCAs that include information exchange between each component / subsystem such as hardware, software, and humans, and processes that require human approval. "Stop too soon / Applying too long" was not extracted in Table 3.

Table 3. UCA identified

(UCA No) CA	(UCA No.) Not Providing [related safety constrains]	(UCA No.) Providing causes Hazard [related safety constrains]	(UCA No.) Too early / Too late [related safety constrains]	Stop too soon / Applying too long
(UCA3) Check status of onboard systems (from Remote operator (captain) to Operational Design Domain confirmation system)	(UCA3-N-1) The remote operator (captain) does not check the status of onboard systems. [SC1][SC3][SC6][SC7][SC8][SC9][SC10][SC11]	(UCA3-P-1) The remote operator (captain) misidentifies the status of onboard systems. [SC1][SC3][SC6][SC7][SC8][SC9]SC10][SC11]	(UCA3-T-1) Too late：Confirmation of onboard systems by remote operator (captain) is delayed. [SC1][SC3][SC6][SC7][SC8][SC9][SC10][SC11]	Nil
(UCA7) Send integrated info. and maneuvering plan (from Autonomous operating system (ROC) to Autonomous operating system (onboard))	(UCA7-N-1) Do not send integrated information even when the transmission time comes. [SC1][SC3][SC4][SC5][SC6][SC7][SC8][SC9][SC10][SC11]	(UCA7-P-1) Send with errors in the integrated information. [SC1][SC3][SC4][SC5][SC6][SC7][SC8][SC9][SC10][SC11]	(UCA7-T-1) Too late：Delayed transmission of integrated information. [SC1][SC3][SC4][SC5][SC6][SC7][SC8][SC9][SC10][SC11]	Nil
(UCA16) Send onboard systems' status, sensor information, etc. (from Onboard systems to Operational Design Domain confirmation system)	(UCA16-N-1) Onboard systems does not send own status, sensor information, etc. [SC1][SC3][SC4][SC5][SC6][SC7][SC8][SC9][SC10][SC11]	(UCA16-P-1) The onboard systems send inappropriate own status, sensor information, etc. [SC1][SC3][SC4][SC5][SC6][SC7][SC8][SC9][SC10][SC11]	(UCA16-T-1) Too late：Transmission of own status, sensor information, etc. from onboard systems is delayed. [SC1][SC3][SC4][SC5][SC6][SC7][SC8][SC9][SC10][SC11]	Nil

3.4. (Step 4) Identify Loss Scenarios
For the UCA identified in Section 3.3, control loop diagrams between components / subsystems were developed by paying attention to the connection and operation between the component / subsystem. The hint word set for HCF (Hazard Causal Factor) [22] were also utilized to identify the HCF. As an example of a control loop diagram, Figure 4 shows a diagram of onboard systems and operational design domain (hereinafter ODD) confirmation system. Table 4 shows a summary of the loss scenarios for UCAs extracted in this way and their countermeasures.

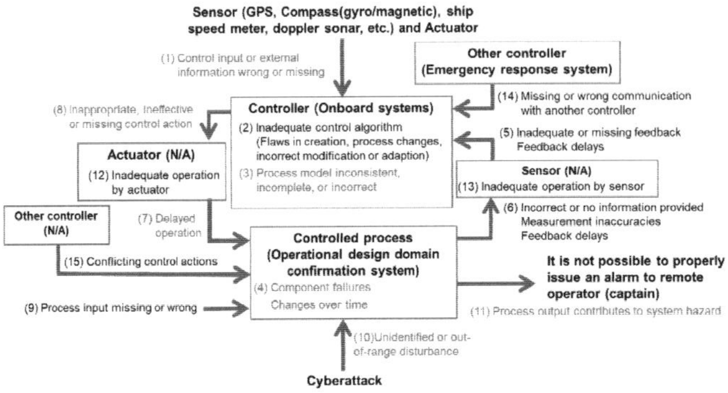

Figure 4. Control loop diagram between onboard systems and Operational Design Domain confirmation system

Table 4(a). Loss scenarios for UCAs extracted and countermeasures (UCA3-N-1)

HCF (HCF No.)	Loss scenarios	Countermeasures
The remote operator (captain) forgets/neglects to check the status of onboard systems periodically. (HCF3-N-1-6)	An error occurs in onboard systems and deviates from ODD. - The ODD confirmation system displays deviation from ODD without alarm. - The remote operator (captain) does not confirm the ODD confirmation system due to other operations, - The remote operator (captain) does not notice any abnormalities and does not give any instruction to the emergency response system. - Improper autonomous maneuvering will continue, and the MASS may approach or collide with other ships or drifting objects.	<Measures to prevent the occurrence of HCF> - Thoroughly pre-instruction to the remote operator (captain) that regular confirmation is required even if the alarm does not sound. - Addition of alarm function to periodically prompt confirmation to the remote operator (captain). - Optimization of the workload of remote operator (captain)
The remote operator (captain) does not notice the alarm. (HCF3-N-1-12)	An error occurs in the onboard systems and deviates from ODD. - An alarm is issued from the ODD confirmation system to the remote operator (captain), but the remote operator (captain) does not notice alarm because the alarm volume setting is low / the alarm is turned off by mistake / physical condition of the remote operator (captain) is bad. - The remote operator (captain) does not notice any abnormalities and does not give any instruction to the emergency response system. - Improper autonomous maneuvering will continue, and the MASS may approach or collide with other ships or drifting objects.	<Measures to prevent the occurrence of HCF> - Optimization of alarm volume. - Visually noticeable alarm <Measures to prevent accidents even if HCF occurs> - Thoroughly pre-instruction to the remote operator (captain) that regular confirmation is required even if the alarm does not sound. - Addition of detection function of remote operator (captain) condition
The ODD confirmation system has not received the information. (HCF3-N-1-15)	An error occurs in the onboard systems and deviates from ODD. - Onboard systems' error cannot be sent due to sensor failure, communication device failure or cyberattack, etc. - Since the ODD confirmation system has not received the error information, the error is not displayed on monitor and alarm is not issued. - The remote operator (captain) does not confirm the ODD confirmation system. - The remote operator (captain) does not notice any abnormalities and does not give any instruction to the emergency response system. - Improper autonomous maneuvering will continue, and the MASS may approach or collide with other ships or drifting objects.	<Measures to prevent the occurrence of HCF> - Ensuring the reliability of communication between the MASS and ROC - Implementation of reliable cyber security measures - Ensuring the redundancy of sensors - Ensuring the reliability of each onboard system <Measures to prevent accidents even if HCF occurs> - Alarm function to communication abnormality independent of the ODD confirmation system

Table 4(b). Loss scenarios for UCAs extracted and countermeasures (UCA7-N-1and UCA7-P-1)

HCF (HCF No.)	Loss scenarios	Countermeasures
The Autonomous operating system (ROC) cannot send integrated information due to poor communication status. (HCF7-N-1-1)	The Autonomous operating system (ROC) cannot send integrated information to the Autonomous operating system (onboard) due to poor communication. - The latest integration information is not sent from the autonomous operating system (onboard) to the emergency response system. - Even if the emergency response system finds the mismatch between information from autonomous operating system and information from onboard systems, its findings cannot be sent to the remote operator (Captain) due to poor communication status. - Because the remote operator (captain) does not pay special attention to the autonomous operating system due to dozing or sudden illness, etc., the remote operator (captain) cannot notice the poor communication status. - Autonomous maneuvering continues with poor communication status, and the MASS may approach or collide with other ships or drifting objects.	<Measures to prevent the occurrence of HCF> - Ensuring the reliability of communication between the MASS and ROC - Implementation of cyber security measures <Measures to prevent accidents even if HCF occurs> - Addition of function to check the current communication status of the autonomous operating system - Installation of a function that can detect and report doze and sudden illness of remote operator (Captain)
Each information of the integrated information is missing or incorrect (HCF7-P-1-1)	Each information of the integrated information (position of own ship / other ships / drifting objects, weather and sea condition, etc.) is missing or incorrect. - The MASS may be maneuvered based on an incorrect decision, and the MASS may approach or collide with other ships or drifting objects.	<Measures to prevent the occurrence of HCF> - Addition of function to verify reliability of each information of integrated information - Development of reliable calculation algorithm <Measures to prevent accidents even if HCF occurs> - Examination of countermeasures when reliability is low

Table 4(c). Loss scenarios for UCAs extracted and countermeasures (UCA16-P-1)

HCF (HCF No.)	Loss scenarios	Countermeasures
Incorrect information is sent to the ODD confirmation system due to improper processing by some of onboard systems. (HCF16-P-1-1)	Incorrect information is sent to the ODD confirmation system due to improper processing by some of onboard systems. - Since the ODD confirmation system processes based on incorrect information, it is not possible to issue an alarm properly. - Even if the MASS status is out of ODD, the remote operator (captain) does not give an instruction to the emergency response system because the alarm is not issued properly. - Improper autonomous maneuvering will continue, and the MASS may approach or collide with other ships or drifting objects.	<Measures to prevent the occurrence of HCF> - Ensuring data quality - Ensuring the redundancy of sensors <Measures to prevent accidents even if HCF occurs> -Addition of function to onboard systems to verify the reliability of each sensor information. And do not send sensor information to the ODD confirmation system if reliability is suspected. -Addition of function to the ODD confirmation system to notify the remote operator (captain) of an error when the information from onboard systems is missing.

From Table 4, we got the awareness from two viewpoints, one that leads to the improvement of the CS diagram and another that leads to the detailed design (concrete functional requirements) of each component / subsystem.

For example, the following items were extracted as common to multiple HCFs that would lead to the improvement of the CS diagram.

Improvement points of the CS diagram for the ROC are shown as follows.

- Review of the number of people performing tasks at ROC (it is necessary to change the composition that one remote operator (captain) must check the status of the onboard systems.)
- Clarification of countermeasure when the remote operator (captain) is in poor physical condition (Establishment of fallback system in ROC)"
- Installation of a component / subsystem that can detect and report doze or sudden illness of remote operator (captain).

Similarly, the following items were extracted as those that lead to the detailed design (materialization of functional requirements) of each component / subsystem.

- For the systems on ROC, "reliable cyber security measures", "audiovisually noticeable alarms", and "alarm function to communication abnormality between the MASS and ROC" are required.
- For onboard systems (own ship position cognitive system, weather and sea condition display system, other ship and drifting object monitoring system, etc.), "function to verify reliability of each information of integrated information" and "development of reliable calculation algorithm" are required.

4. Discussion

4.1. Consideration of system safety design through validity verification of conceptual design from a certification point of view

In the case of MASS, a highly complicated system will be installed, and it is difficult to set uniform requirements for such a large-scale system. This is because even if the system has the same function, the requirements and performance standards to be specified differ depending on the conditions under which it is operated [1],[6],[10].

The policy of the NK guidelines [6] is goal-based. In case of ClassNK, examinations based on the functional requirement specifications of autonomous operation systems will be conducted. For example, the NK guidelines describe the basic elements for ensuring safety that should be considered at the conceptual design stage. The purpose of this is to correctly understand the characteristics of the autonomous operation system by organizing the information based on these basic elements. Table 5 shows a summary of the basic elements for the hypothetical autonomous ship dealt with in this study.

Table 5. Items to be considered at conceptual design phase and the settings in the hypothetical autonomous ship [6], [11]

Basic elements for ensuring safety	Model of the hypothetical autonomous ship
Target task of automated operation on a ship	Ship maneuvering tasks in each phase (planning, berthing and unberthing, and voyage)
Division of roles between humans and automated operation systems	Situation awareness：Onboard systems Decision：Remote operator (captain) Action：Onboard systems
Prerequisite specification for system installation	See the sensors and actuators in Figure 3. Surrounding condition monitoring cameras, LiDAR, distance meters, cargo monitoring camera, and communication antennas for onboard systems to send information to and ROC are notable differences from conventional ships.
Operational design domain (ODD)	Materialized at detailed design
Fallback executor	Emergency response system
Human machine interface (HMI)	Materialized at detailed design
Cyber security	Materialized at detailed design
Reliability of computer systems	Materialized at detailed design

In the actual examination by ClassNK, a document examination and a function verification test are performed to confirm the validity of the design and that the developed system meets the functional requirement specifications. The document examination also includes a review of risk assessment results. In addition to confirming that the validity of the overall system configuration has been properly verified, it is also an important point whether the three relationships between ODD, fallback, and MRC (Minimum Risk Condition) are properly designed. As for the function verification test, the test items are to be decided according to the characteristics of the autonomous operation system. So, it is important to properly organize the functional requirements of the autonomous operation system in advance. That is why, the NK guidelines also mention the design and development process, and the application of the Verification and Validation (V&V) processes is recommended as an example in order to execute this flow effectively. The general procedure of V&V process is as follows.

a) Validation of conceptual design
b) Extraction of functional requirements for each component / subsystem to be considered during detailed design
c) Verification for each component / subsystem
d) Verification for integrated subsystem
e) Validation of the entire system

Although it is possible to receive a review of classification societies with complete evidence after the V&V process has been completed, it is also expected that the time required for examination for certification will be shortened by adding classification societies to the flow of this V&V process and conducting a step-by-step review.

In this study, consideration from a certification point of view using the NK Guidelines [6] for the analysis results of "the hypothetical autonomous ship" by STAMP / STPA were performed. Assuming analysis for conceptual design and keeping Approval in Principle (AiP) in mind, STAMP / STPA was applied to the processes a) and b). At the conceptual design stage, the validity of the entire system is mainly confirmed rather than the details of the subsystem / component.

4.1.1. Validity of conceptual design. As described in Section 3.4, the following notices have been obtained through the analysis, which will lead to the improvement of the CS diagram.

- Review of the number of people performing tasks at ROC (it is necessary to change the composition that one remote operator (captain) must check the status of the onboard systems.
- Clarification of countermeasure when the remote operator (captain) is in poor physical condition (Establishment of fallback system in ROC)
- Installation of a component / subsystem that can detect and report doze or sudden illness of remote operator (captain).

Based on this awareness, the modified CS diagram in which remote operator monitoring system is newly installed is shown in Figure 5.

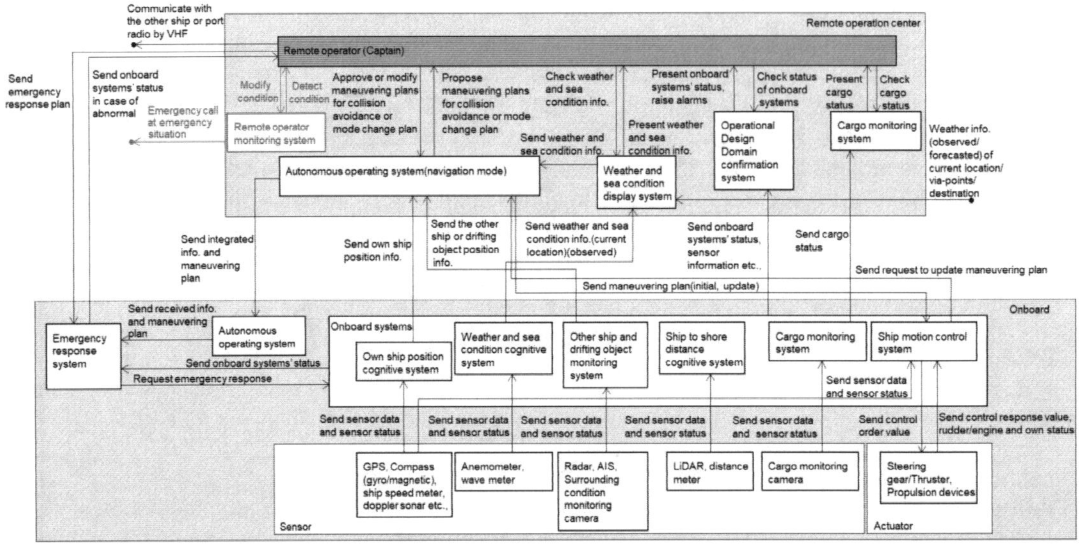

Figure 5. Modified CS diagram

Although not reflected in Figure 5, the analysis in this study clearly showed that there are a large number of functional requirements for each component / subsystem, especially those required of remote operator (captain). With this in mind, we can review the minimum number of people required in ROC at the conceptual design stage. Also, it is important to thoroughly consider the necessary HMI of ROC equipment required at the detailed design stage. These items could be functional requirements for the ROC operations and ROC equipment, respectively.

In this way, by describing loss scenarios and examining countermeasures, the CS diagram could be improved. It can be said that the validity of the conceptual design could be verified by STAMP / STPA. According to the NK guidelines, it is necessary to confirm the safety of the design of the autonomous system by using an appropriate risk analysis method. From the result of this study, it is considered that the verification required by the classification society can be effectively carried out by using STAMP / STPA at conceptual design phase.

4.1.2. Extraction of functional requirements (for component / subsystem) to be considered during detailed design. As described in Section 3.4, applying STPA to the CS diagram (Figure 3) modeled by STAMP, we got the awareness which will lead to the detailed design (materialization of functional requirements) of each component / subsystem.

For the systems on ROC, awareness has been obtained that following items are required:

- reliable cyber security measures,
- audiovisually noticeable alarms, and
- alarm function to communication abnormality between the MASS and ROC

For onboard systems (own ship position cognitive system, weather and sea condition display system, other ship and drifting object monitoring system, etc.), following items are required:

- function to verify reliability of each information of integrated information and
- development of reliable calculation algorithm

By paying attention to the cooperation between components / subsystems, for example, it is possible to realize the necessity of having a function to check the reliability of the received information in the downstream subsystem (the subsystem that receives the information). In addition, it has been possible to extract specific requirements for HMI of components / subsystems that require cooperation with humans. In this way, STPA makes it possible to extract the functional requirements of subsystems / components in consideration of the characteristics of autonomous operating systems.

It will lead to a great reduction in development rework by sharing and agreeing on the functional requirements extracted in this way with the certification side such as classification societies. In addition, the content and method of the test to be carried out in step c) can be efficiently discussed with certification side.

4.1.3. Consideration for test method (Verification). In this study, it was confirmed that the upstream part of V & V process can be effectively carried out even for the hypothetical autonomous ship by analyzing the CS diagram modeled by STAMP / STPA. On the other hand, once the CS diagram can be modeled, it is possible to check or sophisticate it virtually by computer simulation such as 1D simulation, and such efforts are being promoted in the automobile industry, etc., which model-based development has already been adopted. In the shipping industry as well, the Comprehensive Simulation System for MASS is being developed by Minami et al. [24]. In this paper, verification using 1D simulation has not been carried out, but it is considered effective to combine STAMP / STPA and 1D simulation as a method to consider the validity of the MASS design efficiently and effectively.

4.2. Points to keep in mind when applying the STAMP/STPA method to MASS widely

(1) Safety Constraint

Assuming that designers who are unfamiliar with STAMP / STPA will analyze by themselves, there is a concern that considerably disjointed SCs will appear.

Therefore, it is effective to present the market view of SC by presenting analysis examples. Based on such analysis examples, it is important to share the awareness among stakeholders about the safety level required for the system from the initial design stage.

(2) UCA

While the UCA could be extracted efficiently by having the guide words, the work of describing the UCA in combination with the guide words is monotonous and mechanical work continues. If the CS diagram is drawn in too much detail from the beginning, the amount of work will be enormous. By keeping the appropriate size of the component / subsystem to be analyzed in such as information gathering / situation awareness / decision / action (control), it is possible to efficiently define UCA.

(3) HCF and Loss scenarios

The points that we noticed when describing the loss scenarios in this study are shown below.

- When extracting HCF, it is unclear whether to describe the scenarios that lead to all related SC violations or only the most probable scenario.
- The interaction between components / subsystems can be expressed as CA, but the interaction between CAs (dependency, order of CAs in time series, etc.) is difficult to express.
- When extracting the HCF, whether to consider not only the direct causes of UCA but also the indirect causes or not. Combinatorial explosions can be occurred if the cause investigation is traced back too much to.

Since there is no clear explanation for these points in the method, it can be said that it is necessary to set rules when performing analysis.

(4) Consideration throughout

The major feature of the maritime industry is that there are many stakeholders, but the structure becomes more complicated in MASS, where the concept of operation (CONOPS) and the degree of system autonomy must be considered in parallel. Therefore, risk analysis for MASS tends to be complicated as well, but at least at the conceptual design stage, it is possible to carry out risk analysis, which some classification societies can accept, by using STAMP / STPA.

In the case of MASS, it seems that STAMP / STPA is more effective when the system scale is large and the degree of design progress is upstream. On the other hands, as the design goes downstream, conventional methods may be easier to use for stakeholders in the shipping industry. Therefore, it is necessary to select a method according to both the size of target system and the degree of design progress, considering whether the analysis target is a ship or system as well as conceptual design or detailed design.

5. Conclusions

In this study, STAMP / STPA [13] was carried out for the hypothetical autonomous ship by Shiokari et al. (2021), and the effectiveness of STAMP / STPA at the conceptual design stage of MASS was evaluated, together with consideration how STAMP / STPA works in goal-based safety evaluation process from a certification point of view. The findings obtained in this study are summarized below.

a) By using STAMP / STPA for risk analysis at the conceptual design stage of MASS, it is possible to easily share the characteristics of the autonomous operation system with the certification side such as classification societies. In addition, the analysis result can be used as a material for the classification societies' review for AiP and so on. For example, the NK guideline requires "system architecture that gives an overview of the autonomous operation system" at the time of examination, and the CS diagram modelled by STAMP/STPA can be utilized as this material.
b) In this study, we showed that STAMP / STPA extracted the functional requirements for the hypothetical autonomous ship. By sharing and agreeing on the extracted functional requirements with the certification side such as classification societies, it is expected to reduce development rework and help analysts efficiently consider the content and method of the test to be conducted. On the other hand, we analyzed the typical use cases assumed from the system model and extracted the functional requirements this time, but since multiple different uses are possible depending on the system model, it is necessary to organize the use cases and consider the relationship with the functional requirements.
c) Considering that there are many possible forms of autonomous operating systems, it is also necessary to perform risk assessment in a form that is easy for certifiers to review. From that point of view, this paper has showed the possibilities that STAMP / STPA can work effectively as a method of both risk assessment and extraction of functional requirements. On the other hand, STAMP/STPA is still unfamiliar in the maritime industry, especially for ship designer. This means there is a risk that the analysis by ship designer may become analyst-dependent. In order to solve this, it is necessary to present an analysis example like this paper and proceed with standardization.

Acknowledgements
This work was partly supported by JSPS KAKENHI Grant Number JP20K14969.

References

[1] Yamada T: Safety Evaluation for Technologies related to Autonomous Ships, ClassNK Technical Journal No.3, 2021(I)
[2] IMO MSC.1/Circ.1604 (2019), INTERIM GUIDELINES FOR MASS TRIALS
[3] Maritime Bureau, Ministry of Land, Infrastructure, Transport and Tourism: [Guidelines for Safety design of MASS] Jidouunkousen no anzen sekkei gaidorain (2020) (Japanese)
[4] VTMIS, EU OPERATIONAL GUIDELINES FOR SAFE, SECURE AND SUSTAINABLE TRIALS OF MARITIME AUTONOMOUS SURFACE SHIPS (MASS)
[5] Norwegian Maritime Authority, RSV 12-2020: Guidance in connection with the construction or installation of automated functionality aimed at performing unmanned or partially unmanned operations
[6] ClassNK: Guidelines for Automated/Autonomous Operation on ships (Ver. 1.0) (2020)
[7] DNVGL: Autonomous and remotely operated ships, DNVGL-CG-0264 (2018)
[8] Bureau Veritas: Guidelines for Autonomous Shipping, Guidance Note NI 641 DT R01 E (2019)
[9] ABS: ABS advisory on autonomous functionality (2020)
[10] Shiokari M, Itoh H and Yuzui T: Towards the Development of Risk Analysis Method for Autonomous Ships, Conference proceedings of the Japan Society of Naval Architects and Ocean Engineers, No. 30, 393–396, 2020. (In Japanese)
[11] Shiokari M, Itoh H, Yuzui T, Ishimura E, Miyake R, Kudo J and Kawashima S: Application of

Risk Analysis Method with System Modeling to Conceptual Design of Autonomous Ships, Conference proceedings of the Japan Society of Naval Architects and Ocean Engineers, No. 32, 355–366, 2021. (In Japanese)

[12] Itoh H, Yuzui T, Shiokari M, Ishimura E, Miyake R, Kudo J: Risk Assessment of Autonomous Ship Systems, ClassNK Technical Journal No.4, 2021.

[13] Leveson NG and Thomas JP, STPA handbook, 2018.

[14] Yamaguchi S, Shirasaka S: Evaluation of the Sequential Safety Analysis Method Based on the Related Severity of Hazards Using STAMP/STPA, JOURNAL OF JAPAN SOCIETY FOR SAFETY ENGINEERING, Vol. 58 No.2,pp.124-132,2019.

[15] Xiang-Yu Zhou, Zheng-Jiang Liu, Feng-Wu Wang, Zhao-Lin Wu, Ren-Da Cui: Towards applicability evaluation of hazard analysis methods for autonomous ships, Ocean Engineering, 214(October), 2020.

[16] Wróbel K, Montewka J, Kujala P: Towards the development of a system-theoretic model for safety assessment of autonomous merchant vessels, Reliability Engineering and System Safety 178, 209–224 (2018)

[17] Wróbel K, Montewka J, Kujala P: (2018). System-theoretic approach to safety of remotely-controlled merchant vessel. Ocean Engineering, 152(January), 334–345. https://doi.org/10.1016/j.oceaneng.2018.01.020

[18] Børge Rokseth, Odd Ivar Haugen and Ingrid Bouwer Utne: Safety Verification for Autonomous Ships, MATEC Web of Conferences 273, 02002 (2019), ICSC-ESWC 2018. https://doi.org/10.1051/matecconf/201927302002

[19] Hyungju KIM, Odd Ivar HAUGEN, Børge ROKSETH, Mary Ann LUNDTEIGEN: Comparison of Hazardous Scenarios for Different Ship Autonomy Types using Systems-Theoretic Process Analysis, Proceedings of the 29th European Safety and Reliability Conference (2019)

[20] Osiris A. Valdez Banda, Sirpa Kannos, HAZARD ANALYSIS PROCESS FOR AUTONOMOUS VESSELS, Novia Publikation och produktion, serie R: Rapporter 2/2019, ISBN: 978 - 952 - 7048 - 47 - 4 (online), ISSN: 1799 - 4179 (2019)

[21] Meriam Chaal, Osiris A. Valdez Banda, Jon Arne Glomsrud, Sunil Basnet, Spyros Hirdaris, Pentti Kujala: A framework to model the STPA hierarchical control structure of an autonomous ship, Safety Science 132 (2020) 104939

[22] Information-technology Promotion Agency, Japan: STAMP Workbench, https://www.ipa.go.jp/english/sec/reports/20180330.html

[23] Information-technology Promotion Agency, Japan: The introductory guidebook of STAMP/STPA (in Japanese)

[24] Minami M, Kokubun K, Kobayashi M, Hikida K, Yoshimura K, Sato K, Saito E, Sawada R: Development of Comprehensive Simulation System for Autonomous Ships, ClassNK Technical Journal No.4, 2021(II)

Wave-field prediction based on radar snapshots taken on a moving vessel

Yun Zhi LAW,[1*] Harrif SANTO,[1] Kie Hian CHUA[1]

[1]Technology Centre for Offshore and Marine Singapore (TCOMS),

12 Prince George's Park, Singapore 118411

Corresponding author: Yun Zhi LAW; e-mail: law_yun_zhi@tcoms.sg

Abstract. This paper presents a framework that is capable of producing phase-resolved wave-field prediction in real-time based on radar measurements taken off a moving vessel. Phase-resolved wave prediction capability is essential in providing the operating environment that an autonomous vessel would encounter in order to plan for maneuvering and pathway. We use Higher Order Spectral (HOS) method to generate synthetic short-crested wave-field. Based on the snapshots of wave-field as would be obtained by an X-band radar mounted on a moving vessel, we perform wave-field reconstruction and propagation using a predictive model based on linear wave theory to produce predictions of encountered wave-field in real-time. Comparisons are made between the predicted encountered wave-field and the underlying synthetic wave-field, and reasonably good agreement can be achieved considering the use of low-order model. The wave prediction capability can be further evolved into the prediction of vessel motion and structural response.

Keywords: real-time wave prediction, smart and autonomous ship, X-band radar

1. Introduction

A short-term deterministic sea wave prediction (DSWP) capability is essential to provide the operating environment for an autonomous vessel to plan for maneuvering and pathway. Based on the far-field wave measurements by an X-band radar on-board of the vessel, the DSWP capability is able to provide prediction of wave-field to be encountered by the vessel which can be further turned into prediction of vessel motion and structural response.

There have been extensive studies on DSWP from the literature. One notable example is the Environmental and Ship Motion Forecasting (ESMF) programme funded by Office of Naval Research (ONR), where commercial-off-the-shelf (COTS) Doppler marine radar was modified to determine the wave-field surrounding the vessel, non-linear wave theory was used to propagate the wave surface forward in time, and seakeeping theory was utilised to predict future vessel motions, e.g. see Alford et al. (2015), Connell et al. (2015). The ESMF programme subsequently yielded a commercial solution coined FutureWaves™, see Kusters et al. (2016). In addition, there are also solutions available making use of standard non-coherent X-band navigation radars, e.g. Hilmer and Thornhill (2015), which subsequently led to a commercial WaMoS II product.

The vast majority of the solutions make use of 3D FFT (Fast Fourier Transform) methods applied on measurement snapshots to derive characteristics of the underlying wave components (such as frequencies and wavenumbers) as part of wave-field reconstruction. For more details on 3D FFT, see Blondel-Couprie and Naiijen (2012) and Naiijen and Blondel-Couprie (2012).

Once the wave-field reconstruction is available, most of the solutions deploy linear wave propagation model to propagate the wave-field forward in time and produce wave prediction in or ahead of real-time. There are also methodologies which propose the use of non-linear physics-based wave propagation models, see e.g. Simanesew et al. (2017) and Köllisch et al. (2018), among others, as well as data-driven reduced-order model, such as Bukka et al. (2021).

Our objective is to develop a framework that can assimilate measurement of wave-field into a numerical engine to yield wave prediction in real-time. In this paper, we generate synthetic non-linear shorted-crested waves using HOS-Ocean code with varying sea-state steepness values (Bonnefoy et al., 2010; Ducrozet et al., 2016). We then assume a vessel, equipped with X-band radar capable of measuring phase-resolved wave-field, is moving with forward speed following certain path and heading. Subsequently, we make use of 2D FFT in order to obtain wave-field reconstruction from the moving vessel. The use of 2D FFT enables the use of 2-3 successive radar 2D snapshots, thereby avoiding the need of a series of 2D snapshots of wave-field as required for 3D FFT. We then use a linear wave propagation model based on linear wave theory to account for wave propagation, which works reasonable well for sea-states with low steepness. The developed framework can be further enhanced using non-linear wave propagation model and extended to yield vessel motion and structural response prediction in the future.

This paper is organized as follow. A description of our developed framework is presented in Section 2. Then, we describe the problem setup of this study in Section 3, followed by the detail of the implementation of linear wave theory on our developed framework in Section 4. The result and performance of our implementation is presented in Section 5, while the conclusion and the future work of this study is presented in Section 6.

2. Framework's description

Our developed framework consists of three components: (i) wave-field reconstruction, and (ii) wave-field propagation, and (iii) encountered wave-field prediction. Figure 1 shows the flow chart of these operations. Each of the components has a dedicated task to handle, and it should receive a well-defined input and produce a well-defined output. By fixing the type of input and output of each component, it allows us to upgrade or switch the algorithm implemented in each component easily.

It should be noted that our framework is originated from the similarity in the procedure of the works conducted on short-term deterministic sea wave prediction. For instance, the methodology presented in Blondel-Couprie and Naiijen (2012) can be viewed as having two components: (i) wave-field reconstruction where 3D FFT is applied on snapshots of radar images to compute the underlying linear wave components, and (ii) wave-field prediction by propagating these components linearly. Similarly, the work of Köllisch et al. (2018) can also be viewed as having two components: (i) wave-field reconstruction by estimating the surface potential via Jacobian-Free-Newton-Krylov method, and (ii) wave-field prediction by propagating the wave field in time using High-Order Spectral method. Most of the works stop at the second component, as their focus is on the wave-field prediction. We extended it by adding the third components on the encountered wave, as encountered wave is more relevant to the operation of a vessel.

We elaborate each of these components in the following subsections.

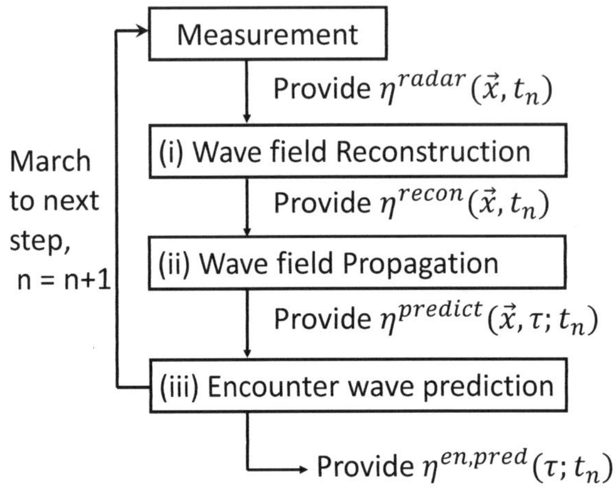

Figure 1. Flow chart of the developed framework.

2.1. Wave-field reconstruction

This component is responsible of providing the reconstructed wave-field $\eta^{recon}(\vec{x}, t_n)$ and its corresponding surface velocity potential $\phi^{recon}(\vec{x}, t_n)$ by processing collected measurement. The measurement can be collected from any sources, such as wave buoy data and radar images. The implementation under this components should filter and process the measurement, then produce a reconstructed wave field and its surface velocity potential. Several methods can be implemented under this components, such as Jacobian-Free-Newton-Krylov (JFNK) method for estimation of surface potential by Köllisch et al. (2018), Ensemble Kalman Filter (EnKF) for wave-field assimilation by Wang and Pan (2021), and 3D FFT method for wave-field reconstruction by Blondel-Couprie and Naiijen (2012). Note that, for 3D FFT, it may not be trivial if the radar is mounted on a moving vessel. This is because all snapshots are captured at different locations and heading, and a direct 3D FFT is not applicable to those snapshots.

2.2. Wave-field prediction

The task for this component is to propagate the reconstructed wave-field forward in time (into the future). The input for this component is the wave-field and surface potential reconstructed in the previous components. Using them as an initial condition, this component is expected to predict the wave-field in the future as a prediction. For the prediction to be practical, it has to be produced faster than real time. For sea state with lower non-linearity, linear wave theory is the choice because of its simplicity and efficiency. For steeper wave with higher non-linearity, non-linear wave propagation models such as High Order Spectral (HOS) method and multi-layer Boussinesq-type models are more suitable. Data-driven models such as Artificial Neural Network (ANN) can be implemented as well by training it with sufficient data.

2.3. Encountered wave prediction.

The task for this component is to predict the encountered wave-field based on the propagated wave-field. The input of this component is the predicted wave field from the previous component and the projected vessel position and heading, while the output is the encountered wave experienced by the vessel. The task of this component is trivial if we assumed that the vessel will follow its planned path and heading, such that the encountered wave is just an interpolation of the predicted wave-field to the vessel position and heading in the future. It becomes non-trivial if we take the effect of the wave on the vessel path into account. This component can be further extended into the vessel motion prediction, where the encountered wave can be served as an input to compute the vessel motion.

3. Problem Setup

We consider a generic scenario where an X-band radar is mounted on a moving vessel as shown in Figure 2. The vessel manoeuvres along an arbitrary path with its position and heading measured. In this paper, we assume the path to be a turning circle without loss of generality. While the vessel is moving, we further assume the on-board radar is measuring phase-resolved wave-field bounded by a square region with length L at a constant time interval δt. The square region is centred at the vessel, with its y-axis aligns with the heading of the vessel. We denote the wave-field measured by the radar at time t_n as$\eta^{radar}(\vec{x}, t_n)$. The vessel's position and heading are denoted as $\vec{x}_{vessel}(t_n)$ and $\theta_{vessel}(t_n)$, respectively. Our objective is to predict the wave-field to be encountered by the vessel, $\eta^{en}(t_n)$, which is an essential information needed to predict the vessel motion.
We choose the following set of parameters and assumptions:

1. Radius of the turning circle, R = 700 m,
2. Time to complete 90 deg turning circle, T = 140 s. (i.e. vessel speed around 15 knots),
3. Effect of wave diffraction due to ship is not considered,
4. Radar measurement area, L = 1.6 km, with the vessel always at the centre of the measurement area, and
5. Radar measurement interval, δt = 2.5 s.

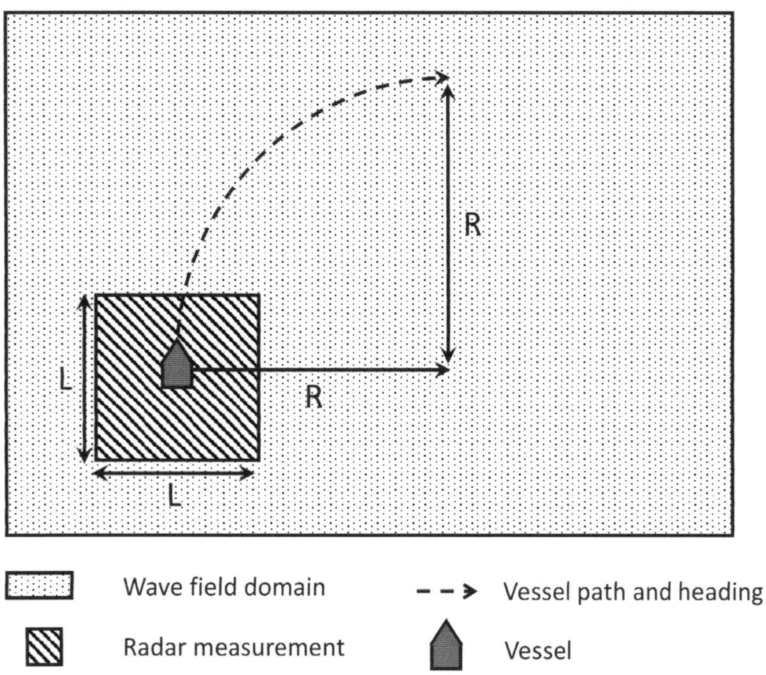

Figure 2. Schematic diagram of the problem setup.

4. Implementation using linear wave theory

For this study, our intention is to demonstrate the implementation of the framework. We considered the scenario where a radar is mounted on a moving vessel, and the radar is measuring a snapshot of the surrounding wave field at a uniform time interval. We have implemented linear wave theory (LWT) based method in both the wave-field reconstruction and wave-field prediction components. LWT is chosen because of its simplicity and efficiency. For encountered wave prediction, we made the assumption that the vessel will follows a specific path and heading. The detail of the implementation for each component is discussed in the following subsections.

4.1. Wave-field Reconstruction

The input of this component is the wave-field measurement collected by a radar mounted on a vessel. A commercial-off-the-shelf X-band radar, such as WaMoS and FutureWaves, is capable of measuring wave-field over a range of 1 km from the radar. During operation, the radar will provide snapshots of sea surface elevation at a uniform sampling rate dependent on its rotation per minute (RPM). If the radar is mounted on a moving vessel, its measuring area will follow the vessel position and heading. A schematic diagram on the wave radar's snapshots and how the wave-field is reconstructed is shown in Figure 3.

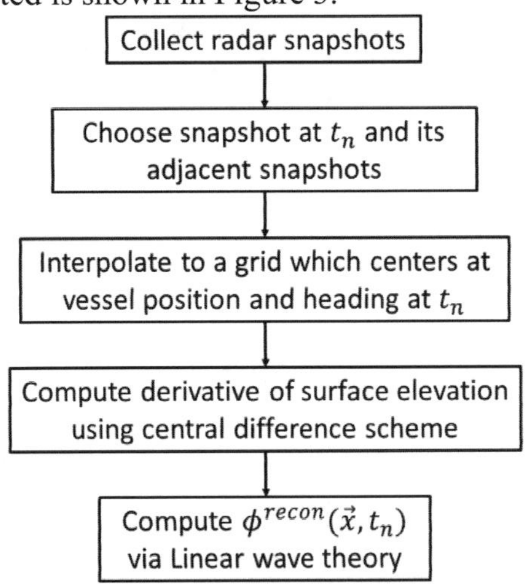

Figure 3. Schematic diagram of wave-field reconstruction using X-band radar.

Assuming the radar is providing snapshots at every δt, the time instant of each snapshot, t_n, can be written as $t_n = n\delta t$. We denote the radar snapshot at t_n as $\eta^{radar}(\vec{x}, t_n)$. In order to estimate the surface velocity potential, we need to compute $\partial\eta/\partial t$. This is achieved by invoking mid-point method (or central difference scheme) to compute the temporal differences between wave-fields at different time instances. To illustrate, assuming that the latest snapshot available is t_{n+1} (i.e. $t_{now} = t_{n+1}$), we choose the last three snapshots (i.e. $t = t_{n-1}, t_n, t_{n+1}$) to interpolate and zero-pad into an identical grid to obtain $\eta^{recon}(\vec{x}, t)$. The grid is chosen to be aligned with the vessel heading at t_n and centred at the vessel position. The size of the grid is 3 km by 3 km. To minimize spectral leakage, a ramping envelope is applied at the boundary of the radar snapshots to smoothen the changes from radar snapshots to the zero-padded region. The time derivative of the wave-field is obtained by the following equation,

$$\left.\frac{\partial\eta^{recon}(\vec{x},t)}{\partial t}\right|_{t=t_n} = \frac{\eta^{recon}(\vec{x},t_{n+1}) - \eta^{recon}(\vec{x},t_{n-1})}{2\delta t}.$$

While the surface velocity potential, $\phi^{recon}(\vec{x}, t_n)$, is obtained by linear wave theory,

$$\phi^{recon}(\vec{x}, t_n) = \Phi^{recon}(\vec{x}, z = 0, t)$$
$$= \sum_{p,q} \frac{Z_{p,q}}{|\vec{k}_{p,q}|} \exp(i\vec{k}_{p,q} \cdot \vec{x}).$$

$Z_{p,q}$ is the complex Fourier coefficient of $\partial\eta/\partial t$ in space, where

$$\frac{\partial\eta^{recon}}{\partial t}(\vec{x}, t_n) = \sum_{p,q} Z_{p,q} \exp(i\vec{k}_{p,q} \cdot \vec{x}).$$

It should be noted that in our implementation to reconstruct the wave-field and the corresponding surface velocity potential at t_n, the radar snapshot at t_{n+1} is required. This introduces a delay of δt in our prediction.

4.2. Wave-field Propagation

LWT assumes the wave-field can be written in the following expression

$$\eta(\vec{x}, t) = \sum_{p,q} A_{p,q} \exp\left(i\vec{k}_{p,q} \cdot \vec{x} - i\omega_{p,q}t\right) + B_{p,q} \exp\left(i\vec{k}_{p,q} \cdot \vec{x} + i\omega_{p,q}t\right),$$

Where both $A_{p,q}$ and $B_{p,q}$ are the complex Fourier coefficients of the $+\omega t$ and $-\omega t$ wave components, which further correspond to wave components travelling in the opposite directions. Their corresponding surface velocity potential will have the following expression

$$\phi(\vec{x}, t) = \sum_{p,q} \frac{-i\omega_{p,q}A_{p,q}}{|\vec{k}_{p,q}|} \exp\left(i\vec{k}_{p,q} \cdot \vec{x} - i\omega_{p,q}t\right)$$
$$+ \frac{i\omega_{p,q}B_{p,q}}{|\vec{k}_{p,q}|} \exp\left(i\vec{k}_{p,q} \cdot \vec{x} + i\omega_{p,q}t\right).$$

We assume that the wave-field, $\eta^{recon}(\vec{x}, t_n)$, and the surface velocity potential, $\phi^{recon}(\vec{x}, t_n)$, computed by the reconstruction component can be decomposed as the linear summation of wave components shown above. Figure 4 shows a schematic diagram of the procedure of wave propagation based on LWT.

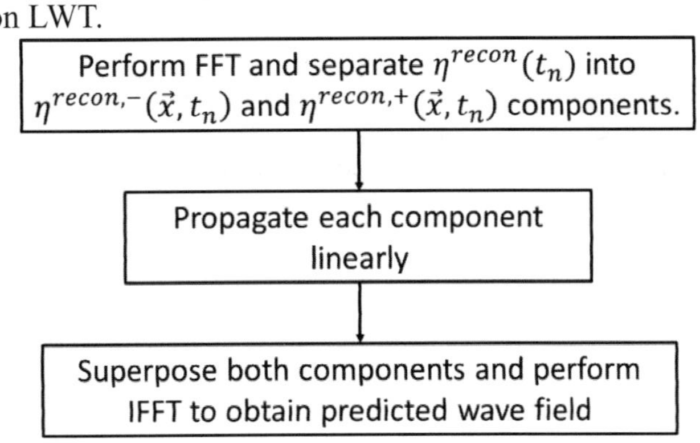

Figure 4. Schematic diagram of wave-field propagation using LWT.

To start with, we decompose the reconstructed wave-field into $+\omega t$ and $-\omega t$ wave components separately. First, we perform spatial 2D FFT to obtain both the wave-field and the surface velocity potential

$$\eta^{recon}(\vec{x}, t_n) = \sum_{p,q} Z^{\eta}_{p,q}(t_n) \exp\left(i\vec{k}_{p,q} \cdot \vec{x}\right),$$

$$\phi^{recon}(\vec{x}, t_n) = \sum_{p,q} Z^{\phi}_{p,q}(t_n)\exp(i\vec{k}_{p,q}\cdot\vec{x}),$$

where $Z^{\eta}_{p,q}(t_n)$ and $Z^{\phi}_{p,q}(t_n)$ are the complex Fourier coefficients of the reconstructed wave-field at t_n. The $+\omega t$ and $-\omega t$ component can be written as

$$\eta^{recons,\pm}(\vec{x}, t_n) = \sum_{p,q} Z^{\eta,\pm}_{p,q}(t_n)\exp(i\vec{k}_{p,q}\cdot\vec{x}),$$

with

$$Z^{\eta,\pm}_{p,q}(t_n) = \frac{1}{2}\left[Z^{\eta}_{p,q}(t_n) \mp \frac{i\left|\vec{k}_{p,q}\right|}{\omega_{p,q}} Z^{\phi}_{p,q}(t_n)\right].$$

The above expression is derived by eliminating $A_{p,q}$ and $B_{p,q}$ respectively via combination of η and ϕ. The wave-field propagated based on the reconstructed wave-field at t_n, denoted as $\eta^{predict}(\vec{x}, \tau; t_n)$, where τ lead time of the prediction, will then be written as

$$\eta^{predict,\pm}(\vec{x}, \tau; t_n) = \sum_{p,q} Z^{\eta,\pm}_{p,q}(t_n)\exp(i\vec{k}_{p,q}\cdot\vec{x} \pm i\omega\tau),$$

$$\eta^{predict}(\vec{x}, \tau; t_n) = \eta^{predict,+}(\vec{x}, \tau; t_n) + \eta^{predict,-}(\vec{x}, \tau; t_n)$$

4.3. Encountered Wave-field Prediction

We made the assumption that the vessel will follow a planned path and heading, such that we can evaluate the predicted wave-field at the projected position to compute encountered wave. We denote $\eta^{en,pred}(\tau; t_n)$ as the vessel encountered wave-field predicted based on $\eta^{predict}(\vec{x}, \tau; t_n)$, and $\vec{x}^{vessel,pred}(\tau; t_n)$ as the projected vessel position, we have

$$\eta^{en,pred}(\tau; t_n) = \eta^{predict}\left(\vec{x}^{vessel,pred}(\tau; t_n), \tau; t_n\right)$$

The predicted encountered wave-field can be further used to predict the vessel motion and structural response.

5. Results and Discussions

To assess the performance of our framework and methodology, we generate numerical wave-field using the HOS method which serve as a reference wave-field. The short-crested wave-field is generated on a 512 x 512 grid, with non-linear order of 5, and based on Pierson-Moskowitz (PM) spectrum $S(f, \theta) = F(f)G(\theta)$ where

$$F(f) = \frac{5}{16}\frac{H_s^2 f_p^4}{2\pi f^5}\exp\left[-\frac{5}{4}\left(\frac{f_p}{f}\right)^4\right],$$

$$G(\theta) = \frac{1}{\beta}\left[\cos\left(\frac{\pi\theta}{2\beta}\right)\right]^2.$$

The directional spectrum $G(\theta)$ follows the definition by Ducrozet et al., 2016. The spreading parameter, β, is set to 45 deg relative to mean wave direction that travels in –y direction of global coordinate. Two different sea-states are simulated: $H_s = 0.65\ m,\ T_p = 8$ s. and $H_s = 1.45\ m, T_p = 12\ s$. Both sea-states have identical steepness of $H_s/T_p^2 = 0.01$. The vessel is assumed to move in a turning circle as shown in Figure 1, where

$$\vec{x}^{vessel}(t) = \begin{pmatrix} -500\cos(\omega^{vessel}t) \\ 500\sin(\omega^{vessel}t) \end{pmatrix},$$

with $\vec{x}^{vessel}(0) = (-500, 0)$, $\vec{x}^{vessel}(140) = (0, 500)$. The reference encountered wave-field is computed based on the vessel position. Radar snapshots are generated at every 2.5 s by casting the reference wave-field into the radar measurement area. We starts our encountered wave-field prediction from $t_1 = 0$ s to $t_{57} = 140$ s at uniform time interval of 2.5 s. Each prediction provides a predicted encountered wave-field of up to $\tau = 90$ s ahead.

We assess the accuracy of encountered wave-field prediction at t_n by comparing with the underlying reference wave-field based on the following normalized error:

$$\epsilon(\tau; t_n) = \frac{\left|\eta^{en,pred}(\tau; t_n) - \eta^{en,ref}(\tau; t_n)\right|}{H_s^{radar}(t_n)/2},$$

where $H_s^{radar}(t_n)$ is computed as four times the standard deviation of the underlying surface elevation measured by the radar at t_n. We also compute the root-mean-square (r.m.s) of $\epsilon(\tau; t_n)$ over the prediction at different t_n,

$$\varepsilon(\tau) = \sqrt{\frac{1}{N}\sum_{i=1}^{N} \epsilon^2(\tau; t_i)}.$$

Figure 5 shows the normalized r.m.s error ($\varepsilon(\tau)$) of the encountered wave-field prediction for two different sea-states. For sea-state with $T_p = 8$ s, the normalized r.m.s error is found to be around 0.25 for the first 80 s, then increases to around 0.35 at 120 s. Meanwhile, for sea-state with $T_p = 12$ s, the normalized r.m.s error is around 0.15 for the first 60 s, and then increases beyond 0.4 after 80 s. We observe that the prediction error is larger for the shorter wave period. This is because the sampling interval of the radar at 2.5 s may not be sufficient to capture the time derivative accurately for the shorter wave period. Besides, it is clear that for sea-state with $T_p = 12.0$ s, the error increases after 60 s. This is because the prediction above 60 s falls beyond the predictable zone. However, it is important to take note that the predictable zone may vary for different sea-states and different projected vessel positions. An indicator of confidence level should be developed to indicate the length of prediction that remains reliable.

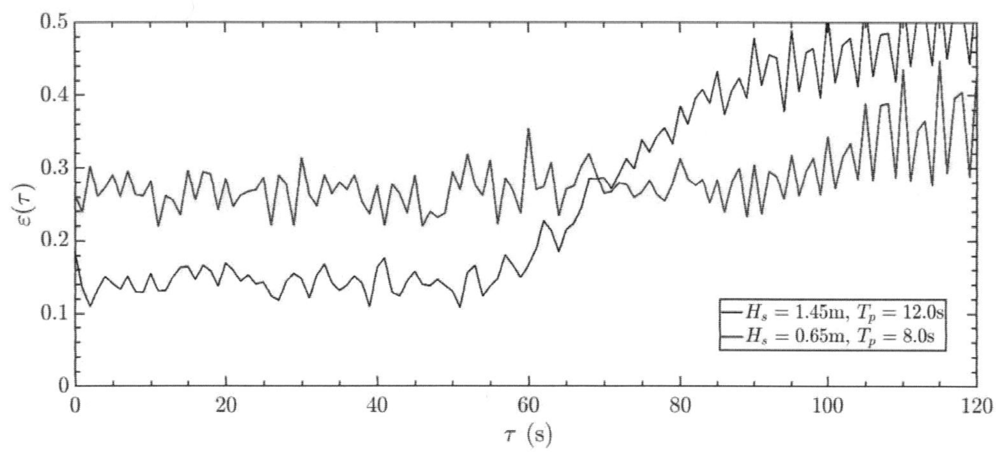

Figure 5. Normalized r.m.s error (ε(τ)) of the encountered wave-field prediction for two different sea-states.

Figure 6 and 7 show the encountered wave-field predictions and the normalized errors for the two sea-states at $t_n = 70$ s, where the vessel is heading at 45 deg (relative to the 0 deg/North). For sea-state with $T_p = 8$ s, it is observed that while the predicted encountered wave-field captures the phase well up to 80 s, it loses its predictability beyond that. Besides, a large deviation occurs around 25 s. This may be due to the larger error in the wave-field reconstructed at $t_n = 70$ s, which later propagates and arrives at the vessel. For sea-state with $T_p = 12$ s, our prediction has better accuracy, with error less than 20%

for prediction up to 80 s. It is also noted that the prediction from 100 s onwards decays to zero, which indicates that the prediction goes beyond the predictable zone.

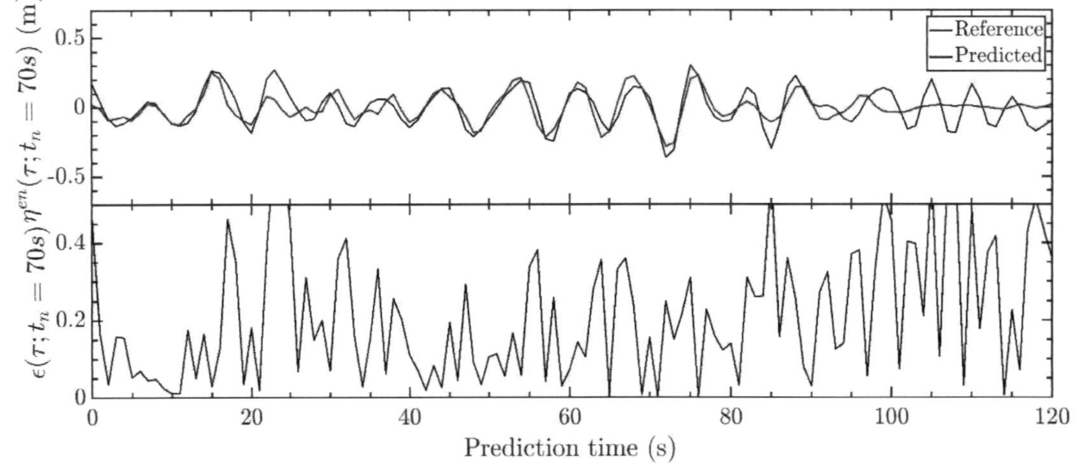

Figure 6. Encountered wave-field prediction and normalized error for $H_s = 0.65$ m, $T_p = 8$ s, $t_n = 70$ s.

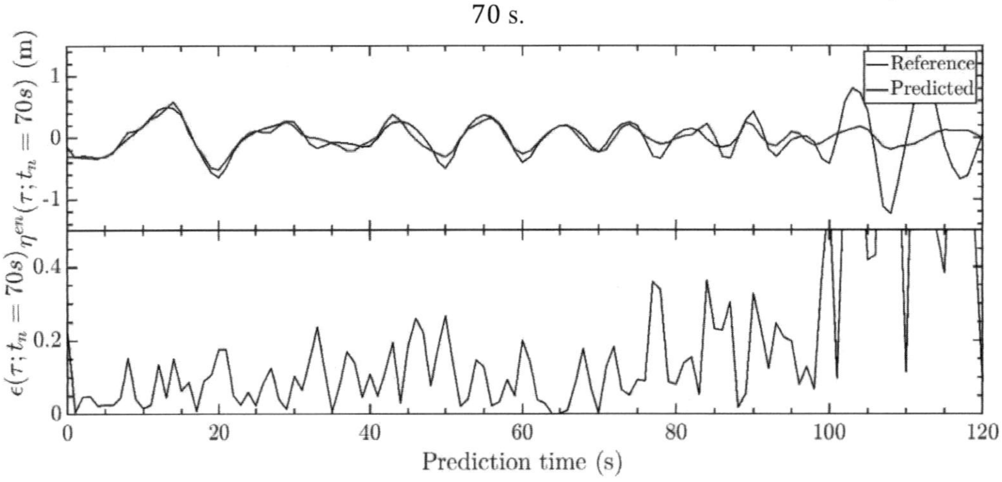

Figure 7. Encountered wave-field prediction and normalized error for $H_s = 1.45$ m, $T_p = 12$ s, $t_n = 70$ s.

We also conduct a test to compare the accuracy of our prediction at different wave steepness values. Figure 8 shows the normalized r.m.s error of the encountered wave-field prediction for $T_p = 12$ s for three different sea-state steepness values, i.e. 0.01, 0.02, and 0.04. Based on the observation, the error does not change significantly with different steepness. This may indicate that the source of error arises predominantly from the wave-field reconstruction. More results are required in order to make a more conclusive remark on the performance of the model. However, since the emphasis of this paper is on developing a framework, a more thorough investigation is left for future study.

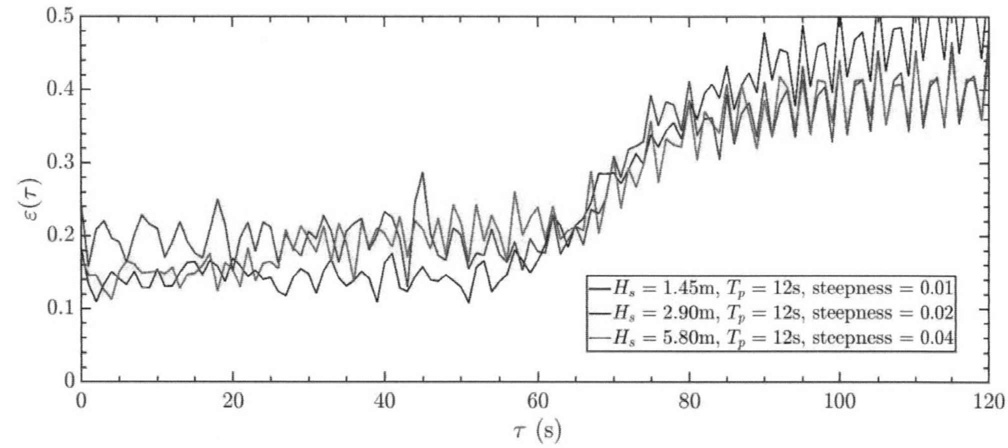

Figure 8. Normalized r.m.s error ($\varepsilon(\tau)$) of the encountered wave-field predictions for $T_p = 12.0$ s at different steepness values.

6. Conclusion

We have developed a framework to predict wave-field encountered by a moving vessel based on measurements taken off an X-band radar mounted on the vessel. The framework is capable of handling a vessel moving in an arbitrary path, with the assumption that the vessel location and heading can be measured. The framework, which consists of three main components (reconstruction, propagation and prediction), is highly modularized, such that each component can make use of different implementation to improve the performance further. Our current implementation of wave reconstruction and propagation is based on linear wave theory, which serves as a baseline performance for our framework. The computation is very efficient, where it takes 2.5 s to produce prediction up to 120 s. In terms of the accuracy, for two directional sea-states considered, a normalized error of 0.25 is achieved for sea-state with $T_p = 8$ s, and 0.15 for sea-state with $T_p = 12$ s. The larger error for the shorter wave period may arise from the inaccuracy in the estimation of ϕ^{recon}, where the sampling interval of 2.5 s becomes comparable to the sea-state with T_p of 8 s.

Possible future study includes assessment of different implementations to each of the components, such as EnKF and JFNK for wave reconstruction, HOS, data-driven and reduced-order model for wave propagation. An indicator of confidence level should also be developed for the prediction, such that only reliable prediction is presented.

7. Acknowledgements

This research is supported by the National Research Foundation, Singapore and the Singapore Maritime Institute under its Maritime Transformation Programme White Space Funding Support for the Establishment and Operation of the Centre of Excellence for Autonomous and Remotely Operated Vessels (CEAOPS) (SMI-2019-MTP-01).

References

[1] G. Wang and Y. Pan, "Phase-resolved ocean wave forecast with ensemble-based data assimilation," *Journal of Fluid Mechanics,* vol. 918, 2021.

[2] A. Simanesew, K. Trulsen, H. E. Krogstad and J. C. N. Borge, "Surface wave predictions in weakly nonlinear directional seas," *Applied Ocean Research,* vol. 65, pp. 79-89, 2017.

[3] P. Naaijen and E. Blondel-Couprie, "Reconstruction and prediction of short-crested seas based on the application of a 3D-FFT on synthetic waves: Part 1—reconstruction," in *International conference on offshore mechanics and arctic engineering*, 2012.

[4] J. G. Kusters, K. L. Cockrell, B. S. H. Connell, J. P. Rudzinsky and V. J. Vinciullo, "FutureWaves™: A real-time Ship Motion Forecasting system employing advanced wave-sensing radar," in *OCEANS 2016 MTS/IEEE Monterey*, 2016.

[5] N. Köllisch, J. Behrendt, M. Klein and N. Hoffmann, "Nonlinear real time prediction of ocean surface waves," *Ocean Engineering,* vol. 157, pp. 387-400, 2018.

[6] T. Hilmer and E. Thornhill, "Observations of predictive skill for real-time Deterministic Sea Waves from the WaMoS II," in *OCEANS 2015-MTS/IEEE Washington*, 2015.

[7] G. Ducrozet, F. Bonnefoy, D. Le Touzé and P. Ferrant, "HOS-ocean: Open-source solver for nonlinear waves in open ocean based on High-Order Spectral method," *Computer Physics Communications,* vol. 203, pp. 245-254, 2016.

[8] B. S. H. Connell, J. P. Rudzinsky, C. S. Brundick, W. M. Milewski, J. G. Kusters and G. Farquharson, "Development of an environmental and ship motion forecasting system," in *International conference on offshore mechanics and arctic engineering*, 2015.

[9] S. R. Bukka, Y. Z. Law, H. Santo and E. S. Chan, "Reduced order model for nonlinear multi-directional ocean wave propagation," *Physics of Fluids,* vol. 33, p. 117115, 2021.

[10] F. Bonnefoy, G. Ducrozet, D. Le Touzé and P. Ferrant, "Time domain simulation of nonlinear water waves using spectral methods," in *Advances in numerical simulation of nonlinear water waves*, World Scientific, 2010, pp. 129-164.

[11] E. Blondel-Couprie and P. Naaijen, "Reconstruction and prediction of short-crested seas based on the application of a 3D-FFT on synthetic waves: Part 2—Prediction," in *International Conference on Offshore Mechanics and Arctic Engineering*, 2012.

[12] L. K. Alford, R. F. Beck, J. T. Johnson, D. Lyzenga, O. Nwogu and A. Zundel, "A real-time system for forecasting extreme waves and vessel motions," in *International conference on offshore mechanics and arctic engineering*, 2015.

Prediction of tidal-driven currents using convolutional neural network

Jian Feng CHOO [1], Jeng Hei CHOW [1], Pavel TKALICH [1,2]

[1]Technology Centre for Offshore and Marine, Singapore (TCOMS), 12 Prince George's Park, Singapore 118411.
[2]Tropical Marine Science Institute (TMSI), NUS, S2S Building, 18 Kent Ridge Road, Singapore 119227

Abstract. Efficient forecasting models of coastal currents are required for the predictive control of smart vessels with autonomous or remote operating capabilities. Accurate physics-based numerical models demand computational resources that often are unavailable onboard ships; moreover, internet bandwidth at remote locations might not be sufficient to exchange the large computational files with onshore servers. In this case, light and efficient Data-Driven Models (DDM) may be developed to be used as a surrogate for a physics-based model.
Currents and surface elevation are hydrodynamic variables driven by astronomic tide potential, allowing for harmonic analysis and synthesis of otherwise irregular solutions of Navier-Stokes equations. Once numerical solutions are obtained and analysed in the Fourier domain, the identified harmonics (tide constituents) can be used as a dataset to train the DDM. One approach to building a DDM is by utilizing a Convolutional Neural Network (CNN).
CNN is mainly applied in image recognition processes, where it learns the spatial features of image matrices of pixels arranged in columns and rows and uses the learned features to predict an output. This approach can be applied to 2D or 3D fields of predominantly tidal-driven coastal currents. This paper extends the CNN framework to approximate and predict coastal hydrodynamic state variables in space and time. The depth-averaged Navier-Stokes equations were used to create the dataset necessary to train the adopted CNN autoencoder architecture.
Two methods of CNN model development are explored. The first method assesses CNN's capabilities for time forward prediction, using 2D time series of hydrodynamic variables. The second method investigates CNN's parameter (using Manning number) prediction ability using 2D spatial plots formed by the top 20 most energetic tidal constituents that were obtained from the numerical solution. The hyperparameters of the CNN model for both methods were carefully optimized and tuned using the loss function, Root Mean Square Error (RMSE). The coefficient of determination, the relative error of the time series, and its amplitude and phase were used to gauge the performance of the CNN model for both methods at the final testing stage.
Once accurate solutions are achieved, the trained DDM is almost instantaneous and uses negligible computational resources as compared to traditional fluid solvers. Therefore, trained DDM can be deployed onboard ships where global forecast information (e.g.,

from Global Forecast System) can be used to provide an accurate prediction of the current forcing. To generate the instantaneous current predictions for navigation and smart ship control, the global data can be pulled or pushed to the ship before the mission.

1. Introduction

The availability of forecasted current velocities is important for all maritime-related activities. One example would be vessel manoeuvring through the coastal oceans with limited access to a met-ocean forecasting facility. With the fast and accurate prediction of currents or other met-ocean variables, the vessel may navigate safely through complex paths with the least power consumption. Other example includes the various sea operations such as ocean oil drilling, laying underwater cable and many other expensive offshore construction projects, which relies heavily on hydrodynamic predictions for safe and efficient planning. Predictions of ocean currents could be generated and disbursed by an onshore met-ocean forecasting centre, using advanced physics-based solvers. However, this method could be prohibitively computationally expensive to be carried onboard vessels. The alternative solution described in this study would be having a light Data-Driven Model (DDM) on-site board using a machine learning algorithm.

Machine learning in recent years has gained a lot of attention due to its ability to approximate the input to output vectors using linear and non-linear functions. The various architectures of machine learning are being widely used commercially in multiple industries such as finance, retail, energy, and defence. In 1989, LeCun et al [1] created a convolutional layer for efficient 2D and 3D image recognition that uses low computational power. These convolutional layers are then used in various image recognition processes, such as predicting tumours from MRI scans, facial recognition applications and image classification in search engines. Applying CNN onto fluid flow is still at its exploratory stage. Morimoto Et al [2] applied CNN on to wake interaction problem for both flat plate and two side by side cylinders, in a steady-state laminar flow to find a practical methodology for fluid flow analysis. Nils Thuerey et al [3] used CNN to predict the 2D horizontal velocity field of unseen air foil geometry in turbulence flow. In the field of turbulent combustion modelling, Lapeyre et al [4] tried to extract complex flame topology to predict subgrid-scale wrinkling. All these examples showed promising results of using CNN to depict complex flow features in a 2D hydrodynamics field.

This paper extends the CNN framework to predict predominantly tide-driven hydrodynamic state variables (currents and water level) in space and time. The depth-averaged Navier-Stokes equations were used to create the dataset necessary to train the adopted CNN autoencoder architecture. The model has been constructed using two methods.

The first method investigates CNN's capability on time forward prediction using 2D time series, obtained from the numerical model. The second method assesses CNN's Fourier prediction ability using Fourier coefficients, extracted from each numerically computed data of varying Manning numbers. In both methods, three domains of increasing complexity were used for capability testing.

2. Methodology

2.1 2D Depth Average Navier Stokes Equation

$$\frac{\partial}{\partial x}[(h+\zeta)\bar{u}] + \frac{\partial}{\partial y}[(h+\zeta)\bar{v}] + \frac{\partial \zeta}{\partial t} = 0 \quad (1)$$

$$\frac{\partial \bar{u}}{\partial t} + \bar{u}\frac{\partial \bar{u}}{\partial x} + \bar{v}\frac{\partial \bar{u}}{\partial y} - f\bar{v} = -\frac{1}{\rho}\frac{\partial p_a}{\partial x} - g\frac{\partial}{\partial x}(\zeta - \bar{\zeta}) + \frac{\tau_{sx} - \tau_{bx}}{\rho(h+\zeta)} \quad (2)$$

$$\frac{\partial \bar{v}}{\partial t} + \bar{u}\frac{\partial \bar{v}}{\partial x} + \bar{v}\frac{\partial \bar{v}}{\partial y} + f\bar{u} = -\frac{1}{\rho}\frac{\partial p_a}{\partial y} - g\frac{\partial}{\partial y}(\zeta - \bar{\zeta}) + \frac{\tau_{sy} - \tau_{by}}{\rho(h+\zeta)} \quad (3)$$

$$\tau_s = C_D \rho_a |W| W \quad (4)$$

$$\tau_{bx} = \frac{g\bar{u}\sqrt{\bar{u}^2 + \bar{v}^2}}{hC^2} \tag{5}$$

$$\tau_{by} = \frac{g\bar{v}\sqrt{\bar{u}^2 + \bar{v}^2}}{hC^2} \tag{6}$$

$$C = \frac{1}{n}R^{\frac{1}{6}} \tag{7}$$

The 2D depth average Navier Stokes equations are commonly used to model the coastal ocean currents. Here, $\bar{u}, \bar{v}$ are the depth average current velocities in the x and y direction. h, ζ, f, ρ, g represents the water depth, surface elevation, Coriolis parameter, density of the ocean and gravity respectively. The quantities τ_{sx}, τ_{sy} are the wind stresses in the x and y direction. τ_{bx}, τ_{by} denotes the bottom shear stresses in the x and y direction. p_a, ρ_a are defined as the atmospheric pressure and density. C_D, W, C, R, n represents the drag coefficient of wind, wind speed, Chezy number, hydraulic radius, and Manning number. The Manning number in equation (7) is used to parameterize bottom shear stress.

2.2 Tides and Harmonic

The dynamics of the coastal regions are often dominated by astronomic tides schematically depicted in Figure 1. Due to its periodic nature, currents and water level driven by astronomic tides can easily be decomposed into individual frequencies of the cosine function, known also as tidal constituents. The harmonic analysis can express surface elevation, u-velocity, and v-velocity in terms of amplitude (a), frequency (ω) and phase (φ).

$$\zeta = \sum_{i=1}^{N} a_i \cos(\omega_i t + \phi_i) \tag{8}$$

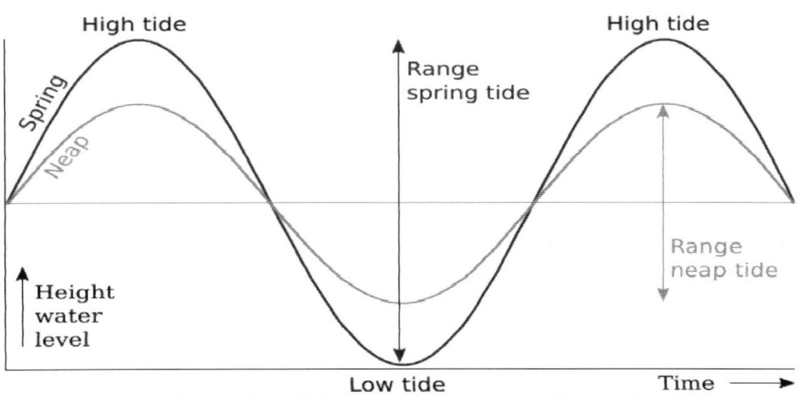

Figure 1 Definitions of astronomic Tides [5]

2.3 Convolutional Neural Network

The general equation for the machine learning model is shown in equation (9), where x is the input, y is the output and f are the machine learning model function. The focus in this section will be on convolutional layers.

$$y = f(x) \tag{9}$$

$$\mathbf{z} = \sum_{\mathbf{j}} \sum_{\mathbf{k}} \mathbf{x}(\mathbf{m} - \mathbf{j},\ \mathbf{n} - \mathbf{k})\ * \mathbf{k}(\mathbf{j}, \mathbf{k}) \tag{10}$$

CNN uses matrix wise multiplication shown in equation (10) to connect one layer to the next shown in Figure 2. The variables z, k, x, m, n, j, k represents the output matrix, filter matrix, input matrix, x and y direction of the input matrix and x and y direction of filter matrix respectively. This configuration reduces the number of trainable weights when computing for 2D and 3D matrices, compared to the traditional multi-layered perceptron shown in Figure 3 which connects layer to layer linearly.

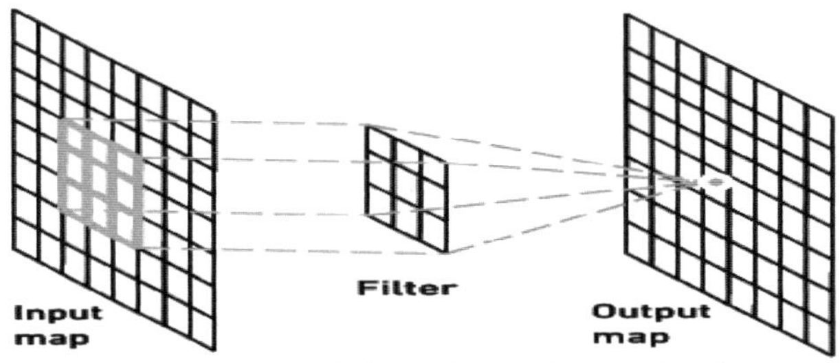

Figure 2 Convolutional layer Connection [6]

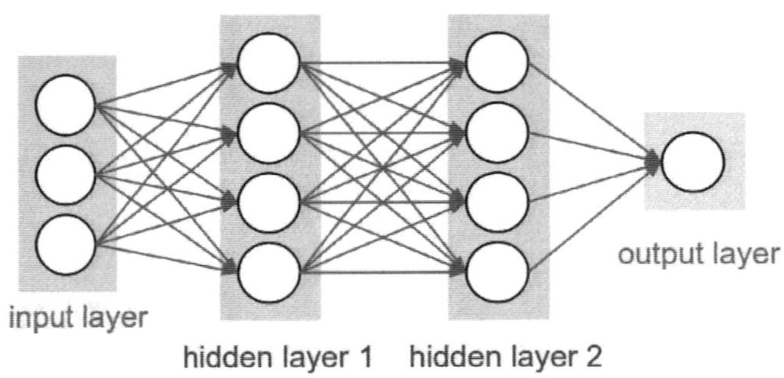

Figure 3 Dense layer connection [7]

Each convolutional layer can define a few important hyperparameters; the number, size, and stride of filters. The filters are 2D matrices that store the trainable weights which captures the features in the 2D input matrix. The activation function is an important feature of a neural network that introduces non-linearity into the neural network equation. This may allow the neural network to approximate non-linear complexities present in the data. The pooling layer shown in Figure 4 is another type of CNN layer that reduces the dimension of 2D matrices without affecting the number of filters present, created by convolutional layers.

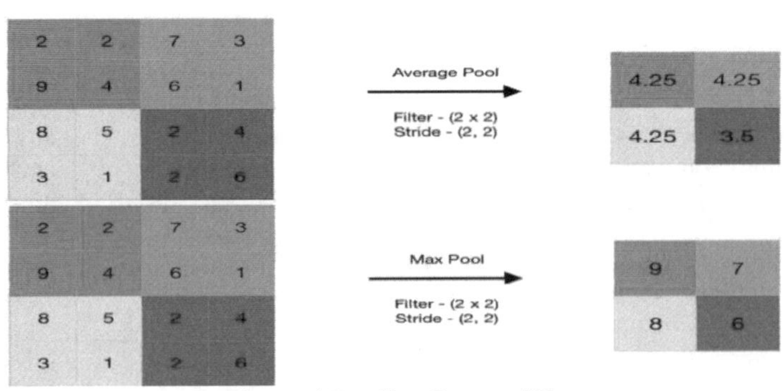

Figure 4 Pooling Layers [8]

Machine learning trains and update its model by the backpropagation method. The summarized training process shown in Figure 5, starts from passing the training dataset forward through the randomly initialized model. Then, the loss function is computed by comparing the output and ground truth data. Right after, the model computes the gradient of the loss function concerning the weights. Finally, the weights are updated using a gradient descent algorithm.

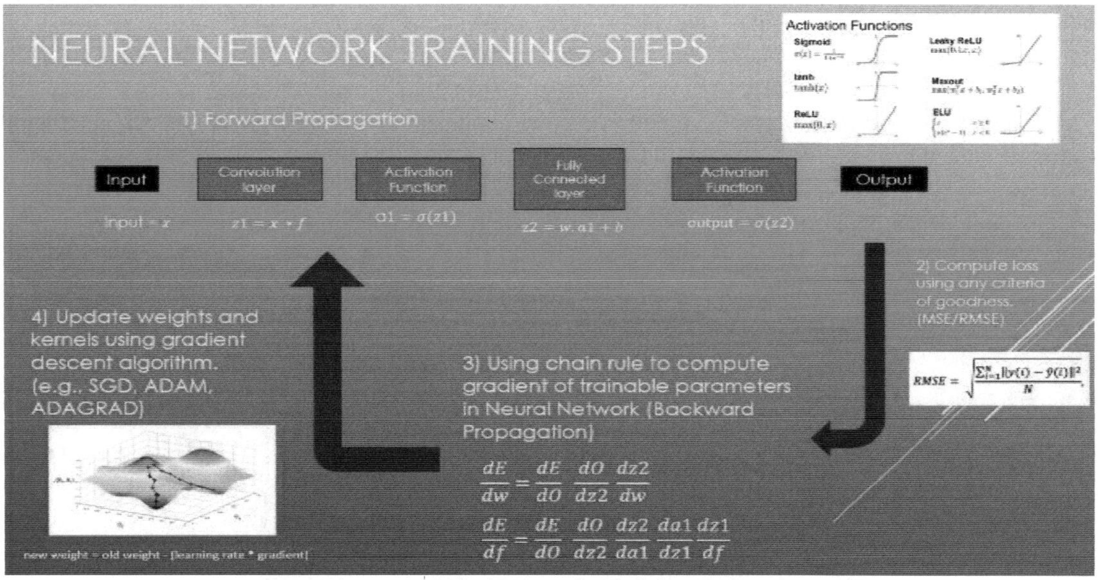

Figure 5 Neural Network Training Process

3 Pre-processing of datasets and experiment setup

To test the application of CNN onto a 2D hydrodynamics field, three domains of increasing complexity were created (Figure *6*). The first geometry is a straight channel that has a size of 150km x 22km x 60m, which approximates the dimension of Singapore Straits. The geometry is semi-enclosed where the open boundary is on the bottom left. The second geometry has the same overall size as the first geometry but with a jetty (12km x 3km) protruding out in the middle of the fluid domain which causes eddies to form near it. The open boundary is located at the same position as the first geometry. Lastly, the method is applied to a real domain located around St John's Island. The geometries will be used to test two methods of CNN. The first method predicts the time series of the hydrodynamic variables at the 2D domain. The second method predicts Fourier coefficients of decomposed time-series of the hydrodynamic variables at the 2D domain, limited to 20 most energetic constituents.

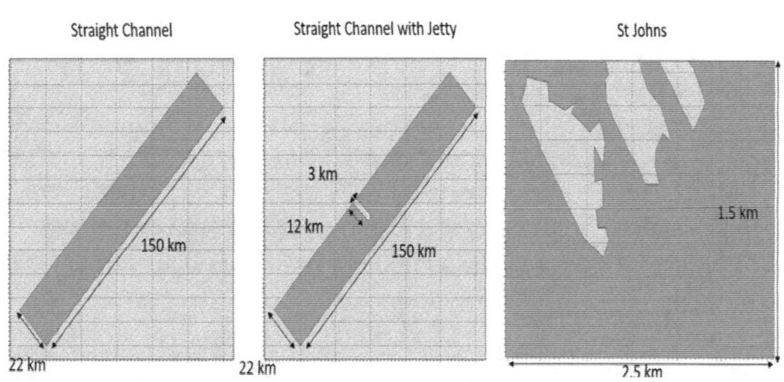

Figure 6 Layout of the 3 configurations: Straight Channel (Left), Straight Channel with Jetty (Middle) and St John's Island (Right)

3.1 Time Prediction Method

The training dataset for the first method is computed for 2 years using the MIKE21-FM model with boundary conditions generated from Global Tide Model (GTM). The output data were converted from an unstructured mesh to a 100 by 100 structured grid using bilinear interpolation in MATLAB. For the convenience of the activation function used in the CNN model, the dataset was normalized using a min-max normalizer. Normalizing the data will also be important when using both surface elevation and currents as inputs together.

The CNN autoencoder structure (Figure 7) that will be adopted is a commonly used CNN structure when applied to fluid problems [9], [10], [2], [4]. It reduces the high dimensional input, and thereafter computes the output using the low dimensional data. This concept is similar to Proper Orthogonal Decomposition (POD) [11]. One of the advantages of an autoencoder is the nonlinear characteristics that can be embedded inside its equation, whereas POD is linear. Hence, CNN might be a better approximation method.

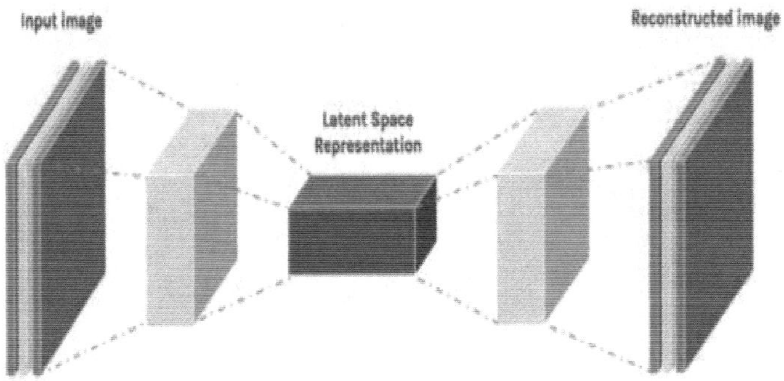

Figure 7 Example of an Autoencoder structure [12]

The CNN model was trained using the first 11 months of the data, from January to November, validated with the first 8 days of December, and tested on the whole of December. The weights of the model were updated by the training dataset. The validation set was simultaneously passed through the model to observe the model's accuracy during the training process. Lastly, the test set which was unseen by the model was used to analyse and gauge CNN's forward prediction performance. The performance was gauged by comparing the predicted output and ground truth data of the following: the relative error time series and the R^2 values of time series, amplitude and phase.

The model was trained using the supervised learning method. The input and output were chosen to be the same. This simplification allows us to isolate the performance solely to CNN capability. The model was trained with Adam optimizer, coupled with the loss function, Root Mean Square Error (RMSE), to gauge the performance of the model. The number of training iterations was chosen based on the step at which RMSE converges.

Multiple models with different hyperparameters such as, number of kernels, number of neurons, activation function, kernel size, number of dense layers, number of convolutional layers and learning rate were trained at fixed random seeds such that the most suitable model is used. Due to the stochastic nature of the gradient descent algorithm, the optimized model is then further tuned by training at the different random seeds.

The CNN model is further tuned by introducing a multiplying layer at the end of the model shown in Figure 9, and additional input to mask out the land values to 0, shown in Figure 8. This additional layer and input eliminate the need for the CNN model to train the land areas, hence the computational time was reduced.

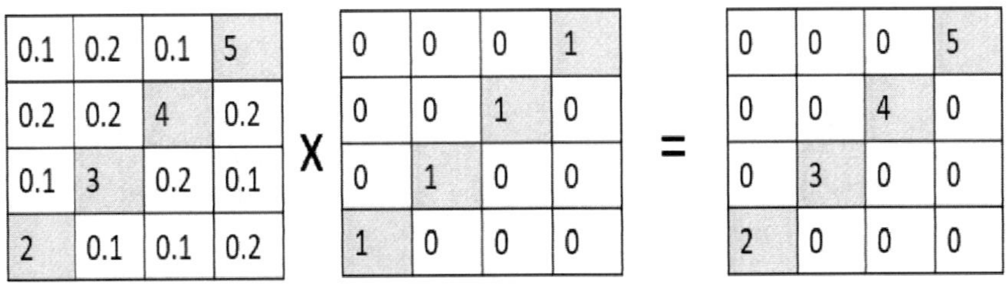

Figure 8 Masking operation in CNN model

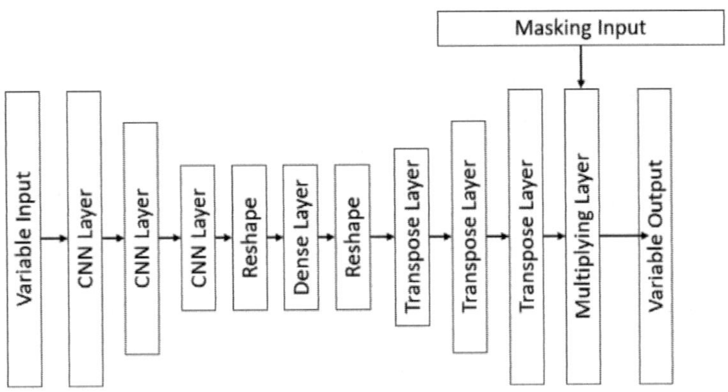

Figure 9 CNN Autoencoder with additional masking input and multiplying layer.

Three variations of CNN models were tested in the time series method: Independent input, partially adjoint input and fully adjoint input shown in Figure 10. The independent variable structure inputs and outputs a single variable. Partially adjoined structure inputs and outputs two variables at the same time. Lastly, the fully adjoined variable inputs and outputs all three variables together. The setup of these models was used to test CNN's ability to capture the continuity relationship between the three variables together.

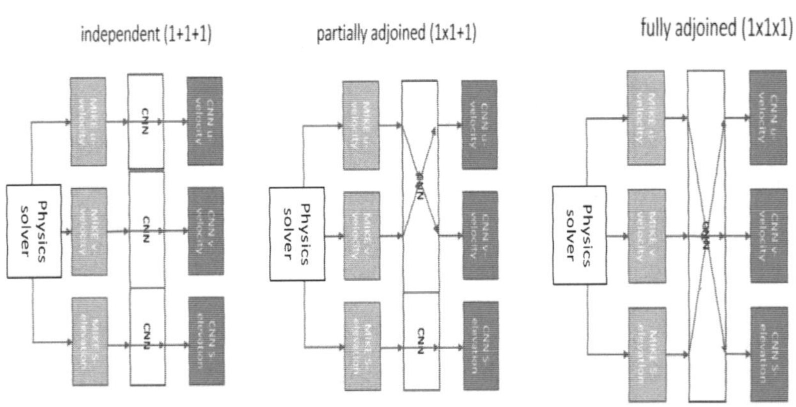

Figure 10 CNN Schematic Diagram for time series method. Independent Variable (Left), Partially Adjoint Variable (Middle), Fully Adjoint Variable (Right)

3.2 Fourier Prediction Method

$$\mathbf{F_k^{i,j} = A_k^{i,j} \cos\left(B_k^{i,j}\, t - C_k^{i,j}\right)} \tag{11}$$

The dataset for the Fourier prediction method was created for 1 year with a constant time step of 10 minutes using MIKE21-FM. The dataset was computed using 36 different cases of varying manning numbers from 10 to 45 with a constant parameter step of 1. The manning number parameterize the bottom shear stress, thus affecting hydrodynamic state variables.

The process of creating the Fourier method dataset starts with converting the 2D unstructured output of each sea state to a 100x100 structured grid. Thereafter, equation (11) is used to compute the Fourier coefficients of each cell in the structured grid using Fast Fourier Transform (FFT), and only the top 20 most energetic Fourier coefficients are extracted per cell. The number of most energetic harmonics is derived using the analysis of Tanjong Pagar's Tide Gauge measurements. The surface elevation measurement was decomposed into its Fourier coefficients and thereafter reconstructed back starting from its most energetic Fourier coefficient. The reconstructed data were compared with the original measurement using RMSE as the criteria of goodness. It is observed that the RMSE value converges around its 20 most energetic harmonics. A similar methodology is applied to the current data from the Buoy measurements provided by MPA (Marine Port Authority). The same observation was made, where the top 20 most energetic Fourier coefficients were sufficient to predict the current data as well. The obtained top 20 harmonics were arranged to their basis vector. After which, the 2D Fourier coefficients (representing amplitudes and phases of tide constituents) of each hydrodynamic variable were concatenated together to form the Fourier prediction method dataset. Finally, the amplitudes and phases were converted from polar to cartesian coordinates because there were no mathematical equations in CNN to capture the phase wraps between 0 to 360 degrees.

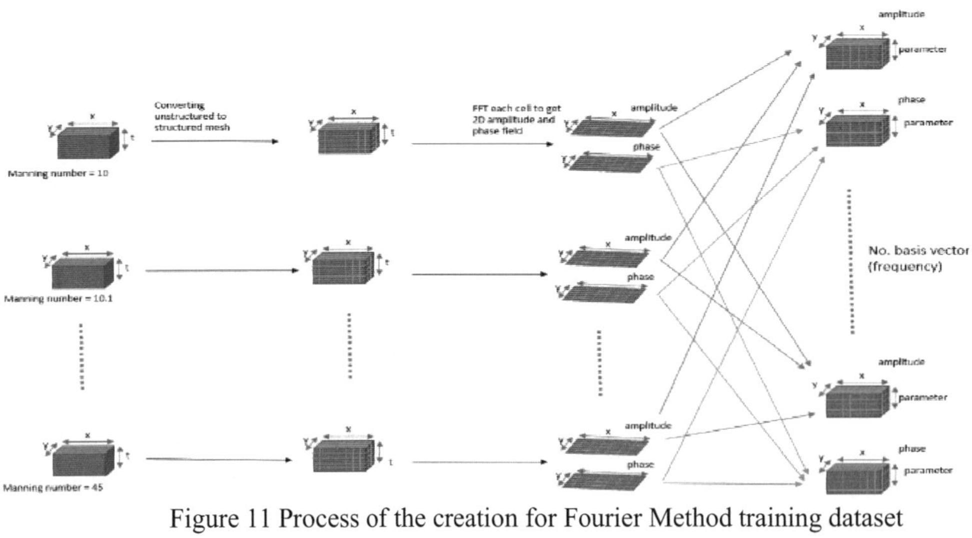

Figure 11 Process of the creation for Fourier Method training dataset

The dataset for the Fourier prediction method consists of 34 training sets, one validation case and one test case. This simple methodology will test CNN's capability in predicting the Fourier coefficients of different variables. Like the time prediction method, the model updates the weights using only the training set. The validation set enumerates the performance of the model simultaneously while training. Lastly, CNN was tested on the unseen test set, where the performance was gauged by comparing the predicted output and ground truth data of the following: time series relative error and the R^2 values of time series, amplitude and phase.

The CNN structure shown in Figure 9 was adopted in the Fourier prediction method. Each x and y component are trained separately shown in Figure *12*. Similarly, the model also uses the same input and output variable to simplify the case. Each frequency vector basis that was obtained is trained independently per model. The same stringent optimizing and tunning of the hyperparameters in the model described in the time series method was also applied in the Fourier prediction method.

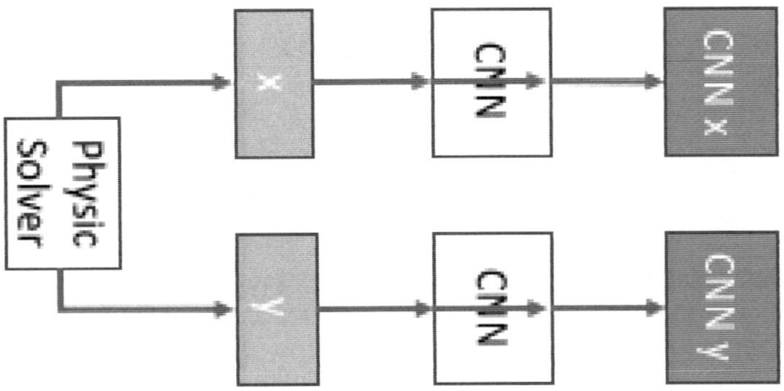

Figure 12 CNN Schematic Diagram for Fourier prediction method.

4 Results

Both CNN methods were tested on the geometries of increasing complexity: the straight channel, the straight channel with jetty, and the St. John's Island domain. Five points shown in Figure 13 for St John's Island domain are sampled to analyse CNN performance at different flow conditions including eddies. All 5 points showed similar model performance; therefore, this section shows the results on point 4 of the St John's Island domain (Figure *13*).

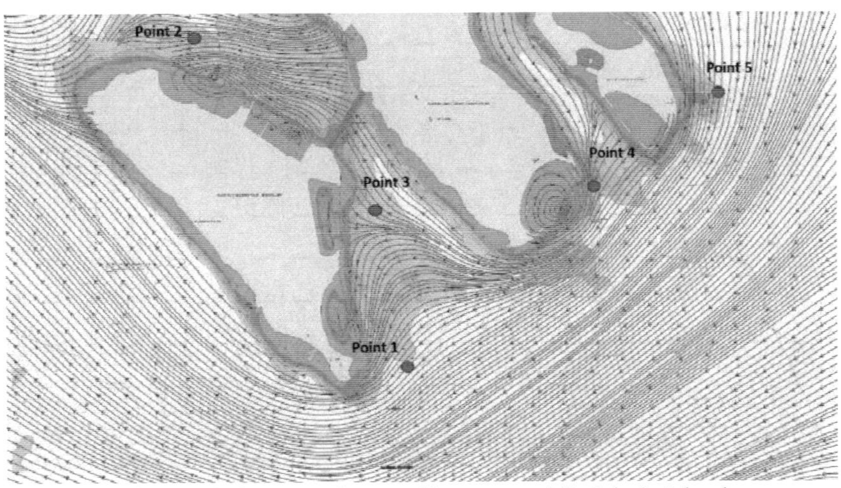

Figure 13 Computed streamlines around St John's Island

4.1 Time Prediction Method Results

Figure *14* compares the streamlines computed using the numerical model and CNN independent input model at the same timestep around St John's Island domain. From this comparison, we observed CNN have a negligible difference in shapes and sizes when compared to the numerical model, CNN performance was very promising in terms of predicting the location and shape of eddies at all time steps.

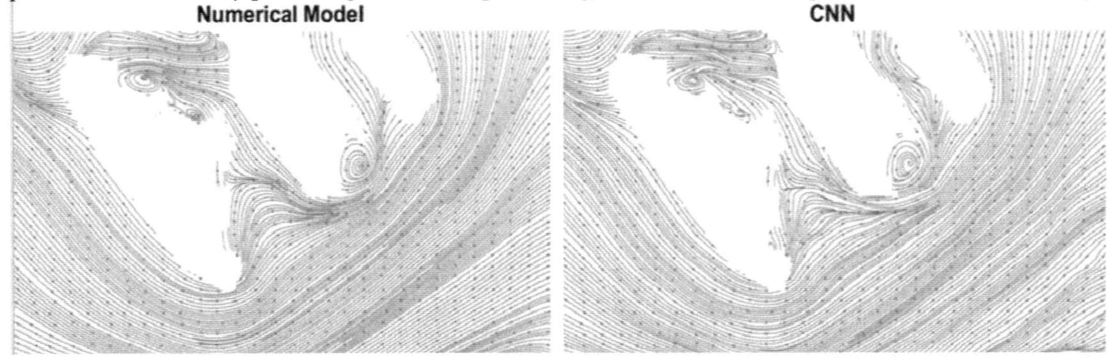

Figure 14 Time prediction method - Streamline comparison of St John's Island domain. Numerical Model (Left), CNN (Right)

When comparing the time series between the numerical model and CNN independent input model at point 4 shown in Figure *15*, we observe that higher error occurs around the same period for all sea state variables. The overall error is usually around 15% to 25% for all points. The R^2 for all sea state variable suggest that the predicted time series from CNN fits adequately with the numerical model. This behaviour was also present in other points taken around St John's Island and the other flow geometry. Figure *16* compares the amplitudes and phases of the time series in its cartesian coordinate. It is observed that CNN fits satisfactorily along the reference line. The R^2 value for all the plots suggests that CNN fits quite accurately with the numerical model. Similar results can be found in the other points around St John's Island and the other two geometries. The results suggest that the high relative error might cause by some phase discrepancy at certain constituents.

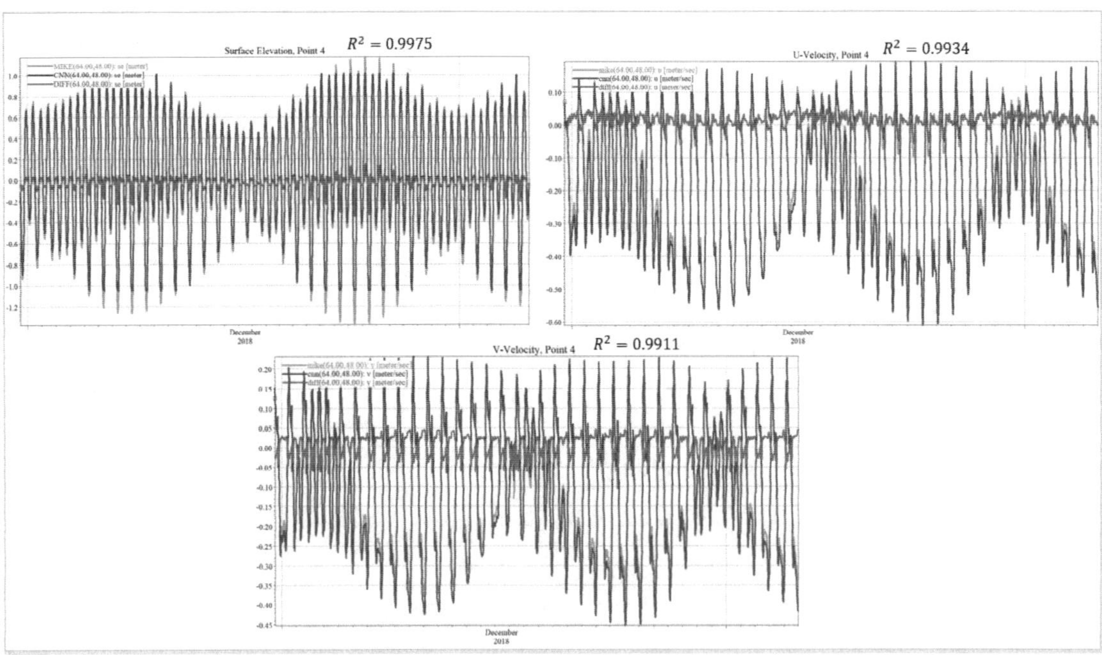

Figure 15 Time prediction method – Time series plots for surface elevation (Top left), u-velocity (Top right), V-Velocity (Bottom) at point 4.

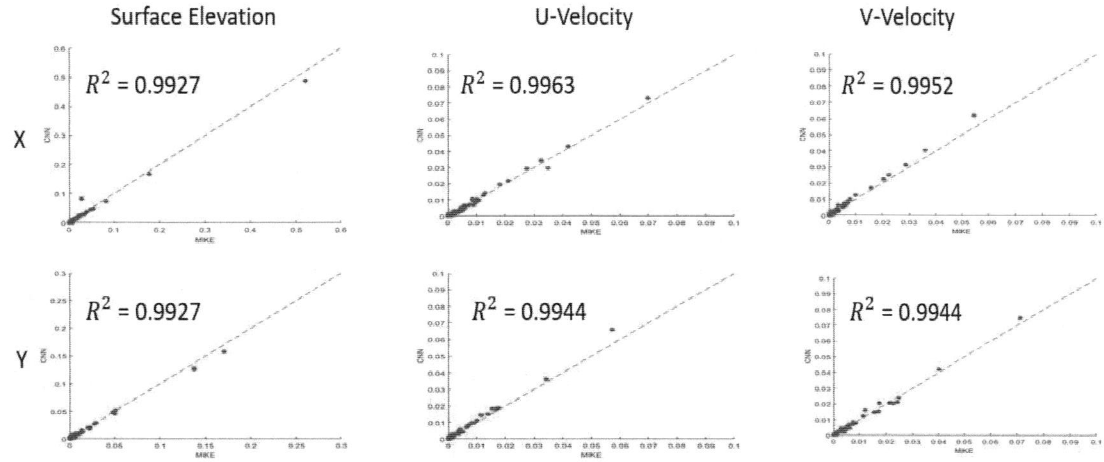

Figure 16 Time prediction method - Comparing numerical model vs CNN using $\boldsymbol{R^2}$ of amplitude and phase in cartesian coordinate.

Table 1 compares the independent, partially adjoint and fully adjoint model using R^2 value for amplitude, phase and time series. It was observed that when surface elevation is introduced, the R^2 value decreases across x, y and time series. This might be due to the inverse phase relationship between surface elevation and current velocities. Hence, CNN might not be able to capture the inverse relationship. However, introducing the v-velocity variable improves the R^2 of u-velocity for x, y and time series. Similar conclusions were also found when looking at the perspective of surface elevation (se) where the R^2 value decreases when currents are introduced as an additional input.

Model	Input/Output	R^2 in x	R^2 in y	R^2 of Timeseries
Independent	u	0.9963	0.9944	0.9911
Partially Adjoint	u,v	0.9983	0.9942	0.9927
Partially Adjoin	u,se	0.8709	0.9413	0.8990
Fully Adjoint	u,v,se	0.9854	0.9624	0.9496

Table 1 U- velocity R^2 values of amplitude and phase in cartesian coordinate and time series for respective CNN models in point 4 of St John's Island domain.

4.2 Fourier Method Results

Figure *17* shows the streamline comparison between the physic-based and CNN independent input models. It is observed that CNN can predict the Fourier coefficient of varying hydrodynamic state variables accurately for all time steps.

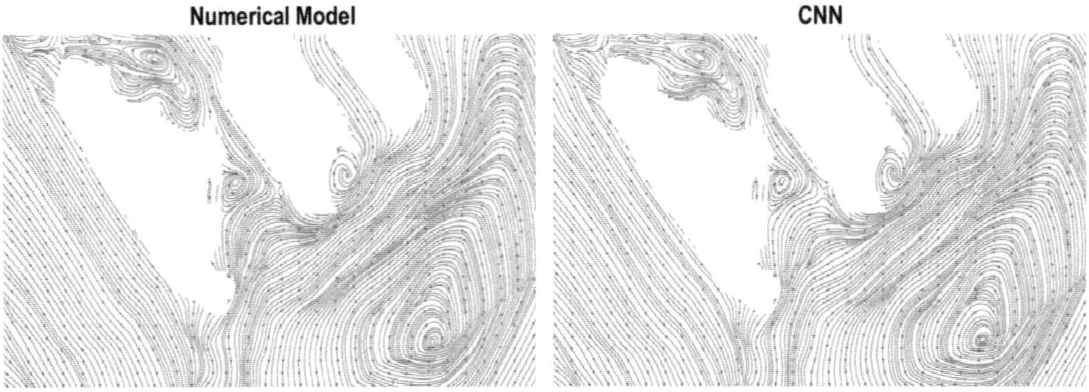

Figure 17 Fourier prediction method - Streamline comparison of St John's Island domain. Numerical Model (Left), CNN Model (Right)

Figure *18* shows the time series plots for St John's Island at point 4. The error between the numerical and CNN independent input model was insignificant, within 5% for all sea state variables. The R^2 value showed that the CNN model fits the numerical model accurately. The same behaviour can also be seen at other points around St John's Island and geometry cases.

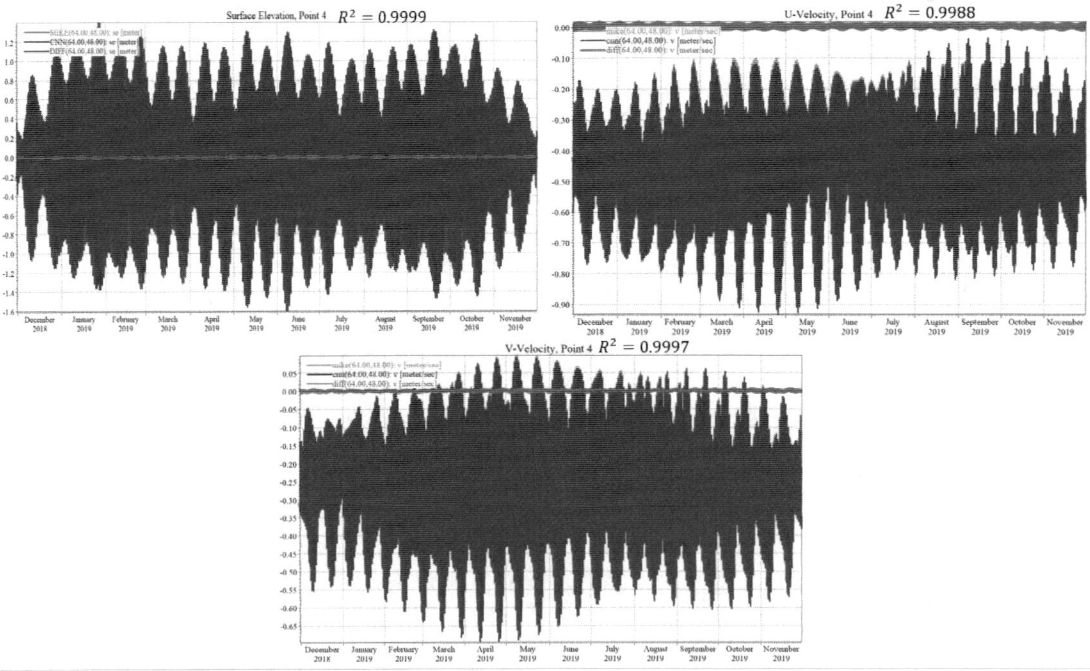

Figure 18 Fourier prediction method - Timeseries plots for surface elevation (Top left), u-velocity (Top right), V-Velocity (Bottom) at point 4.

The R^2 plot shown in Figure *19* compares the amplitude and phase between the numerical model and CNN independent input model. It is observed that CNN can predict the Fourier coefficients of varying hydrodynamic state variables accurately, with an R^2 value close to 1. Similar observations are also found in other points around St John's Island and other test cases as well.

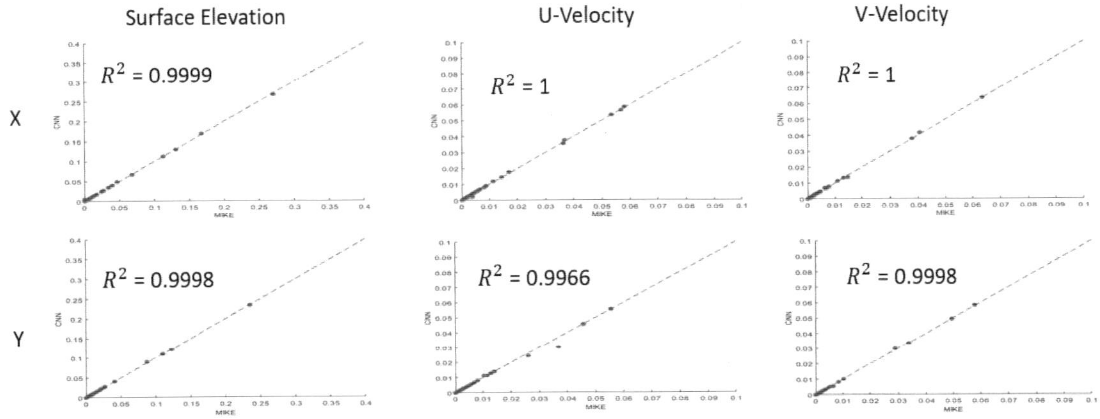

Figure 19 Fourier prediction method - Comparing numerical model vs CNN using R^2 of amplitude and phase in cartesian coordinate.

5 Conclusions

This paper investigates CNN's applicability to coastal hydrodynamics driven by predominant astronomical tides. Three complex hydrodynamic flow fields are investigated using two methods.

The first method tests CNN's capability on time series prediction. Although CNN predicts the complex flows with an acceptable error, it was able to predict the location and capture the eddies satisfactorily. CNN was also able to generally give an adequate prediction of the time series and its Fourier decomposition when applying error measures of relative error and R^2 values. We have also tested on three different variations of CNN model described in Figure 10. When compare the models whose inputs without surface elevation and those with surface elevation, the ones with the latter have higher RMSE and lower R^2 value. This shows that including surface elevation as inputs with currents decreases the model performance and vice versa.

The second method tests CNN's ability to predict Fourier coefficients of varying Manning numbers. It was observed that CNN can capture the spatial features of 2D Fourier coefficients accurately with R^2 values very close to 1 for its time series, amplitude and phase.

In conclusion, CNN performs better when predicting non-sequential data (Fourier method) compared to predicting sequence data (time series). Further development using CNN to predict non-sequence data (Fourier method) should be considered as a promising method to elaborate hydrodynamic forecast in tidal-dominant conditions. As for sequence data (time series), future works may include applying Conv2DLSTM that may give further accuracy increase of time forward predictions. Physics-informed Neural Network would also be another step forward to incorporate Navier-Stokes equations into machine learning models such that the coupling between the hydrodynamic state variables can be improved.

Acknowledgements

This research is supported by the National Research Foundation, Singapore and the Singapore Maritime Institute under its Maritime Transformation Programme White Space Funding Support for the Establishment and Operation of the Centre of Excellence for Autonomous and Remotely Operated Vessels (CEAOPS), SMI-2019-MTP-01.

References

1. LeCun, Y., et al., *Backpropagation Applied to Handwritten Zip Code Recognition.* Neural Computation, 1989. **1**(4): p. 541-551.

2. Morimoto, M., et al., *Convolutional neural networks for fluid flow analysis: toward effective metamodeling and low dimensionalization.* Theoretical and Computational Fluid Dynamics, 2021. **35**.
3. Thuerey, N., et al., *Deep Learning Methods for Reynolds-Averaged Navier–Stokes Simulations of Airfoil Flows.* AIAA Journal, 2019. **58**: p. 1-12.
4. Lapeyre, C.J., et al., *Training convolutional neural networks to estimate turbulent sub-grid scale reaction rates.* Combustion and Flame, 2019. **203**: p. 255-264.
5. willemsen, D. *The astronomical origin of tides.* 2019; Available from: https://www.sailingissues.com/navcourse6.html.
6. Yakura, H., et al., *Malware Analysis of Imaged Binary Samples by Convolutional Neural Network with Attention Mechanism.* 2018. 127-134.
7. inc., D., *Dense Tensor.* 2022.
8. Savyakhosla, *CNN | Introduction to Pooling Layer.* 2021.
9. Murata, T., K. Fukami, and K. Fukagata, *Nonlinear mode decomposition with convolutional neural networks for fluid dynamics.* Journal of Fluid Mechanics, 2020. **882**: p. A13.
10. Bukka, S.R., A.R. Magee, and R.J.a.F.D. Jaiman, *Deep Convolutional Recurrent Autoencoders for Flow Field Prediction.* 2020.
11. Vinuesa, R. and S. Brunton, *The Potential of Machine Learning to Enhance Computational Fluid Dynamics.* 2021.
12. despois, j., *Latent space visualization — Deep Learning bits #2.* 2017.

Model-Based Design and Safety Assessment for Crewless Autonomous Vessel

Takuya Nakashima*[1], Koji Kutsuna[2], Rui Kureta[2], Hisaki Nishiyama[2], Tomoya Yanagihara[2], Jun Nakamura[2], Hideyuki Ando[2], Hideaki Murayama[1] and Satoru Kuwahara[3]

[1] Graduate School of Frontier Science, The University of Tokyo, Kashiwa, Japan
[2] MTI Co., Ltd., Tokyo, Japan
[3] Japan Marine Science Inc., Kawasaki, Japan

E-mail: nakashima@s.otpe.k.u-tokyo.ac.jp

Abstract. The Designing the Future of Full Autonomous Ship (DFFAS) Project conducted the crewless maritime autonomous surface ship long-distance demonstration in congested waters in March 2022. This study shows a model-based systematic design methodology and a safety evaluation method for autonomous ships conducted through this project. A reference model of ConOps (Concept of Operations) for autonomous navigation systems is proposed using the DFFAS system as an example. System Theoretic Process Analysis (STPA) is also applied at the subsystem and component levels according to the actual system development phase to extract appropriate granularity of safety requirements.

1. Introduction

In Japan, the ageing of the seafarers for coastal shipping and the difficulty of securing personnel have been big issues, significantly impacting the local economy and related industries. The introduction of autonomous and crewless vessels is expected to solve this problem.

The Nippon Foundation launched "demonstration tests of The Nippon Foundation MEGURI2040 Fully Autonomous Ship Program" in 2020 to support the development of autonomous technologies. It promotes the transformation of the logistics, economy, and social infrastructure by fostering momentum for further technological development through the world's first successful demonstration of crewless domestic vessels. With the support of this project, the Designing the Future of Full Autonomous Ship (DFFAS) Project was launched, consisting of 30 domestic and foreign companies and institutions.

Based on the concept of "creating the future of crewless maritime autonomous surface ships - a grand design devised by various experts," the DFFAS project aims to develop a crewless vessel with open collaboration. It also aims to develop autonomous navigation and support functions such as monitoring and diagnosis from shore (including a communication system) and remote operation in emergencies.

The DFFAS project successfully carried out the demonstration experiment using a domestic container ship between Tokyo Bay and Ise Bay from February 26 to March 1, 2022. It demonstrated for the first time the use of a comprehensive fully autonomous navigation system (including remote control and land support) for a container ship operating in congested waters.

2. Background and Objectives

2.1. Application of MBSE to Autonomous Navigation System Development

When considering the design of an autonomous navigation system, it can be difficult to achieve an optimal system design by reassembling existing components based on conventional functions, as the function of the autonomous system will be changed by a role distribution between humans and machines. The concept of systems thinking is essential to consider overall requirements and functions based on the purpose and goal of the system. Particularly in development with multiple stakeholders, the concept of Model-Based Systems Engineering (MBSE) is effective describing the envisioned system as a model rather than as a natural language to provide a common understanding of the envisioned system.

In the DFFAS project, the conceptual design of an autonomous navigation and operation system has been conducted using the MBSE approach. There are several examples of conceptual design methods for autonomous navigation systems. As described in [1], the design should be materialised by the submitter and approved step by step for obtaining and maintaining approval of an alternative and/or equivalency. Tools and methodologies for autonomous ship design are being discussed [2], but most of the research focus are mainly on the conceptual phase. There are few examples of applying and breaking down this methodology to the autonomous ship to be built.

INCOSE defines MBSE as “the formalized application of modeling to support system requirements, design, analysis, verification and validation activities beginning in the conceptual design phase and continuing throughout product development and retirement. [3]” In this paper, we focused on the ”modeling to support system requirements, design, analysis” from the perspective of system safety.

2.2. Establishment of Safety Evaluation Method for Autonomous Navigation Systems

It is important to deal with emergent hazardous events in designing complex systems using new technologies. The objective of a ship operation system is to ensure the safe, timely, and efficient operation of a ship in any environment, and particularly, safety seems to be the main perspective for defining requirements.

Several risk analysis methods for the system of systems have been developed to deal with emergent hazardous events caused by interactions of system components. This paper applies STPA (System Theoretic Process Analysis) based on the concept of STAMP (Systems-Theoretic Accident Model and Processes) proposed by [4]. Safety of the Intended Functionality (SOTIF), which has been discussed in the automotive industry as a complement to functional safety [5], is an approach to ensure safety, including emergent hazards and human factors due to the interaction of systems. In [6], STPA is also introduced as one of the analysis methods for emergent hazards.

There are several hazard analysis results regarding autonomous ship [7]. Some references apply STAMP/STPA to autonomous navigation systems. A framework for analysing and discussing uncertainty in mitigation measures obtained through STPA analysis was provided and the initial-stage concept was analyzed in [8]. A framework to model a hierarchical control structure for the STPA analysis of an autonomous ship was presented in [9]. A preliminary concept of safety validation method using STPA was proposed in [10]. A systematic and systemic hazard analysis and management process for a conceptual design phase were provided in [11]. They applied it to two pre-construction ferries, but the extracted safety actions are still general and broad.

This paper attempts to evaluate the safety of the entire autonomous navigation system, to organise the system requirements based on the evaluation, and to reflect them in the design using STPA.

3. Methodology

3.1. Establishment of ConOps reference model for crewless autonomous navigation system

The development of a system begins with its conceptual design. Concept of Operation (ConOps) is a document that defines the requirements of a system by envisioning its users and their usage scenarios. ConOps is also considered essential in the context of Requirements Engineering. ConOps is also used to study new systems in various fields, such as aerospace [12]. There have been many guidelines for autonomous ships in recent years [13][15][14][16], and some refer to the necessity of ConOps.

It seems to be common that ConOps is a document to confirm the purpose, assumptions, and scope of the system, to support requirements elicitation based on use cases etc., and to involve relevant stakeholders to convince them of the necessity of the system to be built. However, the granularity of this document is not clearly defined, especially for the autonomous navigation system. It is necessary that, by using ConOps, the system's requirements are extracted at an appropriate level of granularity and are utilised without reducing the degree of design freedom. Based on the literature survey on ConOps, a reference model of ConOps is presented using the DFFAS system as an example.

3.2. Establishment of safety requirement elicitation scheme by STPA for detailed design

While extracting appropriate safety requirements and reflecting them in the design is an important process, there seems to be no completely established methodology. For example, safety analysis cannot be performed until the system structure has been realised to some extent, but there are no examples to show how much the system should be realised and at what stage safety analysis should be performed. The implementation of STPA during the requirements definition process of the V-model was systematised [17]. However, the proposed process did not address the case where the requirements are repeatedly detailed and hierarchically divided into subsystems and modules.

Therefore, a step-by-step safety analysis for the system to be developed should be needed. This paper discusses how the safety analysis can function in parallel with the development process and the appropriate granularity of the safety analysis with experts' opinions. By doing this, the flow from the concept to the detailed design of the system using ConOps is concretised.

Figure 1 shows the focus of this paper in the system development process. The italic number in the figure corresponds to the section number of this paper.

4. Result

This chapter introduces the results of the study conducted in the DFFAS project.

4.1. Construction of Concept of Operation

Based on the literature review, the elements necessary for ConOps for autonomous navigation systems has been visualised in Figure 2. In addition to the system's requirements, functions, and components, it is necessary to consider the system's background, purpose, use cases, the external environment, and other external constraints. It is also necessary to consider the risk and impact of the system.

ConOps and Operational Concept are described separately in [18]. In this case, the former is a document from the business strategy viewpoint in promoting the system, and the latter is closer to the ConOps to be discussed here. Also, in some documents, it is mentioned that the system's future development should be considered [19]. Therefore, a chapter on the system's future development to clarify the current position of the envisioned system is included.

Based on the contents in Table 1, the Concept of Operation for the DFFAS system is outlined.

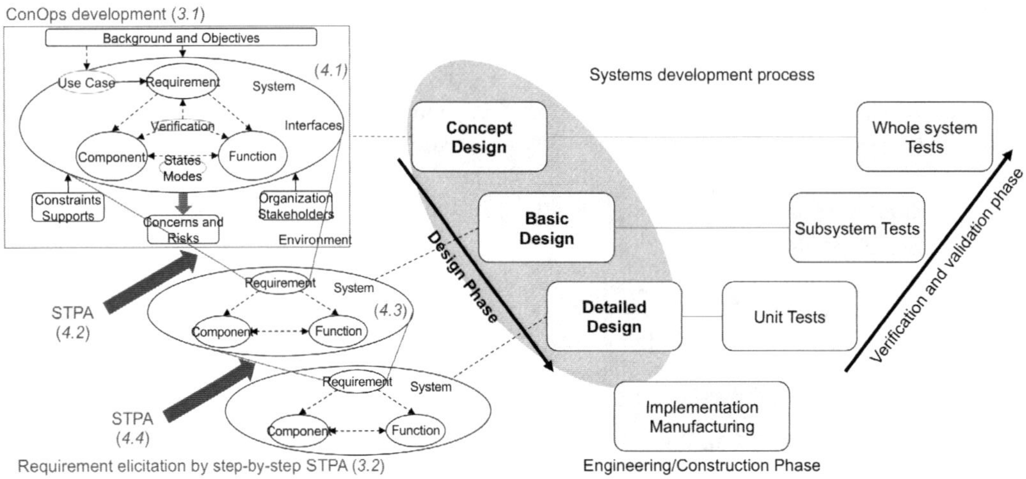

Figure 1. Focus of this paper.

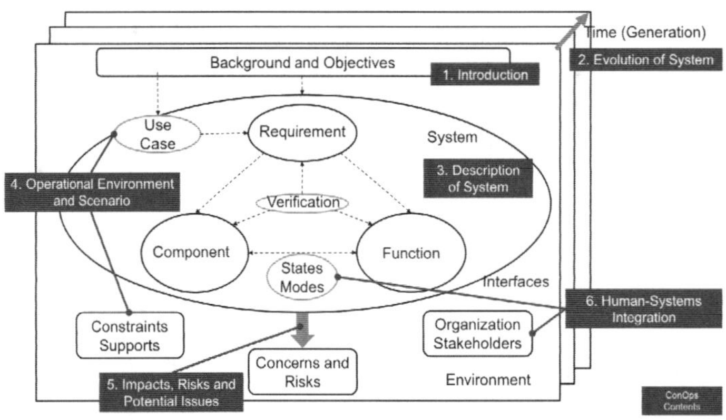

Figure 2. Required elements for system description.

(1) Introduction The background and objectives of the DFFAS project are as described above; to construct a system for unmanned navigation in Japan's coastal shipping lanes and to complete a demonstration experiment.

Table 2 shows the goals and major assumptions for constructing this system. These were determined through discussions within the DFFAS project, in consultation with The Nippon Foundation and other related stakeholders. In addition, the scope of the system to be built has been detailed in Table 3 based on the goals and assumptions.

(2) Evolution of systems There are various possible use cases for introducing unmanned systems, both in terms of time and area. Therefore, developing a general-purpose system that will contribute to the gradual introduction of unmanned navigation systems is necessary to

Table 1. ConOps Contents.

Content	Description
1. Introduction	Background, System Scope, Assumption and Constraints
2. Evolution of System	Justification for changes Future Roadmap and Status of the envisioned system
3. Description of System	Needs, Goals and Objectives of the system Overview Architecture incl. Interfaces (Major System elements and interconnections) Modes of Operation Basic Functions (Proposed Capabilities)
4. Operational Environment and Scenario	Use Cases (Nominal, Off nominal) Actors/Stakeholders Operational Scenario Data flow (input and output of the system)
5. Impacts and Potential Issues	Operational impacts, Environmental Impacts, Organizational Impacts, Scientific/Technical Impacts Regulatory Compliance, How to Implement the system
6. Human-Systems Integration	Human-in-the-loop involvement Human-machine interface etc.
Appendix	Glossary, Acronyms, Reference Documents

Table 2. Goals and assumptions of the DFFAS project.

Goal	- To implement a crewless autonomous vessel based on economic rationality. - To contribute to ensuring maritime safety by reducing the number of accidents caused by human error.
Assumption	- The system should be designed to take advantage of the actual business strengths of various stakeholders, such as shipyards and marine equipment manufacturers. - The system should be implemented at a low cost by utilising existing facilities as much as possible. - The system should be designed to withstand conditions like actual coastal ship operation. (The system should be used in narrow waterways and congested waters.) - Do not aim to make all operations unmanned but focus on the parts that contribute to safe operations and actual business. - The system should be able to be easily backed up by humans in case of emergency.

reduce the workload at sea and ensure maritime safety, regardless of the revision schedule of laws and regulations. This aim is also considered in defining the system's scope.

In addition, open development methodology based on Model-Based Systems Engineering will contribute to developing new concept ships, not just limited to autonomous ships in the future. So, constructing a general framework for many stakeholders to develop and verify novel systems can be a target as well.

Table 3. Scope of the system to be built with DFFAS.

Element	Details
Target ship (Automation functions)	- Retrofit a 749 GRT container ship. (Average size for a Japanese coastal ship). - The actuators are assumed to be standard equipment for general merchant ships.
Internal environment	- A highly autonomous Berth-to-Berth operation system will be established. - Emergency response is assumed to be carried out by the seafarer, and the operator ashore will be in charge of remote operations. - The shipowner and captain will be responsible for maintaining the ship's seaworthiness and formulating long-term voyage plans. - Mooring and cargo handling operations, which are greatly affected by infrastructure development, are outside the scope. - It should be possible to flexibly change the division of roles between machines and humans, and between onboard and ashore, depending on the situation.
External environment	- A situation is assumed where there is a mixture of ships of various autonomous operation levels and existing ships. - Port control is assumed to be existing. No onboard pilot is assumed. - Current navigation rules are to be observed (COLREG, SOLAS, Port regulations and the other domestic laws).

(3) Description of the system Based on the above assumptions, the necessary elements of the system have been organised. First, ship operation and manoeuvring can be divided into seven stages: information acquisition, information integration, situation analysis, planning, plan verification, detailed control order, and execution of control. It is assumed that humans would carry out these tasks with the support of machines and equipment in existing ships. However, in the case of autonomous navigation, the information acquisition, integration, situation analysis and planning would, in effect, be processed continuously by the computer, making it difficult to distinguish between them. Therefore, the function group is divided as shown in Table 4 and Figure 3. The tasks from information acquisition to verification can be divided into strategic and tactical tasks, which may vary depending on the time axis of planning.

It is difficult to operate a vessel with a system in which the roles of man and machine are firmly fixed in a situation in which the ocean, communication environment, and internal system conditions change dynamically. Therefore, it is required to have a system status manager that can appropriately judge the system's status based on the information obtained.

The system structure to achieve these functions has been defined at the subsystem level in Figure 4. Based on the communication environment and the business feasibility, it is assumed that strategic decisions would continue to be made on land. In addition, an executor for the propulsion subsystem is assumed to be on board, as it seems unfeasible to execute maintenance of current machinery room by automation e.g., robots with manipulators.

The architecture has been created based on the scope mentioned in *(1)* and *(2)*. While assuming the use of existing systems as much as possible, the Fleet Operation and the Central Information Management (CIM) subsystems are two new systems for this novel system. The former is responsible for navigation support from shore. The shipowner is expected to be

Table 4. Basic Function.

Function	Description
Object and Environment Detection (LOED)	A function group corresponds to detecting objects and events via integrating information on a wide-range environment.
Short-term Object and Environment Detection (SOED)	A function group corresponds to detecting objects and events via integrating information on a surrounding environment.
Long-term Event Response and Path Planning (LERPP)	A function group corresponds to planning based on LOED and making a navigation plan.
Short-term Event Response and Path Planning (SERPP)	A function group corresponds to planning based on SOED and making an action plan.
Dynamic train control and Actuation	A function group corresponds to the control and actuation of the ship as planned.
System Status Management	A function group corresponds to managing information of system status.
Communication	A function group corresponds to communication between ship and shore (Fleet Operation Centre: FOC).
Propulsion and Power Management	A function group corresponds to managing electricity and controlling the power systems of the ship.

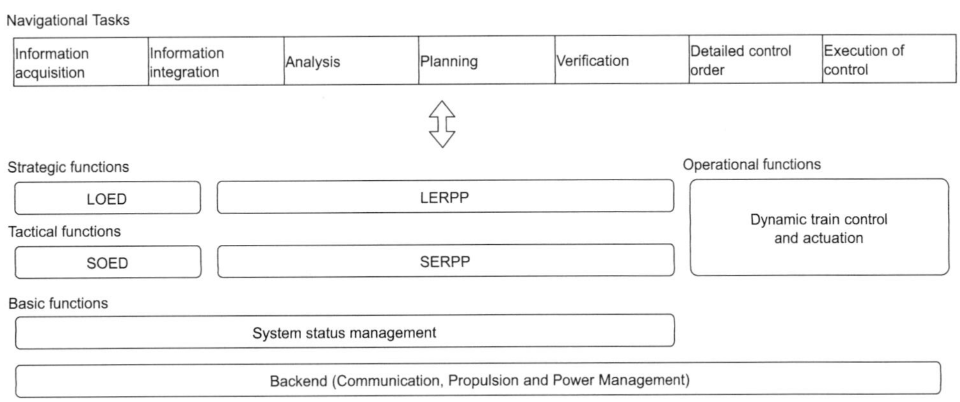

Figure 3. Functions, Tasks, Category of components and Role.

responsible for long-term voyage planning, and the master and crew are also expected to be responsible for monitoring and fallback operations. The latter is responsible for the functions of System Status Management and is designed to change the operational status according to the situation.

(4) Operational Environment and Scenario This section describes the environment surrounding the system, including the use cases, related stakeholders, and data exchanged inside and outside the system. As the assumed scenarios define parts of the functions, it is concretised in an iterative manner with *(3)*.

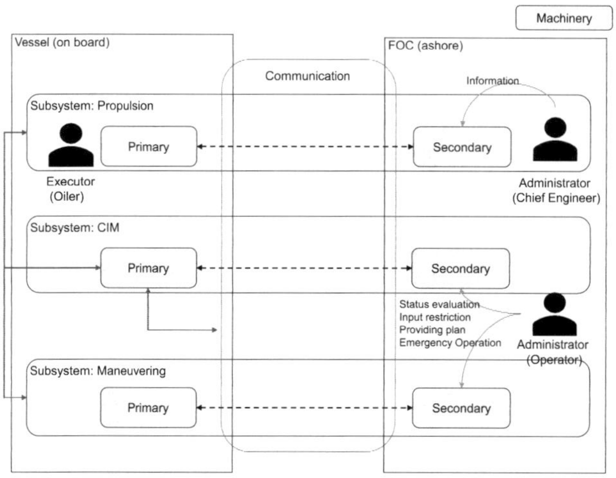

Figure 4. System structure at the subsystem level.

Actors and stakeholders include shipping companies and operators, port authorities, administrations and classification societies, shipyards, marine manufacturers, insurance companies, etc. We worked closely with these stakeholders to exchange opinions and obtain feedback on the design based on their needs.

As for use cases, berth to berth operation is assumed in the nominal case, while remote operation from the Fleet Operation Centre is assumed in the off-nominal case. In addition, the system automatically stops safely in an emergency to ensure Minimal Risk Condition.

As an Operational Scenario, since there are various navigational patterns for berth-to-berth operations, the modes of the operation have been defined. The actual voyage can be divided into seven modes: unberthing, leaving, harbour out, coastal, harbour in, approaching and berthing.

The degree to which humans should contribute varies depending on the internal and external environment. Here, the following four levels of system statuses are defined based on existing literature. The definition of each status is shown in Table 5. In the basic and detailed design stages, the criteria for each mode are refined.

Table 5. Definition of Status.

Status	Description
Normal (N)	Running without any involvement by the operator
Active Monitoring (AM)	Running under monitoring and verification by the operator
Remote Fallback (RFB)	Running under fallback operations by the operator at FOC
Independent Fallback (IFB)	Running under fallback operations by the machinery on the vessel to keep the system at Minimal Risk Condition

The mode change is implemented in the form of embedding it in the waypoint according to the operational status. As a result, the mode changes automatically according to the position information of the own ship. On the other hand, the status is determined by the CIM subsystem. The status decision criteria are divided into two categories: those defined by the system's internal state (internal health level) and those defined by the external influence

(External Operational Design Domain: EODD). The internal health level is defined by the state of the subsystem responsible for the operation task, divided into the Manoeuvring, Propulsion, and Communication subsystem. The state of each subsystem is defined as shown in Tables 6, 7 and 8. The whole system status is defined in Table 9, assuming the system is in EODD.

The EODD represents the range of external conditions that can be expected when a ship is in operation. The appropriate range of rudder angle and speed should be determined based on the magnitude and direction of the external forces, in accordance with the ship's manoeuvrability. Although the ship should be designed to operate automatically without deviating from these conditions, if the ship deviates from these conditions and gets out of EODD, it is designed to immediately switch to Independent Fallback (IFB) Operation.

The basic idea of the status setting is shown in [20]. As the status can define responsibilities between humans and machines, it can be considered as the foundation for the operational envelope proposed in [21].

There are two types for status transition: one is Approval, in which the status is switched with the operator's approval, and the other is Acknowledgement, in which the operator is only notified of the switch. Approval is required for those that require human involvement in the future and those that no longer require active involvement, while Acknowledgement is used for those that require a quick transition to IFB.

The concept of Dynamic Positioning System (DPS) was referred to for the concrete realisation of each subsystem level [22]. The DPS should be designed redundantly, and its functions are defined at different levels according to the situation.

Table 6. Manoeuvring subsystem health level.

Level	Definition
Level1	Possible to design an action plan with sufficient reliability all by oneself
Level2	Possible to design an action plan under monitoring by the operators; The operators complement reliability and integrity
Level3	Possible to design an action plan within a limited scope; the operators complement subsystem function
Level4	Possible to design an action plan within a limited scope; machines on the vessel complement subsystem function
Level5	Unable to design an action plan; Unable to take supplementary operation by either human or machines on the vessel

Table 7. Propulsion subsystem health level.

Level	Definition
Level1	Possible to deliver the power of propulsion and control
Level2	Possible to deliver the power of propulsion and control, but with alert (caution or warning) or on monitoring by the operators
Level3	Possible to deliver the power of propulsion and control within a limited scope, e.g., alert to trigger auto-slow-down is activated
Level4	Possible to deliver the power of propulsion and control within an extremely limited scope, e.g., the subsystem can continue only anchoring
Level5	Unable to deliver the power of propulsion and control

Table 8. Communication subsystem health level.

Level	Definition
Level1	Possible to monitor and control the system
Level2	Possible to monitor and change the status of the system within a limited scope; RFB operations unavailable
Level3	Disconnection or impossible to monitor

Table 9. Whole system status definition.

C: 1	P: 1	P: 2	P: 3	P: 4	P: 5
M: 1	N	N	RFB	IFB	NUC
M: 2	AM	AM	RFB	IFB	NUC
M: 3	RFB	RFB	RFB	IFB	NUC
M: 4	IFB	IFB	IFB	IFB	NUC
M: 5	NUC	NUC	NUC	NUC	NUC
C: 2	**P: 1**	**P: 2**	**P: 3**	**P: 4**	**P: 5**
M: 1	N*	IFB	IFB	IFB	NUC
M: 2	AM*	IFB	IFB	IFB	NUC
M: 3	IFB	IFB	IFB	IFB	NUC
M: 4	IFB	IFB	IFB	IFB	NUC
M: 5	NUC	NUC	NUC	NUC	NUC
C: 3	**P: 1**	**P: 2**	**P: 3**	**P: 4**	**P: 5**
M: 1	N*	IFB	IFB	IFB	NUC
M: 2	IFB	IFB	IFB	IFB	NUC
M: 3	IFB	IFB	IFB	IFB	NUC
M: 4	IFB	IFB	IFB	IFB	NUC
M: 5	NUC	NUC	NUC	NUC	NUC

M: Maneuvering subsystem health level, P: Propulsion subsystem health level, C: Communication subsystem health level, N: Normal Operation, AM: Active Monitoring Operation, RFB: Remote Fallback Operation, IFB: Independent Fallback Operation, NUC: Not Under Command, *: With a time limit

(5) Impacts and potential issues The impacts of this system on various fields are summarised in Table 10 in terms of operation, environment, organisation, technology, legal system, and social implementation. As a result, the important elements for future risk assessment and requirement elicitation are summarised.

(6) Human-Systems Integration The division of roles between humans and machines is practically included in *(3)* and *(4)*. It can be organised again at the subsystem level, as shown in Figure 5.

4.2. Initial safety analysis using STPA

Based on the ConOps, safety analysis using STAMP/STPA has been conducted. The STPA consists of four major steps: two preparation phases (Step 0-1, 0-2), Step 1 and Step 2.

Table 10. Impacts and potential issues.

Category	Description
Operation	The envisioned system reduces the load on the crew and ensures safe and efficient operation (with the effect of reducing accidents caused by human); maintenance requires flexible software updates from remote locations.
Environment	Efficient operation by the envisioned system can reduce impacts.
Organisation	The number of seafarers could be reduced. It is also necessary to have a shore support centre.
Technology	Design and construction require the system integration capabilities to bring together many technologies and achieve efficiency and safety and the ability to manage many stakeholders.
Legal	Although the introduction of the envisioned system does not violate current legal system, it is necessary to lobby the IMO and the national government to change rules such as the lookout to make efficient use of this system.
Implementation	At present, there are hurdles in implementing the envisioned system because of the initial cost.

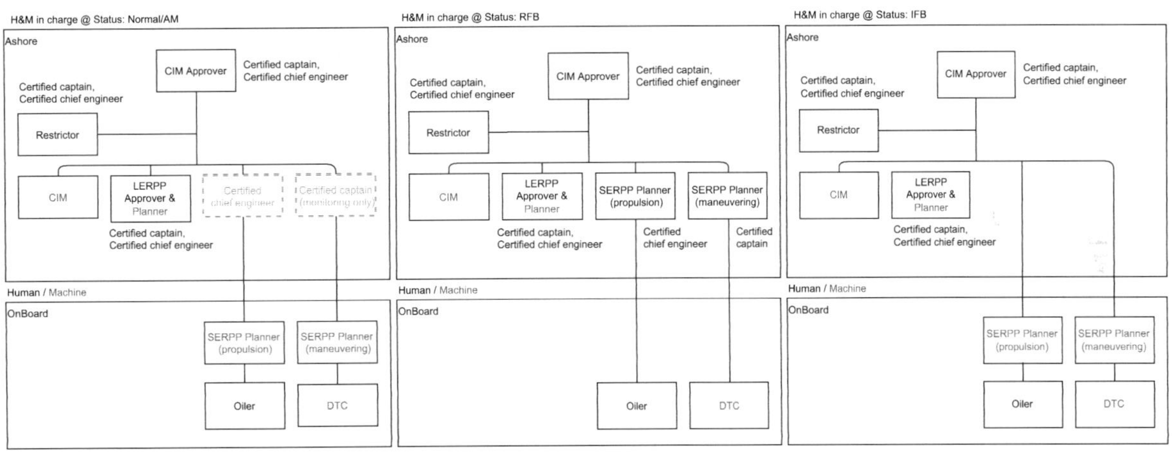

Figure 5. Human Machine Roles in each status. (Left: Normal and Active Monitoring, Centre: RFB, Right: IFB)

In Step 0-1, the accidents and hazards of the system and the Safety Constraints (SCs) that the system should guarantee are defined. Collisions and groundings were set to be our focus, which account for more than half of all accidents. Note that fire and other incidents are excluded from the scope of this analysis. In general, Safety Constraints seem to be defined as close to the final consequences as possible. However, in considering a complex system such as this one, it seems difficult to extract the Unsafe Control Actions (UCAs) exhaustively with abstract Safety Constraints, as some control actions can be too far from the consequences. Therefore, a chain of events that could lead to "close to obstacles' and 'loss of seaworthiness' in operation has been assumed and these intermediate events were defined as Safety Constraints, which are shown in Table 11 and Figure 6.

In Step 0-2, the control structure should be developed, which has already been created in the ConOps phase (Figure 4).

In Step 1, the UCAs are extracted by checking the Control Actions among the subsystems described in the control structure one by one. The following four words are used as guide words referring to [23]: *1. A control action required for safety is not provided or is not followed, 2. An unsafe control action is provided that leads to a hazard, 3. A potentially safe control action is provided too late, too early, or out of sequence, 4. A safe control action is stopped too soon or applied too long (for a continuous or non-discrete control action).* Based on this, the Unsafe Control Actions were extracted. A couple of them are shown in Table 12. The square brackets indicate the Safety Constraints that conflict.

In Step 2, the Hazard Contribution Factors (HCFs) were extracted for each UCA. The guiding words for the HCFs were those used in [23]: *failures related to the controller, inadequate control algorithm, unsafe control input, and inadequate process model.* The obtained UCAs and HCFs were organised in BowTie to increase their visibility, as shown in Figure 7. The countermeasures against the HCFs are shown as the barriers, which function as the requirements for the subsystems.

Table 11. Safety Constraints (SCs).

SC	Description
SC1	Own vessel states must be detected: system conditions and sensor-detected values etc.
SC2	Other vessels and those states must be detected: existence and course, heading, speed and positions.
SC3	Natural environments which affect the system must be detected: wind, wave, tidal stream, temperature, etc.
SC4	Static constraints which are essential to achieve voyage must be obtained.
SC5	Navigation and/or action plan must be established.
SC6	Control signal must be calculated based on navigation/action plan.
SC7	Geographic information to navigate must be detected.
SC8	Seaworthiness including condition of equipment and hull must be analysed and actions must be selected based on own status and surrounding environment.
SC9	Dynamic constraints must be analysed based on static constraints and internal/external environment (e.g., short stopping distance, Turning circle).

Table 12. Examples of Unsafe Control Actions (For interaction between subsystem).

Control Action	Not Providing	Providing	Too early / Too late
Sending constraints from CIM to Maneuvering (LERPP)	The subsystem is unable to work. [SC5]	The subsystem works improperly. [SC5]	N/A (Small delay of providing constraints does not trigger hazardous situation.)
Sending constraints from Maneuvering (SERPP) to CIM	CIM is unable to decide system status. [SC8]	CIM decide improper system status. [SC8]	CIM is unable to decide system status at suitable timing. [SC8]

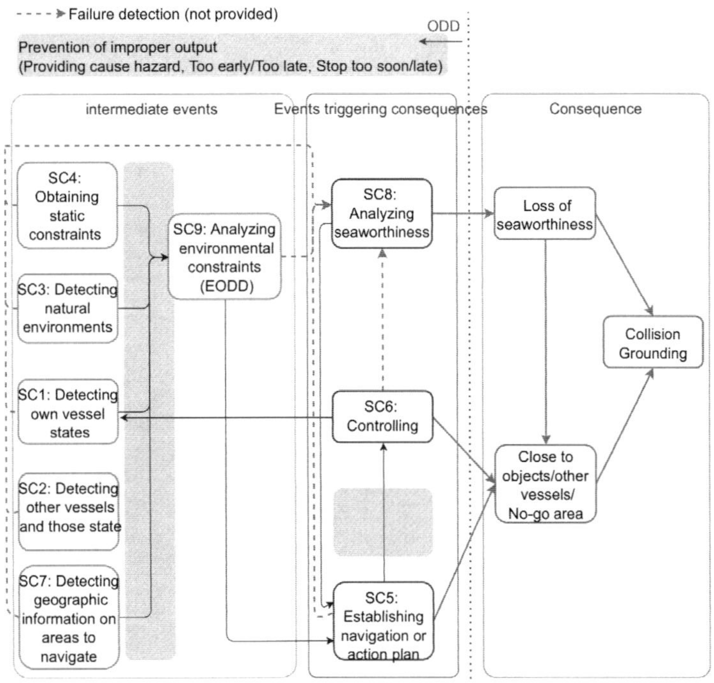

Figure 6. Flow Diagram of Safety Constraints.

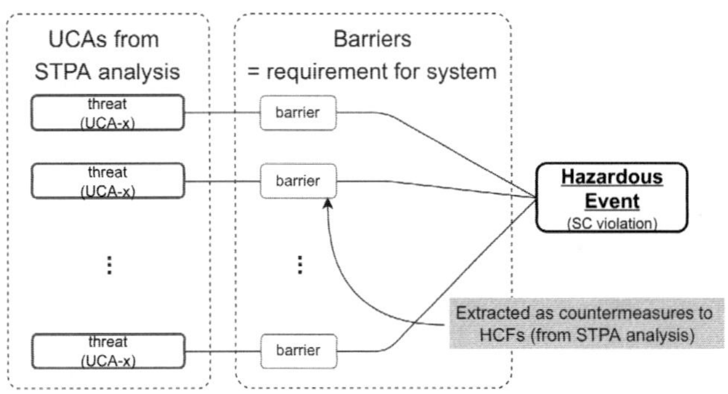

Figure 7. Schematic image of BowTie analysis.

4.3. Requirement Definition, Decomposition and Architecting
Based on ConOps and STPA, and referring to the framework of SysML, the model diagrams of the system have been organised, clarifying the relationships among them. Requirements for the system have been set based on the Hazard Causal Factors (HCFs) described above.

The Block Diagram and N2 Diagram of the CIM subsystem are shown to specify the functions and the system configuration in Figure 8 and 9. The Block Diagram shows the process flow, and the N2 Diagram shows the flow of information generated in each process.

As a result, the system configuration, including the modules and components in each subsystem, has been detailed. The modules in each subsystem correspond to some of the functions in the above diagram.

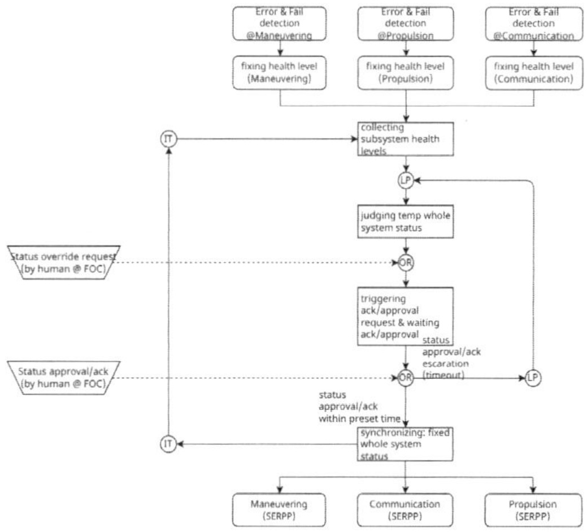

Figure 8. Block Diagram of CIM subsystem.

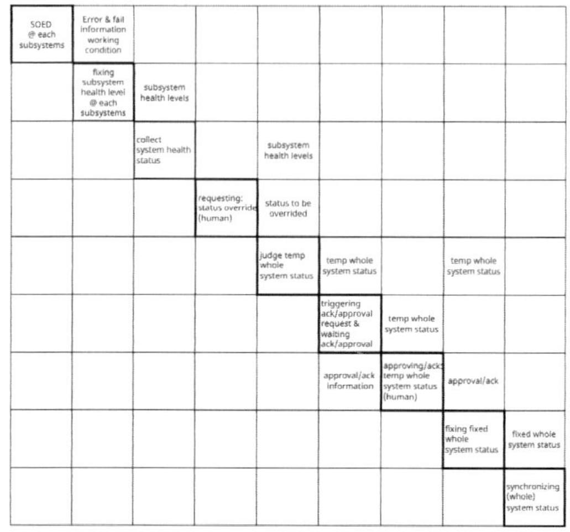

Figure 9. N2 Diagram of CIM subsystem.

4.4. Detailed Safety analysis using STPA

In this phase, the same analysis as in *4.2* is performed on the system that has been concretised to the module level within the subsystem, and more detailed safety requirements are extracted. Part of UCAs in the subsystem are shown in Table 13.

The extracted requirements have been classified into those reflected in the system design, the operational procedures, and the training requirements. The classified requirements have been referred to in the verification phase, i.e., integration test, demonstration test, etc.

Table 13. Examples of Unsafe Control Actions (For interaction inside subsystem).

Control Action	Not Providing	Providing	Too early / Too late
Sending Temporary whole system status	CIM is unable to make output fixed status. [SC8]	CIM chooses an improperly fixed status. [SC8]	CIM is unable to choose fixed status at suitable timing. [SC8]
Sending Long term Re-route request	LERPP cannot establish a navigation plan. [SC5]	LERPP cannot establish a proper navigation plan. [SC5]	N/A (Even if the re-route request is given delay, LERPP has some delay, but not critical.)

4.5. Summary

The overall flow up to this point is shown in Figure 10. In defining the requirements for the safe autonomous system, STPA has been conducted to concrete and design the behaviour and structure step by step. To conduct the safety analysis, it is not possible to extract appropriate requirements unless the system's behaviours and structures are defined to some extent. By introducing the MBSE and STPA concepts to the development of an autonomous vessel, an actual example of the appropriate granularity and methodology has been presented.

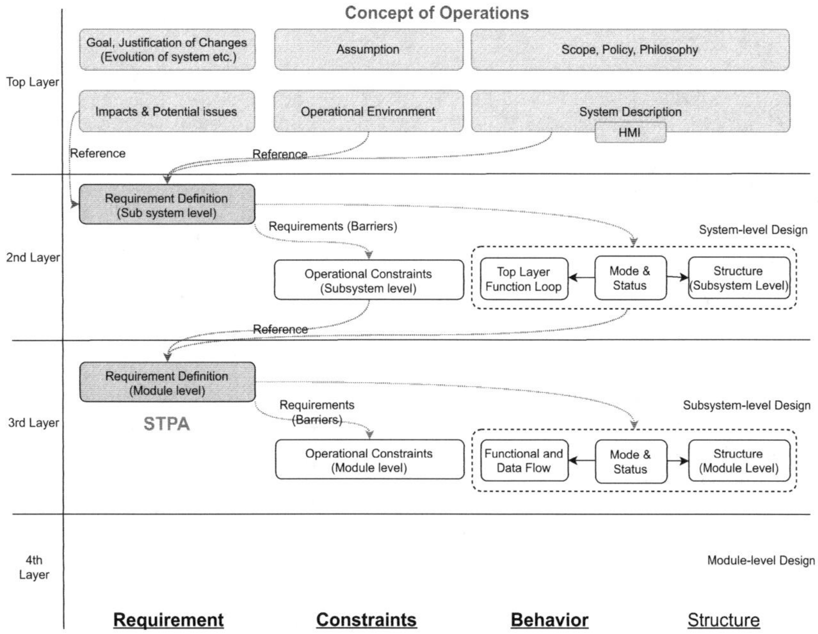

Figure 10. Model-based design for a crewless vessel with STPA.

5. Conclusions

In this paper, a methodology for model-based design of autonomous navigation systems has been described using real system development as an example. The appropriate ConOps elements for the autonomous navigation system have been studied and a set of functions and subsystems is designed in relation to the system's objectives, assumptions etc. This paper also proposes a new system concept for managing the system status.

A step-by-step safety analysis method has been conducted using STPA to implement them at an appropriate granularity according to the actual system development phase to extract appropriate safety requirements. Appropriate Safety Constraints for the autonomous navigation system are also proposed, which will be an effective method of STPA for such a complex system.

One of the future challenges is model-based validation and verification. Safety argumentation methods for autonomous vessels and simulation systems with appropriate scenarios should be considered.

Acknowledgments

This research was supported by The Nippon Foundation's "demonstration tests of The Nippon Foundation MEGURI2040 Fully Autonomous Ship Program". We would also like to express our sincere gratitude to Prof. Yan Jin and his colleagues at Impact Lab, Nippon Kaiji Kyokai (ClassNK), and the National Maritime Research Institute for helpful discussions.

References

[1] IMO, Guidelines for the Approval of Alternatives and Equivalents as provided for in Various IMO Instruments, MSC.1/Circ.1455, 24 June 2013.

[2] L. Andreas and L. Wennersberg, AUTOSHIP D3.2 Autonomous-ship-design-methods-and-test-principles v1.1, 2021.

[3] International Council on Systems Engineering, Systems Engineering Vision 2020, INCOSE-TP-2004-004-02, 2007.

[4] N.G. Leveson, Engineering a Safer World: Systems Thinking Applied to Safety. MIT Press, Cambridge, MA., 2011.
[5] ISO 26262:2018 Road vehicles – Functional safety
[6] ISO/DIS 21448: 2021, Road vehicles – Safety of the intended functionality
[7] C. A. Thieme, C. Guo, I. B. Utne, and S. Haugen, Preliminary hazard analysis of a small harbor passenger ferry-results, challenges and further work, J. Phys. Conf. Ser., vol. 1357, no. 1, 2019.
[8] K. Wróbel, J. Montewka, and P. Kujala, Towards the development of a system-theoretic model for safety assessment of autonomous merchant vessels, Reliab. Eng. Syst. Saf., vol. 178, no. June, pp. 209–224, 2018.
[9] M. Chaal, O. A. Valdez Banda, J. A. Glomsrud, S. Basnet, S. Hirdaris, and P. Kujala, A framework to model the STPA hierarchical control structure of an autonomous ship. Safety Science, 132, 2020.
[10] B. Rokseth, O. I. Haugen, I. B. Utne, Safety Verification for Autonomous Ships., MATEC Web of Conferences, 273, 2019.
[11] O. A. Valdez Banda, S. Kannos, F. Goerlandt, P. H. A. J. M. van Gelder, M. Bergström, and P. Kujala, A systemic hazard analysis and management process for the concept design phase of an autonomous vessel. Reliability Engineering and System Safety, 191, 2019.
[12] NASA, NASA System Engineering Handbook Revision 2, Natl. Aeronaut. Sp. Adm., p. 297, 2016.
[13] American Bureau of Shipping, Guide for Autonomous and Remote Control Functions, 2021.
[14] ClassNK, Guidelines for Automated/Autonomous Operation of ships Design development, Installation and Operation of Automated Operation Systems/Remote Operation Systems , 2020.
[15] Bureau Veritas, Guidelines for Autonomous Shipping, 2019.
[16] DNVGL, Class Guideline DNVGL-CG-0264 Autonomous and remotely operated ships, 2018.
[17] S. Yamaguchi, Evaluation of the Application of the Safety Analysis Method Based on System Theory "STAMP/STPA" to Requirement Development Phase, Keio University, 2019, PhD thesis. (Japanese)
[18] ISO/IEC/IEEE 29148:2011 Systems and software engineering – Life cycle processes – Requirements engineering
[19] The Federal Aviation Administration, Concept of Operations for Urban Air Mobility (UAM) (ConOps 1.0), 2020.
[20] K. Kutsuna, H. Ando, T. Nakashima, S. Kuwahara, and S. Nakamura, NYK's Approach for Autonomous Navigation-Structure of Action Planning System and Demonstration Experiments, J. Phys. Conf. Ser., vol. 1357, no. 1, 2019.
[21] Ø. J. Rødseth, L. A. Lien Wennersberg, and H. Nordahl, Towards approval of autonomous ship systems by their operational envelope, J. Mar. Sci. Technol., no. 0123456789, 2021.
[22] American Bureau of Shipping, Guide for Dynamic Positioning Systems, 2021.
[23] N.G. Leveson and J.P. Thomas, STPA Handbook, 2018.

Development of Recommendations for Digital Testing of MASS Navigation Safety prior to Sea Trials

Kie Hian CHUA[1], Savio COUTINHO[2], Aizad NORAHIM[1], Dimitrios KONOVESSIS[3]

[1]Technology Centre for Offshore and Marine, Singapore (TCOMS), 12 Prince George's Park S(118411),
[2]Singapore Polytechnic, 500 Dover Road S(139651),
[3]University of Strathclyde, 16 Richmond St, Glasgow G1 1XQ, United Kingdom

Corresponding author: chua_kie_hian@tcoms.sg

Abstract. For Maritime Autonomous Surface Ships (MASS), a key area that has seen active development is in the use of autonomous capabilities for vessel navigation and control. This can range from a simple use case of waypoint navigation, which takes into account bathymetry and navigation markers, to complex collision avoidance scenarios where the autonomous systems are required to detect, evaluate and execute evasive manoeuvres based on time and spatially varying dynamic behaviour of other vessels. In Singapore, it has been identified that there is a need to carry out accurate digital testing of MASS navigation safety before sea trials. This is where a vessel developer is required to demonstrate that the MASS is able to carry out the sea trials safely, and to stress test high risk scenarios that may not be practicably tested in the sea trials. A study has been carried out to develop recommendations for the digital testing, which takes into consideration the need for accurate representation of the actual MASS being built, as well as the verification of the autonomous navigation algorithm's capabilities to safely control the vessel in real-world scenarios. Based on the study, a three-stage framework is proposed. Firstly, the accuracy of the digital model in representing the dynamic responses of the actual vessel is verified and any discrepancy with benchmark data is to be quantified. Secondly, tests are carried out to ascertain that the autonomous navigation algorithm is able to control (virtually) the dynamically-accurate vessel from one point to another, taking into account the real-world environmental loads. Lastly, the ability of the autonomous navigation algorithm in carrying out collision detection and avoidance is verified. As part of the study, a review of the current state-of-art and engagement with the industry has been carried out. These details are described in this paper.

1. Introduction

With the advancement of digitalization in the maritime industry, autonomous or remotely controlled operations could undertake mundane and high-risk tasks, thereby enhancing efficiency and safety and efficiency. This has helped increase the interest in autonomous or remotely controlled vessels.

Globally, several companies are actively developing and integrating technologies to automate and autonomise maritime navigation and operations. Since 2017, Kongsberg Maritime has been involved in the development of navigational, power generation and propulsion systems for the Yara Birkeland which has been launched in Jan 2022 [1]. Wärtsilä is also actively developing autonomous solutions, starting from ferries and harbour tugs. In 2020, Wärtsilä and PSA Marine [2] retrofitted an existing tug with a suite of sensing and control systems, and have successfully demonstrated the autonomous capabilities through a series of sea trials, under the IntelliTug project. MPA has also funded other MASS projects, including the retrofitting of 2 other tugs with autonomous capabilities by ST Marine [3] and Keppel Offshore and Marine [4], respectively.

Despite the examples above, the maritime autonomy sector is generally still considered to be in its early stages of development. At the current state-of-the-art, autonomous navigation requires the integration of navigational, propulsion and control systems that require continuous improvements due to the evolving global maritime traffic and unpredictable sea conditions.

As part of the effort to enable MASS trials and operations, a study has been carried out to develop a set of recommendations for digital testing of autonomous navigation capabilities prior to actual sea trials. These recommendations take into consideration the need for accurate representation of the actual MASS vessel being built, as well as the verification of the autonomous navigation algorithm's capabilities to safety control the vessel in real-world scenarios. In addition, any recommendations put forward for MASS testing should not impede technology innovation, and should consider the goal-based processes used by the industry for qualifying MASS.

In our study, literature from current research and technical practice is reviewed, as well as with actual data from initial MASS sea trials that have been carried out in a "sandboxed" manner in Singapore waters. Based on the review and engagement with stakeholders, an underlying framework is first developed, addressing the need for first-principles, verification of vessel model accuracy, as well as the autonomous navigation algorithm's ability to steer the vessel safely (in the virtual representation), including navigation under the dynamic influence of environmental loads, as well as the algorithm's collision avoidance capabilities.

To establish equivalence in terms of navigation safety, the digital testing will gauge whether an autonomous navigation algorithm can achieve safe distances that are comparable to existing manned vessels. Quantitative and qualitative data on the typical closest point of approach (CPA) for various vessel manoeuvring scenarios is collected to establish the existing benchmark for the manned vessels. AIS data of ships plying around Singapore is used to quantitatively establish the CPA between a vessel and another encountered vessel or navigational marker. At the same time, qualitative data is also gathered from mariners, based on their perception of CPA as they navigate through the same scenarios in Singapore waters.

In the next section, an overview of the research and technical literature is reviewed. This is to provide the background of the existing state of the art and the current industry practice relating to testing and verification of autonomous vessel safety. In Section 3, the frameworks established by the various Classification Societies for qualification and certification of autonomous vessels and the associated technological capabilities are also reviewed. Section 4 describes how CPA values that can be used to demonstrate safety equivalence to the existing conventional vessels are determined from AIS data and through a survey of seafarers. Lastly in Section 5, we present the proposed digital testing framework before concluding in Section 6.

2. Literature Review

There have been various developments using smart or autonomous algorithms to provide navigation guidance, including obstacle detection, trajectory control, and collision avoidance. A selection of the relevant literature has been reviewed and are disused below.

2.1. Autonomous navigation and collision risk assessment

In [5], the authors proposed a fuzzy control way point tracking system that calculates the proximity of the first two waypoints individually using distance to closest point of approach (DCPA) and time to closest point of approach (TCPA) and then subsequently plan out the appropriate route. The authors also noted that the differing manoeuvring characteristics of different vessels could pose a challenge for way point tracking systems.

Given the dynamic nature of ship navigation, information relating to positioning, navigation, and timing (PNT) is crucial in ensuring navigation safety due to the need to determine the relative positions between ships and / or obstacles, including the evolution of the trajectories. The use of AIS as a tool to provide PNT information is investigated in [6] by using the signals from different AIS stations. A ship's position can be predicted based on the associated displacement vector.

While the AIS can give real-time information between nearby ships and the operation of encountered ships, it is not able to provide a broad interpretation for a vessel to take any evasive actions to reduce a collision risk. In [7], the authors proposed a new design of AIS-based embedded system for vessel collision avoidance that provides visual display information of the vessel navigation for the mariners and provide advance collision warning and operation plan. A framework for collision warning using a risk model where the ship's system states, and position-related information are used in [8] to determine the level of risk of a ship-to-ship collision.

To ensure the safety of the autonomous ship, online assessment of navigational information should be carried out in real-time, similar to how existing ships maintain constant watchkeeping. The following two categories of collision risk detection method are proposed in [9]: Closest Point Approach Method (CPA - 2D method) and Predicted Area of Danger Method (PAD - 3D method). In the former, the predicted shortest distance between an autonomous ship and an encountered vessel is used to assess the collision risk. In the PAD method, various possible trajectories of the autonomous ships is projected as an inverted cone, while the encountered ship's trajectory is projected as an inverted cylinder. The intersection between the cone and cylinder is considered to be the area of danger where there is a risk of collision.

2.2. *Assessment and Regulation of Autonomous and Remotely Operated Vessels.*

In [10], the potential impact of autonomous and remotely operated vessels on the overall regulatory framework is discussed and a goal-based regulatory approach is proposed to facilitate the use of various technologies. An approach to apply the Systems-Theoretic Process Analysis (STPA) for the holistic safety verification for autonomous ships is proposed in [11], where a case study for an autonomous ship, covering hazard identification for potential scenarios involving collisions into obstacles and loss of navigational control at a high-level is also described. A system developed by DNV GL for testing of autonomous navigation systems using a ship's Digital Twin is described in [12]. Other main components of the system include the Operating environment, Test management system, and the Test interface that could be used for evaluating how well the autonomous navigation system performs based on compliance with COLREGs.

With regard to the generation of scenarios for digital testing, there are also other developments in terms of how scenarios could be generated for testing of collision avoidance capabilities. In [13], the authors developed a data-driven approach based on AIS information to generate traffic scenarios for collision avoidance testing. Machine learning techniques for feature extraction is applied for abstraction and automatic identification of various ship encounter scenarios such as passing, crossing of other vessels. A method for automatic generation of hazardous scenarios for the testing of an autonomous ship's collision avoidance system based on the distance and time considerations between the ship and the encountered vessels is proposed in [14].

2.3. *Quantifying the collision regulations into actionable algorithms*

Since its origins in 1840, the COLREGS have been written in a general manner, requiring interpretation by seafarers taking into account the circumstances, which includes a combination of the specific COLREG rule and on what is commonly referred to as "the ordinary practice of seamen". This term, mentioned in Rule 2, relates to the prevalent culture of the area being navigated. While this rule requires compliance to COLREGS it also allows for deviation when it is necessary to avoid immediate danger. COLREGS does not provide information on the deviations nor definition of what could constitute "ordinary practice of seamen." Thus, the challenge for navigation decision making is to identify the appropriate time horizon and spatial distance before determining it is required to deviate for collision avoidance. This decision is dependent on an assessment of the prevailing circumstances where the following considerations should be taken into account by the navigator:

i. Size, speed and manoeuvrability of the vessels involved.
ii. Available safe sea room.
iii. Location and number of other vessels in the vicinity which could increase the complexity of the situation.
iv. Influence of environmental aspects, such as, visibility, state of sea, current etc.

There are other COLREG rules that have been laid out in a general manner similar to Rule 2. For example, in Rule 16 a give-way vessel is required to "take early and substantial action to keep well clear." The subjectivity of this rule poses a challenge as to what would constitutes "early and substantial" action, which in turn could be any combinations of speed or course change. Rule 17 requires the stand-on vessel to monitor if the other vessel is taking action in accordance with the appropriate rules and monitor of this action is made in ample time and allows for safe passing of both vessels. Such a vessel is required by another part of the same rule to take action when collision cannot be avoided by the action of the give-way vessel alone. With the subjective terms such as "action as will best aid to avoid collision" and "if the circumstances at the case admit", Rule 17 adds further complexity to implementation of autonomous collision avoidance strategies.

In a high complexity situation however such as the Straits of Malacca and Singapore where there is heavy vessel traffic in a constrained sea space, an evasive manoeuvre for one ship may lead into a close quarters situation with another ship, resulting in a cascade of complex and unpredictable interactions and outcomes. In [15], it is suggested that a study of Automatic Identification System (AIS) data for a given area could be undertaken to elucidate the effective outcomes of "early and substantial" actions undertaken in a manner that is consistent with "the ordinary practice of seamen" for the area. By combining learning from recent AIS data with large number of digital simulations, developers would be able to ensure that the limits identified are up-to-date with respect with respect to the "ordinary practice of seamen" in the area, which will evolve as the overall vessel demography changes with increase in adoption of autonomous technologies. This would also pave the way for development of "learn-as-you-operate" capabilities where the navigation knowledge and experience derived from machine learning or artificial intelligence (AI) is improved with more data that is kept updated.

3. Development in Qualification and Certification of MASS

Various Classification Societies, in their role as Recognized Organizations for flag Administrations, have published guidance on the certification of MASS. A selection of these guidance documents is reviewed in our study, in order to elucidate the current approaches on how MASS and the system components can be qualified as fit-for-purpose.

3.1. American Bureau of Shipping (ABS)

ABS defined a five-stage goal-based qualification process [16], which requires comprehensive engineering evaluation and detailed risk assessments to be done at each stage. The five stages of the process are: Feasibility, Concept Verification, Prototype Validation, System Integration, and Operational. The two sets of guidance notes from ABS [16] and [17] can be used for the purpose of classifying ships and offshore vessels integrating new technologies which have yet to be proven and tested in the maritime and offshore industry, such as the novel systems that are used in MASS. In addition, an accompanying procedure [18] provides further guidance on the smart capabilities that can be implemented on existing vessels, that can be used to enhance the safety and efficiency.

3.2. Bureau Veritas (BV)

BV has developed a risk-based guideline to classify new technologies in the marine and offshore industry. The objective of this risk-based systematic approach is to locate the risks associated with the failure modes identified from this new technology [19]. BV has also published a set of Guidelines for autonomous shipping [20] which provides recommendations for the operation and design of the autonomous systems, to help the designers and operators of these autonomous systems better understand the regulatory framework governing this type of ships.

3.3. Det Norske Veritas (DNV)

DNV has published a set of recommended practice (RP) which provides guidance for the qualification of new concepts and technologies [21]. The RP provides a staged qualification process for identifying and minimizing unpredictability of these technologies, through technical evidence demonstrating the performance of associated systems. The process compromises six stages: Qualification Basis, Technology Assessment, Threat Assessment, Qualification Plan, Qualification Execution and Performance Assessment. DNV has also published the Class Guideline (CG) for Autonomous and Remotely Operated Ships [22], providing recommendation for operation, design, as well as the qualification process for certification of MASS. The CG states the rationale for adopting a risk-based assurance process, i.e. that the MASS sector is still in the nascent stages where novel technologies and concepts are being introduced and it "is therefore currently not possible or desirable to provide detailed rules for all areas and combinations of concepts".

3.4. Lloyd's Register (LR)

Consistent with other Classification Societies, LR introduced a structured goal-based approach to the qualification process of unmanned marine systems (UMS) with a defined set of goal, functional objectives, and performance requirements [23 & 24]. LR's rationale for adopting a goal-based approach is similar to DNV's rationale for adopting a risk-based assurance process.

4. Determination of CPA from AIS Data and Survey of Seafarers

Given the above, an exercise to analyse AIS data on actual closest point of approach (CPA) for various types of ships in the Singapore Strait is undertaken. As described earlier, such CPA can be inferred to be 'safe outcomes' that a typical vessel should adhere to ensure safe navigation, given the "ordinary practice of seamanship" in this area. This AIS data was captured using an AIS receiver located at the National University of Singapore (NUS), which provided information on vessel movement in the western part of the Singapore Strait. This is shown in the red circle of Figure 1.

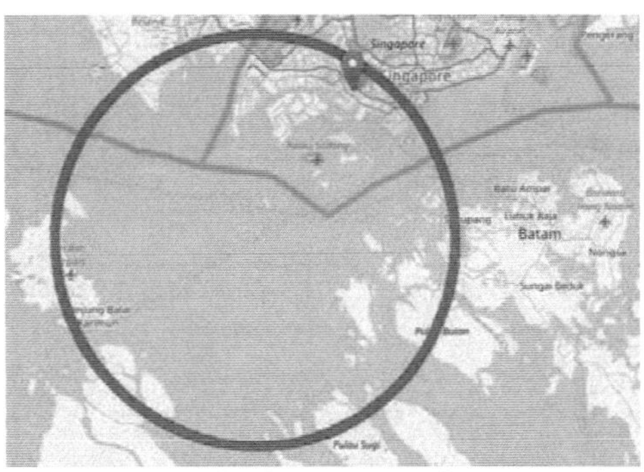

Figure 1. AIS data coverage of NUS receiver

Data over a period of 5 months from April 2020 to August 2020 was collected for the following types of vessels and for the manoeuvres as indicated in Table 1. From the AIS data, it was observed that while the CPA values may vary according to the type of vessel, the range of variation is less than half a nautical mile for each type of manoeuvre. This is likely due to the influence of a restricted waterway dominating over the vessel type. Consequently, a condensed table of the CPA values was obtained and shown in Table 2.

Table 1: Types of vessels and manoeuvres studied using the NUS AIS data

Types of vessels for which AIS data recorded	Manoeuvres for which AIS data recorded
• Bulk Carrier • Oil Tanker • Offshore Vessel • Container Vessel • Passenger Vessel/Ferry • Ro-Ro Vessel	• Passing ahead of another vessel • Passing astern of another vessel • Passing alongside another vessel (head-on or overtaking situation) • Passing a fixed navigation mark

Table 2. Ranges of CPA values according to types of manoeuvres

Manoeuvre Type	CPA range derived from AIS data
Passing ahead of another vessel	0.56 – 0.63 NM
Passing astern of another vessel	0.55 – 0.61 NM
Passing alongside another vessel (head-on situation)	0.61 – 0.72 NM
Passing a fixed navigation mark	1.10 – 1.50 NM

A survey of past and present seafarers with experience navigating the Singapore Strait is also conducted to further ascertain the value of similar safe distances for the manoeuvres is listed in Table 1. The following questions were posed to the participants through the survey:

i. In a risk of collision situation during transit in the Singapore straits, what CPA would you deem to be safe and appropriate when passing ahead of the other vessel?
ii. In a risk of collision situation during transit in the Singapore straits, what CPA would you deem to be safe and appropriate when passing astern of the other vessel?
iii. In a risk of collision situation during transit in the Singapore straits, what CPA would you deem to be safe and appropriate when passing clear of the other vessel in an overtaking or head-on situation?
iv. In a situation during transit in the Singapore straits, what CPA would you deem to be safe and appropriate when passing clear of a fixed navigation mark?

A total of 179 respondents were recorded through the survey, a breakdown of their respective rank at time of completion of survey is shown in Table 3 while Table 4 tabulates the results on the expected CPA obtained from the survey, together with the ranges derived from AIS data.

Table 3. Breakdown of seafarer survey participants by rank

Seafarer Rank	No. of respondents
Master Mariner	72
Chief Officer/First Mate	28
Navigational Officer (Operational)	50
Trainer/Lecturer	15
Sea-Pilot	9
Shore-Based (ex-seafarer)	5

Table 4. CPA ranges obtained from seafarer survey compared with ranges derived from AIS data

Survey Question	CPA Ranges (NM)	
	Seafarer	AIS
1	0.97 – 1.14	0.56 – 0.63
2	0.62 – 0.73	0.55 – 0.61
3	0.68 – 0.75	0.61 – 0.72
4	0.41 – 0.51	1.10 – 1.50

The following observations were made from the CPA ranges obtained from the survey:

i. The CPA values obtained from the survey is expected to be more conservative due to the perception of a navigator exercising prudence, to maintain a higher level of safety, especially for Question 1.

ii. There is general consistency between the two datasets for Questions 2 and 3.

iii. For Question 4, the difference in CPA values is likely due to the way AIS calculates CPA which uses the actual centre of the navigation mark as a reference point. Seafarers on the other hand, would usually consider the edge of landmass on which the navigation mark is situated as the point of reference for determining CPA.

iv. While good seamanship dictates variable safe distances be maintained from passing vessels or objects based on ship length, ship speed and size to take into account special manoeuvring characteristics of vessels, the CPA ranges obtained above from both the AIS data and seafarer survey are in the order of around 0.60 to 0.75NM for most cases. This is attributed to the unique context Singapore Strait where the high volume of traffic navigating in close quarters due the limited sea room at hand play a more dominant influence on the CPA compared to the vessel sizes.

Based on the above study and taking into consideration that the recommendations on digital testing of autonomous ships should be kept as simple as reasonably practicable to facilitate harmonised implementation across different stakeholders, it is recommended that a CPA range of 0.60NM – 0.75NM should be adopted for the various indicated manoeuvres and interactions with other target vessels / obstacles. This would suffice to demonstrate that the autonomous ships are able to achieve safety outcomes that are equivalent to the existing ships navigating in this area. Through the industry and seafarer engagements, there was also a request for simplicity in terms of specifying any recommendations of CPA for MASS. As such, the range of 0.60NM – 0.75NM is also applied to the scenario when a MASS passes a fixed navigation mark. In addition to having a simple and standardised metric for assessing the collision avoidance capabilities of an autonomous ship, this CPA range also provides an additional safety buffer distance compared to the range of CPA values obtained from the seafarer survey, as well as the AIS data (after accounting for the manner in which CPA is computed by AIS systems).

5. Recommended Digital Testing Framework

A three-stage digital testing framework is proposed based on the foregoing review. Descriptions of each of these stages are described below.

The first stage, Model Accuracy, is to first ensure that the underlying numerical model of the ship within the digital model is verified to be reliable and accurate before conducting the simulation test cases using the autonomous navigation algorithm. This is carried out by using the numerical model to simulate turning circle and zig-zag manoeuvres, and comparing against benchmark data of the actual ship. These manoeuvres have been selected from the IMO Standards for Manoeuvrability [25] and are thus familiar to the industry.

In the second stage, Navigation Safety, the digital model will be tested for its ability to autonomously navigate from one point to another, such as along a series of waypoints in various environmental conditions. This is to demonstrate that the autonomous navigation algorithm is able to steer the ship safely, accounting for the external environmental forces the bathymetry and coastline.

In the third stage, Collision Detection and Avoidance, the collision avoidance capabilities of the autonomous ship will be assessed, where the autonomous ship is required to take action in accordance with COLREGS, allowing for safe passing maintaining a pre-determined safe distance. By defining incidents as breaches of a prescribed safe distance and setting the incident tolerance level to be the same as the current level, the framework helps to ensure that the autonomous vessel is at least as safe as the existing ships in operations.

The parameters used for the second and third stages would depend on the relevant context, such as the operating area of the autonomous ship, and considers the need for a goal-based approach as recommended by the industry and IMO. Through our industry discussions, representatives stressed the criteria for such tests to adopt a functional approach, where the autonomous ship is tested for functions and goals it is designed and built to perform. Nevertheless, the advantages for standardization of the tests to ensure a harmonised degree of rigour, to provide a level playing field across industry stakeholders is also recognised, and to minimize the likelihood of differing interpretations by different stakeholders. As such, in the second Navigation Safety stage of the digital testing, a set of prescribed environmental conditions based on conditions normally encountered in this region is proposed. For autonomous ships operating in other areas in addition to Singapore, environmental conditions applicable to that area should be applied in addition, to ensure that the navigation algorithm is able to account for those environmental effects. Considering that there would be a certain level of uncertainty and the need to demonstrate that the navigation algorithm is able to generate consistent results across multiple simulation runs, the point A to point B navigation tests should be repeated 30 times, to quantify the prediction interval of the navigation course travelled by the digital model.

Based on a similar consideration for standardisation, collision avoidance testing of typical COLREGS scenarios in the third Navigation Safety stage should be progressively tested up to a minimum of two encountered target vessels. For these test scenarios the autonomous ship will be required to take avoiding action in a risk of collision situation involving one or two target vessels. The action taken shall be in accordance with COLREGS and allow for safe passing based on a pre-determined safe distance. Based on the simulated outputs, this safe distance will be obtained in the form of the CPA between the autonomous ship and encountered vessels. Corrections to account for discrepancies between the digital model and the actual vessel, as well as uncertainty of the prediction should be applied to the CPA. Thereafter, the corrected CPA will be assessed against the CPA values recommended in the previous section, as a "pass criteria". In addition to the CPA, the actions taken by the autonomous ship should be in accordance with COLREGS. Thus, the digital model should be deemed to have failed a test case if it undertakes an action that is not aligned with COLREGS or where the obtained CPA is less than the recommended values.

The following combinations of ship speed and draught conditions are recommended for the digital tests.

<u>Ship speed</u>

a. 6 knots;
b. 12 knots; and
c. At least 90% of the ship's speed corresponding to 85% of the maximum engine output, as per the IMO Standards for Manoeuvrability [25]

<u>Draught condition</u>

a. Full load or design draught; and
b. Lightly loaded or ballast draught.

The ship speeds have been selected based on the speed limits within the Singapore Port waters, as well as the speed specified in [25]. By testing at both higher and lower speeds, any influence of ship speed on manoeuvrability would be taken into account. Similarly, by testing at the heaviest and lightest loading conditions, the influence of vessel loading (including weight distribution) on the dynamic responses would be taken into account.

5.1. Stage 1: Model Accuracy

In a digital model for motions of a dynamic system, the motion responses of that system under external forces would be captured using a set of mathematical equations. For an autonomous ship, these equations would be implemented as a manoeuvring model, upon which the autonomous navigation and collision avoidance capabilities will be built on. It is thus important to establish the accuracy of the manoeuvring model as inaccuracies could mean erroneous predictions of an autonomous ship's ability to manoeuvre away from shallow waters, obstacles and other vessels, resulting in grounding or collisions. As such, any discrepancies from the actual ship in terms of dynamic responses should be quantified and subsequently accounted for in the overall assessment of navigational safety.

In the proposed digital testing guidelines, the well-established turning circle and zig-zag manoeuvres are adopted from [25]. These are selected to streamline the testing requirements as much as possible, given that at least part of the benchmark data would be available for existing ships. The digital model would be used to simulate these manoeuvres and the various values of longitudinal and lateral distances obtained from these manoeuvres would be compared against corresponding data representative of the actual ship. Possible sources for such data are:

i. Data from prior sea trials, which could be available if the autonomous ship being tested is converted from an existing conventional vessel. In this case, sea trials would have been conducted when the vessel was first delivered;
ii. Data from model tests; or
iii. Data from high-fidelity computational fluid dynamics simulations in which the actual hull form, propulsion and steering systems are modelled.

The turning circle manoeuvres should be carried out for both starboard and port turns in calm water, and for a range of rudder angles, similar to [25]. This would help ensure that the range of turning ability is captured. Similarly, the zig-zag manoeuvres should be carried out at different rudder angles.

5.1.1. Turning Circle Manoeuvres. To simulate the turning circle manoeuvres, the rudder deflection was set at three different rudder angles: 10°, 20° and 35°. The key information from the turning circle simulations that will be obtained and compared against benchmark data are advance, transfer and tactical diameter. Figure 2 illustrates the manoeuvre, as well as how these three values should be measured.

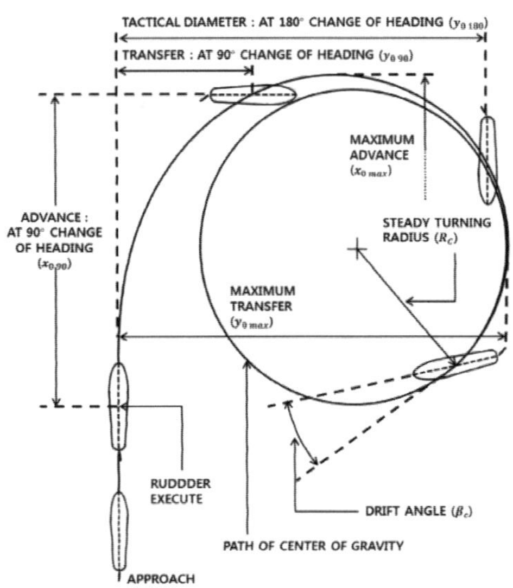

Figure 2. Turning circle manoeuvre from ITTC, 2017 [26]

5.1.2. Zig-zag Manoeuvres. To simulate the zig-zag manoeuvres, from a straight approach, the rudder will then be deflected at a specified degree angle and once the vessel has achieved this specified degree of course change, the rudder angle will be shifted to the opposite side.

The key data from the zig-zag test simulations are the longitudinal distance travelled from the start of the manoeuvre to the point where the ship is at the 2nd overshoot angle, and the maximum lateral displacement. This lateral displacement is the maximum perpendicular distance measured from the base course to the ship's zig-zag trajectory. By comparing these values obtained from the digital tests against benchmark data, the variability in longitudinal and lateral coursekeeping can be taken into consideration. The key information to be obtained from the zig-zag manoeuvres are illustrated in Figure 3.

5.2. Autonomous Navigation

Once the accuracy of the manoeuvring model relative to the actual ship is established, the digital model will be used to simulate autonomous navigation of the ship from one location to another and under various environmental conditions. This is to verify that the autonomous navigation algorithms would be able to steer the vessel along a safe course, with adequate control to account for the ship dynamics, environmental forces and any shallower bathymetry and fixed obstacles.

For an autonomous ship operating in Singapore port waters, it is recommended to apply the local environmental conditions for the autonomous navigation simulations. If the autonomous ship is designed for operations in other areas, environmental conditions corresponding to those areas should also be applied to ensure that the navigational algorithm is able to account for the resulting external environmental forces.

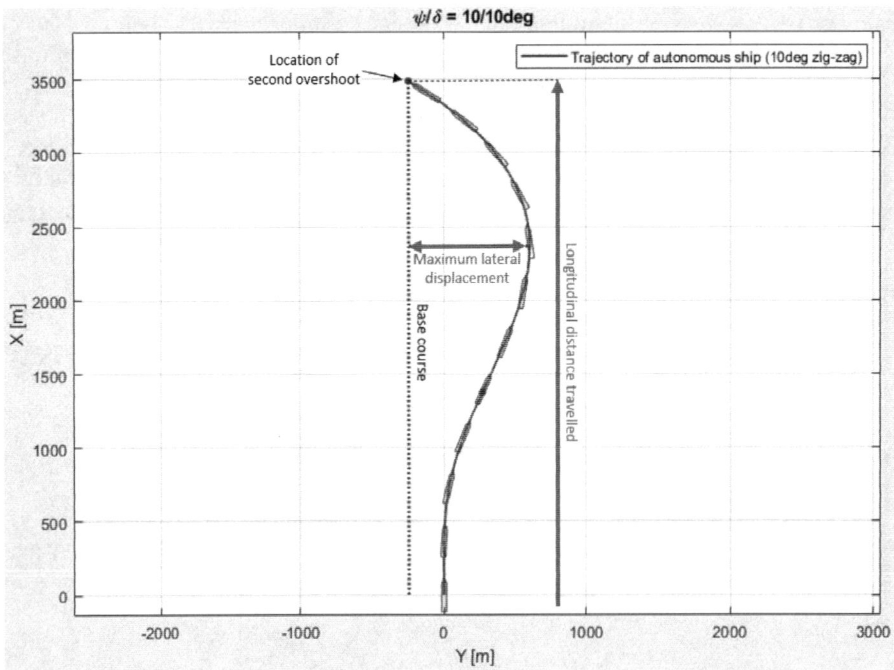

Figure 3. Key information to be obtained from zig-zag manoeuvres

Through our industry discussions, it is recognised that the autonomous navigational capabilities could be based on a diverse range of algorithms, some of which include uncertainty and scatter in terms of the predicted navigational course. It is thus prudent to assume that any numerical prediction of a physical system would include possible scatter within certain confidence limits. As such, it is recommended to undertake at least 30 repeats of each of the various combinations of autonomous navigational scenarios. This is based on the assumption that the scatter of the discrete points along a navigational course follows a Gaussian assumption, which is deemed to be reasonable, based on similar applications in other vessel collision models [27]. The scatter would be used to quantify the prediction interval, which can be calculated from the cross-track error (XTE) as follows:

$$Prediction\ interval_p = 3.983\left(XTE_p\sqrt{1+\frac{1}{30}}\right)$$

where the subscript p represents the combination of ship speed and draught condition being simulated. The value of 3.983 is used based on the assumption that the scatter in predictions can be modelled using a Gaussian assumption and corresponds to a sample size of 30.

By combining the prediction interval with the percentage difference between the digital model and benchmark data, the initial correction factor for the p combination of ship speed and draught can be obtained using the formula below:

$$u_p = prediction\ interval_p \times \Delta_{\max(p)}$$

Lastly, considering that vessels in the real-world environment would navigate in a range of speeds, the largest value of correction factor across the three ship speeds tested would be taken as the final correction factor at a given draught condition. The is expressed below:

$$u_{\text{final}(d)} = \max\left(u_{6\ knots,d}\ , u_{12\ knots,d}, u_{90\%,d}\right)$$

where $u_{6knots(d)}, u_{12knots(d)}$ and $u_{90\%(d)}$ represent the initial correction factor at draught d for ship speeds at 6 knots, 12 knots, and at least 90% of the ship's speed at 85% of the maximum engine output, respectively.

5.3. Collision Detection & Avoidance

The third and final stage of the recommended digital testing guidelines aims to verify that the autonomous ship would be able to take COLREGS-compliant corrective actions and maintain a safe distance to avoid collision when encountering other vessel(s) in a seaway, where risk of collision exists. Various scenarios, derived, based on COLREGS, have been recommended for testing of the digital model. These scenarios comprise one - and two - vessel encounters and should be considered as the minimum level of testing. Tests with more than two vessels would help to further stress test the autonomous collision avoidance capabilities.

It is assumed that these tests are likely to be carried out prior to the certification or qualification of the sensor hardware systems and any sensor fusion algorithms required for the autonomous ship's situational awareness capabilities. As such, it is necessary to assume that the collision detection capabilities relating to the picking up target vessels is working as specified. Thus, if the range for collision detection is found to be lower than what was applied in the digital tests, the tests should be repeated using the new range.

Considering there is a possibility that some autonomous algorithms may have underlying probabilistic behaviour, resulting in multiple possible outcomes for each collision avoidance scenario, at least 5 simulations should be carried out using randomly generated heading angle(s) for the encountered vessels in each scenario.

5.3.1. Single Vessel Encounters

Rule 13 – Overtaking scenario, target ship. In this scenario, the autonomous ship should maintain course and speed and the target vessel, being the overtaking vessel, should be the vessel taking actions to allow safe passing. However, when it becomes apparent that the target vessel is not taking sufficient action, the autonomous ship shall take avoiding actions to pass clear with sufficient safety distance.

Rule 14 – Head-on Scenario with target vessel. Here the autonomous ship should alter course to starboard and pass port-to-port with sufficient safety distance. Noting that the COLREGS do not specify any heading angle values for the target vessel, the guidance provided in [28] has been adopted here, where the target vessel's course should be within "6° of being opposite (+/-180°)".

Rule 15 – Crossing Scenario with target vessel on starboard side. In this scenario, the autonomous ship should alter course to starboard and pass the target vessel from stern where practical with sufficient safety distance.

Rule 15 – Crossing Scenario with target vessel on port side. In this scenario, the autonomous ship should maintain course and speed until it becomes apparent that the target vessel is not taking action, the autonomous ship shall alter course to starboard, make a round-turn and pass clear with sufficient safety distance.

5.3.2. Two Vessel Encounters. In such multiple target vessel scenarios, the autonomous ship should assess the developing situation and risk of collision in accordance with appropriate COLREGS Rules. Avoiding action should be taken based on the prevailing circumstances. The evasive actions recommended herein are suggested manoeuvres – the autonomous ship could undertake different set of actions to pass safely.

One target vessel in head-on scenario & one target vessel approaching from starboard stern quarter. In this scenario, autonomous ship would assess the CPA and TCPA of the two target vessels and take avoiding action based on their course and speed. Sufficient safety distance should be kept relative to both target vessels.

Two target vessels in head-on scenario. In this scenario, the autonomous ship is required to make a bold alteration to starboard and pass port-to-port of both target vessels with sufficient safety distance both target vessels.

One target vessel in head-on scenario and one target vessel crossing on starboard side. In this scenario, the autonomous ship is required to alter course to starboard and pass both target vessels with sufficient safety distance.

One target vessel in head-on scenario and one target vessel crossing on port side. Here, the autonomous ship should assess the CPA and TCPA of the two target vessels and take avoiding action accordingly to maintain sufficient safety distance with both target vessels.

One target vessel crossing on starboard side and one target vessel overtaking from starboard stern quarter. Similar to the previous scenario, the autonomous ship should assess the CPA and TCPA of the two target vessels and take avoiding action accordingly to maintain sufficient safety distance with both target vessels.

One target vessel crossing on port side and one target vessel overtaking from starboard stern quarter. Here, the autonomous ship should maintain course and speed until when it becomes apparent that the target vessels are not taking action. The autonomous ship would then assess the CPA and TCPA of the two target vessels and take avoiding action accordingly to maintain sufficient safety distance with both target vessels.

One target vessel crossing on port side and one target vessel overtaking from port stern quarter. Similar to the previous scenario, the autonomous ship should maintain course and speed until it becomes apparent that the target vessels are not taking actions. The autonomous ship would then assess the CPA and TCPA of the two target vessels and take avoiding action accordingly to maintain sufficient safety distance relative to both target vessels.

One target vessel crossing on starboard side and one target vessel overtaking from port stern quarter. Here, the autonomous ship should alter course to starboard and pass with sufficient safety distance from both target vessels.

One target vessel in head-on scenario and one target vessel overtaking from port stern quarter. Similar to the previous scenario, the autonomous ship should alter course to starboard and pass with sufficient safety distance from both target vessels.

One target vessel overtaking from port stern quarter and one target vessel overtaking from starboard stern quarter. In this scenario, the autonomous ship should maintain course and speed until it becomes apparent that the target vessels are not taking actions. The autonomous ship would then assess the CPA and TCPA of the two target vessels and take avoiding action accordingly to maintain sufficient safety distance relative to both target vessels.

One target vessel crossing on starboard side and one target vessel crossing on port side. In this scenario, the autonomous ship should assess the CPA & TCPA of the two target vessels and take avoiding action accordingly to maintain sufficient safety distance with both target vessels.

5.4. CPA Corrections

For each of the 5 simulations carried out in this stage for each scenario and combination of draught conditions, an initial CPA value (represented as CPA_0) should be obtained as the initial safety distance. Recalling that the final correction factors (u_{final}) have been calculated from the outputs of Stages 1 and 2, to account for the prediction interval of the digital model, as well as the potential differences between the digital model and the actual ship, respectively. These correction factors are subtracted from CPA_0 to give the $CPA_{corrected}$ values for each draught, as per the equation below.

$$CPA_{corrected(d)} = CPA_{0(d)} - u_{final(d)}$$

For all scenarios, it is recommended that the $CPA_{corrected}$ should be at least 0.75NM, based on the findings of this study, as described in the earlier sections. In scenarios where the autonomous ship is unable to achieve this, such as in the unlikely scenario where the CPA of the autonomous ship relative to other obstacles or ships occur at the same time instant, the lowest $CPA_{corrected}$ value should not be less than 0.60 NM. An example of such a scenario is illustrated in Figure 4 below (not drawn to scale), where the autonomous ship is caught between an encountered vessel and another obstacle such that the CPA are attained at the same time instant. The figure also illustrates the relationships between CPA_0, the safety correction (u_{final}) and $CPA_{corrected}$.

5.5. Discussions and Limitations of Proposed Framework

To the authors knowledge, there are no standardised testing framework for assessment of an autonomous vessel's safety using digital means. In the existing practice, testing and assessment plans are jointly developed on a case-by-case basis by vessel developers and the certifying organisations, such as Class and / or Administrations. While this practice enables the development of a bespoke test plan that considers the high-level goals and associated risks to be addressed by each specific autonomous vessel to be certified, it does not facilitate comparison of the vessels' safety with other autonomous vessels nor existing vessels. This can hinder technological development efforts and safety level given that the requirement to be of "equivalent" safety level as existing vessels may not be determined in a consistent and transparent manner.

As noted earlier in this paper, autonomous maritime technologies are still undergoing active development. This means that what is considered as "ordinary practice of seamen" would change as the vessel demography shifts towards increasing level of autonomy. This is an important limitation of the proposed framework – whereby the testing framework will have to be reviewed from time to time to ensure relevant to the prevalent culture of the sea area being navigated. There are two aspects to this: firstly, the CPA values will have to be reviewed at regular intervals. This review will need to include not only AIS data and expert judgement from seafarers, navigational performance data from autonomous vessels will also need to be taken into consideration given the heterogeneous mix of vessel types. Updates of to the CPA values will help to ensure that the outcomes of collision avoidance manoeuvres (in the form of CPA) will be kept up-to-date as the overall level of vessel intelligence increases in a given sea area.

Another important limitation of the current framework relates to COLREG compliance – the encounter scenarios involving one vessel and two vessels have been laid out in a manner that reduces the subjectivity of interpretation. However, it is recognised that complex multi-vessel encounter situations may often require evasive actions that rely on the subjective aspects of COLREGS. In such situations, the appropriate actions to be undertaken will still have to be jointly established by the certifying organisation and the vessel developers. With the operationalisation of more autonomous vessels, information on the appropriate actions can be captured and used to derive metrics to ascertain the navigational safety of autonomous vessels in the future. Concurrently, with further evolution collision detection and avoidance technologies (e.g. from a combination of maritime, as well as from adjacent land-based transportation and robotics sectors), the future capabilities of autonomous vessels with regard to compliance of COLREGS would be enhanced, including in complex situations. This would prompt a further review of the proposed framework.

6. Conclusions

In this paper, a study has been carried out to develop a framework for digital testing of MASS navigation safety prior to sea trails. A review of the existing state-of-the-art, both in terms of technology developments and industry guidance in the form of publications from Classification Societies have been carried out. Based on the review and discussions with industry, a three step digital testing framework has been proposed, to first verify the accuracy of the digital model in representing the dynamic responses of the actual vessel, followed by tests to ascertain that the autonomous navigation algorithm is able to control (virtually) the vessel from one point to another, taking into account the real-world environment and lastly, that the autonomous navigation algorithm is able to carry out collision detection and avoidance.

Through the three stages, the framework takes into account any uncertainties in using the digital model to represent the actual autonomous vessel, potential scatter in terms of predicting a given navigation path, as well as the repeatability when carrying out collision avoidance actions. The subjective nature of COLREGS, as well as the need for a safety level that is equivalent to the existing situation with manned vessels is addressed through the use of CPA data to infer the safe outcomes of collision avoidance actions in Singapore Strait.

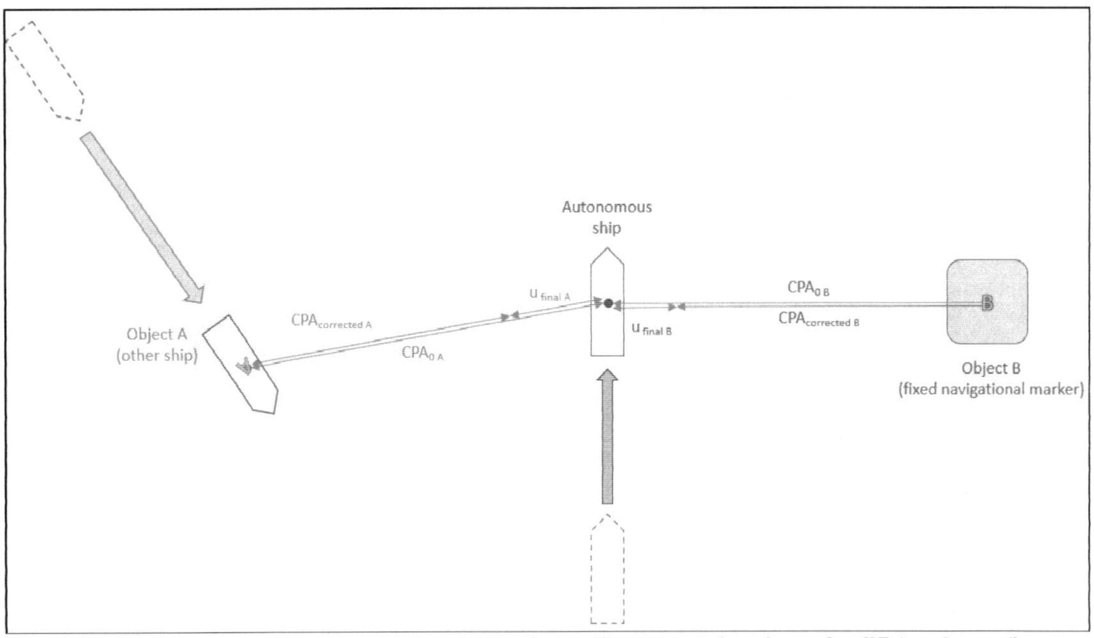

Figure 4. Illustration of the CPA_0, $CPA_{corrected}$ and u_{final} for a scenario where the CPA values of an autonomous ship relative to another ship, as well as a fixed navigational marker occurs at the same instant.

The proposed framework will provide a harmonised system for digital testing to reduce the likelihood of differing interpretations by different stakeholders, and concurrently, account for the need for a goal-based approach to testing.

Acknowledgements

This research is supported by the National Research Foundation, Singapore and the Singapore Maritime Institute under its Maritime Transformation Programme White Space Funding Support for the Establishment and Operation of the Centre of Excellence for Autonomous and Remotely Operated Vessels (CEAOPS), SMI-2019-MTP-01.

References

[1] Yara. (2020). Yara Birkeland press kit. https://www.yara.com/news-and-media/press-kits/yara-birkeland-press-kit [Accessed 17 Feb 22]
[2] Wärtsilä. (2020). Initial sea trials successfully completed by Wärtsilä & PSA Marine's ground-breaking 'IntelliTug' project. https://www.wartsila.com/media/news/13-03-2020-initial-sea-trials-successfully-completed-by-wartsila-psa-marine-s-ground-breaking-intellitug-project-2660851 [Accessed 17 Feb 22]
[3] ST Engineering. (2020). Sea Trials for Autonomous Tug Project https://onward.stengg.com/2020/04/29/sea-trials-for-autonomous-tugs [Accessed 17 Feb 22]
[4] Keppel Corp. (2019) https://www.kepcorp.com/en/media/media-releases-sgx-filings/keppel-to-develop-its-first-autonomous-vessel-for-operations-in-singapore-6564/ [Accessed 17 Feb 22]
[5] Ahmed, Y.A., & Hasegawa, K. (2016). Fuzzy Reasoned Waypoint Controller for Automatic Ship Guidance. IFAC, 49, 604-609.
[6] Jiang, Y., Zhang, S., & Yang, D. (2014). A Novel Position Estimation Method Based on Displacement Correction in AIS. Sensors. 14. 17376-17389. 10.3390/s140917376.
[7] Chen, D., Dai, C., Wan, X., and Mou, J. (2015). A research on AIS-based embedded system for ship collision avoidance, 2015 International Conference on Transportation Information and Safety (ICTIS), pp. 512-517.
[8] Goerlandt, F., Montewka, J., Kuzmin, V., & Kujala, P. (2015). A risk-informed ship collision alert system: Framework and application. Safety Science, 77, 182-204.
[9] Imazu, H. (2017). Evaluation method of collision risk by using true motion. TransNav: International Journal on Marine Navigation and Safety of Sea Transportation, 11(1).
[10] Chircop, A. (2018). Testing International Legal Regimes: The Advent of Automated Commercial Vessels. The German Yearbook of International Law, 1-34.
[11] Rokseth, B., Haugen, O. I., & Utne, I. B. (2019). Safety verification for autonomous ships. In MATEC Web of Conferences (Vol. 273, p. 02002).
[12] Pedersen, T.A., Glomsrud, J.A. & Haugen, O.I. (2019). Towards Simulation-based Verification of Autonomous Navigation Systems. In: Proceedings of the International Seminar on Safety and Security of Autonomous Vessels (ISSAV) and European STAMP Workshop and Conference (ESWC).
[13] Hwang, Taewoong, and Ik-Hyun Youn. 2021. Navigation Situation Clustering Model of Human-Operated Ships for Maritime Autonomous Surface Ship Collision Avoidance Tests. Journal of Marine Science and Engineering 9, no. 12: 1458.
[14] Bolbot, V., & Theotokatos, G. (2021). Automatically generating collision scenarios for testing ship collision avoidance system using sampling techniques.
[15] Porathe, T. (2019) Maritime Autonomous Surface Ships (MASS) and the COLREGS: Do We Need Quantified Rules Or Is"the Ordinary Practice of Seamen" Specific Enough? TransNav, Int. Journal on Marine Navigation and Safety of Sea Transportation 13.3.
[16] ABS (2017a) Guidance Notes on Qualifying New Technologies. American Bureau of Shipping.
[17] ABS (2017b) Guidance Notes on Review and Approval of Novel Concepts. American Bureau of Shipping.
[18] ABS (2020) Guide for Smart Functions for Marine Vessels and Offshore Units. American Bureau of Shipping.
[19] BV (2020) Risk Based Qualification of New Technology – Methodological Guidelines, Bureau Veritas.
[20] BV (2019) Guidelines for Autonomous Shipping, Bureau Veritas.
[21] DNV (2013) Recommended Practice for Technology Qualification. DNV-RP-A203. DNV.
[22] DNVGL (2018) Class Guideline for Autonomous and remotely operated ships. CG-0264, DNVGL

[23] LR (2017a) LR Code for Unmanned Marine Systems. London: Lloyd's Register
[24] LR. (2017b). New code to certify unmanned vessels announced. https://www.lr.org/en/latest-news/new-code-to-certify-unmanned-vessels-announced/ [Accessed 17 Feb 22]
[25] IMO MSC Resolution MSC.137(76). Standards for Ship Manoeuvrability (2002) https://wwwcdn.imo.org/localresources/en/KnowledgeCentre/IndexofIMOResolutions/MSCResolutions/MSC.137(76).pdf [Accessed 17 Feb 22]
[26] ITTC (2017). ITTC Quality System Manual Recommended Procedures and Guidelines Procedure Full Scale Manoeuvring Trial. Retrieved from: https://www.ittc.info/media/8179/75-04-02-01.pdf [Accessed 17 Feb 22]
[27] Pedersen, P.T., (1995). Collision and grounding mechanics. In: Proc. WEMT'95, Danish Society of Naval Architecture and Marine Engineering, Copenhagen, pp. 125–157.
[28] Cockcroft, A.N. and Lameijer, J.N.F. (2012). A Guide to the Collision Avoidance Rules (7th Edition), Butterworth-Heinmann, 2012

On the use of maritime training simulators with humans in the loop for understanding and evaluating algorithms for autonomous vessels

Anete Vagale[1], Ottar L Osen[1,2], Andreas Brandsæter[1], Marius Tannum[2], Christian Hovden[2] and Robin T Bye[1]

[1] Cyber-Physical Systems Laboratory, Department of ICT and Natural Sciences, NTNU – Norwegian University of Science and Technology, Postboks 1517, NO-6025 Ålesund, Norway
[2] Faculty of Technology, Natural Sciences and Maritime Sciences, University of South-Eastern Norway, Postboks 235, NO-3603 Kongsberg, Norway

E-mail: anete.vagale@ntnu.no

Abstract. The prospect of a future where the maritime shipping industry is dominated by autonomous vessels is appealing and gaining global interest from industry majors, research institutions, and academia. Potential advantages include increased operational safety, reduced costs, and lower environmental footprint. However, the transition will not happen overnight and is not without challenges. For example, algorithms for autonomous navigation must take into consideration safety concerns of the own ship, its crew and passengers, other surrounding ships, and the surrounding environment. This raises a need to test and verify safety, performance, and robustness of the algorithms responsible for the autonomous functionality. In addition, the transition towards fully autonomous ships is likely to be gradual and involve remote control centres and ships with varying degrees of autonomy. Hence, humans will inevitably have to interact with autonomous vessels in a variety of scenarios, including overriding own ships from land or on board, as well as communicating with autonomous ships from other fleets. Inevitably, full scale scenario testing involving real vessels and humans is costly, impractical, time-consuming, and potentially dangerous. In this paper, we propose an alternative approach, and explore how maritime navigation training simulators with humans in the loop can be used as a testbed for understanding and evaluating algorithms for autonomous vessels. In the proposed setting, we can directly compare choices made by an algorithm with those of a skilled human navigator for a variety of navigational tasks. Moreover, we can study in real-time the behaviour and decision-making of human navigators in mixed scenarios that also include autonomous ships, whether this is known beforehand or not. Our paper provides an overview of related work, details on maritime simulators and how algorithms can be tested, and some of the technical requirements. To exemplify our approach, we present two example test setups, and provide a brief discussion of our findings. We conclude that using maritime training simulators enables the study of several interesting and vital research questions, including that of the interaction between autonomous and traditional vessels operating side by side.

1. Introduction

With autonomous ships on the horizon that may affect the safety of the own ship and other ships, its crew and passengers, and the surrounding environment, the need to test and verify the safety, performance and robustness of the autonomous functionality is a top priority. Autonomous ships therefore require sophisticated sensors and algorithms to collect and fuse information about the ship and its surroundings to form an adequate situational awareness. Based on the situational awareness, an advanced navigation algorithm[1] must be able to automatically change the ship's route and perform collision avoidance manoeuvres, ensuring safe sailing in compliance with rules and regulations. Lack of unambiguous specifications is also a challenge for an autonomous ship, since current regulations, including The Convention on the International Regulations for Preventing Collisions at Sea (COLREG) [1], use expressions which require interpretations, such as ample time, good seamanship, safe speed, etc. For an overview of relevant intelligent autonomous navigation algorithms, see [2].

Due to the complexity of navigating a vessel, traditional algorithms from control engineering and other classic fields can easily come up short, with algorithms from the field of artificial intelligence (AI) such as deep neural networks, evolutionary algorithms, and reinforcement learning tending to be better choices. However, AI-based algorithms and particularly machine learning algorithms often suffer from a lack of explainability or interpretability, and they may also be inherently complex. As a consequence, testing such algorithms quickly becomes a complex problem in itself. Full scale realistic testing of algorithms employed on a real ship is inevitably going to be costly, dangerous, impractical and probably not even allowed without thorough testing in a safe environment in advance. Two existing testing approaches are: (i) simulation-based testing and (ii) functionality-based testing. In this paper, we discuss and focus on the simulation-based testing approach. A procedure for simulation-based verification of autonomous navigation systems is proposed in [3]. The verification and assurance of machine learning algorithms needs to be fundamentally different from traditional assurance and verification processes based on requirements and physical understanding [4, 5, 6, 7]. Moreover, since autonomous vessels will have to relate to traditional vessels with humans onboard and vice versa, the human dimension becomes very important.

There are several ship simulators, e.g., [8, 9, 10], that could enable us to test path planning and collision avoidance scenarios but as described later in this article, there are many challenges that require more advanced setups. For example, in order to evaluate interactions with humans, there is a need for a very realistic setup. Likewise, algorithms could be tested by simulating data from the automatic identification system (AIS) during realistic navigation scenarios but the lack of a visual model for the AIS-simulated vessel will impact the human navigator's decision-making. It will therefore be of limited value to study the human navigator's actions towards an invisible AIS-simulated "ghost ship."

Due to humans' ability to learn and adapt to new and unexpected situations, humans can play an important role in complex technological systems such as autonomous navigation to ensure safe and efficient operation [11]. Testing how human navigators will react to an autonomous vessel piloted by an algorithm in real life is one of several interesting research questions. It could also be valuable to compare how well an algorithm solves a given set of scenarios compared to a human navigator.

In this article, we argue that the use of maritime training simulators will enable testing and research that is otherwise very difficult to perform, and propose to use these simulators as a testbed to investigate research questions similar to the above. The proposed test setups could aid in evaluating how well algorithms for autonomous vessels perform in general and especially with humans in the loop. We suggest looking both at a setup when an intelligent algorithm is controlling target ship(s) in the simulator, and a setup where an intelligent algorithm is controlling the bridge and executing navigation tasks. Another benefit of using a simulator for these experiments is the ability to control the experiment and making sure that the situations can be replicated.

[1] In this paper, the word 'algorithm' may also refer to several algorithms combined and used together, or an autonomous system as a whole.

The remainder of the paper is organized as follows: Related work is reviewed in Section 2. Section 3 provides an overview of existing maritime simulators, their history and possible simulator setups. Testing of autonomous algorithms is described in Section 4. Section 5 presents technical requirements for the proposed use cases. Different test setups are proposed in Section 6. Section 7 contains a discussion, and finally, some concluding remarks are drawn in Section 8.

2. Related work

Autonomous vessels are already investigated using various simulators. However, using *maritime training simulators* normally used in education of cadets, as proposed in this paper, is a novelty. Performing tests in maritime training simulators enables a means for verifying autonomous navigation systems or path planning algorithms, with humans in the loop, in a safe simulated environment before implementing them on a real vessel at sea.

A maritime training simulator that could fulfill this purpose is the K-Sim simulation platform developed by Kongsberg Digital, certified by DNV and fulfilling Standards of Training, Certification and Watchkeeping (STCW) requirements. Its intended purpose relates to virtual prototyping, testing, and verification of autonomous algorithms for vessels [12]. This includes scenario generation faster than real-time, energy consumption predictions, taking into account environmental conditions, and route verifications. Another use case of the platform is simulation-based prediction, as well as using live data from a ship and its environment.

In another project, Wärtsilä delivered a navigation simulator and specific mathematical models for Intelligent Shipping Technology Test Laboratory (ISTLAB) at the Satakunta University of Applied Sciences (SAMK) in Finland [13]. The goal of that project, launched in early 2019, was to develop a testing environment for remotely controlled, autonomous vessels.

With a focus on simulation-based verification of autonomous navigation systems, Pedersen et al. [3] propose to use a digital twin for testing purposes. A complete test system includes scenario manager, test evaluation module, operating environment, test interface, and digital twin that represents the systems of the own ship. The authors indicate that the Open Simulation Platform (OSP) [14] may potentially be used later for testing purposes.

Finally, we draw attention to research by Vagale et al. [15], who focus specifically on the evaluation of path planning algorithms from the risk and safety perspective.

3. Maritime training simulators

Maritime simulators for training can perhaps be labelled "very serious games." These simulators have a very realistic touch and feel since they typically have a vessel bridge replica with large screens with a wide view angle, all essential handles and equipment found on board a real ship, and often even an audio system that provides realistic sounds. This very realistic physical setup combined with strict rules on expected behaviour, and maybe even combined with use of uniforms, makes it possible to create a theatre-like atmosphere that is immersive and blur the lines between simulator and real life. Indeed, the provided training in these simulators is considered realistic enough that, according to International Maritime Organization [16], cadets are permitted to fulfil part of their training in a simulator.

Although marine training simulators may seem realistic to human navigators, it cannot be assumed that the simulators are sufficiently realistic for testing a particular class of autonomous vessel functionality, namely that of object detection and classification. Hence, this functionality cannot be directly implemented in the simulators. To overcome this challenge, authors in [17] propose utilising Cycle-Consistent Adversarial Networks (Cycle-GANs) to transfer the simulator data to a real-world-like environment before autonomous functionality such as object detection and classification is performed.

3.1. History

Technically sophisticated simulators are in particular found in the training of professionals where one seeks to increase safety, such as aviation, shipping, and healthcare [18]. The earliest ship bridge simulators with visual night scenes were introduced in the 1970s [19] and full bridge simulators have already been commercially available for decades, reaching the point where the international rules and regulations as prescribed by the International Maritime Organization (IMO) mandates the use of simulator training as a basis for certifying professional navigators through its STCW guidelines.

The 2010 amendments of the STCW 1978 code section A-I/12 added requirements for its member states to include simulator-based training with modern technology like electronic chart displays information system (ECDIS), AIS, and dynamic positioning (DP). The updated codes also define that seafarers may demonstrate their competence to handle a ship in all conditions through approved simulator training while at the same time providing the nautical student with limited credit as an equivalency for sea-going service depending on factors comprising the level of simulation and scenarios, time spent, student-teacher ratio, pre-brief and de-brief procedures, and integration with other elements in the approved training program [16]. The STCW requires the approval of simulators used for mandatory training or assessment of seafarers, and classification societies often do the travail of simulation certification. Det Norske Veritas (DNV) for instance has developed a standard for certifying simulator compliance, where they also define four classes of bridge operation simulators, Class A, B, C and S, where only Class A, full mission simulator, meets the IMO requirement of "manoeuvre and handle a ship in all conditions" demanded for approval of sea-time equivalency [20].

3.2. Major vendors

Major vendors of full mission maritime training simulators in accordance with STCW 2010 are:

- K-SIM by Kongsberg Digital [8]
- NTPRO 5000 by Wärtsilä [9]
- NAUTIS by VSTEP [10]
- SIMFLEX by FORCE Technology [21]
- BOREALIS by Poseidon [22]
- REMBRANDT by BMT [23]
- ARI SMS by ARI Simulations [24]
- IS FMBS by Image Soft [25]
- ANS6000 by Rheinmetall Electronics [26]

3.3. Simulator setup

Typically, a simulator setup consists of at least two or more rooms (one-to-many setup). Firstly a room where the main instructor controls the exercise, configures the training scenario, selects vessel types and sizes, adjusts weather parameters, and more. From this room, the instructor may monitor and record the cadets' performance during the exercise. In simpler cases, a one-to-one simulator with only one navigation bridge connected to the instructor station is also possible (see Fig. 1). An example setup of five navigation bridges connected to a main instructor station is given in Fig. 2.

In addition, the simulator can contain one or more vessel bridge replicas. For simple exercises a single bridge is sufficient, however, both in order to train on more complex challenges and to train more cadets at the same time it is convenient to use several bridges. Typically, the bridges are linked with the same simulation environment, thus allowing the different bridges (simulated vessels) to interact with each other. A picture of a typical training bridge is shown in Fig. 3.

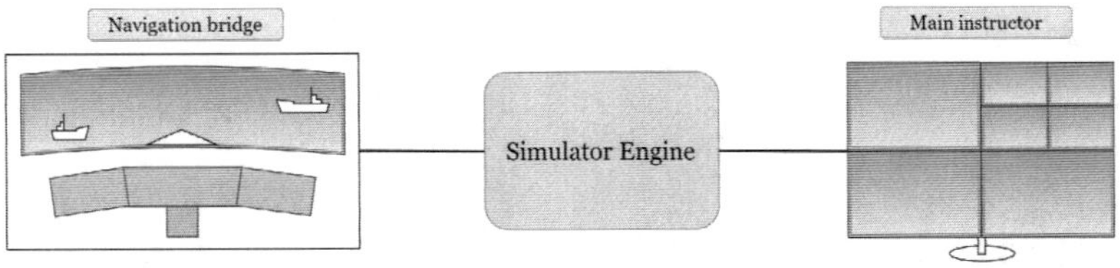

Figure 1. Setup of a one-to-one navigation training bridge connected to the main instructor station through a simulator engine.

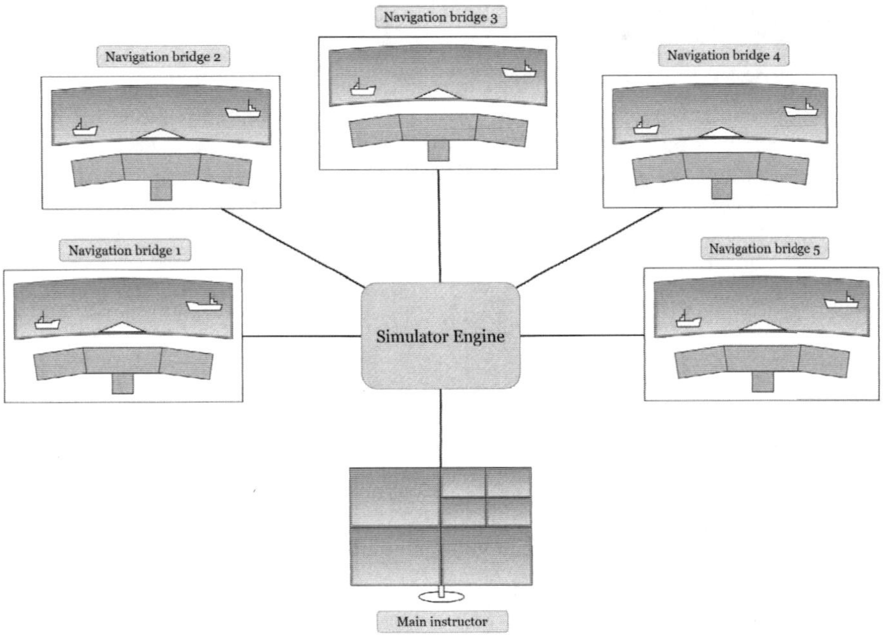

Figure 2. One-to-many setup of the main instructor station (in the centre) connected to five navigation training bridges through a simulator engine.

An example of an instructor station overlooking the actions of the cadet navigating the bridge is shown in Fig. 4. A research training simulator is shown in Fig. 5.

3.4. Target vessels

In addition to one or more training bridges, the operator or instructor has the possibility to control a number of additional vessels (target vessels) in the training scenario. These vessels have no separate simulator bridge, but are merely vessels, chosen from a library of vessels, that are placed into the simulation environment and given a heading and speed. Throughout the exercise, the operator may adjust these vessels' heading and speed. Later in this paper, we present setups where the target vessels are controlled by autonomous path planning and collision avoidance algorithms.

Figure 3. Navigation training bridge at the Department of the Ocean Operations and Civil Engineering at NTNU in Ålesund. Photo: Anete Vagale.

Figure 4. Instructor station at the Department of the Ocean Operations and Civil Engineering at NTNU in Ålesund. Photo: Terje Ole Slinning.

Figure 5. Research training bridge at the Department of the Ocean Operations and Civil Engineering at NTNU in Ålesund. The research bridge consists of the navigation bridge with controls, three screens (on the right), and the instructor's area (on the left). Photo: Anete Vagale.

4. Testing of algorithms

In order to evaluate the performance and capacities of algorithms, extensive testing is required. These tests should be standardized, reproducible, relevant and realistic. As of today, we are not aware of any existing standardized tests for autonomous ships. An attempt at defining requirements for simulator based testing is given in [27]. We expect different actors to develop their own tests before some governing bodies establish some official tests, especially related to safety. However, in addition, there will probably be several other "industry standards" that relate to issues such as efficiency.

4.1. Challenges with assurance of machine learning algorithms

Traditional verification, including the V-model (e.g., see ISO 26262 *Road vehicles - functional safety* in [28]), typically assumes that the requirements of a component are completely specified and that "each refinement can be verified with respect to its specification" [29]. Problems like machine perception and ship navigation cannot be clearly specified. For example, COLREGs [1] that must be complied with when navigating on waters, are in many cases open for interpretation. This has motivated the use of machine learning algorithms which learn from examples rather than being programmed based on a specification [5].

However, understanding and interpreting a machine learning algorithm's reasoning is challenging or even impossible, which makes it challenging to verify the algorithm or to determine when the algorithm's reasoning is in error [30, 31, 32]. Explainable AI (XAI) for algorithms attempts to give a reasoning for the decisions suggested by an algorithm. Additionally, machine learning algorithms are completely dependent on the quality of its training data. Hence, any verification and assurance process needs to include a comprehensive data analysis, documenting if the dataset sufficiently covers the input domain, is representative and complete, particularly regarding corner cases [4]. The choice of resampling strategy, that is, how the data is split into training and test sets using cross validation or bootstrap techniques, must also be evaluated.

The robustness of the algorithms must also be tested and documented. Brandsæter at el. [33] provide an example showing how minor, seemingly insignificant changes like a small image rotation can confuse a classifier. In their example, a one-degree rotation causes a mis-detection of a vessel. It is also shown that when the image is rotated further (three degrees), the algorithm once again correctly detects the vessel, illustrating the unpredictable behaviour of the algorithm. Image manipulations and transformations utilising, for example, Generative Adversarial Networks (GAN) (e.g., [34, 35]) to transform a scene from summer to winter or daylight to night, as well as simple augmentations including rotations, share, blur, and similar (e.g., see [36, 37, 7]) can be utilised to increase the test scope of a limited dataset.

4.2. Metrics

The capability and performance of detecting other ships is an essential part of an autonomous ship's situational awareness capability. Relevant metrics for object detection includes number of correctly detected ships (true positives), but also missed targets (false negatives), as well as true negatives and false positives. These numbers are often summarised in other metrics such as precision, recall / sensitivity (also known as true positive rate), specificity (also called true negative rate), and F1-score. But are all targets equally important? Detecting a ship which is far away is less important than detecting a nearby ship on collision course. As the relevance of the different targets depends on the navigation, this should be taken into consideration. Hence, assessing the ship's situational awareness in isolation is not recommended, although methodology for assessing the situational awareness of humans is available, e.g., see [38, 39, 40].

Øvergård et al. [41] argue that existing research is suggestive of limitations of the reliability of subjective assessments, and propose an initial version (prototype) of an automated assessment algorithm for a specific maritime operation. The following control requirements for the safety of navigation, collected based on open interviews with six subject-matter experts who all held deck-officer certifications, were used in the prototype:

(i) distance to land based on own ship length,
(ii) distance to moving objects (vessels) based on own ship length,
(iii) distance to floating objects based on ship length,
(iv) the deviation between ship heading and heading of dock (meaning that the ship should be parallel to the dock during the last part of docking), and
(v) the minimum depth below the ship's keel (the so-called 'safety depth').

The aforementioned parameters can be considered a part of a greater geometric collision risk assessment (CRA) set. More information on metrics for CRA for autonomous vessels, and a comparison of 45 different path planning and collision algorithms, is given in [42].

On the other hand, Vagale et al. [15] propose introducing additional relevant metrics to evaluate path planning algorithms of autonomous surface vehicles (ASVs), such as:

- efficiency / path fitness (e.g. length of path / time used / fuel used),
- compliance with regulations (COLREGs), and
- good seamanship practice.

The above-mentioned metrics are still not clearly defined and in some cases could be overlapping. Typically, compliance with regulations and safety often overlap. Similarly, good seamanship practice requires following COLREGs and vice versa. It is important to look at these metrics in connection with each other; the safest move might be not to move at all, however, that would obviously affect efficiency negatively.

4.3. Turing test for autonomous ships

In addition to evaluating the above-mentioned metrics, interviewing cadets and instructors after the exercise might be useful to capture their experience with the autonomous ship. In this case, it could also be interesting to consider doing blind tests where the cadets and/or instructors do not know which ships are autonomous (similar to Turing's imitation game).

Usually, the cadets train on scenarios where the other vessels (target vessels) are navigated by the instructor. In such a setup, some (or all) target vessels could be autonomous ships. It would then be interesting to see if the cadets notice any difference. An even more interesting scenario emerges from linking several simulator bridges together (see Figure 2) to share a common scene. For a cadet, any observed ship could either be a target ship placed there by the instructor, another ship operated by another cadet(s), or in our case, an autonomous ship. It would be impossible to know without interacting or studying the ship's behaviour.

Finally, it could also be very interesting to do such an experiment in a way that the instructor played the Turing game. Either by letting him / her navigate a simulator bridge instead of the cadet(s), or by manipulating the instructor panel in such a way that there is no way of knowing which ship is autonomous and which is manually steered. This experiment is of great interest because the instructor have years of experience monitoring cadets' navigational behaviour.

How could an autonomous vessel divulge its nature? For one, by performing actions that are contradictory to good seamanship and/or COLREG rules. For instance, several algorithms that have been proposed, e.g., [43, 44], would change heading very often or continuously. This is against COLREG rule 8 since your intentions should be communicated clearly and your manoeuvres should be predictable. This is typically done by navigating in straight lines and making as few heading changes (way-points) as possible. This kind of navigation will often be in stark contrast to many algorithms that focus on optimization and rely on cost or fitness functions. Another way of identifying vessels is radio communication. This is an integral and important part of the cadet's training. Training without the use of radio would usually be seen as out of the ordinary and reduce the quality and immersiveness of the setup. Hence, it could be better to let the instructor or another human to reply on radio on behalf of the autonomous vessel, but it could be a challenge since it would be impossible for a human to accurately predict the algorithm's intentions and relay these truthfully on radio upon request.

5. Technical requirements

In order to use a maritime simulator as a test bed for autonomous vessels, it is crucial that the simulator provides an interface that enables third party computers to control various settings and controls in the simulation environment. Below, we have listed some interface properties in order of importance:

- controlling target vessel's speed and heading,
- reading position of fixed and moving obstacles (incl. other vessels),
- controlling (emulating) bridge handles and controls,
- reading bridge view (screen), radar, ECDIS, AIS and other instruments,
- controlling other simulation parameters, such as environmental settings, etc.

Since the designers of the maritime simulators probably did not have this use of the simulators in mind when the simulators were made, there is likely no dedicated interface readily available for this use. However, the features listed above correspond quite well with the needs for the operator/instructor's needs and are probably available through some API or similar used by the software for the operator station. However, existing APIs/interfaces are generally not public and most likely internal to the developer and subject to change. Hence, a crucial element in setting up such a proposed testing environment is access to such an interface, and this most likely depends on close cooperation with and goodwill from the simulator vendor.

6. Test setup and research questions

Below, we have listed two main test setups (see Fig. 6) for maritime training simulators and described what kind of interesting research questions these tests could contribute towards answering.

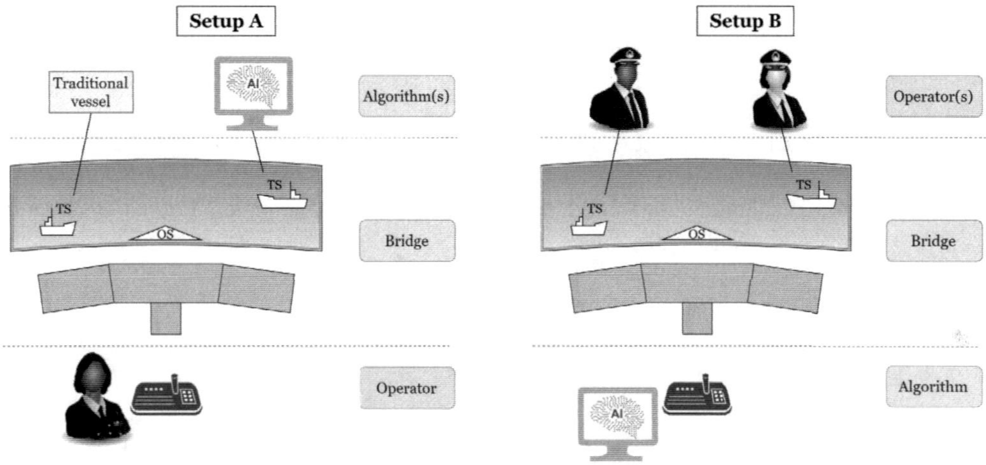

Figure 6. Visual representation of the test setups A and B.

- *Test setup A* — where the algorithm(s) control the target vessel(s) while the bridge(s) is/are controlled by human cadet(s).
- *Test setup B* — where the algorithm(s) control the bridge(s).

6.1. Test Setup A

In the first setup (Fig. 6, left), the algorithm(s) control the target vessel(s) and the bridge(s) is/are controlled by human cadet(s) / operator(s). In a mixed environment where traditional ships and autonomous ships co-exist, it is crucial to understand how they may interact. In this test, it is possible to study both how the autonomous vessel(s) responds to the human actions, and how humans react to the actions performed by the algorithm(s). In addition to evaluating the above-mentioned metrics, the cadets may be interviewed after the exercise in order to capture their experience with the autonomous vessel. This setup is suitable to test algorithms similarly to Turing's imitation game.

Another interesting research question is whether it is beneficial for navigators on other vessels to know that the vessel is autonomous? And further, would it be beneficial to know how the algorithm is "thinking"? With explainable AI, it could be possible for an autonomous vessel to broadcast its reasoning and intentions to other seafarers. Could this information be useful, or would it be a distraction? Could it contribute to information overload on humans? On the other side, would other autonomous vessels be able to perform better if they knew the intentions of other autonomous vessels, or could deadlock situations (more easily) occur?

6.2. Test Setup B

In the second setup (Fig. 6, right), the algorithm(s) control the bridge(s). One series of tests that can be performed is to run basic tests of an algorithm with respect to collision avoidance (avoiding fixed and moving obstacles) and even several vessels with identical or different algorithms. These tests can be done with more "Lo-Fi" simulators also, however, these certified simulators may give more reliable results.

A more interesting test is to give the algorithms the same test scenarios that the instructors give the cadets. The instructors have a set of test scenarios they run as part of their training of cadets. An experienced instructor has an excellent reference base on how well humans (cadets) solve the test scenarios. Hence, the instructors should have no problem evaluating how well the algorithm perform compared to humans using the same metrics as they normally do with the cadets. With some minor modifications, it could also be possible to do this test blindly, but the value of that should be weighted against the increased complexity.

Adding another layer of complexity to this test setup, the instructors could compare how well an algorithm performs compared to humans in a situation that could give humans a cognitive overload. One hypothesis could be that humans will outperform the computer in uncomplicated scenarios, however, the computer would be better at handling very complex situations. In addition to answering the aforementioned research questions, the results could also give some insight on the use of algorithms not only to navigate autonomous vessels, but also as decision support tools or as "smart autopilots" on manned vessels.

6.3. Other setups

The suggested setups and research questions above are of course just examples. It is possible to construct numerous different setups by introducing other elements, combining scenarios, etc. Advanced simulators like the maritime training simulators are a great tool that make it possible to do research with humans in the loop and with varying degree of complexity to validate autonomous ship algorithms.

7. Discussion

In order to do studies with humans in the loop, it is important to make a test scenario that is as close to reality as possible in a safe, controllable and observable environment. Although no simulators are exact copies of reality, maritime training simulators are deemed to be good training environments by international bodies such as International Maritime Organization (IMO) [45, 20]. Likewise, the cadets that participate in training are not experienced sailors, and therefore it is hard to generalise all findings on cadets to experienced sailors. However, this may be rectified by at least running a control group, where experienced sailors are subject to the same tests as the cadets and the algorithms.

In the future, autonomous vessels may be required to identify themselves as "robots" by transmitting some kind of identification to other seafarers. In this paper, we suggest that testing should be done also without mentioned identification. This is motivated by an assumption that in a mixed environment with regular and autonomous vessels, the traffic will flow safer and more efficiently if the autonomous vessels behave like regular vessels. We believe the proposed experiments will help shed light on this research question as well.

Additionally, in order to encourage and speed up the development of simulation-based testing for ship navigation algorithms, sharing open access data, software and results would be an important driver. However, it is also vital to consider data sharing restrictions from partners when they prevail.

All in all, we argue that test setups such as those described above will be able to provide valuable answers to research questions that are hard, if not impossible, to answer otherwise.

8. Conclusions

In this article, we have advocated for using maritime simulators to investigate how well algorithms for autonomous vessels perform in general and especially with humans in the loop. We have identified several interesting research questions that may be hard to answer without the use of the mentioned simulators, and we show that these simulators open up numerous possible research setups that may be explored. We have also identified the need for an interface that allows a third party computer to interact with the simulator. The need for test metrics has also been outlined.

Finally, we strongly argue that it is of great importance that the human-machine interaction between autonomous and traditional vessels is investigated in depth in order to prepare for a future where autonomous and traditional vessels will operate side by side.

Acknowledgments

We wish to thank Terje Ole Slinning and the Department of Ocean Operations and Civil Engineering at NTNU in Ålesund for providing photos and giving us a tour of their maritime training simulator area at the Norwegian Maritime Competence Center.

References

[1] International Maritime Organization 1972 COLREGs - International Regulations for Preventing Collisions at Sea

[2] Vagale A, Oucheikh R, Bye R T, Osen O L and Fossen T I 2021 Path planning and collision avoidance for autonomous surface vehicles I: a review *Journal of Marine Science and Technology (Japan)* **26** 1307–1323 ISSN 09484280

[3] Pedersen T A, Glomsrud J A and Haugen O I 2019 Towards simulation-based verification of autonomous navigation systems *International Seminar on Safety and Security of Autonomous Vessels* (Helsinki) pp 1–13

[4] Wood M, Robbel P, Maass M, Tebbens R D, Meijs M, Harb M, and Schlicht P 2019 Safety first for automated driving

[5] Spanfelner B, Richter D, Ebel S, Wilhelm U, Branz W and Patz C 2012 Challenges in applying the ISO 26262 for driver assistance systems *Tagung Fahrerassistenz, München* **15** 2012

[6] Koopman P and Wagner M 2016 Challenges in autonomous vehicle testing and validation *SAE International Journal of Transportation Safety* **4** 15–24

[7] Brandsæter A and Knutsen K E 2018 Towards a framework for assurance of autonomous navigation systems in the maritime industry *Safety and Reliability–Safe Societies in a Changing World : Proceedings of ESREL 2018* (CRC Press) pp 449–457

[8] Kongsberg Digital 2021 K-SIM® Navigation URL https://www.kongsberg.com/digital/products/maritime-simulation/k-sim-navigation/

[9] Wärtsilä 2022 Wärtsilä Navigation Simulator NTPRO 5000 URL https://www.wartsila.com/voyage/simulation-and-training/ntpro-5000-simulator

[10] VSTEP NAUTIS Navigation Simulator URL https://www.vstepsimulation.com/nautis-simulator/nautis-maritime-simulator/navigation/

[11] Endsley M R 2019 Level of automation effects on performance, situation awareness and workload in a dynamic control task *Hearing on Boeing 737-Max8 Crashes — December 11, 2019*

[12] Heierstad T 2020 Simulation technology – a vital resource for testing autonomous ship operations - Kongsberg Digital URL https://www.kongsberg.com/digital/resources/stories/2020/6/simulation-technology--a-vital-resource-for-testing-autonomous-ship-operations/

[13] Wärtsilä 2020 Making ship systems smarter with simulation URL https://www.wartsila.com/insights/article/making-ship-systems-smarter-with-simulation

[14] Open Simulation Platform 2022 Open Simulation Platform. Towards a maritime ecosystem for efficient co-simulation URL https://opensimulationplatform.com/

[15] Vagale A, Bye R T and Osen O L 2020 Evaluation of Path Planning Algorithms of Autonomous Surface Vehicles Based on Safety and Collision Risk Assessment *OCEANS 2020 MTS/IEEE Global: Singapore - U.S. Gulf Coast* pp 1–8

[16] IMO 2011 International Convention on Standard of training, certification and watchkeeping for seafarers, including 2010 Manila Amendments

[17] Brandsæter A and Osen O L 2021 Assessing Autonomous Ship Navigation using Bridge Simulators Enhanced by Cycle-Consistent Adversarial Networks *Proceedings of the Institution of Mechanical Engineers, Part O: Journal of Risk and Reliability* 1–9

[18] Dahlstrom N, Dekker S, Van Winsen R and Nyce J 2009 Fidelity and validity of simulator training *Theoretical Issues in Ergonomics Science* **10** 305–314 ISSN 1464536X

[19] Barnett M, Gatfield D and Pekcan C 2003 A Research Agenda in Maritime Crew Resource Management *Proceedings of the International Conference on Team Resource Mangement in the 21st Century* pp 1–22

[20] DNV 2017 Maritime Simulator Systems *Standard DNVGL-ST-0033* URL https://rules.dnvgl.com/docs/pdf/DNVGL/ST/2017-03/DNVGL-ST-0033.pdf

[21] FORCE Technology 2014 Ship bridge simulators URL https://forcetechnology.com/en/all-industry-facilities/ship-bridge-simulators

[22] Poseidon Simulation AS Borealis Navigation Simulator URL https://poseidon.no/products/borealis/

[23] BMT BMT REMBRANDT: high integrity ship simulations URL https://www.bmt.org/our-innovations/bmt-rembrandt/

[24] ARI Simulation 2021 Ship Maneuvering Simulators URL https://arisimulation.com/products/marine-simulators/bridge

[25] Image Soft 2022 IS Full Mission Bridge Simulator URL https://imagesoft.fi/product/is-full-mission-bridge-simulator-for-certified-maritime-training/

[26] Rheinmetall AG 2022 Bridge and navigation training URL https://www.rheinmetall-defence.com/en/rheinmetall_defence/systems_and_products/simulation_and_training/civil_training_solutions/merchant_marine_training/bridge_and_navigation_/index.php

[27] DNV 2018 Class Guideline: Autonomous and remotely operated ships

[28] International Standardization Organization 2018 ISO 26262-1:2018(en) Road vehicles — Functional safety

[29] Salay R, Queiroz R and Czarnecki K 2018 An analysis of ISO 26262: Using machine learning safely in automotive software *arXiv preprint arXiv:1709.02435*

[30] Caruana R, Kangarloo H, Dionisio J D, Sinha U and Johnson D 1999 Case-based explanation of non-case-based learning methods. *Proceedings of the AMIA Symposium* (American Medical Informatics Association) p 212

[31] Doshi-Velez F and Kim B 2017 Towards A Rigorous Science of Interpretable Machine Learning *arXiv: Machine Learning* 1–13 URL http://arxiv.org/abs/1702.08608

[32] Lundberg S M and Lee S I 2017 A unified approach to interpreting model predictions *Advances in Neural Information Processing Systems* pp 4765–4774

[33] Brandsæter A, Smefjell G, van de Merwe K and Kamsvåg V 2020 Assuring Safe Implementation of Decision Support Functionality based on Data-driven Methods for Ship Navigation *I: e-proceedings of the 30th European Safety and Reliability Conference and 15th Probabilistic Safety Assessment and Management Conference (ESREL2020 PSAM15)*

[34] Liu M Y, Breuel T and Kautz J 2017 Unsupervised image-to-image translation networks *Advances in neural information processing systems* pp 700–708

[35] Zhu J Y, Park T, Isola P and Efros A A 2017 Unpaired image-to-image translation using cycle-consistent adversarial networks *Proceedings of the IEEE international conference on computer vision* pp 2223–2232

[36] Pei K, Cao Y, Yang J and Jana S 2017 DeepXplore: Automated Whitebox Testing of Deep Learning Systems *arXiv preprint arXiv:1705.06640*

[37] Pei K, Cao Y, Yang J and Jana S 2017 Towards Practical Verification of Machine Learning: The Case of Computer Vision Systems *ArXiv* **1712.01785** 1–16

[38] Endsley M R 1988 Design and Evaluation for Situation Awareness Enhancement: *Proceedings of the Human Factors Society Annual Meeting* vol 32 (SAGE PublicationsSage CA: Los Angeles, CA) pp 97–101

[39] Endsley M R 1988 Situation Awareness Global Assessment Technique (SAGAT) *IEEE Proceedings of the National Aerospace and Electronics Conference* (IEEE) pp 789–795

[40] Endsley M R 2000 Direct Measurement of Situation Awareness: Validity and use of SAGAT *Situation Awareness Analysis and Measurement*

[41] Øvergård K I, Nazir S and Solberg A S 2017 Towards automated performance assessment for maritime navigation *TransNav, International Journal on Marine Navigation and Safety of Sea Transportation* **11** 229–234

[42] Vagale A, Bye R T, Oucheikh R, Osen O L and Fossen T I 2021 Path planning and collision avoidance

for autonomous surface vehicles II: a comparative study of algorithms *Journal of Marine Science and Technology (Japan)* **26** 1307–1323 ISSN 09484280

[43] Naeem W, Irwin G W and Yang A 2012 COLREGs-based collision avoidance strategies for unmanned surface vehicles *Mechatronics* **22** 669–678 ISSN 09574158

[44] Wang N, Gao Y, Zheng Z, Zhao H and Yin J 2018 A Hybrid Path-Planning Scheme for an Unmanned Surface Vehicle *8th International Conference on Information Science and Technology* pp 231–236

[45] Zghyer R and Ostnes R 2019 Opportunities and Challenges in Using Ship-Bridge Simulators in Maritime Research *Proceedings of Ergoship 2019* (Haugesund: Western Norway University of Applied Sciences (HVL)) pp 119–131

Maritime Autonomous Surface Ships: Automation Transparency for Nearby Vessels

Ole Andreas Alsos*, Philip Hodne, Oskar Kristoffer Skåden, and Thomas Porathe

Department of Design, Norwegian University of Science and Technology, Trondheim, Norway

E-mail: *ole.alsos@ntnu.no

Abstract. In the future, maritime autonomous surface ships (MASS) will become more advanced, and mariners will move from the bridge to remote operation centres. To avoid dangerous situations and frequent intervention from the operators, a MASS needs to perceive the environment and calculate the current state and future intention of nearby ships. Another challenge for a MASS is to explain the decisions of its algorithms and make the artificial intelligence's decisions transparent to not only the operators, but also external stakeholders, such as passengers, vessel traffic services, and nearby ships. This is called *automation transparency*. In this paper we focus on how MASS, through various means of maritime communication, can (1) *express* their own state and intention to other ships and (2) *perceive* other ships' state and intention. Towards that, we first analyse current maritime communication methods in terms of how well they can be expressed and perceived by today's MASS in comparison to traditional ships. Next, we analyse which gaps in their communication should be focused on to maximise safety at sea. Finally, we present multiple design recommendations for naval architects, human factors engineers, and interaction designers who are designing and building the next generation of MASS.

1. Introduction

Maritime Autonomous Surface Ships (MASS) are becoming more advanced. In the foreseeable future, mariners (or at least some of them) will move from the ship bridge into remote operation centres (ROC) where they monitor multiple autonomous vessels and intervene if necessary. However, the shift from traditional ships to autonomous ships is not unproblematic. Traditional ships often resolve misunderstandings, poor traffic flow, deadlocks, and dangerous situations by communicating with other vessels in the vicinity through distinct course and speed changes and various tools and equipment for maritime communication. While the sensors onboard an autonomous ship easily can detect other vessels, they have limited capabilities of traditional maritime communication to resolve potentially dangerous situations. Therefore, there is a need for ways to make the current state and future intention (and sometimes even the past states) of autonomous ships visible to vessels with no or low level of automation.

In this paper we want to focus on how a MASS (1) *expresses* its own state and intention to other ships and (2) *perceives* other ships' state and intention with traditional maritime communication. Towards that, we first analyse current maritime communication methods in terms of how well they can be expressed and perceived by today's MASS in comparison to traditional ships. Next, we analyse which gaps in their communication should be focused on to maximise safety at sea. Finally, we present multiple design recommendations for naval architects, human factors engineers, and interaction designers that can help improve the next generations of MASS.

2. Autonomous ships and automation transparency

2.1. The transparency challenge for MASS

Autonomous shipping is high on the agenda within the maritime domain and MASS are expected to soon operate in parallel with traditional ships. A MASS is defined by the International Maritime Organisation (IMO) as "a ship which, to a varying degree, can operate independently of human interaction" [1]. There is, however, varying consensus on the exact definition of autonomy in the maritime domain [2, pp. 37]. Autonomy is usually divided into different levels of autonomy, with the lowest levels consisting of vessels controlled manually by humans and the highest consisting of fully independent vessels (Table 1). IMO divides their definition into four different degrees of autonomy, based on the "presence or absence of onboard seafarers and the establishment of remote control" [1].

Table 1. IMOs definition for degrees of autonomy.

Degree 1	Ship with automated processes and decision support (seafarers on board)
Degree 2	Remotely controlled ship with seafarers on board
Degree 3	Remotely controlled ship without seafarers on board
Degree 4	Fully autonomous ship

Still, the development of MASS is in its infancy, with technical issues dominating its development [3][4]. There are arguably no ships that are fully autonomous today. Indeed, the size of ships' crew has been decreasing for centuries [5], however, with MASS the long-term expectation are vessels without any crew. With humans omitted, the ship's autonomous behaviour when sailing is critical, especially in the interaction between itself and crewed vessels [6]. One of the important research challenges for MASS is to explain the decisions of its algorithms and make the artificial intelligence's (AI) decisions transparent to not only the operators, but also external stakeholders. The next section outlines the concept of Explainable AI and how this may be key to help users understand and interact with the autonomous vessel more efficiently.

2.2. Explainable AI and Automation Transparency

The design of the AI and its components is essential to autonomous systems such as MASS. One of the problems with these sets of algorithms is that their complexity may make their inner workings and decisions mostly invisible for onboard or remote operators, crew, passengers, or nearby ships. To establish trust and avoid dangerous situations, an autonomous ship needs to communicate its current state and future intention to traditional ships in the vicinity. In other words, the AI and automation algorithms must, at least partly, display its inner workings and decisions to its surroundings. The push to develop such systems is often referred to as *Explainable AI* (XAI) or, in the context of autonomy and robotics, *Automation Transparency*.

A common definition of XAI's goal is to make AI understandable to people [7]. Moulding the autonomous system around the user's understanding of how it works, often referred to as the user's "mental model" [8], is foundational to ensure trust and efficient interaction. Liao & Varshney argues that even if AI is made entirely transparent in its workings, it does not mean it is explainable. This trait lies in the perception and reception of the user receiving the explanation and can thus still be a bottleneck for people's trust in autonomy [7].

To help communicate the explainability of an autonomous system, one must utilise the human-machine interface (HMI). An HMI is defined as the means by which humans and computers communicate. From the traditional buttons and diodes to today's gesture-based virtual and augmented reality, the definition may be argued to be broad. In his paper on making autonomous systems transparent, Lyons [9] argues that the machine's interface may exist at several levels, including informational (the use of light, symbols, and text), communicative (the use of sounds and voice), and physical (through movement and gestures). It is through these levels one can design an HMI for MASS that is in line with a user's mental model. Still, there are multiple stakeholders involved in the operation of and interaction with MASS. These stakeholders have different needs. The next section will outline the different stakeholders involved in the environment of MASS, and which ones we will focus on in this paper.

2.3. The stakeholders of MASS

Increasing automation transparency is important for multiple stakeholders in MASS' environment. Veitch & Alsos [10] analysed human-centred XAI in the context of MASS and identified and described 5 main groups of stakeholders that all have different needs in terms of automation transparency; (1) the developer, who is responsible for developing the autonomous system (2) the operator, who is responsible for monitoring and controlling the MASS from a ROC, (3) the passengers and crew onboard, (4) non-SOLAS ships, such as sailboats, leisure boats, and kayaks, and (5) ships sailing under the SOLAS regulations. In their work, they also mention connected organisations and regulators, such as shipping companies, IMO, or national maritime directorates. In this paper we are mainly looking at the two last groups, non-SOLAS ships and SOLAS ships, which both need to interpret the state and intention of MASS and communicate their own intention to the MASS to avoid dangerous situations at sea.

2.4. Automation transparency for SOLAS versus non-SOLAS ships

Non-SOLAS ships usually sail with a crew with varying seamanship training, experience, and safety culture. It is usually smaller vessels below 50 feet and includes sailboats, leisure crafts, smaller fishing vessels, and kayaks. These vessels often have limited equipment for maritime communication, and usually communicate with other ships through analogue means such as distinct course changes, navigation lights, shouting, and – depending on the size and type – radio communication on VHF.

SOLAS ships, on the other hand, have trained and experienced mariners with a strong safety culture. These can generally be regarded as merchant ships, such as vessels transporting cargo or carrying passengers for hire. The ships are usually well equipped with a range of different analogue and electronic tools for maritime communication.

There is a lack of research on how MASS may communicate effectively with both of these vessel types [11]. Indeed, MASS is being employed in an environment and industry of rules and regulations with expected behaviour, such as following the *International Regulations for Preventing Collisions at Sea* (COLREG). These are universally accepted as the "rules of the road" [12] and abiding by these rules will arguably be the primary method for achieving trust in the interaction between MASS and traditional ships. However, these rules are not enough. Navigating is more nuanced and dynamic. For example, deviations from the COLREGs are common. Findings by Rutledal et al. [12] monitoring ferry traffic outside Ålesund, Norway, over the course of 1 month showed that 14% of all evasive manoeuvres done by ferries were not in accordance with the COLREGs. They conclude that other types of communication methods than movement were used to avoid collisions. Hence, an initial overview of today's methods of ship-to-ship communication is relevant to understand how MASS can solve, amongst others, these interactions.

In the following sections we will present how traditional ships communicate, show how these means of communication align with today's versions of MASS, and suggest how they can be adapted to highly automated ships. But first we describe how we used Subject Matter Experts (SME) in the analysis of automation transparency.

3. Method

To help us analyse current methods for maritime communication through a MASS lens, we interviewed five Subject Matter Experts (SME). The five SMEs were recruited using snowball sampling from Norwegian research institutions (N=3) and a large, privately held maritime company (N=2). The recruited participants were educated on a master's or PhD level and were experts in maritime shipping and human factors research within the maritime domain. As outlined in Table 2, they had an average of 15-20 years of experience in their respected field, with 2 SMEs having multiple years' experience as crew on SOLAS ships in the North Sea. The three SMEs from Norwegian research institutions were currently focused on marine autonomy and had educational backgrounds in psychology, marine engineering, and mechanical engineering. The interviews were completed over a 3-month period running from September to November 2021. In the interviews we discussed several research questions: First, what forms of communication are applicable to today's SOLAS and non-SOLAS ships, how these express the ships' own state and intention and perceive other ships' state and intention. Finally, we discussed the gaps in these communication methods when transferred to the AI of MASS, and how these gaps might be resolved.

Table 2. Subject Matter Experts interviewed in this study.

SME	Location	Years of experience	Area of expertise	Experience as crew
1	Western Europe	10-15	Human Factors	No
2	Scandinavia	5-10	Human Factors	No
3	Scandinavia	10-15	Shipping	Yes
4	Scandinavia	20-25	Shipping	Yes
5	Scandinavia	15-20	Human Factors	No

The interviews were performed by a pair of interviewers, each lasting between 30 minutes and 1 hour. We chose unstructured interviews because it allowed for interviews to be inductive and specific to each participant's knowledge. The conversations focused on answering the research questions, each adapted to the participant's field of work.

In addition, we based the insights on our own experiences from controlling non-SOLAS ships, such as sailboats, sail yachts, motorboats, motor yachts, fishing vessels, kayaks, etc. In the next sections we will present a literature review about maritime communication and use the findings from the SME interviews to show how maritime communication can be adapted to MASS.

4. Maritime communication

Communicating a ship's state and intention to other ships is usually done through various means of maritime communication. In this section we will present how traditional ships communicate, both directly and indirectly, and how they do it through analogue and electronic tools and equipment.

4.1. Maritime communication by traditional ships

Traditional ships rely heavily on maritime communication tools and equipment to navigate safely at sea. Today, radio and satellite communication with other vessels and shore stations are widely deployed and have reduced the dependency on visual signalling [13]. In this section we outline these communication methods, and these methods' horizontal range from an origin ship. As the maximum distance one can observe is determined by the observer's height above sea level, we will set the height in this paper to 15 meters, as this is the minimum height requirement for placement of radar devices on a ship by the SOLAS [14]. With humans able to see objects at the horizon about 2.7 nautical miles (NM) away at sea level, the theoretical maximum distance is increased 7.45 NM when moved up to 15 meters. However, the authors argue this is possible only on rare occasions at sea because of atmospheric conditions (waves, rain, fog, and so on), making the estimated distance one can see clearly closer to 2 NM. We do not outline communication methods used in situations of distress or for reporting to sea authorities as this is out of scope for this paper.

4.1.1. Radio (VHF). VHF is short for "Very High Frequency" and can be used to communicate through radio frequencies. VHF is mostly for general communication between ships and shore, allowing ships to solve confusion by communicating intent and perception. Communication is usually done on channel 16, meaning that many ships in the same vicinity share the same communication channel [15, pp. 38-40]. This channel is to be left open for distress signals, so ships may agree to switch to another channel (frequency) in informal cases. VHF radios may also have Digital Selective Calling technology, allowing ships to call up similarly equipped vessels or shore stations without disturbing the common channels. There are established norms for how to communicate over VHF radio, most notably the IMO's Standard Marine Communication Phrases where English should be used [16]. VHF radio waves propagate by line-of-sight, so it is dependent on the height of the antennas and effect of the transceiver. Thus, the range is at least up to 5 NM and, depending on aforementioned variables, up to around 40 NM.

4.1.2. Radio (MF/HF). VHF Marine medium frequency (MF) and high frequency (HF) use transceivers like VHF does. However, the frequency is lower, enabling communication over longer distances since the lower frequency allows the radio waves to bend over the horizon. SOLAS vessels in international waters are required to carry MF/HF radios. The range of medium frequency is reliable up to around 150 NM, while high frequency can reach distances hundreds or even thousands of nautical miles (arguably global). The distances are dependent on, amongst others, frequency, and atmospheric conditions.

4.1.3. Automatic Identification System (AIS). AIS is an automatic tracking system that uses transponder and receiver units to identify and locate vessels by electronically exchanging data with other nearby ships, shore, and satellites. AIS broadcasts information about a ship's identity, position, speed and intended destination on the maritime VHF band. AIS information is normally displayed graphically directly on an electronic chart or radar screen. It is required by the IMO that all ships of 300 gross tonnes or above involved in international voyages and all cargo ships of 500 gross tonnes or above not engaged on international voyages is fitted with an AIS in accordance with the SOLAS convention [17]. Since not all ships are required to use AIS it is important to be aware that the information provided might not give an accurate representation of the traffic. There are separate devices to transmit your position (transponders) and to receive signals from others (receivers). You do not need to have both in your ship for the units to work. All ships have VHF-band based transponders, so a vessel's AIS information is only perceivable for VHF-band receivers in its vicinity. Satellite based S-AIS receivers, however, can receive information from AIS transceivers globally. The range is thus at least 5 NM and up to around 40 NM for ground-based AIS receivers, and global if the receiver is satellite-based [18].

4.1.4. VHF Data Exchange System (VDES). VDES is a two-way communications system encompassing many different maritime communication subsystems. The VDES network allows for data messages to be sent and received, and it supports electronic navigation with access to a cloud-enabled database called the Maritime Connectivity Platform. Ships with VDES-equipment will have higher bandwidth and will be able to transmit route exchange in addition to current position and destination. Maritime infrastructure (buoys, etc.) is also expected to broadcast information towards passing ships as an extra source of information [19]. AIS infrastructure is expected to be upgraded with VDES capabilities over the next 5 to 10 years and can thus be regarded as an extension of AIS. Using satellite technology, the range of VDES is global.

4.1.5. Satellite Phone. Satellite phones communicate through satellites providing voice calls and internet connectivity in most or all geographic locations on the Earth's surface. The most common use-case is providing entertainment and internet connectivity to passengers and crew. However, satellite phones may also be used to call key stakeholders at shore from vessels (such as clients at the next port of call) and enabling the crew on the bridge to be called using their satellite phone number. This is a costly procedure and may thus not be a prioritised communication method. The range of satellite phones is global.

4.1.6. Navigation Lights. All ships on the high seas, independent of size, are required to have navigation lights. COLREG regulates the use of light and signal codes. Some of the most common navigation lights are sidelights, masthead lights, and stern lights. These are used to give other ships information about the status, heading and position of a ship, especially at night, and are essential to calculate the right of way. IMO regulates that navigation lights should be visible up to 1 or 6 NM depending on the type of light and size of the vessel. We estimate the range to be a maximum of 7.45 NM, however, atmospheric conditions may impact this range.

4.1.7. Daylight signalling lamps. Daylight signalling lamps are fixed or handheld devices that are explicitly used for visual signalling using white light. Messages are typically sent using morse code. Since the beam is very focused, daylight signalling lamps is suited for long distance communication, and can, as the name suggests, be used in daylight. IMO regulates that the daylight signalling lamps must be visible up to at least 2 NM [20], but the range is estimated to be around 7.45 NM in optimal weather conditions. The range is affected by obstructing objects and atmospheric conditions.

4.1.8. Searchlights. Searchlights are fixed light apparatuses used to make the surrounding environment, including objects such as other vessels and docks, more visible in harsh weather and low-lit environments. They are not inherently a communication method besides making the vessel itself more visible at night. However, some searchlights may be pointed at and used similarly to daylight signalling lamps to get in contact with crew on other vessels. The range of searchlights are estimated to be 2 NM, but this is affected by atmospheric conditions.

4.1.9. Flag signals. Flag signalling is not as common today as it was before but is still used to a certain extent because of regulation. There are several flags that can be combined to show the state and intention of a ship. For example, flags are used to signal that a ship has divers in the water or that the ship is carrying dangerous cargo, which other ships may take as a sign to deviate course slightly. The range is estimated to be 2 NM with the use of binoculars, but this is affected by obstructing objects and atmospheric conditions.

4.1.10. Day shapes. Day shapes are masthead signals that indicate status to other vessels. As the name suggests, day shapes are used at daylight hours instead of navigation lights. Day shapes consist of four different geometric shapes (Ball, Cylinder, Cone and Diamond) that can be combined to communicate different states. The meanings of the shapes are defined by the COLREGs. Day shapes are visible up to 2 NM with the use of binoculars, but this is affected by obstructing objects and atmospheric conditions.

4.1.11. Ship horn. Ships using their ship horn can communicate intended manoeuvres and attract attention. There are rules defined, amongst others in COLREGs, to help differentiate different sound signals. For example, how many times you use the horn in a row communicates the intent. Ship horns are often used in tandem with a light to emphasize the intention and is by regulation, compulsory in heavy fog. IMO regulates that the ship horns on the largest ships needs to be audible at 2 NM.

 2311 (2022) 012027 doi:10.1088/1742-6596/2311/1/012027

4.1.12. Hand movements. Hand movements can also be used to communicate with other ships. Cases where hand movements are relevant are mostly informal; for example, pointing out areas and items of interest to crew members on other ships. Hand movements are estimated to be visible up to a maximum of 2 NM with binoculars. The range is affected by obstructing objects and weather conditions.

4.1.13. Shouting. Verbal communication through shouting is mainly relevant for close-quarter communication. Included in this definition is the use of a loudspeaker or whistle to attract attention. In a multi-ethnic industry such as merchant shipping, language barriers may be argued to be a challenge in all forms of verbal communication and a source of misunderstandings and ambiguity [21]. Shouting is estimated to be understandable up a few meters away, while the use of megaphone and whistle is estimated to be maximum 0.05 NM. At sea, shouting does not travel far as it is affected by background noise.

4.1.14. Ship movement. Clear and distinct alterations in the ship's heading and speed are used to communicate its state and intention to other ships, both visually and by radar (COLREG). With radar equipment the range where ships can perceive objects and their movement is from 0.05 NM up to 96 NM depending on equipment and the size of the other objects. When observing visually, the range is estimated to be maximum of 7.45 NM, given optimal weather conditions. Deciphering visual ship movement at longer distances is done over time, checking if the vessels heading has changed since the last observation.

4.2. Characteristics of maritime communication
When we analyse the list of maritime communication above, we see that the different mediums have different ranges, as seen in Figure 1. The ranges are also split; even though the perceivable range is in theory longer, the definitive range may differ because of equipment or the environment. Theoretical range can thus also be understood as uncertainty. The authors argue this must be taken into account when designing the HMIs of future MASS vessels.

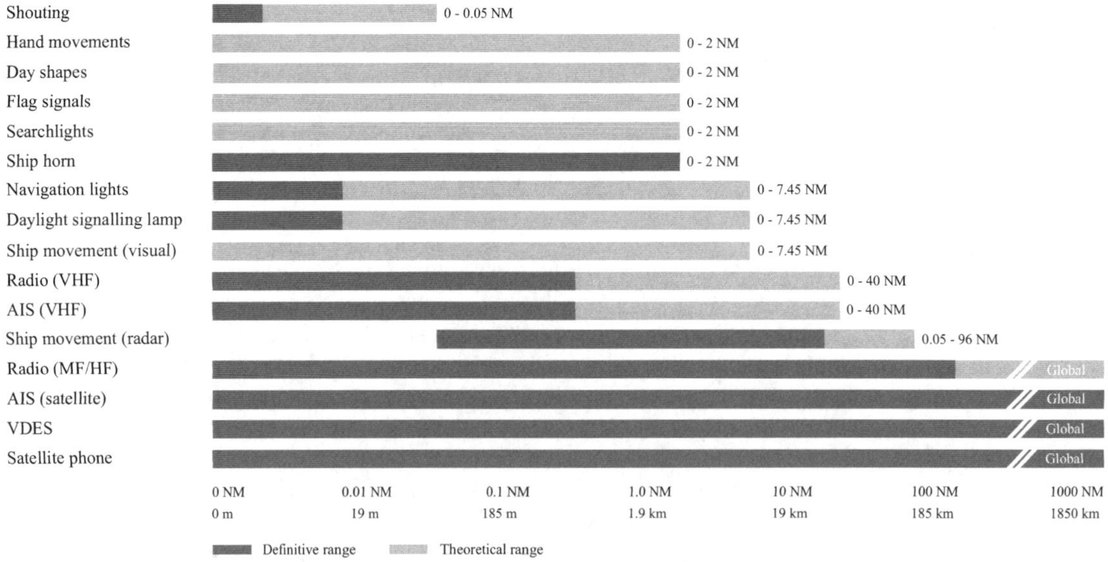

Figure 1. SOLAS ship-to-ship communication methods by distance in NM (logarithmic steps).

The communication methods are also very heterogenous and take different forms. It can be *expressed*, that is, a ship sending information through a medium to one or many ships, or *perceived*, that is, a ship receiving information from another ship. Further, the communication can be *analogue* or *digital* – that is, directly controlled by humans or not. Information can be *broadcasted* to all the ships within the range of the communication medium or communicated *directly* to a specific ship. Finally, information can be communicated *continuously,* or *occasionally*. In Table 3 we list all the communication mediums and map them to the different dimensions mentioned above. We intentionally leave out perceive and express in the table because a modern SOLAS ship can both express and perceive all communication forms. Some forms of communication, such as VHF, have both checked as they can be used in multiple ways. As we will discuss later, the characteristics of the communication can make it difficult for MASS to express or perceive it.

Table 3. The different characteristics of maritime communication between traditional ships.

	Analogue	Digital	Broadcast	Direct	Continuous	Occasional
Shouting	•		•			•
Hand movements	•		•			•
Day shapes	•		•		•	
Flag signals	•		•		•	•
Searchlights	•		•	•		•
Ship horn	•		•	•		•
Navigation lights	•		•		•	
Daylight sign. lamp	•			•		•
Ship mov. (visual)	•		•		•	
Radio (VHF)	•		•	•		•
AIS (VHF)		•	•		•	
Ship mov. (radar)		•	•		•	
Radio (MF/HF)	•		•	•		•
AIS (Satellite)		•	•		•	
VDES		•	•		•	
Satellite phone	•			•		•

5. Maritime communication by MASS

As presented above, maritime communication is essential for automation transparency, since it currently is the only way a MASS can communicate its state and intention to other ships. However, current MASS do not yet master all the mediums of maritime communication due to the limitations in technology. In particular, the autonomous ships have problems expressing and perceiving most of the analogue communication forms. In this section we present the maritime communication methods of MASS, identify gaps and development potential, and present new ways for MASS to communicate its state and intention in line with the concept of automation transparency.

5.1. Current status of MASS maritime communication

Current MASS perceive objects and other ships, and calculate their relative changes in position, course, and speed. This is done with a combination of sensors, as listed in Figure 2. Data from these sensors are combined in a process called sensor fusion [22] and is used to give the automation algorithms awareness of the state and intention of other ships and objects in the vicinity. Today's versions of MASS currently express its own state and intention mainly through its motion as this is the most important method of communication on sea [23]. Since autonomous ships are not particularly good at interpreting the intention of other ships, nor resolve misunderstandings through maritime communication, they often make a very restricted use of right of way. MilliAmpere2 [24], for example, stops for any ship, even if it has the right of way.

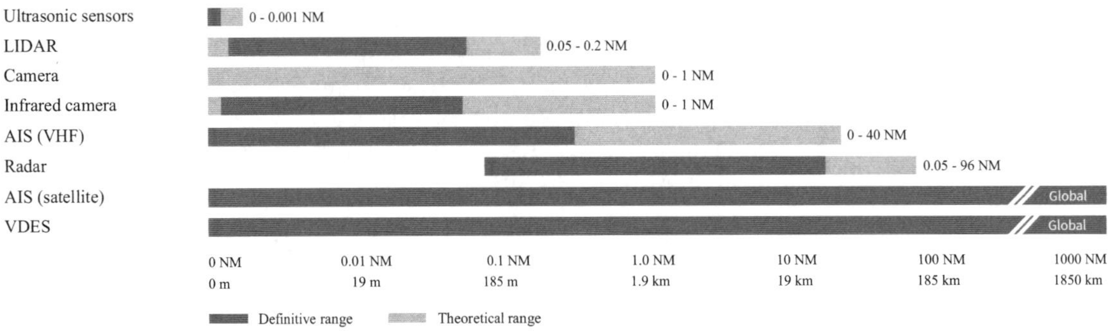

Figure 2. Sensors giving MASS awareness of the state and intention of its environment.

Today, the technical development for MASS is not focused on the analogue means of maritime communication, but on better sensor fusion and improved algorithms for situational awareness. The blindness to analogue means of communication is still an unsolved challenge for MASS. With input from the SMEs, we analysed the communication methods in terms of ability for MASS, such as milliAmpere2 and Yara Birkeland, to effectively perceive and express its state and intention. We also compared it with SOLAS and non-SOLAS ships (Table 4). When analysing MASS, we looked at the vessels' ability to perceive and express autonomously – that is, without any input from a human operator in a ROC.

Table 4. The ability for non-SOLAS ships, SOLAS ships and MASS to perceive and express different means of maritime communication.

	Non-SOLAS vessels		**SOLAS vessels**		**MASS today**	
	Perceive	Express	Perceive	Express	Perceive	Express
Shouting	•	•	•	•		
Hand movements	•	•	•	•		
Day shapes			•	•		
Flag signals			•	•		
Searchlights	•		•	•		
Ship horn	•	•	•	•		
Navigation lights	•	•	•	•		
Daylight signalling lamp			•	•		
Ship movement (visual)	•	•	•	•	•	•
Radio (VHF)	•	•	•	•		
AIS (VHF)			•	•	•	•
Ship movement (radar)		•	•	•	•	•
Radio (MF/HF)			•	•		
AIS (Satellite)			•	•	•	•
VDES			•	•	•	•
Satellite phone			•	•		

While SOLAS ships are equipped and able to perceive and express themselves through all communication methods, non-SOLAS and MASS are not. For non-SOLAS, a key factor is the lack of the same electronic equipment as SOLAS ships. For most leisure crafts, the extent of equipment is usually limited to a VHF radio. They generally rely heavily on observing visual movement and using other human senses such as hearing and speech. MASS perceives the world through technologies such as AIS, radar, LIDAR, and computer vision. Indeed, it can perceive more digital communication methods than non-SOLAS ships, but there are many analogue ones that it is unable to perceive and express. These include shouting, hand movements, day shapes, flag signals, searchlights, daylight signalling lamps, ship horns, VHF, MF/HF, and satellite phones. This divide in communication seen in Table 4 encapsulates part of the challenge with automation transparency for MASS.

5.2. Adapting maritime communications to MASS

We asked the SMEs to reflect on and analyse the cost-benefit of developing technology for MASS so that it can express and perceive analogue communication methods. The SMEs reported that adapting technology to shouting and hand movements would require considerable cost, with poor return on investment. This was driven by the estimated large effort needed to implement equipment, computer vision and natural language processing that understand every contextual meaning of waving and shouting in multiple languages and under difficult weather conditions. The adaption of radio communication by VHF, MF/HF, and satellite phone was considered equally difficult for many of the same reasons above, but with a very high value for radio communication because it is an important tool to quickly align and resolve misunderstandings between ships. Satellite phones was estimated to have little value being adapted, as the key stakeholders can contact each other directly on shore through the ROC using cheaper methods. With ships communicating with each other virtually always through radio, the voice calling use-case for satellite phones is believed to be superfluous.

The SMEs argued that there exist products that may be used to detect the direction and decipher the meaning of a ship horn sequence. High-quality cameras and computer vision algorithms may also be used to detect navigation lights, day shapes, flag signals, as well as the morse code of searchlights, and daylight signalling lamps faster and better than the humans. These communication tools were considered important and relatively easy to adapt to autonomous ships. To display the results from the SME analysis, we plotted them into a two-dimensional diagram (Figure 3).

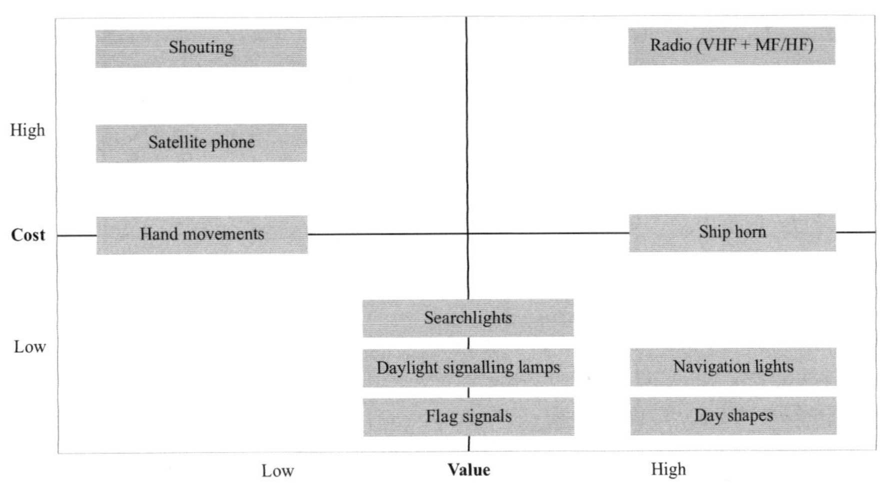

Figure 3. Estimated cost-benefit of automating maritime communication for MASS.

The SMEs also pointed out the challenges of implementing maritime communication for MASS. One expert expressed that fully autonomous vessels would be near impossible in the foreseeable future

– or perhaps forever – and that there would be a need for a ROC operator to take control and communicate with other vessels through radio. In that case, incoming radio communication would be relayed to the ROC. Another expert pointed out that it was a process of development. In the beginning, the MASS might need a human operator to take care of the analogue communication. However, with technological development, new methods of machine-to-machine communication, and more autonomous ships in service, the need for a human operator to communicate would be reduced. As shown above, to make autonomous ships communicate in the same way as a traditional ship is a complex technological challenge. Are there other ways of communicating the state and intention of a MASS? Next, we will propose solutions to help increase automation transparency.

6. Discussion

In this article we have, with the help of five subject matter experts, analysed current maritime communication methods between traditional ships and found that some means of communication poses a problem for autonomous ships, which can give rise to unsafe interactions between manned and unmanned vessels. There is a need for additional ways or methods of communicating the autonomous ship's current state and future intention to nearby ships to ensure safety at sea. We propose a focus on automation transparency to alleviate these problems. We suggest three main strategies for increasing automation transparency that should be done in parallel: (1) to develop the MASS' sensors, algorithms and technology so that they can understand and communicate with ships within existing frames of maritime communication, (2) to evolve existing communication so that it is easier for MASS to use it, and (3) to develop fundamentally new ways of communicating with other ships that are better adapted to the way computers and artificial intelligence perceive and communicate with the world. Below, we will show some examples.

6.1. Automation Transparency within existing frames of communication

First, the principal factor for ensuring safety at sea for all types of vessels is ship movement. This permeates through multiple analogue and digital communication methods. Developing autonomous systems conforming to the expectations of humans is thus the most important development that can be done to communicate the status and intent of MASS. This development is dependent on improved sensor fusion. As seen in Figure 2, current versions of MASS almost exclusively perceive its environment – and makes decisions - through movement, using electronic sensors. However, many misunderstandings and potentially dangerous situations at sea are avoided through more analogue means of communication that autonomous ships are blind to. Our findings indicate that the AI of MASS need to be able to perceive and express itself through more communication methods to be able to sail as safely as a crewed vessel. These include, but are not limited to, radio (VHF and MF/HF), ship horns, navigation lights, day shapes, searchlights, daylight signalling lamps, flag signals, satellite phone calls, shouting, and hand movements. Indeed, as found through the discussion with the experts, existing technology may be adapted to decipher these methods using computer vision and installing microphones that algorithms listen to. However, as seen in Figure 3 this may be extremely costly and not give a high yield. Within existing frames of communication, we propose that MASS should be able to perceive and express ship horns, day shapes, flag signals, daylight signalling lamps and searchlights autonomously without the help of an operator in the ROC. The authors suggest that the other communication methods are not developed within the existing frames of communication. Rather, we suggest changing the expectation of crewed vessels by adapting and further developing today's technology. It is to this proposal we now turn.

6.2. Automation Transparency by improving existing communication

When existing maritime communication is not possible to adapt to autonomous ships, we must improve them. Findings showed that adapting the communication done through radio may be very difficult to achieve. However, they are crucial to ensure safety when interacting with crewed vessels. One expert argued that radio communication would always need to be relayed to the ROC. However, if an operator is to monitor multiple MASS radio channels simultaneously, this may put notable pressure on the person. A system of filtering this communication would be needed. One example is maritime messaging, a new communication tool that could be easily adapted to MASS. Analogue verbal communication on radio could be moved over to a messaging system. One advantage of such text messages over radio communication is that they are lasting and not only interpretable in real time – they can both be read out loud by a synthesized speech as they arrive and be read on a screen later. The use of satellite phone calls would also not need to be automated by the MASS if one could adapt the technology. Without crew, their use case would realistically be handled by the ROC using cheaper communication methods. Moving towards fully autonomous ships we argue that it is realistic to develop AI that can both perceive and express its state and intention using text. If this is through existing or new communication technology may be researched. Another possible improvement is to revise the AIS standard by adding new classes of AIS messages that can be used by MASS to communicate their state and intention. That way, crewed vessels can understand that the vessel is a MASS and may form more correct mental models.

6.3. New forms of automation transparency

There already exist ideas that go beyond current maritime communication. As found in Figure 3, even if MASS can communicate with SOLAS ships without radio communication, it would perhaps not be visible to non-SOLAS ships, as they may lack the needed electronic equipment. This discrepancy may pose a challenge to the deployment and utilization of MASS. Is it possible to make crew on both SOLAS and non-SOLAS vessels implicitly understand that the ship is autonomous? Porathe & Rødseth [25] have proposed separating MASS from conventional ship traffic, creating traffic lanes only to be used by autonomous vessels. Making it clear that a ship is autonomous could also perhaps be done through the physical design of the ship. An example is Porathe [26] which suggests a large display to be integrated into the superstructure of the ship that show text messages and symbols to nearby ships. An urban passenger ferry prototype with LED light strips and panels around the superstructure of the ship to display different states, such as distress, vessel detection or power on thrusters before the acceleration is visible was suggested by Veitch & Alsos [10]. These would be optimal in constrained coastal waters and could potentially work as a communication channel to humans that are shouting or using hand movements. However, these ideas may not be perceptible in all weather conditions and would need appropriate regulation. They are currently concept ideas which need to be tested and explored in future research.

The authors argue that further research into conversational HMIs may yield promising results. HMIs enabling direct conversations with the AI on its terms can help increase automation transparency. New solutions could be developed to access the thoughts of a MASS through onboard electronic equipment such as the upcoming VDES system or perhaps through novel interfaces such as VR and AR. These HMI proposals for MASS have also been theorized in other papers such as Porathe [6] and Porathe & Ekskog [27].

In this paper we have mainly discussed the current states and future intentions of autonomous ships. We also believe that being able to see the past states and intentions of an autonomous ship can help traditional ships understand the current and future intention of it. For example, through digital technology one can (1) display the MASS' past course line in an electronic chart display on board a traditional ship, (2) show when the traditional ship was discovered by the MASS, and (3) reveal when the MASS was in an autonomous mode or manually taken over mode by a ROC operator or onboard crew.

One challenge tied to these developments is how the coming generations of highly automated or autonomous ships are all going to have the same improved capability to perceive others and express themselves. To solve this challenge, we can learn from car producers, who equip new cars with various sensors that prepare them for autonomous driving. As new automation algorithms are being released, the cars become autonomous overnight through automatic software updates. The shipyards and their subcontractors should think similarly: the ships need to be equipped with a range of sensors and equipment, such as microphones and cameras, that potentially can be used to perceive and express more forms of maritime communication. This way, the shift to a safer, more efficient, and more environmentally friendly shipping future can be as efficient as possible.

6.4. Future research
The authors argue that there is a need for further research into developing technology that enables MASS to perceive and express its own and other ships' status and intent. Current versions are blind to multiple communication methods, inhibiting their automation transparency. We argue that some new methods of communication that increases automation transparency is developed. This should be developed into an industry standard for MASS communication. Currently we have ongoing research studies where we look at automation transparency for various stakeholders, such as developers, ROC operators, onboard crew, passengers, nearby ships, and observers. We also experiment with various channels for making automation transparent to these stakeholders, ranging from light, sound, haptic feedback, onboard displays, ship movement, course changes, gestures, and electronic messages. In addition, we also look at how MASS through its motion can provide automation transparency, and how non-SOLAS ships can get a confirmation whether it has been detected by an approaching MASS. A natural expansion of this research is to look at different encounter situations (overtaking, head-on or crossing), different ship type combination (MASS/SOLAS/non-SOLAS) and various number of ships (two ships versus multiple ships). We finally argue that there is a need to develop technology that enable crewed ships to understand that a ship is autonomous, and that they can probe it through a conversational interface. This may be key to improve automation transparency for MASS.

7. Conclusion
In this study we analysed maritime communication for traditional and autonomous ships. We found that current versions of MASS have limited ability to perceive the state and intention of nearby ships and have challenges communicating their own state and intention to these ships, especially concerning analogue communication methods.

To improve automation transparency, we recommend (1) to develop MASS' sensors, algorithms, and technology so that they can understand and communicate with ships within existing frames of maritime communication, (2) to evolve existing communication so that it is easier for MASS to utilise it, and (3) to develop fundamentally new ways of communicating with other ships that are better adapted to the way automation and artificial intelligence perceive and communicate with the world. We also suggest that all three recommendations are developed in parallel. In the future, we expect autonomous ships to be able to transparently communicate its past, current, and future states to other ships, ranging from traditional to fully autonomous ships, making seafaring safer.

Acknowledgements
The authors thank Erik Veitch and the anonymous experts for valuable contributions to this study. The research was funded by the Norwegian Research Council under the projects SFI AutoShip, grant number 309230, and MIDAS, grant number 331921. The authors acknowledge the support of the Department of Design.

References
[1] IMO 2021 *Outcome of the regulatory scoping exercise for the use of maritime autonomous surface ships* Online: https://www.imo.org/
[2] Wright G R 2020 *Unmanned and Autonomous Ships: An Overview of MASS* (New York: Routledge)
[3] Chae C J, Kim M and Kim H J 2020 A Study on Identification of Development Status of MASS Technologies and Directions of Improvement *Appl. Sciences* **10** 4564

[4] Fonseca T, Lagdami K and Schröder-Hinrichs J U 2021 Assessing innovation in transport: An application of the Technology Adoption (TechAdo) model to Maritime Autonomous Surface Ships (MASS) *Transport Policy* **1** 182-195
[5] Levander O 2017 Autonomous ships on the high seas *IEEE Spectrum* **54** 26-31
[6] Porathe T 2018 Interaction Between Manned and Autonomous Ships: Automation Transparency *1st Int. Conf. on Maritime Autonomous Surface Ships (Busan 08.11.2018- 09.11.2022)*
[7] Liao V Q and Varshney K R 2021 *Human-Centered Explainable AI (XAI): From Algorithms to User Experiences* Online: https://arxiv.org/abs/2110.10790
[8] Norman D A 2013 *The Design of Everyday Things: Revised and Expanded Edition* (New York: Basic books)
[9] Lyons B J 2013 Being Transparent about Transparency: A Model for Human-Robot Interaction *AAAI Spring Symp.: Trust and Autonomous Systems* 45-53
[10] Veitch E and Alsos O A 2021 Human-Centered Explainable Artificial Intelligence for Marine Autonomous Surface Vehicles *Journal of Marine Science and Engineering* **9** 1227
[11] Porathe T 2021a Autonomous Ships: A Research Strategy for Human Factors Research in Autonomous Shipping *Int. Conf. on Applied Human Factors and Ergonomics: Advances in Human Aspects of Transportation (USA 25.07.2021- 29.07.2021)*
[12] Rutledal D, Relling T and Resnes T 2020 It's not all about the COLREGs: a case-based risk study for autonomous coastal ferries *3rd Int. Conf. on Maritime Autonomous Surface Ship (Ulsan 11.11.2020- 12.11.2020)*
[13] Craddock D 2021 *What Ship, Where Bound?* (Barnsley: Seaforth Publishing)
[14] IMO 2004 *Adoption of the revised performance standards for radar equipment* Online: https://www.imo.org/
[15] Fletcher S 2011 *Reeds VHF/DSC Handbook* (London: Adlard Coles Nautica)
[16] IMO 2001 *IMO Standard Marine Communication Phrases* Online: https://www.imo.org/
[17] IMO 2015 *Revised guidelines for the onboard operational use of shipborne automatic identification systems (AIS)* Online: https://www.imo.org/
[18] EESA 2022 *SAT-AIS - Satellite Automatic Identification System* Online: https://artes.esa.int/satais-satellite-automatic-identification-system
[19] Lázaro F, Raulefs R, Wang W, Clazzer F and Plass S 2019 VHF Data Exchange System (VDES): an enabling technology for maritime communications *CEAS Space* **11** 55-63
[20] IMO 2000 *Adoption of Performance Standards for Daylight Signalling Lamps* Online: https://www.imo.org/
[21] Porathe T, Eklund P and Goransson H 2014 Voice and Text Messaging in Ship Communication *5th Int. Conf. on Applied Human Factors and Ergonomics* (Kraków 19.07.2014-23.07.2014)
[22] Brekke E F et al. 2022 milliAmpere: An Autonomous Ferry Prototype *IOP Conf. Ser.: The 4th Int. Conf. on Maritime Autonomous Surface Ship (Singapore 06.04.2022- 07.04.22)*
[23] Porathe T, Hoem Å, Rødseth Ø, Fjørtoft K and Johnsen S O 2018 At least as safe as manned shipping? Autonomous shipping, safety and "human error" *Safety and Reliability – Safe Societies in a Changing World* ed S Haugen et al (London: CRS Press) chapter 52 pp 417-425
[24] Alsos O A, Veitch E, Pantelatos L, Vasstein K, Eide E, Peterman F and Breivik M 2022 *IOP Conf. Ser.: The 4th Int. Conf. on Maritime Autonomous Surface Ship (Singapore 06.04.2022- 07.04.22)*
[25] Porathe T and Rødseth Ø J 2019 Simplifying interactions between autonomous and conventional ships with e-Navigation *Journal of physics: Conf. Ser.* **1357** 012041
[26] Porathe T 2021b Human-Automation Interaction for a Small Autonomous Urban Ferry: A Concept Sketch *European Safety and Reliability Conf.: 31st European Conf. (Angers 19.09.2021- 23.09.2021)*
[27] Porathe T and Ekskog J 2018 Egocentric Leisure Boat Navigation in a Smartphone-based Augmented Reality Application *Journal of Navigation* **71** 1299-1311

From captain to button-presser: operators' perspectives on navigating highly automated ferries

Erik Veitch,[1*] Kim Alexander Christensen,[2] Markus Log,[3] Erik Thule Valestrand,[4] Sigurd Hilmo Lundheim,[4] Martin Nesse,[4] Ole Andreas Alsos,[1] Martin Steinert[2]

[1]Department of Design, NTNU, 7491 Trondheim, Norway

[2]Department of Mechanical and Industrial Engineering, NTNU, 7491 Trondheim, Norway

[3]Department of Civil and Environmental Engineering, NTNU, 7491 Trondheim, Norway

[4]Department of Sociology and Political Science, NTNU, 7491 Trondheim, Norway

erik.a.veitch@ntnu.no

Abstract. Teaming with Artificial Intelligence (AI) is changing the way seafarers work. We show that a new kind of seafaring is emerging, characterized by cooperation with AI systems. In this format of seafaring, navigation tasks are controlled automatically while human operators manage the automation, always "in the loop" and ready to take over control if necessary. Ideally, this arrangement sees improvements in overall system performance and safety. However, little is known today about how this format of work will unfold in real-world operations. We investigate this topic by interviewing operators (n = 5) on ferries recently outfitted with state-of-the-art automated navigation technology. We used a mixed-methods approach to analyze the case study interviews, combining quantitative text analysis with Grounded Theory qualitative analysis techniques. The results show that operators perceived a shift towards a backup role coincident with increasing agency of machine autonomy. This role shift was characterized by "button-pressing" to start the machine autonomy and subsequently intervening to stop it when things go wrong. We observed that this shift led to boredom, de-skilling, stretched resources, and compromised organizational harmony – effects running counter to the intention of improved system performance and safety. Synthesizing the findings, we present (1) effects across three operational dimensions: (i) tasks, (ii) human-computer interaction, (iii) organization; and (2) a definition of collaborative (human-AI) seafaring. By identifying issues in the early implementation of highly automated ship navigation, we hope to guide designers of Maritime Autonomous Surface Ships (MASSs) away from potential pitfalls and towards development more in tune with real-world demands of collaborative work.

1. Introduction

The work of seafarers is changing with the development of increasingly intelligent machines used for ship navigation. Meanwhile, Maritime Autonomous Surface Ships (MASS), are evolving from concept designs to real-world implementation, representing the next major shift in seafaring work. MASSs, despite containing high levels of machine autonomy, still require human supervision and remote intervention when things go wrong. Currently, little is known about how this format of seafaring will take place. One approach to addressing this knowledge gap is by examining current changes in how seafaring work is changing aboard highly automated ferries. While not MASS per say, highly automated ferries share similar features. This includes the following precepts for cooperation with the automation: (i) the operators can turn on and off the automation, (ii) the operator must intervene to take preventative actions, and (iii) sufficient situation awareness is needed to "get in the loop" for that preventative action to be appropriately executed. Like MASSs, current highly automated ships also share waterways with predominantly manned vessels, resulting in subtle shifts in expectations for communication and collision avoidance. Developers of MASS have for almost a decade outlined the need for a Shore Control Center, where operators can manage MASS fleets, always "in the loop" and ready to intervene. In this vision, the benefits of machine autonomy are enabled by close cooperation with skilled human operators.

Cooperation with automation is far from straightforward, however. Human factors experts have catalogued a multitude of complications arising from work with intelligent machines, foremost of which is the effects automation can have on their ability to pay attention. Explanations of how operator performance is affected by automation, while still not universally accepted, range from theories of situation awareness [1] to vigilance [2] and mental workload [3]. Yet, experts are unanimous in their agreement that increased automation, if not designed correctly, leads to mental underload among operators – a state of boredom that is potentially hazardous if one's work involves swiftly taking over control to avoid disaster. Dramatic examples have already been reported from highly automated cars. In two separate National Transportation Safety Board (NTSB) reports, overreliance on Tesla's *AutoPilot* system led to fatal accidents when the inattentive driver was unprepared to take over control [4,5]. This led to recommendations directed towards technology companies that included "minimizing driver disengagement, preventing automation complacency, and accounting for foreseeable misuse of the automation" [6]. Looking ahead at MASS operators, the line blurs between seafarer, technologist, and level-headed emergency responder – a new breed of ship operator whose lofty skill requirements may exclude just about any applicant [7].

Seafarers themselves are ambivalent about the development of MASS. Recent studies that have surveyed seafarers show that, while open to technology change, they believe operation of autonomous ships will require significant changes to the skills and competences currently at the core of standardized seafaring curricula [8,9]. Others have reported specific concerns on the part of seafarers contending with automation, which included sensor over-reliance, compromised situational awareness, and increased complacency (while paradoxically providing little reduction in fatigue levels) [10]. The history of ship navigation has drawn a parallel arc to that of modern technology development, punctuated by the advent of tools like the sextant, compass, radio, radar, electronic charts, and Dynamic Positioning (DP). There is a long precedent in academic interest on this topic, both in the context of automation [11] and in the context of collaborative work [12]. Recent advances in Artificial Intelligence (AI), however, challenge current paradigms of technology interaction. Machines capable of completing tasks on their own largely supersede the manual interaction traditionally associated with navigation tools. A sextant cannot locate a star by itself, nor can a DP system hold station without the fine-tuning of an operator. An autonomous ship can, by contrast, sail from port to port by itself, avoiding collisions and even optimizing fuel consumption along the way. John McCarthy, who is credited with coining the term "Artificial Intelligence," defined intelligence as the "computational part of achieving goals in the word" [13]. While this definition is not universally accepted, we adopt it here by virtue of its simplicity. By this definition, the auto-crossing and auto-docking systems that we feature in our case study can be considered AI systems, even if they currently function with rule-based algorithms. Although, if current research trends are any indication, we can expect to see more sub-symbolic techniques commonly associated with contemporary AI applied to future ship navigation (e.g., [14]).

A milestone in autonomous ship development came in 2018 when the International Maritime Organization (IMO) first formally acknowledged MASSs [15]. At that time, it was announced that a scoping exercise would get underway aiming to highlight where MASSs fell outside of existing regulations. The results of the scoping exercise listed "high-priority issues," of which the top three items related to the operator whose assumed role it was to supervise control of the MASSs [16]. Fundamental questions were raised, not least of which questioned the meaning of master, crew, and responsible person in such a system. Around the same time, Stanford University published its influential *AI100 Report*, summarizing state-of-the-art technology and research [17]. "Perhaps the most inspiring challenge," they wrote, "is to build machines that can cooperate and collaborate seamlessly with humans' (p. 19). If the early optimism surrounding intelligent machines was inspired by a technology-driven approach to computational systems, recent thinking elevates human presence and the question of how best to extend attributes of collaborative work to these intelligent machines.

Our case study examines auto-crossing and auto-docking, two recent technologies implemented on Norwegian coastal ferries [18]. The functionality is straightforward (see Figure 1), with one panel containing a series of four buttons: Auto-cross, Manual, Auto-dock, and Autopilot. A typical ferry crossing starts by pressing "Auto-dock," which takes the ferry from dock. "Auto-cross" then initiates an automated crossing by following pre-set waypoints, whereupon "auto-dock" is pressed again upon approaching the arrival dock. Currently, auto-crossing does not detect nor avoid other objects. "Manual" is for taking over control and "Autopilot" is for conventional autopilot. A tablet-sized screen relates system information like the thrusters' power and orientation. For simplicity, we refer to auto-crossing and auto-docking collectively as the "auto-systems." The motivation for interviewing operators of these auto-systems was inspired by the need for "nitty gritty" detail about operators' experiences working with increasingly intelligent systems on ships. Much of the literature is dedicated to the development side, ranging from conceptual frameworks and methodologies to descriptions of new techniques and prototypes. An important question remained unanswered, one that formulates our research question: how does AI technology used to navigate ships affect real-world seafaring work? We set out to design a case study that would shed light on this question. Our aim was to provide more than just an expository account of operators' begrudging views on technology. Rather, it was to describe a phenomenon of growing relevance for MASS operations, and to identify pitfalls on the path from concept design to real-world implementation.

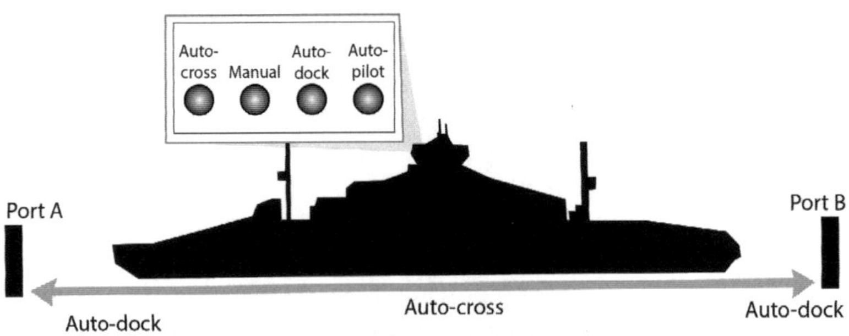

Figure 1*:* Illustration showing a Ro-Ro ferry with auto-crossing and auto-docking.

2. Method

We interviewed captains and chief mates (n = 5) who worked aboard two Roll-On Roll-Off (Ro-Ro) ferries using state-of-the-art automated navigation technologies. One of these ferries, the MF *Korsvika*, was equipped with both auto-crossing and auto-docking. The other ferry, the MF *Vikhammer*, was instrumented with just auto-crossing. Both ferry names are pseudonyms to protect anonymity of the interviewees. The interviews were conducted by the first and second authors in June 2021 and translated from the Norwegian with the help of an expert in marine terminology. Table 1 presents information about the two ferries and the operators' experience levels using the automation technology. The interviews were semi-structured, focused around four main questions: (i) Can you tell us about your background and experience? (ii) Do you have a name you use for the new automated navigation systems? (iii) What changed, if anything, since this technology was introduced? (iv) Have you become a better navigator as a result of the automation? Informed consent was obtained for all interviewees. All interviews were transcribed verbatim.

Table 1. Characteristics of the two Ro-Ro ferries and their operators we interviewed.

Ferry[a]	MF *Korsvika*	MF *Vikhammer*
Length overall	Approx. 140 m	Approx. 105 m
Capacity	Approx. 600 pax; 200 cars	Approx. 350 pax; 120 cars
Crossing length	Approx. 6 nm; highly trafficked	Approx. 1 nm; sparsely trafficked
Operators[b]	**Ola, Lars & Robin**	**Henrik & Magnus**
Auto-crossing	1-2 years' experience	1-2.5 years' experience
Auto-docking	1-2 years' experience	N/A
Avg. experience	26 years	9 years

[a] Names of ferries are pseudonyms.
[b] Names of operators are pseudonyms. Operators held D1 certificates: the highest navigation license in Norway, which reserves the title of "captain."

2.1. *Mixed-methods approach*
We combined qualitative case study analysis and quantitative analysis techniques in what is referred to as a "mixed-methods" approach [19]. We describe the two methods below.

2.1.1. Quantitative analysis. The quantitative approach leveraged the text analysis software Leximancer [20] to identify salient themes, patterns, and relationships in the interview transcripts [21,22]. Leximancer executed a content analysis, where "coding" (i.e., highlighting relevant parts of text) is automated using latent semantic analysis indexing [23]. This resulted in a network of related emergent concepts, as depicted in Figure 2. The interview transcripts were pre-processed into a format that distinguished speakers and their respective ferries. The analysis excluded content from the interviewer. The "concept map" in Figure 2 was generated from linked concepts (words) and thematic clusters (circles). The clusters represented analytical insights, as they were based on patterns and relationships in the interview data. Several interview excerpts were supplied for each theme, providing a rationale for the clustering. In Figure 2 we labelled each cluster with a theme label, each of which we describe in the Findings.

2.1.2. Qualitative analysis. In addition to the quantitative text analysis, a qualitative approach was adopted based on Grounded Theory in the tradition of Glaser & Strauss [24] and Morse [25]. Grounded Theory techniques allow researchers to minimize bias in theory development by virtue of the techniques' transparent grounding in observations. One useful technique we adopted was that of "memos," described by Corbin & Strauss [26] and Glaser & Strauss [24]. In memos, descriptions of thoughts and impressions are immediately written down upon first interaction with the data. In the end, we had 101 such memos, ranging from short notes to descriptive passages. Memos served two practical purposes: (i) they provided a record of early insights at a critical time when the researcher's mind was free from the "pull" of early explanations, and (ii) it served to gauge the extent to which our sampling had reached saturation. Since we sampled as we went along in the analysis (so-called "theoretical sampling"), we reached a sampling saturation when the memos revealed enough findings that we considered it not worthwhile to interview more operators. From the early insights gained in the memos, we proceeded to code the transcripts in the text analysis software NVivo [27] and subsequently aggregate concepts into an overarching narrative structure.

Our aim was to describe a phenomenon that applied in equal measure to the population of ship navigators as to the sampled individuals. Here converging evidence (or "triangulation") played an important role. The data were independently analyzed in the context of our research question, whereupon results were compared and iteratively discussed prior to collating them in the Findings. Triangulation was accomplished across three groups: (i) the first and third authors (qualitative case study analysis) and (ii) the second author (quantitative analysis), and (iii) fourth, fifth, and sixth authors (qualitative analysis on the *Korsvika* subset of the data).

3. Findings

In this section, we present the results of the mixed-methods analysis, identifying and describing effects of AI technology on seafaring work. The quantitative analysis is presented first, before continuing with the qualitative analysis. We conclude the section by synthesizing the findings into a table classifying effects into three operational dimensions (Table 2): (i) tasks, (ii) human-computer interaction, and (iii) organization

3.1. Quantitative analysis

The thematic analysis is visualized in the concept map (Figure 2). Presented below are five themes (named by the researchers) along with corresponding excerpts identified by Leximancer as particularly relevant. The themes are ordered from most significant to least significant based on the amount of text associated with it.

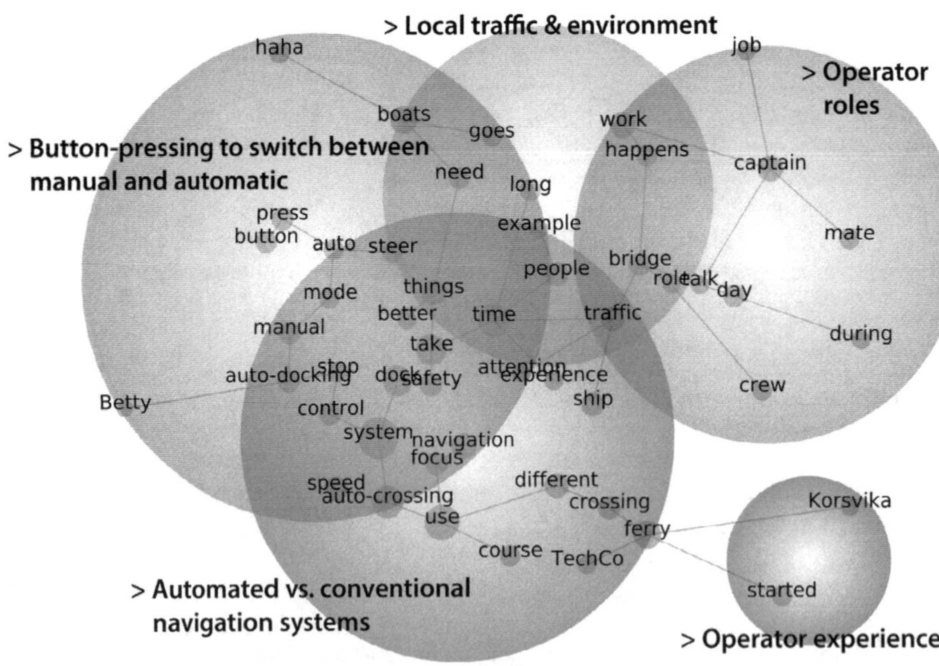

Figure 2: Leximancer concept map. The following are pseudonyms: "Betty" (nickname for auto-system), "TechCo" (developer of auto-system), and "Korsvika" (one of the two Ro-Ro ferries).

3.1.1. Automated versus conventional navigation systems. Conventional navigation consoles required the operators to focus on manually maneuvering the ferry. The new systems, by contrast, could do much of the maneuvering automatically, meaning the operators were free to focus their attention elsewhere. They could keep lookout for other vessels using the chart plotter, radar, or simply by looking out the windows. They could also handle unexpected events or catch up on administrative work. The automated systems were even expected to save fuel compared to conventional navigation.

> "We have gotten more technologically demanding systems, which requires a different focus than before in certain areas. Before, you used an old diesel ferry and pushed it into the dock, and no one cared about fuel consumption." – Lars

3.1.2. Local traffic and environment. The operators on *Korsvika* needed to intervene often (up to once per day) with the auto-crossing system because of small boats and other ships in the area. *Vikhammer* sailed mostly as planned, without navigator intervention, due to a scarcity of local marine traffic. This difference is illustrated below, with one quote from each crossing.

> "I check for small boats that [Vessel Traffic Services] doesn't pick up. They miss a few sometimes, so you have to pay attention." – Henrik

> "I'd love to try another crossing with more ship traffic. You still have to keep an eye out, of course, but work on the bridge has changed." – Magnus

3.1.3. Pressing buttons to switch between manual and automatic. This theme was related to the process of switching between automatic and manual modes and what happened if the correct procedure was not followed. The interplay between the operator and the control panel, which consisted of four buttons, featured heavily in the informants' accounts. (This can be seen in Figure 2, where the concepts "button," "press," and "screen" appeared as linked. These were also linked to "haha," indicating informants' laughter, and "Betty," the nickname two of the five informants assigned the auto-system.)

> "When you press it, you see that it's about to auto-dock. But if you don't see it, it beeps again that you have to take over." – Henrik

> "It's been a half year now since fog and all that. But you could just press the button and [auto-docking] went great." – Robin

3.1.4. Operator roles. The operators provided rich details about their roles. The result was a picture of how work typically transpired aboard the two highly automated ferries.

> "On the bridge, the captain and chief mate shift roles throughout the day. Typically, it's two hours on the bridge, two hours for other tasks." – Lars

3.1.5. Operator experience. This theme was about the operators' background experience. Although it was the least significant of the themes, it served to emphasize how much operators valued experience as part of their work.

> "I started as a deck hand [when I was a teenager]." – Ola

3.2. Qualitative analysis

In this section we expand on the five themes above using qualitative analysis techniques, completing the mixed-methods analysis. We dig deeper into the underlying meaning of the emergent themes, uncovering relevant excerpts as supporting evidence. The analysis is presented in three parts: (i) analysis of the auto-systems' agency, (ii) list of three benefits of the auto-systems, and (iii) list of three drawback of the auto-systems.

3.2.1. Agency of the auto-systems. When asked to describe the automated navigation systems in their own words, two of the captains aboard the *Korsvika* explained that they had taken to calling it by a nickname: "Betty" (a pseudonym). Robin explained:

> "My captain and I, if we're auto-docking, we say that 'Betty's taking care of it.' Then he knows that auto mode is on, haha. If we have normal autopilot on, then I say that 'Betty's not taking care of it,' haha."

Ola also calls the auto-systems Betty. This nickname, traditionally a woman's name, is based on the voice announcements of the system, which are recordings of a woman's voice. Ola reveals how this voice makes it easier to assign the system human-like characteristics:

> "...she can nag a bit, haha. The system works well as it is set up, but it needs some fine-tuning because the system does some weird things that people do not do. She has an arrival that's not logical. Those who've made it certainly don't think so, but for us who steer the ship, we steer it better than the system does."

Here Ola refers to the auto-systems by the pronoun "she" which is characterized as doing "weird things." The attribution of human qualities to physical objects is called personification. Personification is a type of metaphor whose power, as described by Lakoff & Johnsen [28], is to allow one to make sense of phenomena in the world in human terms. Personification lends the AI system agency – the ability to actively influence the world. The agency affords the auto-systems its own "elbow room" on the bridge. As we will see in the remaining analysis, this agency underpins both the benefits and drawbacks we observed in human-AI teaming.

3.2.2. Benefit: Attention is freed up. All ferry operators described the freeing effect the auto-systems had on their capacity for attention. As illustrated by Henrik, this was generally perceived as a benefit:

> "... [when I drive] over with auto-crossing, it's just so I can pay attention the traffic. It [driving manually] is one less thing I need to pay attention to. When the ferry comes to a waypoint and is about to turn, it does so. It follows the planned route. We can pay attention and monitor ship traffic more. There's lots of info you need to process there. So, you have one thing less to think about. As soon as something is wrong, you give alerts. When there's nothing wrong, it's good – then it's an assistant." – Henrik

The term "assistant" is a relevant one, because it indicates a hierarchy whereby the operator can override the auto-system. Lars described this phenomenon in similar terms: "If you free up work one place, now you can focus on another." Lars explained that the natural place to re-allocate attention was on navigation duties. "You make one task easier by not having to control speed and energy consumption, but you can't take away attention to the traffic picture."

Small vessels (recreational craft and sailboats) were highlighted as focal points of re-focused attention. Because auto-crossing does not detect nor avoid other objects, the operators were attentive to potential collision scenarios. While larger ships behaved largely as expected, following the "rules of the road" and communicating intentions over radio, small vessels could behave erratically. All *Korsvika* operators interviewed noted an increasing trend in the number of small vessels and remarked on their limited understanding of marine traffic regulations. "The biggest problem is our summer with small boats and sailboats," related Ola. "These boats don't know how to behave at sea." Manual takeover was evidently more likely for smaller vessels and was reported by one *Korsvika* operator to occur as often as once per day during peak summer season.

3.2.3. Benefit: Docking is easier in poor visibility. All three ferry operators aboard the *Korsvika* reported independently that the auto-docking system made docking in poor visibility significantly more effective. "In thick fog, you spend a very long time on radar and [looking for] fog lights," explained Ola. "In auto you can sit back and let the system do it." Henrik, taking a pragmatic view of the auto-system's benefits, claimed that "It's very nice to use in certain situations... In fog, for example, it's absolutely fantastic." Similarly, Robin explained:

> "...the system is ingenious. We see absolutely nothing in the fog... We see it hits the mark just as well as during clear weather... it's super useful."

3.2.4. Benefit: Maneuvering is more consistent with less "human errors." Several operators pointed out that unlike the navigation crew, each of whom possess their own idiosyncratic style of operating, the auto-system does the same thing, every time. Henrik explained:

> "It [the auto-system] is much more stable, in a way. We have eight different navigators [on this ferry], and all of them do their job a little differently. The system always does the same. There are always some variables for us humans that the system does not have."

Henrik implies that the natural variation among operators is levelled by the auto-system. It also suggests that circumventing "human error" makes maneuvering safer. Lars pitted the operators' ability to maintain low fuel consumption directly against that of the auto-system: "On single trips, the navigator beats the system. But over time, on average, auto-crossing will beat most navigators. A navigator can't handle 100% focus on [optimal energy use] all the time.

3.2.5. Drawback: Resources can be "stretched." Although freed-up attention was mostly framed by operators as a benefit, tensions arose when it was re-allocated in a way that stretched their capacities. Magnus illustrated this with a telling example:

> "I think that technology sometimes trumps safety, trumps people on board... I also run a kiosk, selling chocolate and things. They [management] feel that when auto-crossing is there, they can give him [the operator] other tasks."

Here Magnus reported that following implementation of the auto-crossing system, they are told to periodically leave navigation duties and run the kiosk for passengers during crossing. A latent tension is manifested in the mate's subtle criticism of "they" (presumably the ship's management) for re-allocating resources in a way that was perceived as compromising safety. Lars raised a similar concern, stating that "...auto-crossing is more about energy saving than safety." The benefit of increased efficiency enjoyed by one group (ship management) is cast as the drawback of hazardously stretched resources for another (the operators moonlighting on other tasks).

3.2.6. Drawback: Operators may resent increased automation of their tasks. Friction emerged between operators and developers of the auto-systems when the technology was perceived by the former as part of a misguided endeavour to replace human operators. As Ola explained,

> "The engineers must have something to invent... [and] since they have the shipowners on their side, they make sure that they can get rid of people and hope that the system can be more economical."

Ola continued, resenting developers' short-sightedness in assessing technology benefits, while at the same time feeling powerless to stop them:

> "We call it 'progress' don't we, and the world is moving forward and there are a bunch of people from the university involved. They just have to come up with something. Much of our jobs can be automated... I would call this natural progress."

In a less heavy-handed tone, Robin conveyed a sense that developers did not fully appreciate operators' tacit knowledge for ship maneuvering gained from many years' experience at sea. They related that in some weather conditions, they take over control for the sole reason of making the passage more comfortable for passengers:

> "If it's too windy we run in manual mode. We haven't tested so much in the wind, but these days we say: 'if it's blowing, we steer manually.' I think auto-crossing can be used at any time, but manual mode is more comfortable for passengers... for example if you have rolling, people can fall and hurt themselves. Instead of rolling all the way over, I sail a little North and then a little South to go across the waves."

3.2.7. Drawback: Passivity can lead to boredom and de-skilling. Whereas freed-up attention could be re-allocated in a way that stretched resources beyond their optimal capacities, the opposite was also observed. Attention re-allocated to nothing resulted in boredom. Boredom in the operator's role is clearly dangerous if it leads to inattentiveness and de-skilling over time. The dimension of time, conveyed by words describing past and future, emerged as closely linked to boredom.

> "The working day of the captain will get boring. You press three buttons. There's no challenge. You become an operator who monitors the system and is ready to press a button if there's a bug in the system." – Ola

Here Ola conveys a sense that boredom is a direct threat to ferry operators, whose manual ship maneuvering duties are ceding to automated machines. The sense of safety responsibility also seems undermined, as well as the ebbing away of the high status traditionally with ferry operators. For Magnus, this sentiment is manifested in a desire to transfer to a crossing with more ship traffic, if only to have something to which to pay attention.

> "Now you just press the button and wait for the beep. You become more unfocused on the tasks. You look at the radar occasionally, but here there's no boats. I'd love to try another crossing with more ship traffic. You still have to keep an eye out, of course, but work on the bridge has changed. You get a little unfocused, you know that the computer does the job." – Magnus

In this vein, the topic of land-based control of MASSs emerged, despite never being expressly mentioned by the interviewer:

> "Looking further into the future, I see fewer people. If you eventually sit at the land station and monitor, it'll be a completely different working day." – Henrik

One of the questions we asked was whether operators felt they had become better navigators because of the auto-systems. Surprisingly, no responses came back positive, despite the clear benefits reported. "... No, I wouldn't claim that." responded Ola. "You become complacent. You forget to pay attention. You just leave everything up to the autonomous." Henrik responded to the question similarly: "No, I haven't [become a better navigator], but I'm not any worse either." Magnus answered:

> "No. You become better at understanding technological equipment, but you don't become a better navigator by just sitting there. Okay, the boat goes by itself, so you can have more focus on radar and things like that. But where I am, without ship traffic, you get bored."

Meanwhile, past technologies were occasionally romanticized. Musing on the constant upgrades to navigation technology, whether it be electronic chart display or automated systems, Robin quipped, "It's the oldest stuff that works." Magnus wistfully described his past work, saying, "I experienced a lot of the world with ancient navigation, using astronomy and a bearing compass... Before I started with [my current employer]."

De-skilling was linked to increased passivity. Asked how often Magnus used auto-crossing aboard *Vikhammer*, he responded, "We only use auto-crossing now. Every day, every trip." Following up by asking if there was ever need for manual takeover, Magnus responded:

> "We might occasionally drive if we have an ambulance dispatched… with an ambulance it's life and health. Apart from that… it's been one-and-a-half years since I stopped doing it [driving manually] myself."

Hearing this, one might ask, is Magnus prepared to "get in the loop" and drive manually in such an emergency? Accounts from operators aboard the *Korsvika*, by contrast, whose operators regularly took over manual control, suggested the answer is likely "No." Robin elaborated on how manual steering is now regular practice on *Korsvika*, where 50-70% of crossing "go on auto:"

> "We've set it up so you'll sail it yourself during the day to maintain your driving. That's written in our procedures now. If you've had a holiday, you're allowed to steer the whole shift, there's no one that says you must use auto-crossing… when I come in on Monday, I usually steer for the first crossing, it's nice to feel the weather a bit, the conditions on the water."

Despite the differences between the two vessels, (who have different owners, different automation technology, and sail on crossings with different traffic conditions), all interviewees experienced a fundamental shift in their role as operators. The shift was best described as one towards a "backup" role to the automated navigation system. Or, in the language of the operators (and as indicated in the concept map in Figure 2), the operators were becoming "button-pushers," arbiters of when to start and stop the automation. As Ola put it, "You only become an operator who monitors the system and is ready to press a button if there's a bug in the system."

3.3. Summary

Table 2 contains a synthesis of the findings, where the effects of AI technology on seafaring work have been classified into three dimensions, or "orders." Starting with the first-order benefits of introducing the auto-systems ("Tasks"), we can proceed to second-order interaction effects ("Human-Computer Interaction") and finally third-order effects among groups of people ("Organization").

Table 2. Effects of implementing automated navigation aboard *Korsvika* and *Vikhammer.*

1st-order effect (Tasks): **Efficiency improved**
Immediate benefits: Fuel optimization; Docking made easier in poor visibility; Lower frequency of human errors in control; Higher consistency of operations
2nd-order effect (Human-Computer Interaction): **Attention freed up**
If refocused on too little: Boredom; De-skilling *If refocused optimally: Operator "in-the-loop"* *If refocused on too much: Stretched resources*
3rd-order effect (Organization): **Operators' shift to "button-pressing" backup role**
Resentment towards developers: Perceived undervaluing of tacit knowledge gained through experience; Undermining captain status *Resentment towards owners:* Safety perceived as compromised when resources are stretched; Foreboding of job automatization

The benefits of the auto-systems appear exclusively as first-order effects in Table 2. The drawbacks emerge later, and with less obvious manifestations. How appropriately the system is implemented hangs in the balance of how newly freed-up attention is re-focused. The shift towards a "button-pressing" backup role resulted in feelings of resentment towards both developers and ferry management, whose short-sightedness in prioritizing first-order benefits they perceived as undermining safety, their most fundamental duty as ferry operator.

4. Discussion

Here we explore how the findings can help advance implementation of machine autonomy. In this aim, we compare our findings with state-of-the-art research, drawing especially from the growing communities of Explainable AI (XAI), machine behaviour, and AI safety.

We begin by acknowledging the need for a working definition for the shifting format of seafaring work marked by collaboration with highly automated navigation technology. Such a definition serves to distill the observations presented in the Findings into a from useful for further discussion and construction of the concept. Below is presented a working definition of the concept of collaborative human-AI seafaring (or "collaborative seafaring" for short).

> **Collaborative seafaring:** a format of seafaring whereby tasks are coordinated between machine agents and human operators in such a way that improves the ability to collaborate with and support seafarers' primary control duties and safety responsibilities.

A key property of agency is that it is definable only in relation to those whose actions are influenced by it. Only recently are we beginning to understand the implications of introducing machine agents into our working lives. In our case study, the has focus of these implications has been on ferry navigation. For MASS in general, progress has been made towards establishing design frameworks laying the foundation for shared control between humans and machine agents (e.g., [29]). Recent work has extended this framework specifically for supporting approval of MASS designs, based on design guidelines of major classification societies and maritime authorities [30]. They showed that tasks assigned to machine agents and human operators form a critical part of the "operational envelope" on which design approval hinges. The operational envelope also intro introduced the amount of time required to take over control from the autonomous systems and get "in the loop" as a key system property. This time is specified as "maximum response time [T_{MR}]," which in a given operational scenario is defined as the "maximum time the operator will need to reach the control position, gain situational awareness, and be ready to perform actions to maintain safety" (p. 4). In our case study interviews, we observed accounts of real takeovers. Most often, this included collision avoidance with small vessels. Other non-critical takeovers involved intervening for increased comfort of passengers in certain weather conditions. Building on the work of Rødseth et al. [30], our findings indicated that T_{MR} is not a fixed value. Depending on how freed-up attention was re-allotted, boredom or stretched resources would likely increase T_{MR}. Additionally, operators aboard *Korsvika*, who engaged in regular manual maneuvering practice, could also reasonably be expected to have lower T_{MR} than those aboard *Vikhammer*, who did not. Testing is needed to confirm these effects and whether they should be considered in the operational envelope.

It is also interesting to discuss the difference between how developers perceived work versus how it was actually carried out. To shed light on this topic, we examine a seemingly paradoxical result: despite expressly listing benefits of using the auto-systems, no operator perceived that it made them a better navigator. On the contrary, operators communicated their resentment towards a group they saw as undermining their primary control duties and safety responsibilities. As shown in the Findings, a group that we labelled "developers" (defined by one of the operators, Ola, as "Those who made the system") were antagonized and even resented. The gulf between operators and developers has been investigated before in the field of organizational science. Hollnagel & Woods [31] framed the issue as one of an abstract distance between "work-as-imagined" (development of operational work) and "work-as-done" (how that operational work is carried out). According to this explanation, only through better understanding of operator work can this distance be bridged, leading to improved organizational safety.

Multidisciplinary efforts may also be needed to build effective collaborative seafaring systems for MASSs. In Rahwan et al. [32] the authors argue that insofar that AI agents exhibit "machine behaviour," human interaction with these agents will inevitably influence human behaviour. We observed this with "Betty," the nickname two operators assigned to the auto-system, whose behaviour, among other things, was called "strange" and "nagging." Rahwan et al. propose that incorporating disciplines traditionally outside of those that produce intelligent machines can serve to incorporate scientific methodologies and frameworks that lead to a better understanding of how this behavioural influence is taking place. In this vein, the authors underscore the need for techniques traditionally used in the quantitative behavioural sciences, like randomized experiments and observational inference.

System safety must also be considered. Leveson et al. [33] defines safety as an "emergent system property," whose accidents can be understood as "the result of flawed processes involving interaction among people, societal and organization structures, engineering activities, and physical systems components" (p. 97). In the field of risk science, approaches inspired by systems engineering have been incorporated to address system safety. On such example is System Theoretic Process Analysis (STPA), which has emerged one of the foremost approaches for risk-based design of MASSs with human operator collaboration (e.g., [34–37]). Although these are promising directions, there is still a lack of concrete cases allowing for evaluation of risk-based approaches in real-world MASS implementations.

In the work presented herein, we have described the new concept of collaborative seafaring that is emerging in tandem with highly automated ship navigation systems. We have also highlighted some of its potential pitfalls in implementation. More work, however, is needed to explain its mechanisms. Among other things, future research will be motivated by the need to address mental underload caused by automation. Can this be counter-balanced by increasing operational complexity, for example by having a single operator responsible for multiple MASSs? We have also witnessed the effects on organizational structures, with evidence suggesting that procedures for regular manual practice may provide an effective safety control against automation-induced de-skilling. These are indications of a community advancing beyond considerations of first-order technology benefits and towards clear-eyed orchestration of these benefits in collaborative work.

5. Conclusion

This work aimed to identify and describe the emerging phenomenon of collaborative human-AI seafaring. Our motivation lay in uncovering the "nitty gritty" details of operators whose daily work has become influenced by the agency of intelligent machines. In our case study, ferry operators using recently developed auto-crossing and auto-docking technologies provided an opportunity to investigate this topic. Applying a mixed-methods approach, we identified emergent themes in the data that did not fit conventional paradigms of ship navigation. The study highlighted potential pitfalls when introducing machine agents into real-world seafaring work. To be sure, our study demonstrated that, at least aboard Ro-Ro ferries, tangible positive outcomes are indeed arising from implementation of highly automated navigation systems. The paradox, though, was that despite the benefits, operators collaborating with the auto-systems perceived that their abilities to safely navigate were unchanged or even hampered. The problem, we observed, lay in AI-interaction effects. Attention of operators, when freed up from manual tasks, was often re-focused on little else. Or, on the opposite spectrum, attention re-assigned to new roles served to stretch available resources to the resentment of operators. Among the latter group, there was an underlying belief that exclusive focus on immediate benefits of auto-systems could undermine overall system safety, serving the opposite of their intended purpose. These are pitfalls that, if avoided, will lead to safer implementation of MASSs and the lowering of barriers impeding further introduction of machine intelligence in ship operation.

Acknowledgments
The authors gratefully acknowledge the financial support provided by the Norwegian Research Council under the MAROFF-2 project "Land-based Operation of Autonomous Ships (LOAS) directed by project leader Kongsberg Maritime AS and partnered with the Institute for Energy Technology (IFE). The authors also acknowledge the support of Fosenregionen IKPR, TrollLabs NTNU, and the Department of Design at NTNU. Finally, we acknowledge the support all the ferry operators we interviewed.

References

[1] Endsley M 1995 Toward a theory of situation awareness in dynamic systems *Hum. Factors* **37** 32–64
[2] Molloy R and Parasuraman R 1996 Monitoring an Automated System for a Single Failure: Vigilance and Task Complexity Effects *Hum. Factors* **38** 311–22
[3] Young MS and Stanton NA 2002 Attention and automation: New perspectives on mental underload and performance *Theor. Issues Ergon. Sci.* **3** 178–94
[4] National Transportation Safety Board 2020 *Collision Between a Sport Utility Vehicle Operating With Partial Driving Automation and a Crash Attenuator, Mountain View, California, March 23, 2018* [Internet] Washington, D.C., USA: National Transportation Safety Board [cited 2021 Apr 26] Report No.: NTSB/HAR-20/01 Available from: https://www.ntsb.gov/investigations/AccidentReports/Pages/highway.aspx
[5] National Transportation Safety Board 2020 *Car Operating with Partial Driving Automation and Truck-Tractor Semitrailer Delray Beach, Florida, March 1, 2019* [Internet] Washington, D.C., USA: National Transportation Safety Board [cited 2022 Jan 26] Report No.: NTSB/HAB-20/01 Available from: https://www.ntsb.gov/investigations/AccidentReports/Pages/highway.aspx
[6] National Transportation Safety Board 2020 *Tesla Crash Investigation Yields Nine NTSB Safety Recommendations* [Internet] Washington, D.C., USA: National Transportation Safety Board [cited 2022 Feb 1] Available from: https://www.ntsb.gov/news/press-releases/Pages/NR20200225.aspx
[7] Lützhöft M, Hynnekleiv A, Earthy JV and Petersen ES 2019 Human-centred maritime autonomy - An ethnography of the future *J. Phys. Conf. Ser.* **1357** 012032
[8] Sharma A and Kim T 2021 Exploring technical and non-technical competencies of navigators for autonomous shipping *Marit. Policy. Manag.* DOI: 10.1080/03088839.2021.1914874
[9] Saha R 2021 Mapping competence requirements for future shore control center operators *Marit. Policy Manag.* DOI: 10.1080/03088839.2021.1930224
[10] Hannaford E and Hassel EV 2021 Risks and Benefits of Crew Reduction and/or Removal with Increased Automation on the Ship Operator: A Licensed Deck Officer's Perspective *Appl. Sci.* **11** 3569
[11] Lützhöft MH and Dekker SWA 2002 On Your Watch: Automation on the Bridge *J. Navig.* **55** 83–96
[12] Hutchins E 1995 *Cognition in the Wild* (Cambridge, MA, USA: The MIT Press)
[13] McCarthy J 2017 *What is Artificial Intelligence?* [Internet] Stanford, CA, USA: Stanford University [cited 2021 Sep 27] Available from: http://jmc.stanford.edu/articles/whatisai/whatisai.pdf
[14] Gjærum VB, Strümke I, Alsos OA and Lekkas AM 2021 Explaining a Deep Reinforcement Learning Docking Agent Using Linear Model Trees with User Adapted Visualization *J. Mar. Sci. Eng.* **9** 1178

[15] IMO 2018 *IMO takes first steps to address autonomous ships* [Internet] London, UK: International Maritime Organization [cited 2020 Sep 9] Available from: http://www.imo.org/en/MediaCentre/PressBriefings/Pages/08-MSC-99-MASS-scoping.aspx
[16] IMO 2021 *Outcome of the Regulatory Scoping Exercise for the Use of Maritime Autonomous Surface Ships (MASS)* [Internet] London, UK: International Maritime Organization [cited 2021 Sep 21] Report No.: MSC.1/Circ.1638 Available from: https://www.imo.org/en/MediaCentre/PressBriefings/pages/MASSRSE2021.aspx
[17] Littman ML, Ajunwa I, Berger G, Boutilier C, Currie M, Doshi-Velez F, Hadfield G, Horowitz MC, Isbell C, Kitano H, et al 2021 *Gathering Strength, Gathering Storms: The One Hundred Year Study on Artificial Intelligence (AI100) 2021 Study Panel Report* [Internet]. Stanford, CA, USA: Stanford University [cited 2021 Sep 20] Available from: Doc: http://ai100.stanford.edu/2021-report
[18] Kongsberg 2020 *Ferry solutions: Automated operations improve safety and efficiency* [Internet] Kongsberg Maritime AS [cited 2021 Oct 12] Available from: https://www.kongsberg.com/maritime/solutions/pax/ferry/
[19] Yin RK 2018 *Case study research and applications: Design and methods* (Thousand Oaks, CA, USA: SAGE Publications, Inc.)
[20] Leximancer 2021 *Leximancer Version 5* Leximancer Pty Ltd
[21] Leximancer 2021 *Leximancer User Guide (Release 5.0)* [Internet] Leximancer Pty Ltd [cited 2021 Nov 8] Available from: https://static1.squarespace.com/static/5e26633cfcf7d67bbd350a7f/t/61a846dd28d1c55296d527b4/1638418188623/LeximancerUserGuide5.pdf
[22] Leximancer 2021 *Sense Making - The Goal of Any Leximancer Analysis* [Internet] Leximancer Pty Ltd [cited 2022 Feb 1] Available from: https://www.leximancer.com/blog/sense-making-the-goal-of-any-leximancer-analysis
[23] Dumais ST 2004 Latent semantic analysis *Annu. Rev. Inf. Sci. Technol.* **38** 188–230
[24] Glaser BG and Strauss AL 1999 *Discovery of grounded theory: Strategies for qualitative research* (Hawthorne, NY, USA: Aldine de Gruyter)
[25] Morse JM, Bowers BJ, Charmaz K, Corbin J, Clarke AE and Stern PN 2009 *Developing grounded theory: The second generation* (Walnut Creek, CA, USA: Left Coast Press Inc.)
[26] Corbin J and Strauss A 2015 *Basics of qualitative research: Techniques and procedures for developing grounded theory* (Thousand Oaks, CA, USA: SAGE Publications, Inc.)
[27] NVivo 2020 *NVivo Release 1.0* QSR International
[28] Lakoff G and Johnson M 2003 *Metaphors We Live By* (London: University of Chicago Press)
[29] Rødseth ØJ 2017 *Definitions for Autonomous Merchant Ships* Norwegian Forum for Autonomous Ships
[30] Rødseth ØJ, Wennersberg LAL and Nordahl H 2021 Towards approval of autonomous ship systems by their operational envelope *J. Mar. Sci. Technol.* DOI: 10.1007/s00773-021-00815-z
[31] Hollnagel E and Woods DD 1983 Cognitive Systems Engineering: New wine in new bottles *Int. J. Man-Mach. Stud.* **18** 583–600
[32] Rahwan I, Cebrian M, Obradovich N, Bongard J, Bonnefon J-F, Breazeal C, Crandall JW, Christakis NA, Couzin ID, Jackson MO, et al 2019 Machine behaviour *Nature* **568** 477–86
[33] Leveson NG, Dulac N, Zipkin D, Cutcher-Gershenfeld J, Carroll J and Barrett B 2006 Engineering Resilience into Safety-Critical Systems ed E Hollnagel, DD Woods, et al. *Resilience Engineering: Concepts and Precepts* (Boca Raton, FL, USA: CRC Press, Taylor & Francis Group) pp 77–94
[34] Utne IB, Rokseth B, Sørensen AJ and Vinnem JE 2020 Towards supervisory risk control of autonomous ships *Reliab. Eng. Syst. Saf.* **196** 106757
[35] Valdez Banda OA, Kannos S, Goerlandt F, van Gelder PHAJM, Bergström M and Kujala P 2019 A systemic hazard analysis and management process for the concept design phase of an autonomous vessel *Reliab. Eng. Syst. Saf.* **191** 106584

[36] Wróbel K, Montewka J and Kujala P 2018 System-theoretic approach to safety of remotely-controlled merchant vessel *Ocean Eng*. **152** 334–45
[37] Zhou X-Y, Liu Z-J, Wang F-W and Wu Z-L 2021 A system-theoretic approach to safety and security co-analysis of autonomous ships *Ocean Eng*. **222** 108569

milliAmpere: An Autonomous Ferry Prototype

Edmund F Brekke[1], Egil Eide, Bjørn-Olav H Eriksen, Erik F Wilthil, Morten Breivik, Even Skjellaug, Øystein K Helgesen, Anastasios M. Lekkas, Andreas B Martinsen, Emil H. Thyri, Tobias Torben, Erik Veitch, Ole A Alsos, Tor Arne Johansen

[1] Corresponding author, e-mail: edmund.brekke@ntnu.no, affiliation/address: NTNU, O. S. Bragstads plass 2D, 7034 Trondheim, Norway. All authors were staff or students at NTNU while performing the work reported in this paper.

Abstract. In this paper, we summarize the experiences with the autonomous passenger ferry development prototype milliAmpere, which has been used as a test platform in several research projects at the Norwegian University of Science and Technology (NTNU) since 2017. New algorithms for motion planning, motion control, collision avoidance, docking, multi-target tracking and localization have been developed and verified in full-scale experiments with milliAmpere. The infrastructure surrounding milliAmpere includes several sensor rigs supporting research on multi-sensor fusion and situational awareness, and a shore control lab which can be used to study the interaction between human operators and the autonomous ferry. Building upon the experiences with milliAmpere, the full-scale autonomous ferry milliAmpere2 was recently launched.

1. Introduction

Among the many potential use cases of autonomy on the sea surface, autonomous ferries represent both a unique opportunity and a unique challenge. Typically, a ferry will only be operating in a limited area, so that the tasks that the autonomy must handle can be more restricted than for a vessel designed operate in more diverse environments. For example, there exists several cable ferries, which could be replaced with autonomous ferries doing exactly the same tasks but without the cable. However, autonomous ferries will often have to operate in environments characterized by heavy and unpredictable maritime traffic. Since they will be used for passenger transport, safety requirements must be high. Therefore, systems for sensor fusion, collision avoidance, navigation and docking must be reliable and of high precision. This also makes autonomous ferries an ideal test case for the development of autonomous surface vehicle (ASV) technology.

Autonomous ferries have received increasing attention during the last 5 years [19], [24]. In 2018, the car ferry Falco performed a fully autonomous transit between Parainen and Nauvo in Finland, using technology by Rolls Royce Commercial Marine, since 2019 a part of Kongsberg Maritime. Since then, a handful of Norwegian ferries operated by Torghatten and Fjord1 have started using auto-crossing functionality in regular operations. The Roboat project [37] aims at developing a fleet of autonomous vessels for transportation and constructing dynamic floating infrastructure such as bridges in the city of

Amsterdam. A prototype was demonstrated in October 2021. Also in the Netherlands, an autonomous water taxi called the Greycraft was demonstrated during 2021 [8].

At the Norwegian University for Science and Technology (NTNU), the autonomous passenger ferry milliAmpere was constructed in 2017, as a half-scale prototype ferry. Subsequently, the full-scale ferry milliAmpere2 was constructed in 2020. The milliAmpere, henceforth known as milliAmpere1, has gradually been equipped with increasingly advanced elements of autonomy, and it has been used extensively in research on marine autonomy by MSc and PhD students at NTNU.

The purpose of the present paper is to give an overview of milliAmpere1, its autonomy systems and research results that have involved milliAmpere1. In Section 2, we describe the background and motivation of milliAmpere1. Section 3 gives a technical description of the ferry. Sections 4 and 5 summarize research results in motion control and automated situational awareness. The Shore Control Lab is described in Section 6, followed by milliAmpere2 in Section 7. A brief conclusion follows in Section 8.

2. Background and motivation

Trondheim, where the largest campuses of NTNU are situated, is one of many cities in the world that is located at a river outlet. The river runs around the city centre, making it a peninsula. Furthermore, the city centre is divided into a southern part ("Midtbyen") and a northern part ("Brattøra") by a canal connected to the river, known as "Kanalen". Municipal plans to build an additional bridge between Brattøra and Midtbyen near the historical fish market "Ravnkloa" were announced in 2016. Such a bridge will become an obstacle for boat traffic in Kanalen. As a response to this, researchers at NTNU began to explore the concept of making an autonomous passenger ferry as a potentially more cost-effective alternative with a smaller environmental footprint. The milliAmpere1 was conceived as a prototype for the future autonomous ferry intended to operate between Ravnkloa and Brattøra. A picture of milliAmpere1, in front of historical buildings along Kanalen, is displayed in Figure 1.

Figure 1. The operational environment of milliAmpere1 (left) and a picture of the ferry (right).

3. The ferry

The milliAmpere1 ferry is built as a mono-hull vessel using aluminium as construction material. The 5 meter long and 2.8 meter wide ferry is capable of carrying 6 persons, although it is not certified for commercial passenger transportation. Instead, the ferry has been built to serve as a platform for developing and testing system components such as motion control system, autonomy system, and various sensor configurations. Figure 2 shows an illustration of milliAmpere1 with the sensors located on the roof. Table 1 contains the major technical data for the ferry. Key components of the system architecture of milliAmpere1 are visualized in Figure 3.

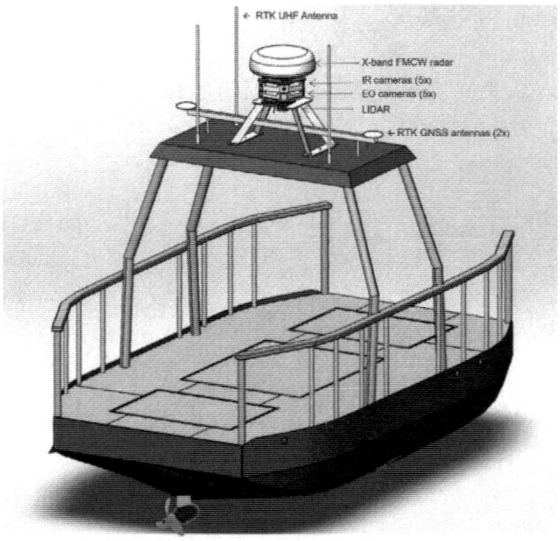

Figure 2. The milliAmpere1 hull with sensor rig mounted on top of the roof. Batteries, chargers, thruster drives, and azimuth servos are located below the deck. Computers, sensors and other electronics are located below the roof.

Table 1. Technical specifications for milliAmpere1.

milliAmpere1	Technical specifications
Length (LOA)	5.0 m
Beam	2.8 m
Draught	0.2 m
Air draught	3.3 m
Light weight	1.8 tons
Max passengers	6
Propulsion	2 azimuth thrusters (2kW each)
Operation speed	3 knots
Max speed	5 knots
Energy	Electric, 24V DC system
Batteries	Lead-Acid VRL, 24 kWh
Navigation sensors	RTK GNSS-compass, IMU
SITAW sensors	IR/EO cameras, X-band radar, lidar

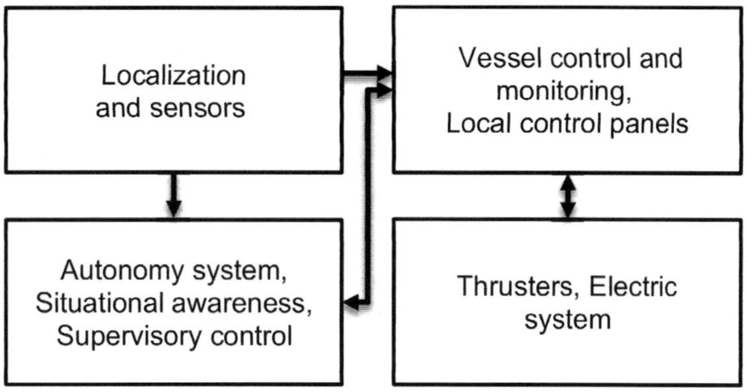

Figure 3: System architecture of milliAmpere1.

The propulsion is provided by two electrical azimuth thrusters located at each end of the hull. This allows the ferry to be highly maneuverable either by using joystick steering from a remote control unit or by the autonomy system. A 24V Lead-Acid (VRL) battery bank with 24 kWh capacity provides electrical power for the propulsion and the on board computers allowing the ferry to operate 6 hours without charging.

We categorize the sensors of milliAmpere1 as proprioceptive and exteroceptive. The former are used to provide information about the vessel itself. These include a Real Time Kinematic (RTK) Global Navigation Satellite System (GNSS) receiver consisting of two antennas together with an internal gyro stabilizer, and a separate Inertial Measurement Unit (IMU), which has integrated GNSS-receiver, magnetometer and barometer. The exteroceptive sensors are used to provide information about the environment around the vessel. All the exteroceptive sensors on milliAmpere1 are placed above the roof, midway between aft and fore, and port and starboard. These include a marine Frequency-Modulated Continuous-Wave (FMCW) X-band radar, a Velodyne VLP16 lidar, and a custom-made sensor rig containing 5 optical Point Grey Blackfly cameras and 5 infrared FLIR Boson cameras.

4. Motion planning and control

Motion planning is an integral part of an autonomous ferry, where the objective of the motion planning is to ensure collision free maneuvering from dock to dock in compliance with the relevant regulations, while ensuring safety, efficiency and passenger comfort. Furthermore, the motion planning relies on a precise vessel motion control system to realize its intentions.

4.1. Motion control system

Some common motion control functionality has been implemented on millliAmpere1 to enable rapid testing of high-level functionality. The motion control system is implemented using the Robot Operating System (ROS), where components are divided into atomic nodes which are connected using publish/subscribe communication. A functional overview of the main components in the motion control system is given in Figure 4. Actuator control nodes provide functionality for automatic control of thruster speed and azimuth angle. The thruster drives are given speed setpoints from the motion control computer, while the azimuth servos are operated in closed-loop with an angle encoder providing azimuth angle feedback and a proportional-derivative (PD) controller driving the servo motor speed. The mapping from desired control forces to thruster speeds and angles are achieved by the thrust allocation node. A novel thrust allocation algorithm has been developed for double-ended ferries which is able to provide a quick solution to the nonlinear control allocation problem [32].

The dynamic positioning (DP) node takes as input the desired and actual pose of the ferry, and issues control force commands to the thrust allocation node in order to track the desired pose reference signal. The DP controller is implemented using a nonlinear proportional-integral-derivative (PID) controller with reference feedforward [23], and a model of milliAmpere1 that was obtained through optimization-based system identification [18]. This controller is capable of tracking a desired trajectory, in addition to traditional stationkeeping. The ferry also supports manual control through a handheld RC controller which controls thrust forces via two joysticks. The milliAmpere1 has a flat keel, which gives the ferry a high degree of maneuverability, but also presents challenges from a tracking control viewpoint [31].

The navigation system estimates the pose of the ferry. This is achieved by an RTK-GNSS receiver with two antennas providing high-precision position and heading. In addition, the ferry is equipped with an IMU providing linear acceleration and angular velocity measurements. The navigation system combines these measurements in an alpha-beta filter to produce smooth and accurate six degree of freedom pose estimates [4].

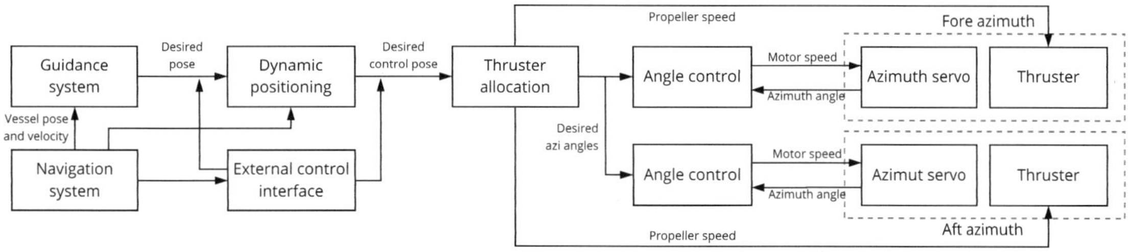

Figure 4: The milliAmpere1 motion control system.

The guidance system takes as input a setpoint for position and heading and produces smooth reference signals for pose, velocity and acceleration. This is achieved by a third-order reference filter with velocity and acceleration saturation. The motion control system is also interfaced through an external control interface which supports both direct force control, and specification of desired pose, velocity and acceleration for the dynamic positioning node.

In [16], a hybrid approach combining reinforcement learning and model predictive control was developed and tested on the milliAmpere1 while performing the four-corner test. The method also demonstrated fault-tolerant capabilities, when an unplanned azimuth thruster failure occurred during testing.

4.2. Planning and collision avoidance

Several motion planning and collision avoidance methods have been developed for autonomous ferry operation and tested on milliAmpere1. In this section, we present three of these methods.

One of the earliest approaches to collision avoidance specifically designed for milliAmpere1 and autonomous ferries is the Single Path Velocity Planner (SP-VP) algorithm [27][28]. The SP-VP algorithm is based on path-time decomposition. In this concept, the ferry is constrained to be located somewhere on a fixed nominal path. Given this nominal path, obstacles parameterized as moving polygons in the Euclidean space can be transformed into the path-time space. A search tree spanning the path-time space, with the current ferry position as the root node and the end of the nominal path as the goal node is then constructed. Collision avoidance is ensured by requiring the edges in the search tree to not intersect with obstacles. Furthermore, a cost dependent on the speed and the closeness to obstacles is associated with each edge, making it possible to find the optimal path-time trajectory using Dijkstra's algorithm. Finally, the optimal path-time trajectory is transformed to a time-expanded Euclidean space as time-parameterized waypoints that the ferry can track.

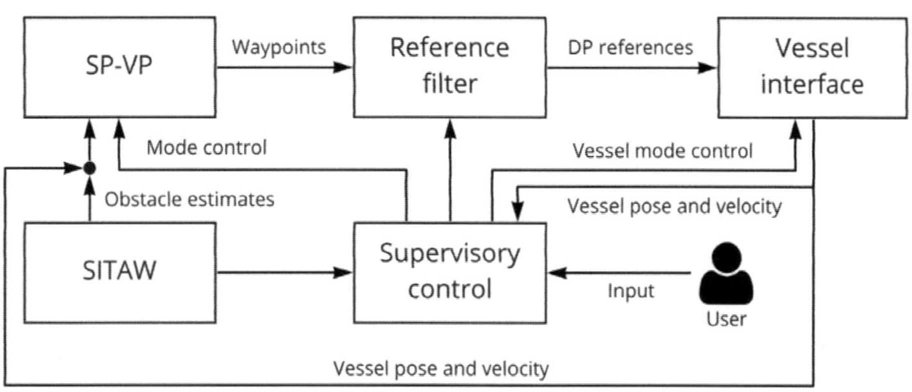

Figure 5: Autonomy system using SP-VP as the core collision avoidance algorithm.

The SP-VP algorithm has been used as part of the basic milliAmpere1 autonomy system (video: www.youtube.com/watch?v=Ry3-yxVaDuE). In this system, the SP-VP algorithm was used together with a tailor-made reference filter to produce continuous pose, velocity and acceleration references for the motion control system. Furthermore, a supervisory control system implemented as a Finite State Machine (FSM) controls the system as a whole to enable autonomous crossing, including docking and undocking. The supervisory control system also implements risk management and several Minimum Risk Conditions (MRCs) serving as fallback solutions in case of unexpected errors. The system can be configured to both run automatically, continuously crossing the canal back and forth, and operate on demand using a button panel to start the ferry crossing.

While the SP-VP is an effective method for collision-free transit across narrow canal-like areas, it does not scale well to a general ferry transit use case. All autonomous vessels must adhere to the maneuvering principles of the International Regulations for Preventing Collision at Sea (COLREGs) when in sight of another vessel, hereafter denoted target ships. This set of rules holds course change maneuvers over speed change maneuvers as the preferred action to avoid collision, and therefore, the fixed path of the SP-VP method makes adherence to these rules problematic. To improve upon this, a more reactive method was developed. The concept was first described in [29], where it was showed that control barrier functions (CBFs) along with encounter-type-specific target-ship domains allowed for maneuvering compliant with COLREGs rules 13-15 and 17. The method ensures collision avoidance with target ships as follows:

1. Classify each vessel-to-vessel encounter with respect to the COLREGs.
2. Assign rule-specific domains to each target-ship. The domains are designed so that if the ferry maneuvers such that it does not violate the target-ship domain, it is also maneuvering in compliance with the relevant COLREGs rule.
3. Formulate CBFs as a function of the distance to and velocity towards each target-ship domain.
4. Apply the CBFs as inequality constraints in a quadratic program that aims to find a generalized force close to the generalized force dictated by the DP controller.
5. Realize the optimal generalized force through the thrust allocation to ensure collision-free maneuvering.

The method also considers static obstacles in a similar way, by formulating CBFs as a function of the distance to and velocity towards each obstacle. The method was verified through experiments in Kanalen with milliAmpere1 and a leisure vessel as target-ship (Figure 6). In the experiments, the target-ship was tracked by lidar and radar, while static obstacles were considered by a combination of electronic nautical charts and lidar.

Works on motion planning among static obstacles include [7], where a two-stage energy-optimised trajectory planning method under polygonal constraints was developed and tested using milliAmpere1. The first stage involved a hybrid A* search, which resulted in a piecewise-linear and collision-free path that was used to warm-start the optimal control solver for further path refinement under dynamic, kinematic and actuator constraints.

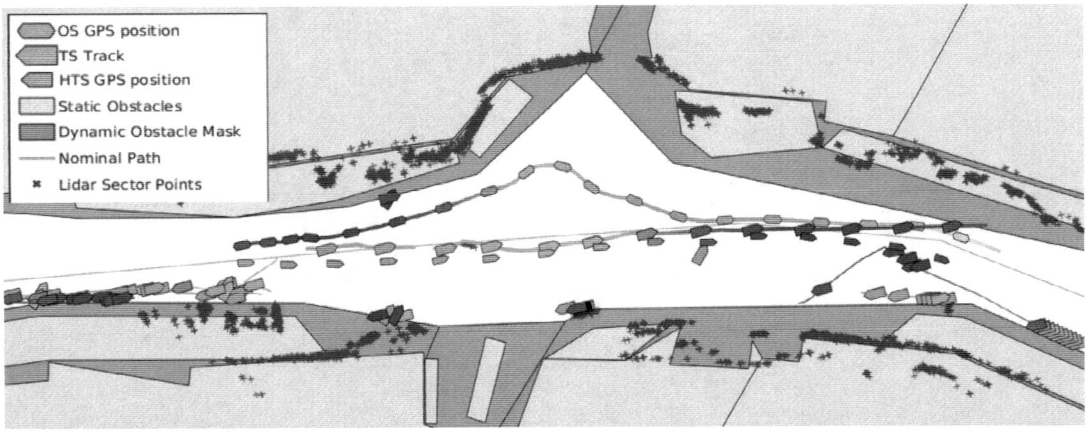

Figure 6: A head-on encounter. The milliAmpere1 tracks a path at the center of the canal and avoids collision by making a starboard maneuver in compliance with COLREGs rule 14. The target ship's tracked position and GNSS position are given by the large and small vessel shapes respectively. The color scale from green to purple represents time. The red crosses give a subset of the lidar points.

4.3. Automated docking

Contrary to motion control scenarios involving path following, trajectory tracking and collision avoidance during the transit phase, automated docking has received less attention in the ASV literature until recently. This is especially the case for works involving field trials and may be attributed to the fact that docking involves intricate maneuvering at low speeds, potentially resulting in high sideslip angles, under the influence of environmental forces. In such conditions, controlling an ASV accurately so as to bring it close to the jetty while avoiding collisions with port infrastructure can be a challenging task even for experienced captains. Moreover, the shape of the ASV has to be considered explicitly to be able to account for collisions between its hull and the infrastructure.

In [14], the finding of docking trajectories was formulated as an optimal control problem (OCP) that incorporates the vessel dynamics and kinematics via a mathematical model, and the harbor and vessel hull layout via appropriate constraints. The OCP trajectory planner was constrained to finding solutions within a static convex set that includes both the ASV in its initial position and the final docking pose.

In [6], this framework was implemented onboard milliAmpere1, with GNSS positioning as the only feedback when maneuvering milliAmpere1 toward the quay from an initial distance of approximately 30m. The previously mentioned PID controller of milliAmpere1 was used to track the trajectories. This docking approach has also been used for undocking in [5]. Then both phases were connected to a transit phase in order to implement a more complete mission.

In [15], the docking algorithm from [6] was further extended to include exteroceptive sensors in the form of lidar and ultrasonic distance sensors. In this way, it was possible to account for unmapped obstacles such as other vessels in the docking area, as well as GNSS positions errors and inaccuracies in the harbor map. An additional contribution was that the convex set, within which the OCP planner computes its solutions, is updated online. Results from the experiments in the Trondheim harbor can be seen in Figure 7, and a video is available at www.youtube.com/watch?v=AyaWlJvI6K8.

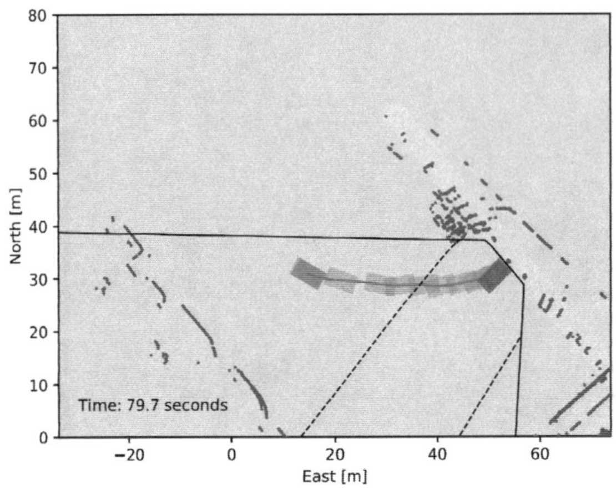

Figure 7: Docking experiments from [15]. The optimization problem plans trajectories within a dynamically-updated convex set. The lidar data (blue cloud) help detect unmapped objects.

5. Situational awareness

Key functionalities for automated situational awareness include detection and tracking of moving obstacles, and localization relative to the surroundings. The latter can be addressed by several approaches, including GNSS-based localization combined with maps, by localization relative to known markers, and by means of Simultaneous Localization And Mapping (SLAM). In this section we summarize the main developments on target tracking and SLAM systems for milliAmpere1.

5.1. Core target tracking system

The basic autonomy system on milliAmpere1 uses RTK-GNSS for navigation, and a combination of radar and lidar for obstacle detection and tracking. Radars have been used in the maritime industry to provide decision support to human navigators long before autonomy, and are especially useful to detect objects far away, at night and in adverse weather conditions such as fog. For these reasons, it is also useful on autonomous vessels. The radar is complemented by a lidar, which is particularly useful for detecting smaller objects that do not have the electromagnetic reflectivity required by a radar. This includes kayaks and small fiberglass/wooden boats, which frequently pass by in urban environments.

The raw sensor data are processed in several steps to extract the information that is used for tracking. In the operational environment of the ferry, most of the data from the sensors are from land-based objects such as buildings and floating piers. In order to remove the reflections from these, the data is transformed to a local Cartesian reference frame. Maps from the Norwegian mapping authority (Kartverket) is used, and only the data from objects on the water is processed further by the detector. This step is highly dependent on the accuracy of the RTK-GNSS. However, in cases with loss of the RTK functionality, the land filtering is still useful to remove most of the detections from land, which reduces the data processing of the subsequent steps. The next step is to cluster point detections from the same target. Points are clustered together via a single linkage clustering algorithm, such that points that are within a defined maximum distance is assumed to originate from the same target. The convex hull of all points in the same cluster is calculated, and this polygon along with its centroid is passed on to the tracker.

Up until this stage, the output of each perception sensor has been processed individually. The tracker fuses measurements from all the available sensors and is based on the integrated probabilistic data association filter (IPDA) [17], which associates the cluster centroids to either existing targets, or creates new targets. It also calculates the existence probability of the target. If the existence probability is sufficiently high, the target is confirmed and used in the motion planning. Tracks with a low existence probability are terminated.

Figure 8: The blue dots are position estimates by means of georeferencing of the ship detections from the image. From the data investigated in [12].

5.2. Active-passive multi-sensor fusion

Research on multi-sensor fusion for milliAmpere1 has focused on measurement-level fusion, where detections from several sensors are fused by a tracking method, as opposed to track-to-track fusion, where separate tracking methods are used for different sensors and then fused. Furthermore, the focus has been on multi-sensor multi-target tracking using variations of the joint IPDA framework. This work is to a large extent based on [10][12], where both radar, lidar, optical cameras and infrared cameras were combined and compared in different fusion configurations (e.g., camera+radar+lidar, pure radar, camera+radar, etc.).

In contrast to the radar and lidar point clouds, the cameras do not supply sensor data in an easily processed format. State-of-the-art techniques for detection and classification in camera images utilize deep learning methods such as convolutional neural networks (CNNs). In the optical camera pipeline used for multi-sensor tracking with milliAmpere1, raw Bayer format images are transmitted via gigabit Ethernet link, converted to color, undistorted to correct for lens distortion (utilizing the camera calibration parameters of each individual camera), and then processed by a CNN such as Yolo v4 to obtain detections (i.e. bounding boxes) as well as class information (e.g. motorboat, kayak, sail boat, etc.). Infrared images are processed in black and white, and otherwise following the same pipeline. The detectors have been trained on images from various littoral environments in Norway.

The passive nature of the imaging sensors only allows detections from images to explicitly encode the direction of the target relative to the camera. This lack of range information can pose a challenge in harbour environments such as milliAmpere1's operational domain because boats moored along the marinas (see top of Figure 8) may flood the system with detections. A solution to this is to use georeferencing, where range is estimated from triangulation by means of the cameras' height above the water [11]. Then, the same land filtering techniques as were used for radar and lidar can also be used to exlude the moored boats. See bottom part of Figure 8.

Notwithstanding these issues, the detection performance of the optical cameras is on level with lidar (above 0.75) and their maximum detection range is also greater. In datasets recorded outside Brattøra in the Trondheimsfjord, an area without moored boats, fusing lidar with optical bearing measurements yielded several performance benefits such as increased tracking accuracy and reduced periods of time when a target was untracked.

Another custom-made sensor rig has also been made for research on stereo vision for milliAmpere1. This rig has two point grey cameras with a baseline distance of 1.75m, enabling reliable distance estimation for vessels up to about 80m away from the ferry [3] [25].

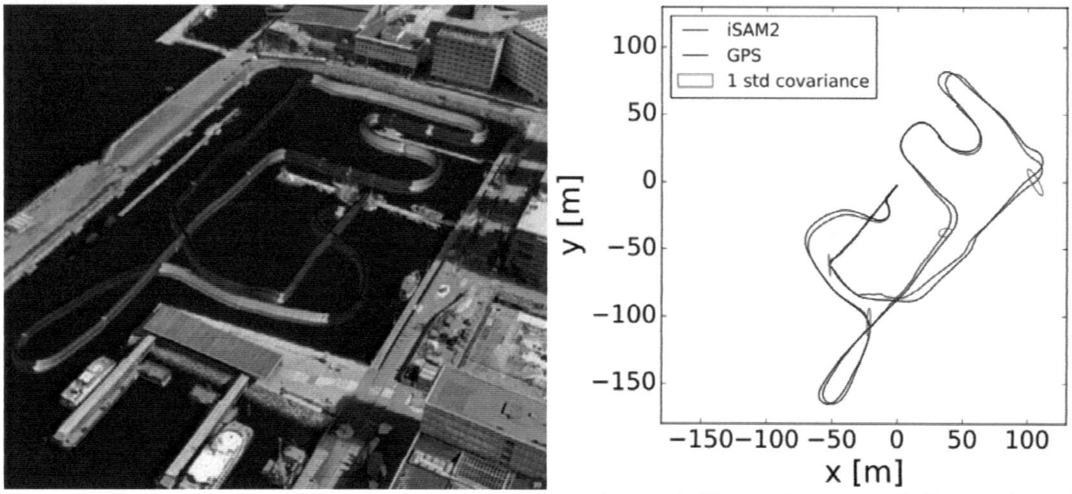

Figure 9: Trajectory of milliAmpere1 during SLAM experiments in Brattørabassenget (left), together with experimental results (right) [22].

5.3. Lidar-based SLAM

Research on SLAM for milliAmpere1 has focused on the lidar. A main advantage of the lidar compared to other sensors is that it both provides high resolution (in common with cameras) and range information (in common with radar) without any need for intrinsic calibration (in contrast to stereo vision). The most mature results so far on SLAM for milliAmpere1 were reported in [22], where a feature-based factor graph solution was used, building upon Incremental Smoothing And Mapping 2 (ISAM2) [13]. Factor graphs is a convenient approach to SLAM, both because they facilitate efficient computational techniques such as ISAM2, and because they are highly flexible, allowing the user to define suitable factors for different information sources such as IMU, GNSS and loop-closure.

In feature-based SLAM it is crucial to be able to extract reliable and stable features from the lidar data, so that the SLAM system can estimate own motion by observing the relative motion of the particular feature points over multiple scans. In this work, the Intrinsic Shape Signatures (ISS) feature extractor was used to extract features, while the Signatures of Histograms of OrienTations (SHOT) descriptor was used to recognize the features between the different keyframes.

The SLAM system was initially implemented as a pure odometry system, which only estimated motion between successive sensor scans. A detailed description of this system, including a comparison of several feature extractors and descriptors, can be found in [21].

The full SLAM system integrates IMU, RTK-GNSS and loop-closure in the factor graph to reduce the drift and allow for more accurate positioning. To enable real-time performance, keyframes are used, and features are only added to the map when milliAmpere1 has moved more than 3 meters since the last keyframe. In this map, the SHOT descriptor is also stored to allow for quicker loop closure detection. To detect the loop closures, a SHOT descriptor is first used, before RANSAC is used to ensure no false positives are detected.

A trajectory used for validation of the lidar-based SLAM system can be seen in Figure 9. This trajectory is approximately 1060 meters and was recorded over 636 seconds.

Results of the system can be seen in the right-hand panel of Figure 9. For this run, the RTK-GNSS signals was disconnected from the system after 323 seconds. The SLAM system is then running with the lidar odometry, the IMU measurements, as well as the loop closure detection ability for the final 323 seconds. During this run, the loop closure is detected at the dock, which is the start and end point, and the whole trajectory. As seen in the figure, the full trajectory is within the 1 std covariance. The average 2D positional error for the second half of the run is 2.3 meters, while the end-to-end error is 0.9m.

5.4. Fiducial SLAM
Non-GNSS localization is especially useful during docking scenarios, when high-precision navigation relative to a floating quay is needed. Due to tidal water the exact location of the jetty may be slowly changing and cannot be pre-programmed. Furthermore, if the lidar is placed on the top of the vessel, as is the case for milliAmpere1, the lidar may not be able to detect the jetty, neither at long nor at short range. An alternative can then be to place fiducial markers such as april-tags at known locations on the jetty. The pose of the ferry relative to the markers can then be estimated by means of its optical cameras. For milliAmpere1, this approach to localization has been investigated in [9]. Again, the ISAM2 framework was used as a back-end, while the segmentation-based AprilTag detector of [36] was used in the front-end. A similar investigation, utilizing stereo vision and machine learning (Yolo v3), was reported in [34].

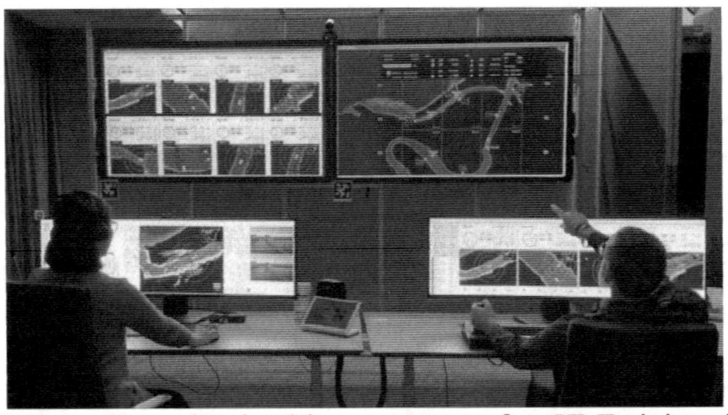

Figure 10: The NTNU Shore Control Lab with a prototype of an HMI giving situation awareness and automation transparency to operators.

6. Shore control lab
The NTNU Shore Control Lab is a testing infrastructure built for the purpose of investigating land-based operation of autonomous vessels. The aim of developing autonomous vessels is not to remove human operators altogether – rather, the intention is to coordinate the actions of the autonomy system and operators. The human remains "in the loop," and is always able to take preventive action if needed. The shore control concept is based on three underlying principles: (1) the operator always understands what the autonomy is "thinking," (2) the operator always has sufficient situation awareness to be able to take over control at a moment's notice, and (3) the operator can turn on and off the autonomy at any time. The physical layout and instrumentation of the NTNU Shore Control Lab is based on resilience engineering and human-centered design methods [33]. Details on the lab's research aims and its integration with the rest of the NTNU autonomous ferry infrastructure are presented in [1].

The milliAmpere1 served as a catalyst in the design of the shore control infrastructure, making abstract requirements become specific requirements for land-based supervision. At the time of writing, both milliAmpere1 and its successor milliAmpere2, are being connected to the NTNU Shore Control Center. From the three principles of shore control needs listed above, data transmission was a key design requirement. The vessels transmit data to the Shore Control Lab via a private 5G network with a bandwidth of approximately 10 Gbps. A rooftop base station receives the signals relayed from the vessels. This includes all four mast-head sensors (360-degree optical and infrared video, radar, and lidar), as well as conditional monitoring information related to the power and propulsion system. Navigation information such as GNSS position, speed over ground, and heading is also relayed. Upon entering the control center, video data is converted into the AVoIP protocol for managing the various inputs. A network-connected DP joystick controls the system aboard ferry. Two-way audio transmitted by radio-over-IP will also be enabled for communication between the ferries. The cyber-security of the network is described in [2].

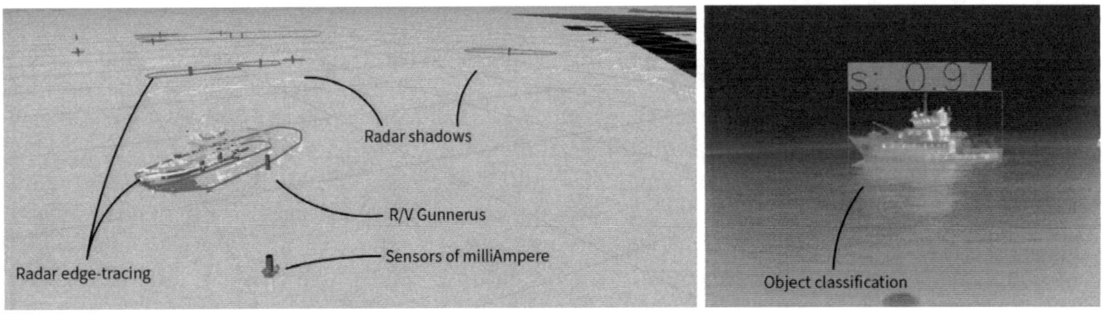

Figure 11: Sensor fusion of radar and lidar (left) and infrared camera (right), which can support human situation awareness and decision making at the NTNU Shore Control Lab.

Some of the questions for investigation at the NTNU Shore Control Lab include:
1. How can we display sensor data in a way that enhances automation transparency?
2. How do we display sensor data in a way that maintains situation awareness?
3. How do we design tasks, procedures, and alarms that ensure that the human controller always knows when to take preventive action?

The sensor data, when transmitted from the ferries to the Shore Control Lab and displayed in a human-computer interface, can be used to test the extent to which human situation awareness is supported. Critical in this regard is maximum response time, T_{MR}. Maximum response time is the amount of time it takes to detect a critical situation, take over control, and gain sufficient situation awareness to appropriately handle a given situation. Besides serving as a core measure in the approval of MASS designs [20], T_{MR} is also an objective measure for comparative assessment of different interface designs.

In such efforts as these, the use of a simulator is crucial, as it enables controlled and repeatable experiments. For this reason, a custom simulator featuring the vessels has been developed in Unity based on the open-source Gemini platform described by [35]. Researchers can build their own traffic scenarios and run trials with participants, varying interface variables and measure the effect on T_{MR}. Collection of operators' biometric data during simulator testing will also allow insights into several important human factors with sound theoretical precedents. Biometric data and their human factor proxies include eye-tracking (visual attention), pupil dilation (cognitive load), as well as heart-rate variability and galvanic skin response (stress). This can allow objective measures that collectively shed light on the extent to which the "traffic picture" can be re-created digitally for interpretation in the mind of an operator. A multi-disciplinary approach is key to such testing efforts, combining randomized human trials experiments and methods traditionally associated with the quantitative behavioural sciences with those disciplines that develop autonomous functionality. The contribution is better design of the human-computer interface (HMI) for autonomous ship systems in the short term, and safer, more robust systems in the long term.

7. milliAmpere2

The full-scale pedestrian ferry milliAmpere2 was launched in June 2021, and has since then completed several tests, including continuous autonomous operation in excess of three hours and autonomous mode with passengers. While milliAmpere1 was designed to enable research on autonomy, the design of milliAmpere2 is to a larger extent governed by the reliability needed for an autonomous ferry that is to be used in continuous operation. The autonomy system of milliAmpere2 builds upon an industrial DP system, and has extensive redundancies to support fail-safe mechanisms. Furthermore, the fragile configuration of two rotating azimuth thrusters is replaced by a configuration of four azimuth thrusters with limited rotation, which control the ferry through their relative forces. Also for milliAmpere2, all closed-loop experiments have made use of radar-lidar fusion. A hazard analysis relevant to both milliAmpere1 and milliAmpere2 has been conducted by [26].

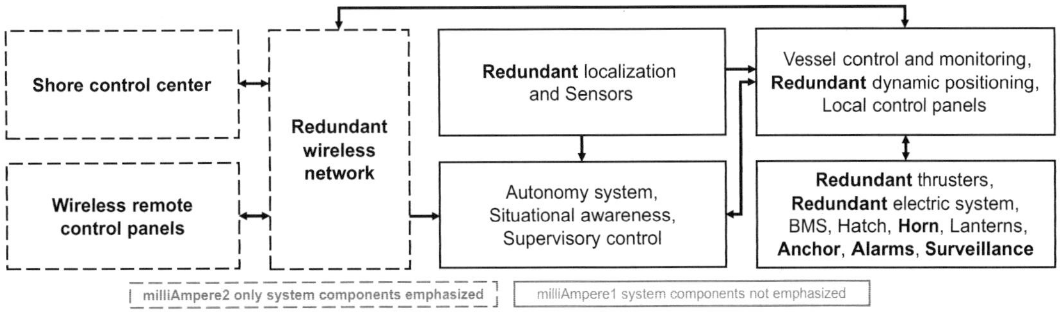

Figure 12: System architecture of milliAmpere2.

8. Conclusions

In this article, we have looked upon several elements of the autonomy systems of the autonomous ferry prototype milliAmpere1. Ultimately, autonomy demands levels of safety that make the system trustworthy. The comprehensive body of experimental research results that milliAmpere1 has been a part of will be valuable when future autonomy systems for other autonomous ship applications like passenger transport are to be designed according to strict safety requirements.

9. References

[1] Alsos O A, Veitch E, Pantelatos L, Vasstein K, Eide E, Peterman F and Breivik M 2022 NTNU Shore Control Lab: Designing shore control centres in the age of autonomous ships J. Phys.: Conf. Ser. Forthcoming 2022

[2] Amro A, Gkioulos V and Katsikas S 2021 Communication architecture for autonomous passenger ship Proc. Institution of Mechanical Engineers Part O: J. of Risk and Reliability

[3] Auestad K 2021 Depth Estimation and Object Detection using Stereo Vision for Autonomous Ferry [Master's thesis] (Trondheim: NTNU)

[4] Bar-Shalom Y, Li X-R and Kirubarajan T 2001 Estimation with Applications to Tracking and Navigation (New York, NY, USA: John Wiley & Sons, Inc.)

[5] Bitar G, Eriksen B-O H, Lekkas A M and Breivik M 2021 Three-phase automatic crossing for a passenger ferry with field trials Proc. European Control Conf. (29 June-2 July 2021 Delft, Netherlands: IEEE) pp 2271–7

[6] Bitar G, Martinsen A B, Lekkas A M and Breivik M 2020 Trajectory planning and control for automatic docking of ASVs with full-scale experiments IFAC PapersOnLine 53 pp 14488–94

[7] Bitar G, Martinsen A B, Lekkas A M and Breivik M 2020 Two-stage optimized trajectory planning for ASVs under polygonal obstacle constraints: Theory and experiments IEEE Access 8 pp 199953–69

[8] Dragan O 2021 This Solar-Powered Electric Ferry is the First Maritime Robotaxi in Europe autoevolution [Internet] ImpactLab [cited 3 March 2022] Available from: https://www.impactlab.com/2021/08/04/this-solar-powered-electric-ferry-is-the-first-maritime-robotaxi-in-europe/

[9] Gerhardsen M 2021 Fiducial SLAM for Autonomous Ferry [Master's thesis] Trondheim: NTNU

[10] Helgesen Ø K, Brekke E F, Helgesen H H and Engelhardtsen Ø 2019 Sensor combinations in heterogeneous multi-sensor fusion for maritime target tracking Int. Conf. On Information Fusion (2-5 July 2019 Ottawa, ON, Canada: IEEE)

[11] Helgesen Ø K, Brekke E F, Stahl A and Engelhardtsen Ø 2020 Low altitude georeferencing for imaging sensors in maritime tracking IFAC-PapersOnLine 53 pp 14476–81

[12] Helgesen Ø K, Vasstein K, Brekke E F and Stahl A 2022 Heterogeneous multi-sensor tracking for an autonomous surface vehicle in a littoral environment Ocean Engineering Forthcoming

[13] Kaess M, Johannsson H, Roberts R, Ila V, Leonard J and Dellaert F 2012 ISAM2: Incremental smoothing and mapping using the Bayes Tree Int. Journal of Robotics Research 31 pp 216–235

[14] Martinsen A B, Lekkas A M and Gros S 2019 Autonomous docking using direct optimal control IFAC PapersOnLine 52 pp 97–102

[15] Martinsen A B, Bitar G, Lekkas A M and Gros S 2020 Optimization-based automatic docking and berthing of ASVs using exteroceptive sensors: Theory and experiments IEEE Access 8 pp 204974-86

[16] Martinsen A B, Lekkas A M and Gros S 2022 Reinforcement learning-based NMPC for tracking control of ASVs: Theory and experiments Control Engineering Practice 120 105024

[17] Musicki D, Evans R and Stankovic S 1994 Integrated probabilistic data association IEEE Trans. on Automatic Control 39 pp 1237–1241

[18] Pedersen A 2019 Optimization Based System Identification for the milliAmpere Ferry [Master's thesis] (Trondheim: NTNU)

[19] Reddy N, Zadeh M, Thieme C, Skjetne R, Sørensen A J, Aanondsen S, Breivik M and Eide E 2019 Zero-emission autonomous ferries for urban water transport: Cheaper, cleaner alternative to bridges and manned vessels IEEE Electrification Magazine 7 pp 32-45

[20] Rødseth Ø J, Wennersberg L A L , and Nordahl H 2021 Towards approval of autonomous ship systems by their operational envelope J. of Mar. Sci. Technol.

[21] Skjellaug E, Brekke E F and Stahl A 2020 Feature-based laser odometry for autonomous surface vehicles utilizing the point cloud library Int. Conf. On Information Fusion (6-9 July 2020 Rustenburg, South Africa: IEEE)

[22] Skjellaug E 2020 Feature-Based Lidar SLAM for Autonomous Surface Vehicles Operating in Urban Environments [Master's thesis] (Trondheim: NTNU)

[23] Sørensen A J, Sagatun S and Fossen T I 1996 Design of a dynamic positioning system using model-based control Modeling, Identification and Control 17 pp 135–51

[24] Tannum M and Ulvensøen J 2019 Urban mobility at sea and on waterways in Norway J. Phys.: Conf. Ser. 1357 012018

[25] Theimann L and Olsen T 2020 Stereo Vision for Autonomous Ferry [Master's thesis] (Trondheim: NTNU)

[26] Thieme C, Guo C, Utne I B and Haugen S 2019 Preliminary hazard analysis of a small harbor passenger ferry – results, challenges and further work J. Phys.: Conf. Ser. 1357 012024

[27] Thyri E H 2019 A path-velocity decomposition approach to collision avoidance for autonomous passenger ferries [Master's thesis] (Trondheim: NTNU)

[28] Thyri E H, Breivik M, Lekkas A M 2020 A path-velocity decomposition approach to collision avoidance for autonomous passenger ferries in confined waters IFAC-PapersOnLine 53 pp 14628–35

[29] Thyri E H, Basso E A, Breivik M, Pettersen K Y, Skjetne R and Lekkas A M 2020 Reactive collision avoidance for ASVs based on control barrier functions Conf. on Control Technology and Applications (24-26 August, Montreal, QC, Canada: IEEE) pp 380-7

[30] Thyri E H and Breivik M 2021 A domain-based and reactive COLAV method with a partially COLREGs-compliant domain for ASVs operating in confined waters Field Robotics Forthcoming 2022

[31] Thyri E H, Bitar G and Breivik M 2021 A 3DOF path-following controller for a non-directionally stable vessel with slow thruster dynamics IFAC PapersOnLine 54 pp 288–94

[32] Torben T, Brodtkorb A and Sørensen A J 2020 Control allocation for double-ended ferries with full-scale experimental results Int. J. of Control, Automation and Systems 18 pp 556–63

[33] Veitch E, Kaland T and Alsos O A 2021 Design for resilient human-system interaction in autonomy: The case of a shore control centre for unmanned ships Proc. Design Society 1 1023-32

[34] Volden Ø, Stahl A and Fossen T I 2021 Vision-based positioning system for auto-docking of unmanned surface vehicles (USVs) Int. J. of Intelligent Robotics and Applications 6 pp 86–103

[35] Vasstein K, Brekke E F, Mester R and Eide E 2020 Autoferry Gemini: a real-time simulation platform for electromagnetic radiation sensors on autonomous ships IOP Conf. Ser.: Mater. Sci. Eng. 929 012032
[36] Wang J and Olson E 2016 AprilTag 2: Efficient and robust fiducial detection Int. Conf. Intelligent Robots and Systems (9-14 October, Daejeon, South Korea: IEEE)
[37] Wang W, Shan T, Leoni P, Fernandez-Gutierrez D, Meyers D, Ratti C and Rus D 2020 Roboat II: A novel autonomous surface vessel for urban environments Int. Conf. Intelligent Robots and Systems (24 Oct-24 Jan, Las Vegas, NV, USA: IEEE)

Acknowledgments

The construction of milliAmpere1 and 2 was funded by NTNU. Furthermore, NTNU has funded research activities related to milliAmpere1 and 2 through the Digital transformation project Autoferry. This work was also supported in part by the Research Council of Norway through projects 296694 (aFerry), 223254 (AMOS) and 309230 (SFI Autoship). Several researchers, postdoctors, PhD candidates, MSc students, BSc students, internship students and technical employees have taken part in the development of milliAmpere1 and related research, and the authors are deeply grateful for all their contributions.

NTNU Shore Control Lab: Designing shore control centres in the age of autonomous ships

Ole Andreas Alsos[1], Erik Veitch[1], Leander Pantelatos[1], Kjetil Vasstein[2], Egil Eide[3], Felix-Marcel Petermann[1], and Morten Breivik[2]

[1]*Department of Design*
[2]*Department of Engineering Cybernetics*
[3]*Department of Electronic Systems*
[1,2,3] *Norwegian University of Science and Technology, Trondheim, Norway*

ole.alsos@ntnu.no

Abstract. As highly automated ships become unmanned, their operators will move into shore-based control centers. In the last years, NTNU has built an advanced and flexible research infrastructure for performing research on autonomous ships and the monitoring and control of these. The infrastructure comprises of (1) the *milliAmpere1* and *milliAmpere2*, which are two all-electric autonomous urban passenger ferries equipped with advanced sensors and equipment for autonomous navigation, (2) the *Shore Control Lab*, a flexible shore control center, where operators can monitor and control a fleet of autonomous ships, (3) a lab section for researchers to give instructions to operators in the control center, and to record, observe, and analyze their behavior, (4) an observation room adjacent to the control room for stakeholders to observe ongoing experiments in the control room, (5) the ferry simulator *Autoferry Gemini* that allows researchers to create challenging or high-risk scenarios where operators can be stress tested without being a danger to ship, crew and passengers, (6) the mixed reality lab *MRLAB*, where we can test physical designs of urban autonomous passenger ferries in a virtual environment, and (7) a dock for passenger handling and with inductive charging capabilities. In this paper, we first describe the research infrastructure's purpose and scope of operation, as well as the technical design, physical setup, and equipment. Secondly, we present a roadmap for the development of the research infrastructure to meet the future research challenges for autonomous ships and the supervision and control of these. Thirdly, we present a number of research questions that are going to be explored in the lab in the years to come.

1. Introduction

As highly automated ships are becoming unmanned, their operators are expected to move into shore-based control centers. In these centers, a team of operators, trained in navigation, can monitor the status of a fleet of autonomous ships and take over their control when intervention is necessary. There is much knowledge on how to design control rooms for remote control of ships and drones. However, in the age of self-driving ships, there are still gaps in our understanding of how to design control rooms.

In 2020, we started the process of building the *NTNU Shore Control Lab*, which is a flexible research infrastructure for developing and testing remote monitoring and control of autonomous ships, including different physical setups and human-machine and graphical user interfaces for such control centers. The infrastructure comprises of:

- ***milliAmpere*** (hereafter called milliAmpere1) and its sibling ***milliAmpere2***, two all-electric autonomous urban passenger ferries equipped with advanced sensors and equipment for autonomous navigation and for providing the data necessary for giving operators superior situational awareness
- a flexible **shore control centre** where a number of operators can monitor and control a fleet of autonomous ships, including *milliAmpere1* and *2*
- a **lab section** for researchers to give instructions to operators in the control center, and to record, observe, and analyse their actions and behavior
- An **observation room** for developers, product owners, product managers, and other stakeholders to observe experiments, field tests, and usability evaluations of different control room setups and user interface designs
- *Autoferry Gemini*, a **ship simulator** based on the open-source Gemini platform allows us to create challenging or high-risk scenarios where operators are stress tested and new user interfaces are evaluated without jeopardizing the ship, crew, passengers, or equipment
- a **mixed reality lab** where physical designs of urban autonomous passenger ferries, as well as the passenger experience, can be tested and evaluated in a safe, virtual environment.
- **ferry docks** with induction charging and land-based terminals for passengers.

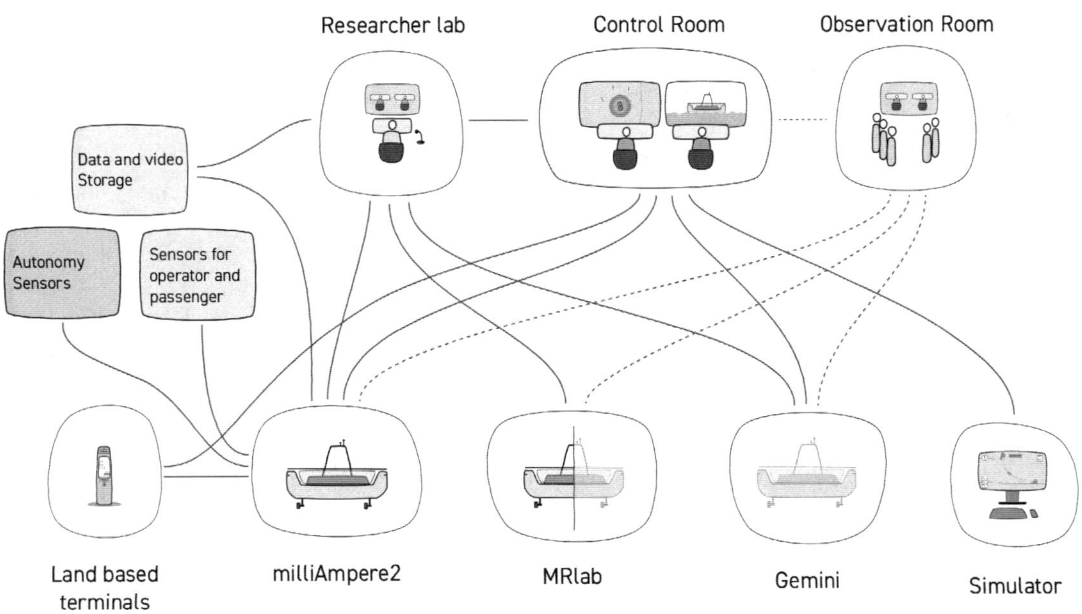

Figure 1: The different parts of the NTNU Shore Control Lab infrastructure and how they are connected

The rest of the paper is organised as follows: In Section 2, we present a review of the current state-of-the-art and trends in remote control centers, both in the maritime domain and other domains, such as vessel traffic service VTS, air traffic control, roads, railways, etc. In Section 3, we describe the research infrastructure, its purpose and scope of operation, as well as a detailed description of the technical design and physical setup. This description also includes the sensors on the ferry, the equipment in the shore control center, and the tools and instruments that the researchers have available in the lab section to measure and analyse the operators' stress levels (e.g. biometrical sensors, eye-tracking, recording equipment). In Section 4, we present a roadmap for the development of the research infrastructure to meet the future research challenges on autonomous ships and the supervision and control of these, together with a number of research questions that are going to be explored in the lab in the years to come.

1.1. State-of-the art in research

Autonomous ships are developing rapidly from concept to creation. Several categories have emerged, including short-sea shipping (e.g., MV Yara Birkeland [1] and SFI AutoShip Project [2]), open-ocean shipping (e.g., MUNIN [3]), urban ferries (e.g., Roboat [4] and milliAmpere [5]), and autonomous surface vessels for data collection [6]. So-called "smart ships" envision a coordinated AI-operator teaming on the bridge [7, 8]. These examples encompass a "Maritime 4.0" research community [9] that combines automation in ships and ports for safer, more effective shipping.

Academic interest in shore control of autonomous ships has increased in tandem with the growth of "Maritime 4.0." In the field of safety and risk science, researchers have examined the role of humans as safety controls in the operation of autonomous ships, applying advanced modelling techniques to predict how critical scenarios like collision avoidance will be coordinated [10-14]. Others have raised the challenges associated with human factors and training requirements of Shore Control Center Operators [15-20]. Meanwhile, applications have expanded from coastal "short-sea" shipping applications to shore control for urban autonomous passenger ferries [21] and inland waterway vessels [22].

In 2021, the International Maritime Organization, the global regulatory agency for international shipping, weighed in. In their much-anticipated "Regulatory scoping exercise for Maritime Autonomous Surface Ships (MASS)," they highlighted where current regulations fall short of addressing the needs for autonomous ship operation [23]. Among the top three "high-priority items," all were related to the Shore Control Center (or the "Remote Control Center" as the IMO calls it) and especially to the roles and responsibilities of the operators working in them. For the autonomous ship as a concept, the coordinating efforts of the Shore Control Center represent its link to real-world operation. In the real world, it is necessary for humans to complete the picture, chaperoning the autonomous ship's conduct in traffic with manned vessels both big and small, maintaining the vessels' integrity preemptively, and appropriately handling unexpected events that no autonomous agent acting alone could adequately address. The Shore Control Lab, in combination with the milliAmpere prototypes, is probably one of the first research platforms dedicated to investigating the relevant questions arising from the vision of safe autonomous ship operation.

1.2. Control rooms in other domains

There is much to be learned from other industries in the field of control rooms embedded in automated systems. In the 1980s, research associated with the digital transformation of control rooms used in the process industries (especially nuclear power plants) addressed issues with how humans interact with technology and led to the inception of whole fields like cognitive engineering [24]. The multi-disciplinary *Computer-Supported Collaborative Work* arose in the 1990s from seminal work focused on line control rooms in the London Underground [25] leading to the use of ethnographic methods in the study of control rooms. Air Traffic Control (ATC) has also been a subject for applied research in safety-critical interaction design [26]. More recently, remote ATCs like the ones in Bodø, Norway [27] and at the German Aerospace Center in Braunschweig are addressing research on the implementation of remote tower centres (ROT). One thing they investigate, e. g. is using additional camera viewing angles and augmenting video feeds with information to enhance the ROT's situational awareness. Additionally, in this ROT, they conduct cognitive work analysis and research with various sensing technologies and interface enhancements [28, 29]. Similar principles to those explored in the aforementioned research study have been utilised to a greater extent in Norway, most notably by the airport operator Avinor, to effectively monitor the airspace in less-operated areas from a distance.

Stanton [30] discuss the design of a future flight deck in which an aircraft has just one pilot who, if required, may get assistance from a ground-based "second pilot" who can operate the aircraft in an emergency like an unmanned aerial vehicle (UAV). This approach is meant to assist in achieving manpower savings while still adhering to the greatest safety requirements. Personnel on the ground must possess the same specialised qualifications as those on an aircraft's flight deck.

Additionally, control rooms are utilised by autonomous underwater vehicles that do a variety of scheduled tasks. Roberts [31], examine the future architecture of submarine control rooms. Consideration is given to factors such as team composition and control room layout, as well as the sensors employed, which were also implemented in a summation environment and serve as a valuable research tool.

Similarly, in the area of powerplants, particularly nuclear powerplants, various research organisations, one of which is the Norwegian Institute for Energy Technology, are examining the design and structure of a control room in more depth (IFE) [32]. Other research organisations are also interested in this subject, and it's worth noting that simulators are often used to study human aspects and the consequences of novel design ideas [33]. This is a very advantageous method for safety-critical situations, as accidents can be fully prevented while new design ideas are being tested. This supports the creation and usage of a simulator in the environment described in this article.

In summary, we may transfer information from other areas to the maritime domain to a certain degree, for example, by using wide-angle camera views, which was researched in aviation. The differences in the technologies used, the dimensions and environmental factors involved, and the knowledge required are too different, which is why shore-based monitoring and control of MASS is a critical research facility for advancing the automation of the Maritime Domain.

1.3. Ongoing projects

The NTNU Shore Control Lab is currently utilised in a number of research projects in collaboration with academic and industry partners, where we have listed some key projects in Table 1.

Table 1: Ongoing research projects utilizing the NTNU Shore Control Lab

Project	Purpose and research aim
Autoferry *2018-2023*	To develop all-electric, autonomous passenger ferries for urban waterways.
SFI Autoship *2019-2027*	A centre for research-based innovation focusing on the development of autonomous ships for safe and sustainable operations.
LOAS 2019-2023	To design and test interaction solutions for “Land-based Operation of Autonomous Ships”
TRUSST *2020-2023*	To develop knowledge and solutions to assure trustworthy, safe and sustainable autonomous passenger transport
MIDAS 2022-2027	To strengthen research and innovation capacity in design and business development for technology companies that develop autonomous solutions for the ocean space
SCAPE *2021-2024*	To develop new innovations that will revitalize urban waterways with zero emission, autonomous passenger ferries
LASH FIRE *2019-2023*	To significantly reduce the risk of fires on board ro-ro ships.
MAS *2021-2025*	To improve the safety level in the petroleum sector through the implementation of human-centered autonomous/digital solutions in safety critical operations

2. Development methodology

After the development of *milliAmpere*1, one of the world’s first prototype of an urban autonomous passenger ferry, the team behind it realised that a more multi-disciplinary approach was needed. The development team, which was a collaboration between the Department of Engineering Cybernetics,

Department of Electronics Systems, and Department of Marine Technology, now contacted the Department of Design, with expertise in human factors, human-centered design, and industrial design, to participate in the development of the successor, *milliAmpere2,* and a shore-based control room to monitor the ferries.

The different parts of the research infrastructure were developed using a range of different methods. The details of the methods used are out of the scope of this paper, but a combination of engineering methods and human centred methods were used. In Table 2, the methods used are listed.

Table 2: The methods used in developing the research infrastructure

milliAmpere1 (2017)	Gemini simulator (2019)	milliAmpere2 (2020)	Shore Control Lab (2021)	MRLAB (2021)	Docs and terminals (2021)
Engineering Goal-based Risk-based approach Nordic boat standard	Software development Game development Interaction design	Engineering Goal-based Risk-based approach Nordic boat standard Human-centered design Industrial design	Human-centered design Prototyping Interaction design	Human-centered design, Prototyping, Software development,	Human-centered design Prototyping Interaction design

3. Research infrastructure

The research infrastructure consists of two main parts. The first part is four prototypes of an urban autonomous passenger ferry and its docks and land-based terminals (see Figure 1):

- **milliAmpere1**, which is described in detail in [5] and not further described here.
- **milliAmpere2**, which is the successor of milliAmpere1.
- **milliAmpere2 simulator**, which is based on the open-source platform Gemini [35].
- **MRlab**, A full scale mixed reality simulator of milliAmpere2.
- **Dock** with inductive charging capabilities and land-based terminals.

The second part of the research infrastructure is the NTNU Shore Control Lab, which is a research facility for the design of shore control centres of the future. The lab consists of four rooms (Figure 2):

- **A reception area** where test participants are welcomed and briefed (not further discussed in this paper)
- **A shore control centre** currently set up for two operators
- **An instructor station** where researchers can instruct and observe the operators during experiments
- **An observation room** where stakeholders and the development team can observe experiments.

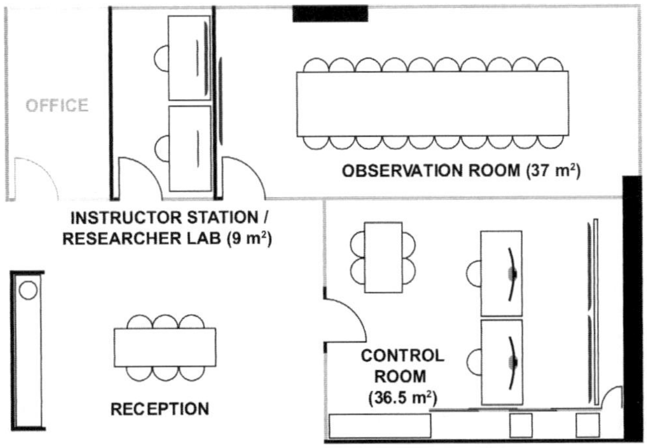

Figure 2: The floor plan of the Shore Control Lab, which consists of a control room, an observation room, an instructor/researcher lab, and a reception area for test participants

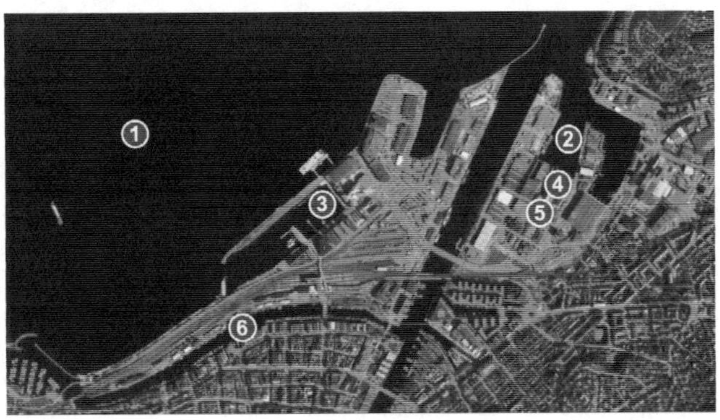

Figure 3: The harbour of Trondheim with (1-3) test areas for autonomous ships, (4) NTNU Shore Control Lab, (5) MRLAB, and (6) milliAmpere2 crossing (Photo: Norgeskart)

The infrastructure is in various locations near the harbour of Trondheim (Figure 3). In the following subsections, we present the details of each of these parts, together with a description of how they are linked together.

3.1. Shore control centre

The control room is designed to let operators (currently two) monitor or control autonomous ships (Figure 4). It can also be used to monitor and control other autonomous or remotely operated systems in various domains, such as ships, drones, or ROVs. The main purpose is to do research related to control room design for autonomous or remote operations and to design innovative solutions for improving the operators' situational awareness.

The main difference between the Shore Control Lab and other remote-control centres is the possibility to run controlled experiments with recordings of sensor data from autonomous ships and lab equipment. It also allows for detailed sound and video recordings of the operators and their interactions with the systems and each other, as well as collecting biometric data from the operator's attention and stress levels. This is further described in the next subsections.

The control room has been designed for flexibility, making it ideal for different setups and experiments as new needs arise. Along the front wall and in the ceiling, there is mounted a rigid grid system that allows for easy mounting or reconfiguring of equipment such as screens, cameras, microphones, etc., as seen in Figure 4. Technical and audiovisual equipment is hidden behind sliding doors. Along the left wall are glass windows with a two-way view to the observation room that can be sealed off with curtains. Care has been given in the design of the room's lighting, acoustics, and cooling to improve the operators' working environment.

The current configuration has two operator stations, each equipped with a 49-inch curved wide screen. On the front wall, two 75-inch touch screens are mounted on a height-adjustable rack so that their working positions can be adjusted both for being viewed at a distance or for touch operation. The operators have access to necessary input devices and a microphone and headset for communicating with crew or passengers onboard the MASS, other ships, VTS or harbour authorities, or the researcher in the instructor station. Every camera or computer display signal can be routed to any screen in any of the rooms in the lab (using AV over IP), adding up to the flexibility of the infrastructure.

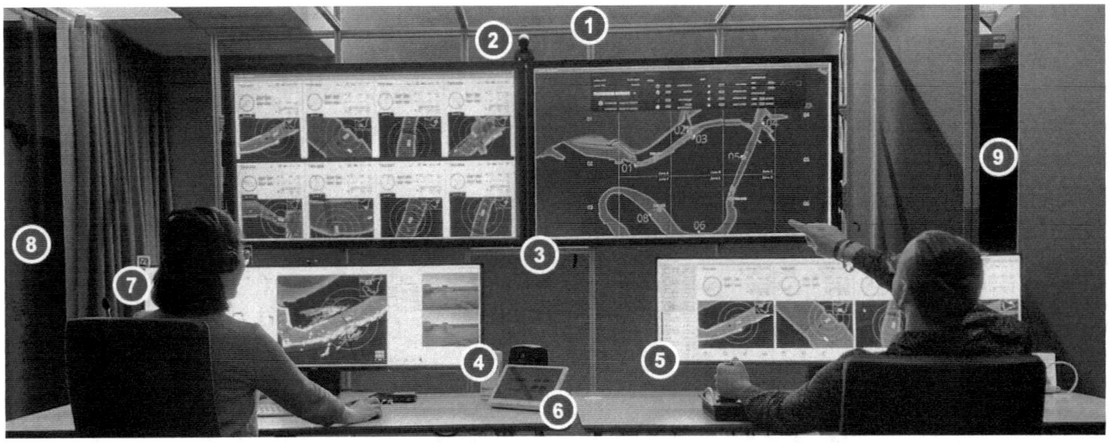

Figure 4: The control room with (1) Aluflex grid for flexibility, (2) cameras, (3) two height adjustable touch screens, (4-5) operator stations, (6) control panel, (7) microphones and loudspeakers, (8) view to observation room, (9) technical equipment (Photo: Erik Veitch)

3.2. Observation Room

The observation room is designed to let researchers, observers, developers, product managers, project managers, and other stakeholders observe experiments or ongoing operations in the control room without disturbing or interfering with them (Figure 5). It can also be used as an emergency room during exercises, real accidents, or extraordinary events.

The observation room seats up to 20 people. It has windows along three of the walls, giving a good view of the harbour area. The room is equipped with a large screen, loudspeakers, microphones, and cameras connected to the system. This allows any video or camera signal in the system to be routed to this screen so that the people in the room can observe what is going on in the lab or on the ship.

Figure 5: The observation room. (Photo: Ole Andreas Alsos)

3.3. Instructor station and researcher lab

The instructor station/researcher lab is designed for researchers to (1) give the operators assignments or scenarios, (2) to monitor experiments or ongoing operations, and (3) to analyse video and sensor data from experiments and evaluations (Figure 6). With the lab, researchers can do a real-time or delayed analysis of the collaboration, individual behavior, and stress level of the operators and discover usability problems in the user interface.

The instructor station/researcher lab is installed in an office space with two workstations. Here, the instructors and researchers have access to one wall-mounted 42-inch screen that shows sensor data and several video streams from the lab. In addition, they have access to a 49-inch widescreen that is connected to the lab computer. The cameras in the control room can be remotely controlled from the lab. It also has an audio system that allows the researchers to communicate with the operators.

Figure 6: The instructor station / researcher lab. (Photo: Nicholas Lund)

3.4. Urban Autonomous Passenger Ferries, milliAmpere 1 & 2

The milliAmpere2 (mA2) is an all-electric, urban autonomous passenger ferry that is also a research platform for research on autonomous ships (Figure 8). It was based on knowledge from its predecessor, milliAmpere1 (mA1), which was built and launched in 2017 [5] (Figure 7). It was developed as a joint project between several departments at NTNU; Department of Design, Department of Electronic Systems, Department of Marine Technology, and Department of Engineering Cybernetics, were all involved in designing and building it. The ferries' main purpose is to investigate research questions related to sensor fusion, autonomy, situation awareness, passenger flow, and trust. Specifications for the two ferries are presented in Table 3.

Figure 7: The all-electric urban autonomous passenger ferry milliAmpere1 (Photo: Egil Eide)

Figure 8: The all-electric urban autonomous passenger ferry milliAmpere2 (Photo: Egil Eide)

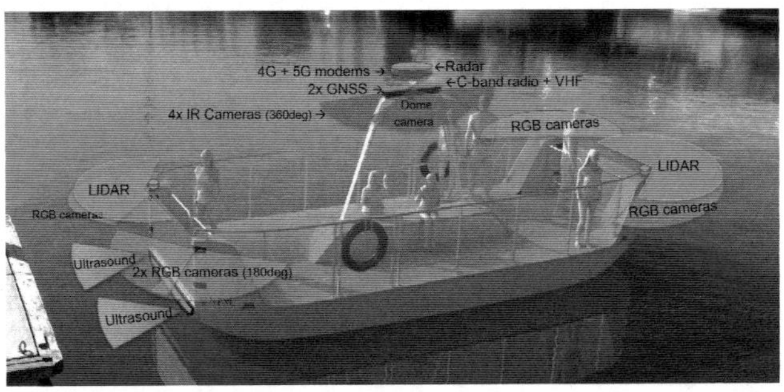

Figure 9: The autonomy sensors onboard milliAmpere2

The milliAmpere2 is equipped with sensors and computer equipment for autonomous operation cameras, ultrasound distance sensors, IR cameras, LIDAR, radar, and GPS (Figure 9). In addition, the ferry is equipped with sensors for operator situation awareness and passenger safety and information, such as a 180-degree dome camera, 3D microphones, loudspeakers, and IMU sensors for registering ferry motion and acceleration. The main purpose of these sensors is to give the operator a feel of the ship so that they can get insight into the ship's state and future intention, in addition to monitor the passengers and surroundings.

The ferry is also equipped with mandatory safety equipment and a user interface for passengers to interact with it. The user interface consists of buttons to start the crossing and to contact the operator, and two touch screens (one on each side of the mast) for information to the passengers.

Table 3: Specifications for milliAmpere1 and milliAmpere2

Specification	milliAmpere1	milliAmpere2
Length (LOA)	5.0 m	8.5 m
Beam	2.8 m	3.50 m
Draught	0.2 m	0.3 m
Air draught	3.3 m	3.5 m
Max pax	6	12
Propulsion	2 azimuth thrusters	4 azimuth thrusters
Operation speed	3 knots	3 knots
Max speed	5 knots	5 knots
Energy	Electric 24V DC	Electric 48V DC
Batteries	Lead-Acid VRL 24 kWh	Lead-Acid VRL 48 kWh
Power	2 x 2 kW	4 x 10 kW
Sensors	IR camera, camera, RADAR, LIDAR	IR camera, camera, RADAR, LIDAR, ultrasonic

3.5. Gemini

Gemini is an open-source project based on the Unity game engine [36] with the purpose of exploring the role game engines have in autonomy, focusing on maritime applications (Figure 10). Due to the lack of maritime alternatives to simulator frameworks such as *Carla* [37] and *AirSim* [38] and the tight integration these have respectively to cars and drones, Gemini went for a toolbox design that allows the designer to develop different applications untied from vehicle specific frameworks. This allows Gemini to be used in various projects.

One of the projects is to simulate the use of *milliAmpere2* and passenger transport on the Trondheim Canal using a simulator referred to as *Autoferry Gemini*. In the Shore Control Lab, Autoferry Gemini has several important purposes: It allows us to (1) test the autonomy system and the sensors [35], actuators, and control systems in a simulated environment, (2) study the relationship between human operators and autonomy [39], (3) expose both the autonomy and the human operators to rare tail events in a safe environment without the risk of people and equipment, and (4) develop, test, and evaluate different human-machine interfaces for monitoring autonomous ships. The former uses several technologies related to video games, high-performance communication, and autonomous systems. Camera modelling is done using Unity's *High-definition render pipeline* designed for real-time high-fidelity simulations. This is also used to model range sensors by attaching *compute shaders* to the pipeline system. These shaders function as general-purpose GPU programs from which sensor models such as radar and lidar are created. Both sensor and generic Unity data are accessible by external programs using *Google's remote procedure calls*. For Autoferry Gemini, this is used, among others, to emulate the sensory system on *milliAmpere2* to test the situational awareness system on the ferry [5].

Journal of Physics: Conference Series **2311** (2022) 012030 doi:10.1088/1742-6596/2311/1/012030

Figure 10: Autoferry Gemini, a simulator for testing and developing HMIs for the operator

In relation to the Gemini simulator, we have started work on a *Scenario Builder* that allows us to quickly create various scenarios that can give the operators missions of varying difficulty and cognitive workload [40]. The same builder can induce rare tail events, where operators are forced to take control of the ferry.

External parties outside NTNU are also using Gemini in their development: (1) The *Njord student challenge*, [41] with the purpose of creating an autonomous ship challenge educating and motivating students in a COVID friendly arrangement. (2) *Zeabuz* simulator [42], with the purpose of verifying their autonomy system for safe operations in urban waterways.

3.6. Mixed Reality Lab

The Mixed Reality Lab (MRLAB) is a 190 m^2 area where we have built two full scale low-fidelity models of the milliAmpere2 superstructure in wood (Figure 11). Its purpose is to test different physical layouts and functionality of urban autonomous passenger ferries in virtual reality. The ferry mockup allows passengers to experience a virtual and tangible ferry trip where they can board the ferry, touch the handrails or mast, and sit on the bench under various weather and traffic conditions. Here, the whole user journey, or parts of it, can be simulated in a highly immersive environment to gather rich feedback from the passengers on their experience and their trust towards the ferry. This allows researchers, industrial designers, and naval architects to test out different physical designs and interaction designs on the ferry and dock in a cheap and risk-free way. Relevant research questions from other parts of the infrastructure can be investigated here in simulations with a high level of realism.

In addition to the two full scale mock-ups of milliAmpere2 mentioned above, the lab consists of VR headsets, hand controllers, simulation computers, and a data collection rig that captures high quality video and sound from the experiments for later analysis.

Figure 11: The MRlab is a mixed reality lab to test different physical layouts and functionality of urban autonomous passenger ferries in virtual reality (Photo: Nicholas Lund).

3.7. Land-based terminals and dock
The ferry docks and land-based terminals are important parts of the research infrastructure that allows for the safe usage of the ferry. The purpose is to do research on (1) induction charging of small-scale ferries, (2) efficient passenger flow on and of small-scale autonomous ferries, (3) passenger interaction with a self-service station for information about the ferry and on-demand booking, and (4) passenger trust in autonomous ferry services.

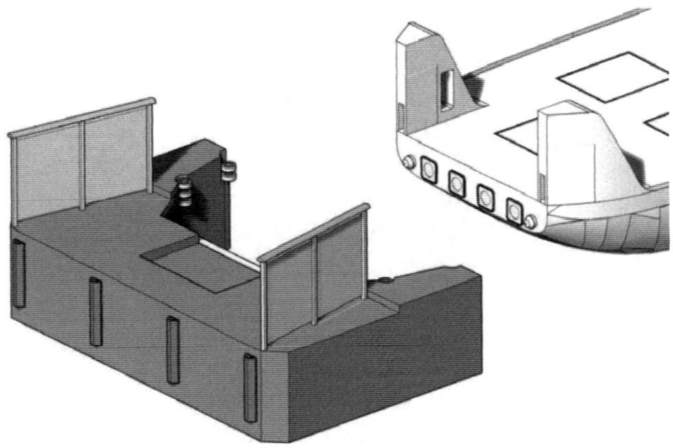

Figure 12: The dock with induction charging (Illustration: Innovation JBA)

The dock, which currently are under construction, consists of two floating docks with automatic induction charging capabilities on the south side (see Figure). In addition, we are currently designing and building a self-service station on each side of the canal where passengers can get information about the ferry service and call it to them if it is located on the other side. A local Global Navigation Satellite Systems (GNSS) reference base station is installed in the area that increases the accuracy for the GNSS position to about 1 cm. Two high-resolution cameras are currently being installed on each side of the canal so that operators can monitor the docks and area where the ferry is operating.

4. Discussion
By combining and connecting different research labs and platforms into a coherent and interconnected research infrastructure, we have created a powerful and flexible basis for investigating a broad range of research challenges related to autonomous ships and the monitoring of these.

4.1. Multidisciplinary
The researchers involved in the investigation of these challenges comes from a broad spectrum of fields and domains, ranging from technology (maritime technology, engineering cybernetics, electronics, cyber security, computer science), design (graphic/information design, interaction design, industrial design), social sciences (cognitive psychology, social anthropology) and economy. This multidisciplinary approach allows us to look at a broader set of research questions and through a range of different lenses. We presume that this form of triangulation will let us conduct research of higher quality.

4.2. Flexibility
The flexibility of the research infrastructure allows for research on more than autonomous passenger ferries. Many other sub fields of autonomous ships and shipping in general can be explored in the labs and on the ferries. For example, we currently conduct activities in the shore control lab in the EU-funded LASH FIRE project, which looks at solutions for fire prevention and firefighting on board

traditional ro-ro ships. In addition, other domains can also utilise the infrastructure. We currently (1) develop design specifications for a space control room for the small-scale satellite HYPSO2 [43], (2) implement human-centered autonomous digital solutions in safety critical operations in the petroleum domain, (3) establish a collaboration with the aquaculture industry on automation and remote operations, and (4) bridge technology, design, and business development in the MIDAS project.

4.3. Education

The infrastructure also serves as an excellent education platform where students (and operators and seafarers) can learn, build, experiment, and test out prototypes, HMIs, software, training procedures, etc. It has already been the basis for several student projects and master theses and is currently utilised in a broad range of courses from several departments, such as *Experts in teams*, *Information visualization*, *prototyping physical user experiences*[1], and many more. Master students and PhD candidates from different fields have used the ferries and the lab, such as engineering cybernetics, industrial design, electronics, marine technology, product development, computer science, and cognitive psychology. This makes the infrastructure a hub for multidisciplinary education and research. It is also the location of the Njord Student Challenge [41], where international student teams design systems to control autonomous ships and compete against other student teams.

4.4. Innovation

The research infrastructure also serves as a platform for innovation and commercialisation. The infrastructure is essential for the research-based innovation centre, SFI AutoShip, and has already resulted in several patents. One example is how the technology from milliAmpere 1 and 2 is now commercialised in the spin-off Zeabuz. Two innovation manager associated with the domain, as well as the local technology transfer office, NTNU TTO, is helping faculty and PhD students to take care of innovation opportunities and intellectual property rights that emerge, and to create innovation into the existing industry.

4.5. Dissemination

The labs and ships are powerful dissemination platforms. The tangibility and visibility of the infrastructure makes it ideal to use for disseminating research results. It has already attracted much media attention, both nationally and internationally, which allows us to advocate the benefits and opportunities of autonomous technology. In addition, the video equipment in the lab allows us to broadcast live and professionally produced presentations[2] of our research from the control room to conferences and events all around the world.

4.6. Future development

For future development of the research infrastructure, our current focus is on improving the different parts and connecting them into an integrated whole. Below, we present the roadmap for future development of the research infrastructure, as well as the research challenges we will undertake in the future.

MilliAmpere1 and milliAmpere2: Future plans for the ferries include installing more on-board sensors for operator situation awareness. This includes cameras, ambisonic microphones, and other sensors that allow the land-based operator to "feel the ship". In addition, we are currently improving connectivity between the ferries and the Shore Control Lab with 5G technology. We are also building prototypes for improved onboard passenger interaction, as well as prototypes for improved automation transparency for nearby ships.

---

[1] With the following course codes at NTNU: TTK4851, TPD4167 and TPD4126

[2]https://www.youtube.com/channel/UCvIJvommzc5kFg08MA0dRaA

Shore Control Lab: Recent purchases of infrastructure equipment facilitate new technologies that have been in development for the past years. This equipment will soon allow us to address several issues we have had in the development process.

The creation of virtual environments requires huge amounts of 3D content that is often prone to change. Moreover, the data generated from both simulated and real experiments with autonomy creates a massive demand for data storage. To address these issues, we are currently building a storage server to facilitate version control and storage of 3D assets using Omniverse's Nucleus as a backend system. Furthermore, we aim to automate the data gathering process from both the real and synthetic ferries using 5G's area network in conjunction with special-purpose virtual machines or docker containers running on the server. We are also improving the resilience by installing UPS systems and a 5G connection for communication with the ferries.

Another important extension of the lab is to connect it to other control rooms at NTNU, such as the Autonomous Underwater Robotics Laboratory, and other autonomous and traditional ships, such as the research vessel R/V Gunnerus. This will enable research on control room handover.

Gemini simulator: Since Gemini was released, technologies that allow real-time raytracing to have disrupted both the movie and game industries with regards to graphics. The potential use cases that follow are, however, not restricted to graphics, meaning applications that were thought to be impossible for the past 30 years are currently being challenged in multiple fields. This new technique of depth measuring in conjunction with specialised hardware offers new possibilities for running sensor models with higher realism than before. We are therefore in the development process of renewing the sensor models to utilize this new technique.

As Gemini is a toolbox untied from applications, the request for having a more "tangible" project to show the community arises. With the development of a simulation builder tool [40] and the recent development of the MRLAB, a job going forward is to merge the Autoferry and Shore Control Lab applications into a united simulation tool. This would allow us to run the actual autonomy system of the milliampere ferries in relation to both the operator and human passengers in different scenarios. The strength of simulations would in this case not be just for verification purposes of the autonomy system and to avoid accidents, but rather to study how humans and machines interact with each other, strengthening the design process.

MRLAB: Currently, the MRLAB is a hard coded setup which gives future passengers an opportunity to do a realistic virtual ferry trial, thereby capturing their perceptions, feelings, and concerns. The possibility for full-scale immersive simulations can, for instance, be used to explore different driving patterns and what is considered a safe journey in the relation between safety and effectiveness. Communication between autonomous vessels, different user groups, and objects is also an interesting topic to explore through mixed reality. However, there could be a huge potential unleashed when Gemini is attached. Real sensor data and the autonomous system identical to the one aboard the milliAmpere2 could then add further realism to the setup. The synergy will be a powerful tool to both experience and further develop the autonomous ferry system. Applications for this lab can span from practical prototyping of new designs to novel research on abstract themes like trust.

Land-based terminals and docks: The future of the docks is to test out small scale inductive charging, and to build and evaluate the land-based passenger terminals so that we can study passenger trust, flow, and behavior.

With the presented research infrastructure and with our plans for future development, we will have a good platform for investigating a broad set of research questions related to the monitoring and control of autonomous ships in shore-based control rooms. We will also learn much about the design of both the autonomous passenger ferries and the control rooms. A short selection of our long list of short-term research plans includes to:

- study the interactions between humans and technology, such as the ferry and its operators, passengers, and nearby ships
- find out how to equip the ferry with sensors that improve the operator's awareness
- understand how operators, passengers, crew, nearby ships, and other stakeholders trust autonomous technology and how we can *design for trust*
- learn how we can design for *automation transparency* so that operators, passengers and nearby ships can understand the current state and intention of autonomous ships [44]
- find out how the tangible VR experience in the MRLAB can be used in the design process of autonomous ships
- explore HMI designs for a team of operators monitoring a fleet of ships
- design HMIs that reduce the reaction time when manual takeover is required
- understand how videogames can aid HMI for autonomous purposes, such as sensor modelling

The nature of research implies that these questions will only lead to new questions, which we look forward to exploring.

5. Conclusion

In this article, we have described an infrastructure for conducting multidisciplinary research on the monitoring and control of autonomous ships. The infrastructure consists of several prototypes of urban autonomous passenger ferries, both real, virtual, or semi-virtual, which are connected to a shore-based control centre with the possibility to conduct controlled experiments. We believe that this infrastructure will be important in contributing to knowledge on how to design shore control centers in the age of autonomous ships.

Acknowledgements

We thank NTNU for funding the design and construction of the Shore Control Lab, and milliAmpere 1 & 2 through the internal AVIT program. We thank the Norwegian Research Council for funding the MRLAB through TRUSST (grant number 331921) and for funding research studies under SFI AutoShip (309230) and MIDAS (331921) and LOAS (296527). The authors acknowledge the support of the Departments of Design, Electronic Systems, Engineering Cybernetics, and Marine Technology. We also thank Thomas Kaland and Mikael Røsbak Hansen for being important contributors in the design and construction of the Shore Control Lab.

References

[1] Kongsberg 2020 autonomous ship project, key facts about yara birkeland Retrieved November 8, 2021 (https://www.kongsberg.com/maritime/support/themes/autonomous-ship-project-key-facts-about-yara-birkeland/)

[2] NTNU 2021 SFI AutoShip - NTNU Retr. March 21, 2021 (https://www.ntnu.edu/sfi-autoship)

[3] Burmeister H C, Bruhn W, Rødseth Ø J and Porathe T 2014 Autonomous unmanned merchant vessel and its contribution towards the e-navigation implementation: The MUNIN Perspective. *International Journal of E-Navigation and Maritime Economy* **1**: pp 1–13

[4] Wang W, Gheneti B, Mateos L A, Duarte F, Ratti C and Rus D 2019 Roboat: An autonomous surface vehicle for urban waterways pp 6340–47 in *2019 IEEE/RSJ International Conference on Intelligent Robots and Systems (IROS)*

[5] Brekke E F, Eide E, Eriksen B-O H, Wilthil E F, Breivik M, Skjellaug E, Helgesen Ø K, Lekkas A, Martinsen A B, Thyri E, Torben T, Veitch E, Alsos O A and Johansen T A 2022 milliAmpere: An autonomous ferry prototype" *(forthcoming) IOP Conference Series: The 4th International Conference on Maritime Autonomous Surface Ship (ICMASS 2022) 6-7 April 2022* Singapore

[6] Dallolio A, Agdal B, Zolich A, Alfredsen J A and Johansen T A 2019 Long-endurance green

energy autonomous surface vehicle control architecture in *OCEANS 2019 MTS/IEEE* Seattle pp 1–10
[7] Wu B, Li G, Zhao L, Johansen Aandahl H-I, Hildre H P and Zhang H 2021 navigating patterns analysis for on-board guidance support in crossing collision avoidance operations
[8] Wu B, Li G, Wang T, Hildre H P, Zhang H 2021 Sailing status recognition to enhance safety awareness and path routing for a commuter ferry *Ships and Offshore Structures* **1**–12
[9] Sullivan, Brendan P.; Desai, Shantanoo; Sole, Jordi; Rossi, Monica; Ramundo, Lucia; Terzi, Sergio. 2020 "Maritime 4.0 – Opportunities in Digitalization and Advanced Manufacturing for Vessel Development." *International Conference on Industry 4.0 and Smart Manufacturing (ISM 2019)* 42:246–53. doi: 10.1016/j.promfg.2020.02.078
[10] Cunlong F, Montewka J and Zhang D 2021 Towards a framework of operational-risk assessment for a maritime autonomous surface ship *Energies* **14**(13):3879
[11] Ramos M A, Thieme C A, Utne I B and Mosleh A2020 Human-system concurrent task analysis for maritime autonomous surface ship operation and safety *Reliability Engineering & System Safety* **195**:106697. doi: 10.1016/j.ress.2019.106697
[12] Ramos M A, Thieme C A, Utne I B and Mosleh A 2020 A generic approach to analysing failures in human – system interaction in autonomy *Safety Science* **129**:104808. doi: 10.1016/j.ssci.2020.104808
[13] Thieme C A and Utne I B2017 Safety performance monitoring of autonomous marine systems *Reliability Engineering & System Safety* **159**:264–75
[14] Kim T and Mallam Steven 2020 A delphi-ahp study on stcw leadership competence in the age of autonomous maritime operations *WMU Journal of Maritime Affairs* **19**:163–81
[15] Sharma A and Kim T-e2021 Exploring technical and non-technical competencies of navigators for autonomous shipping *Maritime Policy and Management* doi: 10.1080/03088839.2021.1914874
[16] Hannaford E and Van Hassel E 2021 Risks and benefits of crew reduction and/or removal with increased automation on the ship operator: A licensed deck officer's perspective *Applied Sciences* **11**(8):3569
[17] Kari R and Steinert M 2021 human factor issues in remote ship operations: lesson learned by studying different domains *Journal of Marine Science and Engineering* *9(4)* doi: 10.3390/jmse9040385
[18] Wróbel K, Gil M and Chae C-J 2021 On the influence of human factors on safety of remotely-controlled merchant vessels *Applied Sciences* **11**(3):1145
[19] Yoshida M, Shimizu E, Sugomori M and Umeda A 2021 Identification of the relationship between maritime autonomous surface ships and the operator's mental workload *Applied Sciences* **11**(5):2331
[20] Guo C, Haugen S and Utne I B 2021 Risk assessment of collisions of an autonomous passenger ferry *Proceedings of the Institution of Mechanical Engineers, Part O: Journal of Risk and Reliability* 1748006X211050714
[21] Goerlandt F and Pulsifer K 2022 An exploratory investigation of public perceptions towards autonomous urban ferries *Safety Science* **145**:105496 doi: 10.1016/j.ssci.2021.105496
[22] Bačkalov I 2020 Safety of autonomous inland vessels: An analysis of regulatory barriers in the present technical standards in europe *Safety Science* **128**:104763. doi: 10.1016/j.ssci.2020.104763
[23] Peeters G, Yayla G, Catoor T, Van Baelen S, Afzal M R, Christofakis C, Storms S, Boonen, R and Slaets P 2020 An inland shore control centre for monitoring or controlling unmanned inland cargo vessels *Journal of Marine Science and Engineering* **8**(10):758
[24] IMO 2021 *Outcome of the Regulatory Scoping Exercise for the Use of Maritime Autonomous Surface Ships (MASS)*. MSC.1/Circ.1638. London, UK
[25] Vicente K J 1999 *Cognitive work analysis: toward safe, productive, and healthy computer-based work CRC Press*

[26] Heath C and Luff P 1991 Collaborative activity and technological design: task coordination in london underground control rooms in *Proceedings of the Second European Conference on Computer-Supported Cooperative Work ECSCW '91*, edited by Bannon L, Robinson M and Schmidt K Dordrecht: Springer Netherlands pp 65–80

[27] Pop V, Stearman E, Kazi S and Durso F 2012 Using engagement to negate vigilance decrements in the nextgen environment *International Journal of Human–Computer Interaction* **28**(2):99–106. doi: 10.1080/10447318.2012.634759

[28] Avinor 2021 Remote towers: The technology of the future at norwegian airports *Avinor* Retr. March 23, 2021 (https://avinor.no/en/remote).

[29] Van Schaik F J, Roessingh J J M, Lindqvist G and Fält K 2010 Assessment of visual cues by tower controllers, with implications for a remote tower control centre. *IFAC Proceedings Volumes* ***43***(13) pp. 123-128

[30] Schmidt M, Rudolph M, Papenfuß A, Friedrich M, Möhlenbrink C, Kaltenhäuser S, and Fürstenau N 2009 remote airport traffic control center with augmented vision video panorama *2009 IEEE/AIAA 28th Digital Avionics Systems Conference* pp 4-E IEEE

[31] Stanton N A, Harris D and Starr A 2016 the future flight deck: modelling dual, single and distributed crewing options *Applied Ergonomics* **53** pp 331-342

[32] Roberts A P J, Stanton N A and Parnell C 2019 Optimising the design and functionality of submarine control rooms from a sociotechnical systems perspective. *UDT* 2019

[33] Skjerve A B and Bye A 2011 Simulator-based human factors studies across 25 years *Springer* London

[34] dos Santos I J A, Luquetti T D V, Ferraz F T and Carvalho P V R 2008 The use of a simulator to include human factors issues in the interface design of a nuclear power plant control room *Journal of Loss Prevention in the Process Industries* **21**(3) pp 227-238

[35] Vasstein K 2021 A high fidelity digital twin framework for testing exteroceptive perception of autonomous vessels *NTNU*

[36] Unity 2022 Automotive, transportation & manufacturing Retrieved February 6, 2022 from https://unity.com/solutions/automotive-transportation-manufacturing

[37] Team C. A. R. L. A. 2022 Carla CARLA Simulator Retrieved February 5, 2022 from https://carla.org/

[38] Microsoft Research 2022 Welcome to airsim# home - airsim Retrieved February 5, 2022, from https://microsoft.github.io/AirSim/

[39] Hoem Å S, Veitch E, andVasstein K 2022 Human-centred risk assessment for a land-based control interface for an autonomous ferry *Submitted to WMU Journal of Maritime Affairs (in review).*

[40] Ek S 2022 Design of simulator for researching autonomous marine vessels *Master thesis* NTNU

[41] Njord 2021 Njord - the autonomous ship-challenge Retrieved February 5, 2022, from https://www.njordchallenge.com

[42] Zeabuz 2022 Zeabuz autonomous & electric vessels Retrieved February 5, 2022 from https://www.zeabuz.com

[43] Gulliksrud S and Jacobsen L 2020 Expandable ground control system for satellite *Master thesis* NTNU

[44] Alsos O A, Skåden O K, Hodne P, Veitch E and Porathe T 2022 Maritime autonomous surface ships: Automation transparency for nearby vessels (forthcoming) *IOP Conference Series: The 4th International Conference on Maritime Autonomous Surface Ship (ICMASS 2022)* 6-7 *April 2022* Singapore

Development of an advanced, efficient and green intermodal system with autonomous inland and short sea shipping - AEGIS

S Krause[1*], L Wurzler[1], O E Mørkrid[2], K Fjørtoft[2], H N Psaraftis[3], M R Vilanova[3], T Zis[3], N F Coelho[4], J van Tatenhove[4], J Raakjær[4], K Kloch[5], M B Billesø[5] and J N Kristiansen[6].

[1] Institut für Strukturleichtbau und Energieeffizienz, Chemnitz, Germany
[2] SINTEF Ocean, Trondheim, Norway
[3] Technical University of Denmark, Kopenhavn, Denmark
[4] Centre for Blue Governance, Aalborg University, Denmark
[5] DFDS, Kopenhavn, Denmark
[6] Aalborg University Business School, Aalborg, Denmark

*Corresponding author: s.krause@institut-se.de, +49 371 3380015

Abstract. The European maritime transport policy recognizes the importance of the waterborne transport systems as key elements for sustainable growth in Europe. A major goal is to transfer more than 50% of road transport to rail or waterways within 2050. To meet this challenge waterway transport needs to get more attractive and overcome its disadvantages. Therefore, it is necessary to develop new knowledge and technology and find a completely new approach to short sea and inland waterways shipping. A key element in this is automation of ships, ports and administrative tasks aligned to requirements of different European regions. One main goal in the AEGIS project is to increase the efficiency of the waterways transport with the use of higher degrees of automation corresponding with new and smaller ship types to reduce costs and secure higher frequency by feeders and provide multimodal green logistics solutions combining short sea shipping with rail and road transport.

1. Introduction

The European Union [1] and the Norwegian Government [2] recognized the importance of waterborne transportation systems for sustainable growth in Europe. The main goal of the AEGIS Project is to shift road transport to waterborne transport thus reducing road congestions and pollution.

In general, waterborne transportation can be very energy efficient especially when using new vessel types with zero or low emission propulsion systems to reach an emission, dust and noise free transport system.

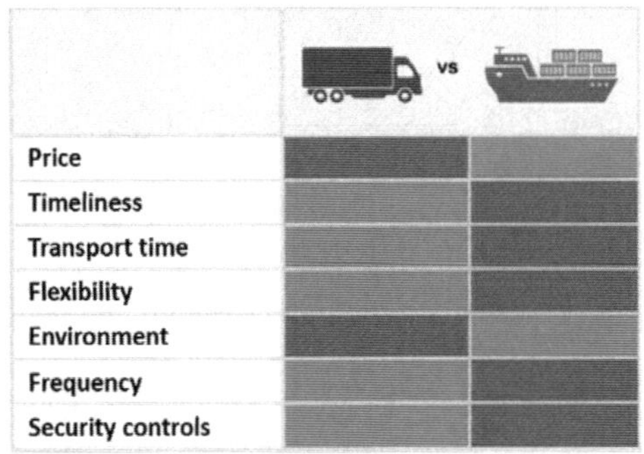

Figure 1 Why Norwegian short sea transport fails to compete with trucks. [3]

Figure 1 summarizes the conclusions on why short sea waterborne transport today cannot compete efficiently with trucks, from the Auditor General in Norway [3]. Ship transport is generally cheaper, is able to transport heavy loads and more tonnage compared to trucks and is friendlier to the environment. On the other side, truck transport offers more service quality and is more flexible on short routes.

The AEGIS project develops a transport system with more and smaller vessels to increase frequency, differentiate speeds, reduce terminal costs and reducing port times. Furthermore, smaller vessels increase reliability and resilience and automated cargo handling as well as standardized cargo units will reduce problems (e.g., human errors) and transhipment costs. The advantage by using and establishing remotely controlled vessels, is the possibility to operate a fleet of vessels from a remote control centre and therefore be more flexible and efficient.

This paper mainly focuses on the development of the new advanced, efficient and green vessel designs (section 4), beforehand the three real life use cases of the project (subsections 2.1-2.3), policy measurements (subsection 2.4) and key performance indicators (KPI) for the vessel development (subsection 2.5) are introduced. In addition, the definition of vessel autonomy for this project as well as autonomous vessel operations are discussed (section 3). In the end, a conclusion and an outlook is presented (section 5).

2. Requirements and Objectives

2.1. Use-case A

Use case A in the AEGIS project refers to a maritime transport corridor from the west coast of Norway down to the continent, that in our case is the port of Rotterdam as destination.

It also includes local transport in Norway, serving the smaller ports in the Trondheimsfjorden region where a transshipment terminal is planned to be at Sandstad in the island Hitra. The use case has three important elements to consider, which is the baseline in this paper:

1. Local transport between smaller ports in the Trondheimsfjorden region operated by a daughter vessel
2. Terminal activities and operations, which differ with different terminal sizes, spanning from the smallest self-served ports to medium-size ports and the large ports in Europe, such as in Rotterdam
3. Short-sea transport from Norway to the large ports in Europe (Rotterdam)

From the studies we consider that there will be different autonomy levels, where we are following the four-degree levels defined by IMO [4], as presented in section 3.2.

When investigating the autonomous transport system, the focus has been mainly on the following parts:

- The autonomous ship with different autonomy levels. In the use case there are two different ship concepts; a mother vessel (see subsection 4.1.1) sailing the short-sea segment and daughter vessels (see subsection 4.1.2) operating locally in Norway.
- There is also a particular focus on crane operations, and land-based autonomous solutions such as terminal tractors and reach stackers.
- An onshore Remote Control Centre (RCC) and its role in supporting the ships and terminals.
- Interaction between ships and automated port services, and specifically regarding solutions for safe berthing and mooring.
- Communication between ships, control centres and crew in logistic operations.

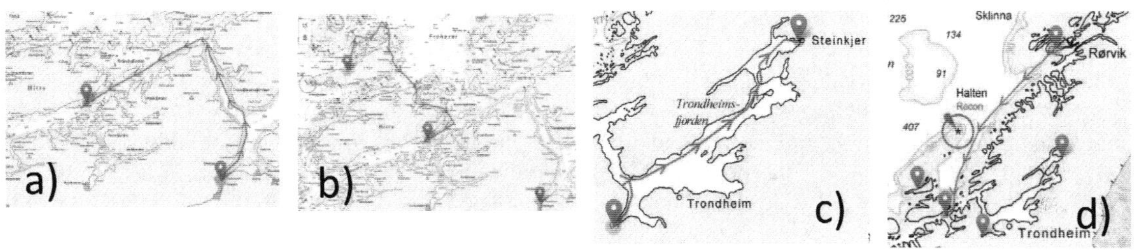

Figure 2 Proposed daughter scenarios for use case A.

Use case A has proposed to look into four different daughter scenarios, as shown in Figure 2. Each of the four scenarios has different cargo volume potential and requirements related to weather conditions and terminals to be visited. While scenarios a), b) and d) have constant cargo flows, the scenario c) will be more varying and maybe also on-demand. Different vessel requirements have been proposed for the four scenarios. For scenario a) the daughter vessel with capacity 60 TEUs is relevant, for scenario b) and c) the barge solution will be the best suited alternative, as these also can be used as temporary storage of cargo. The last scenario, d), might need a bigger and more seaworthy vessel as the daughter with capacity 160 TEUs.

Scenario b) relates to transport of goods from the island Frøya to Hitra with barges and further to Rotterdam with a short sea container liner.

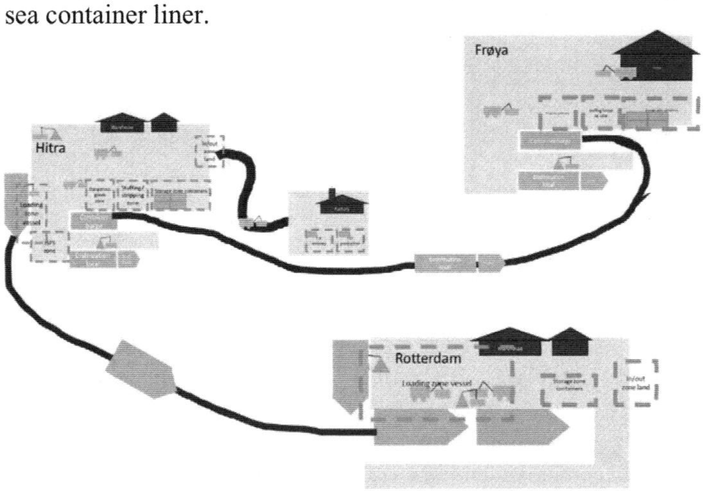

Figure 3 Proposed transport chain in use case A.

As can be read from Figure 3, there are three main parts in this value chain: 1) the production site at Frøya, 2) the terminal operations at Sandstad (Hitra), 3) the arrival terminal in Rotterdam. Regarding autonomy there can be different levels implemented in the individual parts, varying from level 1 to 3 (4 is probably far into the future). The mother short-sea vessel is realistically at a level 1 (maximum 2), whereas the barges will be at level 3. Cargo handling and other terminal related equipment will also be automated and should aim at an autonomy level of at least 2.

2.2. Use-case B

For the purposes of AEGIS Use Case B, with its focus on the inland shipping interface in Belgium and Netherlands, a few important mental framings must be considered before engaging with developing a prospective green, autonomous waterborne transportation system on the inland waterways in Europe in general and in Belgium and the Netherlands in particular.

First, it is worth noting that the purpose of the AEGIS Use Case B autonomous transport system and its focus on RoRo transportation by the inland waterways is not to directly compete with existing LoLo services in the region, but instead to convert existing trucking services unto the inland waterways in the area. This becomes evident when looking at the existing modes of inland freight transportation. According to Eurostat [5], of all the inland freight transport in Europe, the inland waterways constitute a mere 6% of inland freight – a number that has been decreasing yearly from 2013 to 2018 (see Figure 4). In contrast, inland freight transportation by road constitutes more than 75% of the total inland freight – a number that has been increasing yearly from 2013 to 2018.

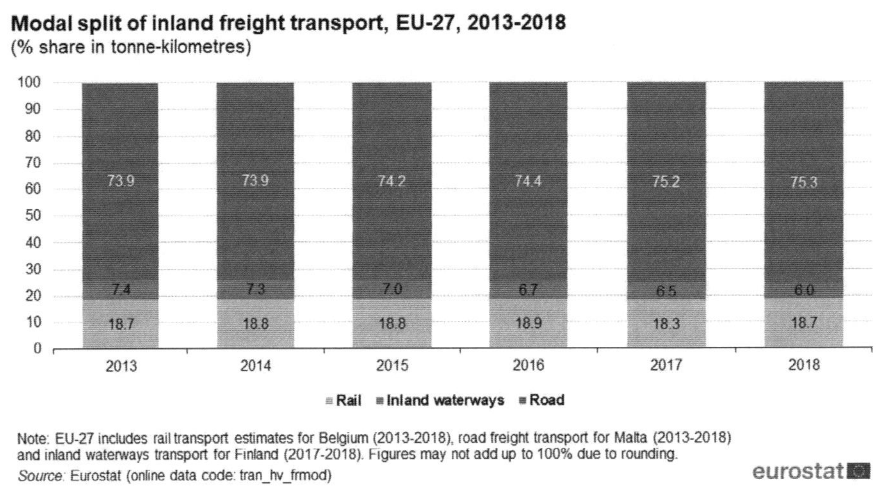

Figure 4 Modal split of inland freight transport, 2013-2018. [5]

Nevertheless, the trucking sector is facing some significant challenges that may see the increase in alternative modes of transport. One of these challenges is the growing congestion on the road infrastructure which negatively influences both speed, reliability, cost-effectivity, sustainability of landborne freight transport as well as the social mobility in the area in general. For the Netherlands, a steady increase in congestion over the past years has been registered and a 35% increase in travel time loss on roads from 2017 to 2023 is predicted [6]. This would place an even heavier burden on the road transport infrastructure in the area and consequently the distribution of cargo out of the Dutch ports. The same problem of congestion is present in Belgium also, where especially the areas around the ports of Antwerp, Ghent and the capital Brussels are facing 83%, 64% and 23% increases in hours lost in congested traffic respectively (see Figure 5) [7].

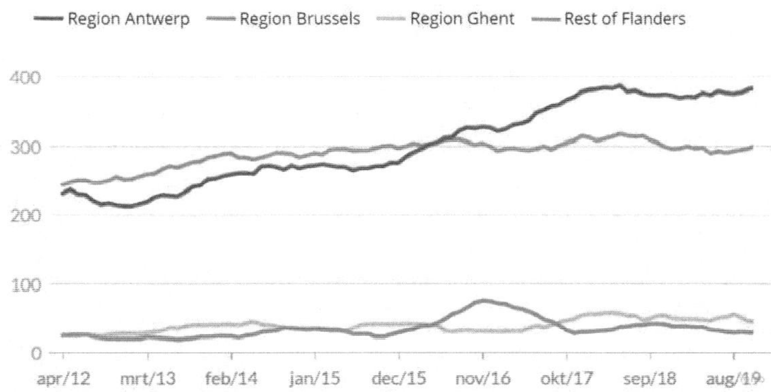

Figure 5 Lost vehicle kilometrehours per working day in the Flemish regions (rolling annual average). [7]

Based on the above, the AEGIS Use Case B predicts an ample opportunity to convert some of this roadborne freight into waterborne freight regardless of whether this comes through the modes of either LoLo or RoRo. This leads back to the point of departure, i.e. that the primary focus is not to convert freight from a LoLo based system unto a RoRo based system, but instead to convert the roadborne freight system unto a waterborne freight system.

Second, when looking at transport systems in general and on multimodal transportation in particular one has to differentiate between the so-called direct runs (A-B), or milk runs (A-B-C...). A direct run is, as the name may indicate, a direct route between two points with no intermediate stops. For this type of route, the vessel will be completely unloaded of cargo at the destination port and potentially reloaded for cargo to travel with the vessel back to the departure port. This means that the cargo onboard the vessel would not necessarily have to be stowed in a particular commercial fashion, as all inbound cargo will be unloaded and subsequently all outbound cargo will be loaded at every single port stay. A milk run is a route with two end points and one to multiple stops along the route between the two end points. For this type of route, the vessel will not necessarily be completely unloaded at every intermediate stop nor will there necessarily be loaded any cargo back on the vessel at the given stop. In the end points, however, the unloading of cargo often follows that of a direct run, but the loading of the cargo must obey by strict commercial planning orders with regards to the destination of every single cargo unit. If one imagines a route from Port of Rotterdam to the Port of Ghent, the two types of routes can be exemplified shown in Figure 6.

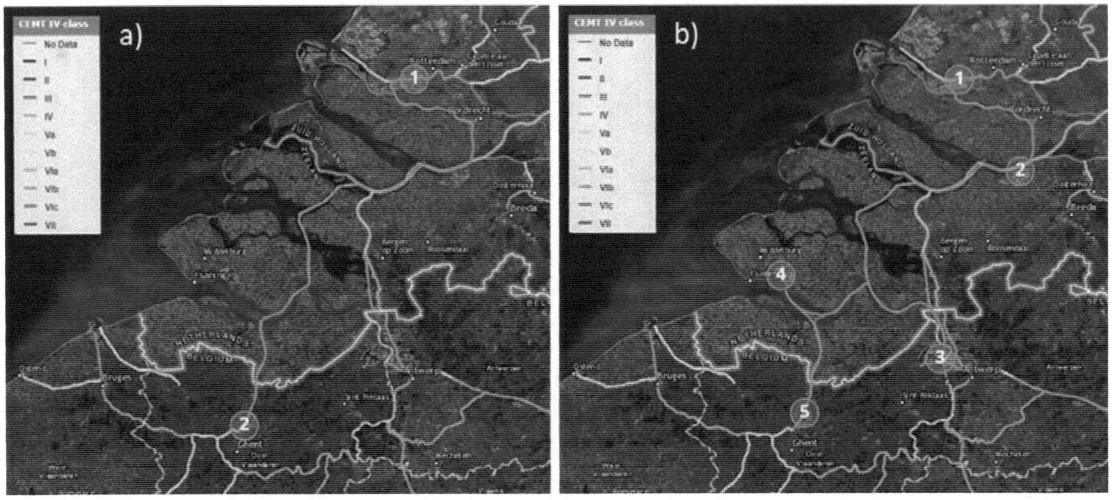

Figure 6 Hypothetical route between port of Rotterdam and Port of Ghent on the inland waterways depicted as a direct run a) and a milk run b).

The two types of routes each offer different advantages and challenges. A direct run offers high cargo volumes to be transported between two given points as fast as possible, but in turn requires that a sufficiently high cargo volume exist between the two points. A milk run instead offers connectivity to terminals with smaller cargo volumes that in themselves could not substantiate a direct run, but at the cost of longer transportation time between the end points due to the multiple stops. A milk run further requires complex planning of the loading and unloading séance in order to have access to the given cargo units destined for unloading at the given port without spending excess resources and time on restoring the cargo units destined for later ports. The direct run does not have this issue. This later point is of utmost importance with regards to establishing a RoRo transport system for the purposes of AEGIS Use Case B. As RoRo vessel normally load and unload through an access point at either the vessel's stern or bow, it becomes difficult to access every cargo unit separately. For a RoRo vessel to be able to engage in effective milk run solutions, new vessel design giving access to as many sperate cargo units as possible must be invented.

One potential way of achieving this is through side loading and unloading of the vessel, instead of stern or bow loading. This requires that the RoRo cargo is places in a transversal direction on the vessel as opposed to the normal longitudinal direction that is common on both RoRo and LoLo vessels of today. Requiring transversal stowage of cargo does however place some rather stringent requirements on the width of the vessel – a requirement that is one of the main restrictors on the inland waterways. A standard, standalone DFDS box trailer measure around 14 meters in length. Including space for various cargo lashing appliances, this would potentially require that a barge with transversal stowage of trailers should have a minimum width of approximately 16 meters. Consequently, for a RoRo barge to be installed in a milk run route and provide the ability to have direct access to as many single cargo units onboard as possible, it would only be able to travel on inland waterways of CEMT Class of VIa or above. In contrast, a RoRo barge installed on a direct run would not be restrained by this width requirement, as there would be no need for side loading and unloading and the trailers could thus be positioned in a longitudinal fashion. To this extend, a standard, standalone DFDS trailer is around 2,5 meters wide and with an additional allowance for lashing would possibly take up to 3 meters transversally allowing for multiple trailers to stand side by side.

Third, it is worth noting the concept of autonomous transport systems as opposed to autonomous transport legs, where making autonomous legs of the transportation instead of an entire autonomous system would likely create significant issues further down the transportation chain. An example of such could be establishing an autonomous LoLo barge which would only be able to visit terminals which would allow for its visiting and more importantly its cargo operation, i.e. terminals with a given level of infrastructural investments. It is true that LoLo vessels could bring their own cargo cranes, by which they could load and unload cargo themselves, but investments must still be made with regards to replacing pure containers unto trailers, i.e. by using so-called reachstackers. It is hard to imagine a solution where these could travel with the vessel itself as trucks often arrive at different times in a terminal and would not be able to load the containers unto their trailers themselves. Looking at the DFDS terminals in the region, it is observed that the cargoes for a given route arrives at the terminal rather sporadically over multiple hours and potentially days. Consequently, an autonomous LoLo barge that carries its own reachstackers would potentially have to wait prolonged periods of time for the trucks to arrive or alternatively leave the containers as they are on the terminal floor. If the former is chosen, a LoLo solution would be extremely slow and cost inefficient. If latter is chosen, the continuous hinterland transportation of the container is potentially hindered by the need to somehow reload the container unto a trailer for a truck to drive off with (see Figure 7).

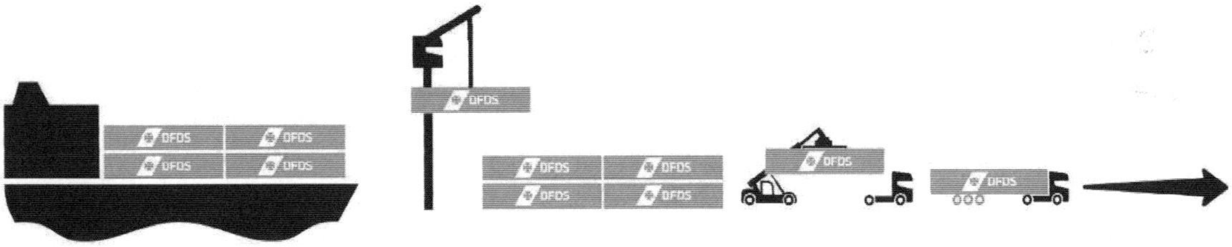

Figure 7 LoLo cargo operations (simplified).

In contrast, RoRo vessels could also bring their own cargo equipment, a so-called tugmasters, that could live onboard the vessel and undertake the cargo operation whenever the vessel is moored in a terminal. This means that upon the unloading of the vessel, the cargo would already be standing on a rolling platform ready for last mile distribution by trucks that merely need to connect to the trailer and drive off into the hinterland. The same logic applies when a truck is coming to a terminal with a cargo, where the truck merely has to disconnect the trailer with the cargo. Below, Figure 8 shows this simplified in comparison to cargo operations for LoLo transportation systems in Figure 7.

Figure 8 RoRo cargo operations (simplified).

To summarise, the continued work in the AEGIS Use Case B and the autonomous transport system that it aims to create builds on the following mental frames:

1. The autonomous inland RoRo transportation option is first and foremost about creating an alternative to inland truck transportation – not directly to inland LoLo transportation. The increasing congestion in the Dutch and Belgian regions may create an ample opportunity to convert roadborne transportation to waterborne transportation.
2. An autonomous inland RoRo barge would encounter width requirements if installed on a milk run as direct access to each individual cargo would necessitate a vessel width of 16 meters and hence an inland waterway of at least CEMT Class of VIa – a CEMT Class which is not widely available in the Benelux region. An autonomous inland RoRo barge would not face same

width requirements if installed on a direct run, which in turn require a certain minimum of cargo volumes to be transported.

3. It is important to consider the full transport system and not only a single transport leg. In this regard, RoRo transportation offers better opportunities to call smaller terminal with no to limited infrastructural investments than LoLo transportation, which does require some level of infrastructural investments. This also creates an opportunity for RoRo vessels to visit infrastructures which would not traditionally be considered a port or a terminal.

The vessel designs for use-case B are required to be able to operate on west European inland waterways (IWW) which are defined by CEMT (Conference of European Ministers of Transport) classes. In addition, the vessels are supposed to only transport trailers as RoRo. There will be no truck drivers or crew on-board as the autonomy level is required to be very high and to safe space and weight on the vessels.

Regarding the different scenarios and the existing IWW, different sizes are proposed to be developed. The first size is to fit the CEMT class II waterways with a capacity of about 10 Trailers, the second to fit CEMT class IV waterways with a capacity of about 20 trailers and the third vessel is required to be able to store the trailers transversal to the sailing directions which results in a vessel fitting the CEMT class VIb.

2.3. Use-case C

The main purpose of the two Danish ports is to develop Blueprints for 1) a Multipurpose terminal in a small port (Vordingborg) and 2) an Intermodal Automatic Green Terminal in a medium size port (Aalborg) as mean to move cargo from road transport to short-sea shipping. This is a way to support the fulfilment of the EU 2030 and 2050 targets for reduction in CO2 emissions.

To move goods from one modality to another is not straightforward: Road transports have historically been highly competitive on price, flexibility, and time. Thus, the challenge for short sea shipping is as far as possible to match road transport on price and set-up systems that through clever planning, can compensate for the need for high flexibility. This entails a focus on goods types where speedy deliverables are less important, as well as focusing on how to reduce the last mile deliverable costs. Collectively, this is to ensure that the short sea shipping transport mode is not outcompeted by road transport. Short-sea shipping (or rail for that matter) has a major advantage to road transport in relation support the green transition and facilitate a carbon neutral transport sector in Europe by 2050. To achieve this highly ambitious target it is important in parallel to optimize and green every step in the value chain. The ports can play an import role in improving terminal design and goods flow on land to increase efficiency. Nevertheless, without also reducing shipping costs, it is unlikely that short sea shipping will become a major player in the green transformation. Thus, new vessel designs, and increased automatizations will be cornerstones for short-sea shipping to increase its market share of movements of goods in Europe.

We conducted a detailed study and estimation of the gross volume of goods that can be shifted from road transport to short-sea-shipping within Denmark more precisely between Northern Jutland and Zealand (other regions not included due to distance, cf. Figure 9). Currently the total goods between these regions (road) are approximately 1 million tons, where the conversion potential is estimated to be around 177.500 tons, which equals approximately 10.000 trips per year.

We combined these findings with international goods, and, based on all the analyses (scenarios with reduced conversion probabilities), it was estimated that the potential gross volume of goods that can be shifted from road transport to Short-sea-shipping within and to/from Denmark is approximately 5 million tons yearly, or about 18% of the relevant goods by truck. It is again important to note that this is provided that any short-sea-shipping solution would be on par or cheaper than a competing direct road solution.

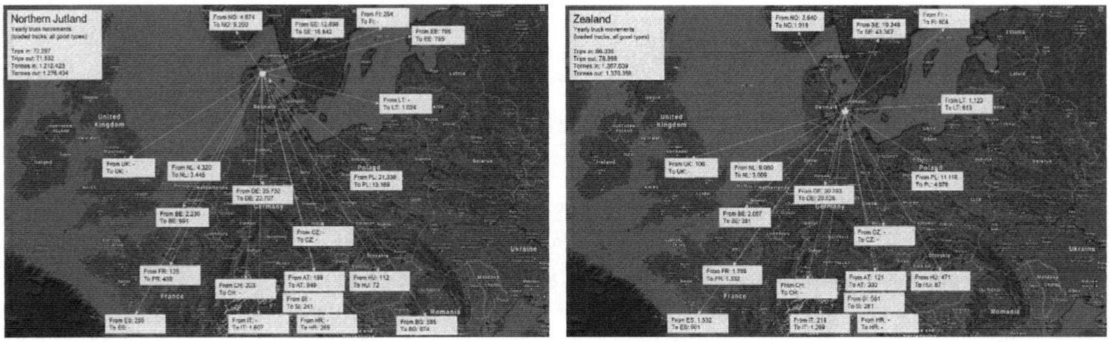

Figure 9 International road transport to/from the regions of Northern Jutland and Zealand. [8]

Presently, Port of Aalborg is called by a feeder line connecting Aalborg to Bremerhaven and Hamburg on a weekly basis, whereas the Port of Vordingborg presently is not connected to the container network. The feeder vessels have a capacity of 1000-1200 TEU and this vessel size is considered optimal but need to be improved in terms efficiency and autonomy to be competitive to road transport. As a Gantry crane constitutes a significant part of CAPEX costs for a medium size container terminal, it will be particularly important to explore onboard crane solutions.

Neither Aalborg nor Vordingborg are on a regular basis called by RoRo vessels. We conducted analyses of relevant activities, potential capital expenditures, operating expenditures, potential goods volumes, as well as developed new costing and pricing models for terminal operators and port authorities respectively. Initially, 12 scenarios have been considered showing that the minimum threshold of an economically viable port call to any of the ports would be 2 calls per week with a 50-trailer RoRo vessel (80 % utilization). We additionally examined 10 scenarios considering the notion of a multipurpose terminal in a small port, where the same terminal area could be used for multiple purposes throughout a week. A RoRo connection would be economically viable at even 1 call per week. The national calling (Aalborg-Copenhagen) showed substantial economic benefit of short-sea-shipping solutions over the road alternative. The international calling (Klaipeda-Vordingborg) similarly showed economic promise. However, the example similarly showed that increased vessel transport time (48 hours), offsets the economic benefit, and also potential market attractiveness, of having seabound solutions that stretch over longer distances, with only a 50-trailer vessel capacity. This finding gives interesting perspectives in min-max thresholds of sea distances and provide important guidance for potential vessel designs for UC-C.

Regarding the different challenges in use case C, the vessel designs are supposed to be as different as them. Therefore, the vessel designs are proposed to be medium sized and combine SSS and IWW shipping. The first design is required to transport only containers with a capacity of about 120-130 TEU with low draught and an on-board handling system. A second design is required to be more flexible in the kind of cargo it is transporting. Therefore, a vessel able to transport bulk as well as containers with the same properties as the first vessel design is proposed. Additionally, a third vessel design is required to be able to fit the RoRo transport needs of the customers operating in the area of use case C. Therefore, synergies are considered from vessel designs from use case B. The capacity is required to be about 50-60 trailers stored transversal to the sailing direction.

2.4. Governance and policy support of the Areas

Shipping is a global industry which is subject to regulatory standards developed under the International Maritime Organization (IMO), a specialized agency of the United Nations. The rules and standards developed at the IMO apply to design, construction, equipment, operation and manning of vessels. The IMO also plays a regulatory role with respect to environmental protection, namely by determining pollution standards. Furthermore, the IMO also impacts on the efficiency of ships, for example via the mandatory Energy Efficiency Design Index (EEDI) and the Ship Energy Efficiency Management Plan (SEEMP). Under international law, states do not find many avenues to depart from IMO regulations, as they are expected to be the result of a consensual compromise. However, there are some avenues for coastal states to develop more stringent norms, especially when enforcing them at port.

Although the IMO has been working on the introduction of digital technology into shipping, regulating Maritime Autonomous Surface Ships (MASS) is a relatively recent endeavor. In 2017 the IMO begun scoping ways to determine how safe, secure and environmentally sound MASS operations might be addressed in IMO legal instruments. The challenge is to identify all existing treaties affected by the introduction of different levels of autonomy and to either redraft them or interpret them consistently. In this early scoping exercise, four degrees of autonomy were considered: crewed ship with automated processes and decision support (Degree One); remotely controlled ship with seafarers on board (Degree Two); remotely controlled ship without seafarers on board (Degree Three); and fully autonomous ship (Degree Four). Each one of these degrees would require different regulatory work, and eventually Degree Three and Four would require a revision of more structural treaties on the law of the sea which did not consider the possibility of unmanned vessels.

A few examples may showcase the gaps between existing instruments and forthcoming technical developments. UNCLOS requires all ships to be in charge of a master and officers who possess appropriate qualifications. SOLAS obliges the master to render or provide assistance to persons in distress at sea. FAL obliges the master to ensure the "security, general health, welfare and safety of the stowaway while he/she is on board, including providing him/her with adequate provisioning, accommodation, proper medical attention and sanitary facilities". Without a human on board, would the flag state be violating international law? Furthermore, COLREG also presuppose a human involvement. It remains to be decided how to interpret the phrases ""not under command" or "restricted in her ability to maneuver" if MASS is introduced without adequate legal revisions.

The AEGIS concept focuses on SSS and IWT, which in principle allows for some more leeway from the existing regulatory scenario. Test sites were created by national maritime authorities, for example in Norway and Denmark, to assist in the development of these technical solutions, and this was made possible within the international legal framework. However for a capital intensive industry such as shipping, taking risks with regulatory unknowns is not an option. And thus, at least for international journeys, more clarity is required.

The AEGIS project is seeking to identify these regulatory avenues to propose regional approaches that fit the use-case scenarios. Considering that the project seeks to develop routes that are not mere cabotage but that link up small terminals to larger transshipment hubs, international law would still apply. That notwithstanding, it is a hypothesis that the EU may seek to facilitate navigation of vessels with advanced technology which is not yet regulated under the IMO.

A less experimental route is the development of policy measures to support SSS without considering the introduction of MASS. The EU has failed in the past to make maritime transport more appealing and it does not seem that lack of IMO regulation was the cause of it. Policy support in this case may have to do with looking at the incentive structures provided to road and rail transport. Environmental policy has developed since the early days of EU Marco Polo programs designed to foster maritime transport and thus it is possible that new avenues are made existent today to support private actors in shifting from road to sea. Yet for that it is important for policy makers to take into consideration the need of private actors for clear and foreseeable rules so that long-term management strategies can emerge.

All in all, the challenges associated with AEGIS from a law and policy perspective relate to the international nature of the maritime space. This makes it relatively easier for UC-C and UC-B to move forward, as port development and inland waterway transport are more heavily ingrained in national jurisdiction. With what regards SSS, and the development of new vessels, there remains a structural challenge which requires political will and legal imagination.

2.5. Key performance indicators for the vessel design

Whatever solutions are contemplated in AEGIS, it is imperative to assess them in a holistic fashion, so as to capture the effects of all conceivable cross-linkages and inter-dependencies and hopefully obtain what we call "win-win" solutions. This analysis and evaluation are carried out in a comprehensive Cost Benefit Analysis (CBA) and environmental assessment including also social aspects. To that effect, we have developed a comprehensive set of Key Performance Indicators (KPIs) in order to assess any solutions further contemplated in AEGIS. KPIs represent the criteria under which the set of solutions developed under AEGIS will be evaluated. They include criteria grouped under the following classes:

- Economic KPIs, including cost, profit, logistical efficiency and others
- Environmental KPIs, including GHG and other emissions
- Social KPIs, including safety, security, externalities and others

The KPIs are aimed to be measurable, concise and compact, relevant, unambiguous and easily understood. Appendix A presents a list of the most important KPIs.

3. Autonomy in vessel designs

3.1. Definition of autonomy for this project

Autonomous operations are widely studied and various defined. The range of defined autonomy levels vary from four autonomy levels introduced in subsection 2.4 [9] to seven [10] defined by the Lloyd's register. Further, Ø. Rødseth defined in one of his publications [11] two different areas of automation on ships, operational and bridge autonomy level. This results into a 4x3 matrix to describe different scenarios. By the definition of Lloyd's Register [10], the different autonomy levels are also categorised in groups to describe manned, remote, automated, and fully autonomous ships in a more detailed way.

This project will discuss the level of autonomy for new designed vessel concepts regarding the definition by the International Maritime Organisation (IMO). Therefore, four levels of autonomy are defined as followed [9]:

- Degree one: vessel with automated processes and decision support and crew on board to control vessel systems and functions. Operations may be automated and at time be unsupervised.
- Degree two: vessel remotely controlled and operated from another location but with crew available on board to take control.
- Degree three: vessel remotely controlled and operated without crew on board.
- Degree four: vessel operates fully autonomous and is able to make decisions and determine actions by itself.

3.2. Autonomous vessel operation

For the developed vessel designs, the degree of autonomy will be developed towards a higher level. The benefits of autonomous operations will take effect at low levels but become more important at high levels.

Autonomous sailing and vessel operations can be described with several actions that need to be performed for secure and safe sailing. First steps are condition detection and condition analysis. Further steps are action planning, action execution and action control. After executing these steps, the circle of actions is reapplied. For these actions, appropriate sensors and fast decision algorithms are needed. The more and divers the sensors are installed on the vessel, the better and more secure will be the condition detection and action control. Appropriate sensors are optical sensors, e.g. camera and LIDAR (light imaging, detection and ranging) systems as well as other sensors, e.g. radar, etc. Additionally, sensors and technology for GNSS (global national satellite system) are essential for autonomous sailing.

The main benefit for autonomous vessel operations is the increase of safety by eliminating the human error, which causes over 75% of accidents on ships, according to Bureau Veritas [12]. Simultaneously, the operational costs are lower due to less or no crew on board, following that during vessel design it is possible to use the space more efficiently without needing any crew accommodation. Another benefit for autonomous vessels is the efficient use of fuel considering environmental influences like weather conditions and waves as well as optimized routeing and efficiency of main engine. According to MUNIN (Maritime Unmanned Navigation through Intelligence in Networks), dependent on the ship type it is possible to save up to $7m over a 25-year period in fuel consumption, crew supplies and salaries [13].

The main challenges for autonomous vessel operations are regulatory issues (see subsection 2.4) which have to be defined by policy measures and by the involved stakeholders. Another challenge are the large capital costs in initially investing due to the new applied technology in an early development stage [10]. Further investments are expected besides the vessel for setting up and operate an onshore centre for monitoring fleet movements [10]. In addition, current infrastructure and unmanned vessels need to communicate and therefore use compatible systems which is a high risk for cyber security. Currently, vendors use their own system, like Kalmar with KalmarOne, which needs to be able to communicate with several other systems on vessels probably designed by different contractors. A lot of work needs to be done create common standards for autonomous vessels.

4. Developed vessel designs

In this section, the developed vessel designs for each use case regarding the conclusions from subsections 2.1 to 2.3 will be presented.

The descriptions of the vessel designs include general main dimensions, like length over all (L_{OA}), breadth (B), and draught by design (T_{Design}) as well as information about the propulsion system, fuel concepts, and on-board cargo handling. Further, the implemented autonomy concept is described.

First the developed vessel designs for use case A are presented in subsection 3.1, where the mother concepts are described in subsection 3.1.1 and the daughter concepts in 3.1.2. Second, in subsection 3.2 the developed inland waterway RoRo vessel designs for use case B are presented. Third, subsection 3.3 depicts developed vessel concepts for use case C.

4.1. Mother and daughter concepts for Use-case A

As mentioned above, this subsection presents the current state of the vessel development for use case A. Note, that these designs are in concept state and not fully developed or detailed.

4.1.1 Mother vessel designs. According to 2.1, different scenarios are defined within the use case, thus requiring more than one vessel design for the mother vessel.

New Design of 1000 TEU vessel

This vessel design is developed to cover the transshipment from Rotterdam to the Trondheim region. Therefore, a container vessel with on-board handling systems and a green propulsion system is developed and shown in Figure 10 and the main dimensions are summed up in Table .

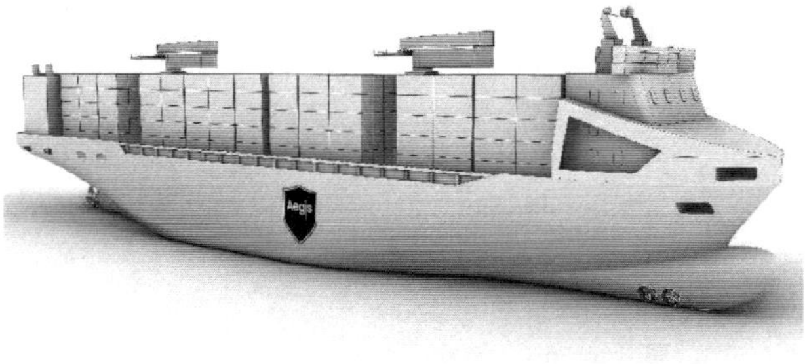

Figure 10 New design of mother vessel.

Table 1 Main dimensions for new design.

L_{OA}[m]	B[m]	T_{Design}[m]
143.8	25.5	8.2

This vessel is designed as an open top concept with a total capacity of 1046 TEU, however cargo places for 40', 45' and reefer containers are considered in the design.

The propulsion system is designed for the route between Rotterdam and the Trondheim region, and for the harsh and windy conditions at the North Sea. Therefore, a methanol combustion engine is chosen (in the first state "methanol ready" to be switched to full methanol operation when the infrastructure is existing) which is expected to reach an average speed of 15 kn. The vessel is driven by a controllable pitch propeller (CPP) which is built with separated blades which move "simultaneously through an arc to change the pitch angle and therefore the pitch" [14]. This system results in optimized propeller efficiency for different loading and weather conditions which leads to a reduced fuel consumption. Two electrically driven bow thrusters enable a high manoeuvring capability. In further developing steps, a hybrid solution with battery support will be considered, which might be needed in highly restricted fjord and near coastal areas. Furthermore, wind-assisted propulsion systems, such as Wing sails or Flettner Rotors [15] might be possible options due to the windy conditions along the Norwegian west coast.

As shown in Figure 10, the new designed mother vessel is equipped with two triple-joint cranes – a development of the AEGIS partner Cargotec/MacGregor (CT/MCG). Each crane has an outreach of 32 m and a safe working load (SWL) of 45 t. The benefit of the triple-joint crane is its high internal stability, space saving design, smooth operation with anti-sway system, and precise movements. It will be operated highly autonomous. CT/MCG offers special autonomy solutions for their cranes, such as anti-collision systems and automated cargo grabbing. Although, in international hubs such as Rotterdam the use of their port-side cranes is obligatory, however an on-board cargo handling system enables a high flexibility at the Norwegian ports. The vessel is not restricted to the availability of port equipment and can use its own green energy for the transhipment process. Also, the option of using green electricity provided by the port will be considered.

The first developed design of the mother vessel is considered to operate on an autonomy level of one to two. There will be crew on board and the captain's bridge and accommodation are part of the vessel concept. During operation an increase of autonomy is expected (due to learning effects and adopted regulations). Therefore, the superstructure has a modular design and can be removed partly as needed. Additionally, automated mooring [16] and automated cargo handling will save time and will allow for transshipments independent of the port operation hours. A digital twin enables customers and shipowners to observe and compare the vessel status and to predict the arrival time.

<u>Design of a coastal feeder vessel</u>

Investigations on cargo volumes in the area of use case A yield to the idea of a coastal feeder vessel design to call ports along the Norwegian west coast before small, autonomous daughter vessels cover the last mile transport within the fjord. Figure 11 depicts the initial design and Table lists the developed main dimensions.

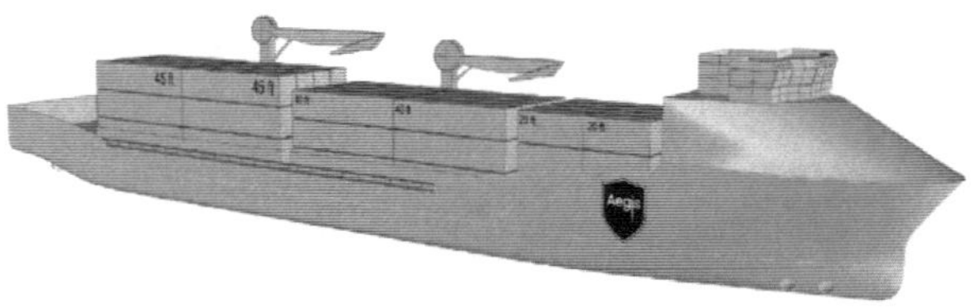

Figure 11 Coastal feeder vessel.

Table 2 Main dimensions for coastal feeder vessel.

L_{OA}[m]	B[m]	T_{Design}[m]
103.5	13.4	2.7

As shown in Figure 11,different container sizes will be transported. The vessel is designed for each 32 TEU, FEU and 45 ft containers, respectively. This results in an overall capacity of 160 TEU. Further developments will consider storage space for high cube and refrigerated container, according to the required data from subsection 2.1.

The coastal feeder vessel travels short to medium distances along the Norwegian west coast with multiple stops along the route which offers multiple opportunities to refuel or recharge. Therefore, propulsion systems like batteries can be used. Hence, no funnel or exhaust system need to be designed. In general, the use of batteries requires more storage place and the loss of payload since the weight of batteries is higher than other fuels by the same covered operating distance. By calling many ports along the route, the operating distances are reduced and batteries could be recharged during the process of loading and unloading the cargo. The required energy to recharge those batteries should be provided by green energy, e.g. hydroelectric power or wind power, to reach the main AEGIS goals. This green energy is used to drive electric engines within two azimuth thrusters at the stern of the vessel. Further, two bow thrusters are used to gain high manoeuvrability and flexibility.

The coastal feeder vessel is equipped with two on-board cranes. Each of them with 35 t SWL, one with 18 m beam and the other with 25 m beam. This ensures the accessibility of each cargo container on the ship. These on-board cranes are designed highly autonomous and are located on portside of the vessel. This ensures the possibility to use terminal cranes in bigger ports. By locating the cranes on portside of the vessel it is assured that these on-board cranes do not interfere with the terminal cranes during loading and unloading.

This concept is developed with an autonomy level of two to three which requires a bridge and crew accommodation. However, the bridge is designed to be modular and therefore, could be dismantled by increasing the autonomy level in the future. Additionally, automated mooring and automated cargo handling as well as the implementation of a digital twin is considered for this developed design.

4.1.2. Daughter vessel designs. This subsection describes two different approached for daughter vessel designs for cargo distribution within the Trondheim region.

<u>Self-propelled Shuttle</u>

The self-propelled shuttle is conceived as a fully autonomous vessel with green propulsion for transporting container only and depicted in Figure 12. In addition, the vessel is equipped with an on-board handling system. With respect to the expected cargo volume along the scenario described in subsection 2.1, the main dimensions were calculated and are listed in Table 3Table .

Figure 12 Self-propelled shuttle.

Table 3 Main dimensions for self-propelled shuttle.

L_{OA}[m]	B[m]	T_{Design}[m]
61.0	12.4	3.3

Required input data are assumed with a capacity of maximum 60 TEU stored longitudinal in four rows and two levels and an operation speed of 10 kn.

The main routes cover short distances for this vessel concept and provide opportunities to stop and recharge or refuel along the route. Therefore, a fully electric propulsion system powered by batteries is considered. As mentioned before, batteries have less range and higher weight than current liquid fuels. This needs to be considered by investigating storage places, stability issues, and safety aspects like fire protection. Another approach for the shuttle concept is the use of fuel cell technology driven by green hydrogen. As for the batteries, hydrogen requires large storage solutions and because of its properties it is suitable for shorter distances and refuelling regularly. This technology uses hydrogen, stored in cooled and/or high-pressure tanks, which is converted in fuel cells to generate electrical power. The biggest challenge in using hydrogen as a fuel is the wide flammability range and the easy ignition. Regardless of which fuel is used, they have to be produced or generated from renewable energy sources like wind, solar, or waterpower. These requirements meet the conditions for green transport systems where zero CO_2 emission is aspired.

This shuttle vessel is equipped with a fully automated, lightweight built gantry crane with telescopic legs. The crane ranges from starboard to portside and can move on guide rails from stern to bow to ensure each container can be reached and long crane booms can be avoided, which otherwise would cause stability problems. The usage of such a mobile crane is beneficial for smaller ships as storage loss can be avoided. During sailing the crane will be secured at the stern. Furthermore, a telescopic spreader for 20 to 45 ft container is used to load and unload cargo.

The shuttle vessel is designed to be highly autonomous. As shown in Figure 12, the vessel has no superstructure since no crew is on board. Dependent on the level of autonomy the vessel could be operated by a remote-control centre on shore (level 3) or sail fully autonomous (level 4). Additionally, automated mooring and automated cargo handling as well as the implementation of a digital twin is considered for this developed design.

Barge convoy

The second vessel concept contains small, autonomous, and self-propelled barges which will be pushed by a green, advanced, and highly automated push boat. As found in the analysis of use case A, the main cargo types are containerized cargo and bulk cargo. Therefore, universal barges are developed with a capacity of 36 TEU stored longitudinal in three rows and in three layers. It is also possible to load bulk or a combination of bulk and containerized cargo as seen in Figure 13. These barges are designed as additional storage space off-shore or as an expansion of port-side storage space for possible customers. Resultant from these assumptions the main dimensions are listed in Table .

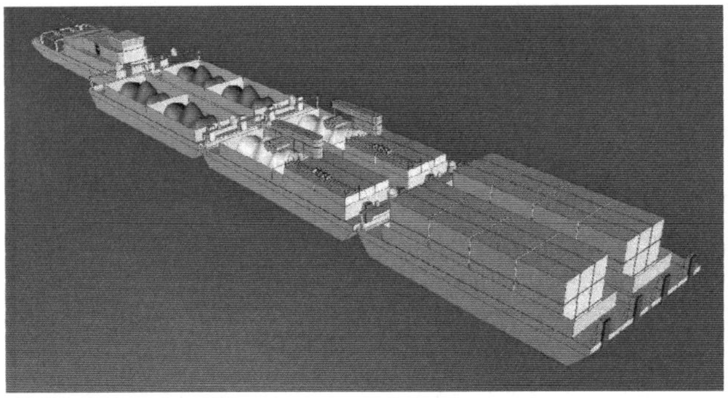

Figure 13 Push convoy with three different barge concepts.

Table 4 Main dimensions for universal push barge.

L_{OA}[m]	B[m]	T_{Design}[m]
33.0	9.8	2.0

The barges can be un-propelled for pure push operation or equipped with a short-range propulsion system. The self-propelled barges are designed to be able to operate small distances. They are equipped with a centred foldable rudder propeller which folds into a hull niche while the barges are part of the pushed convoy. It is also possible to operate this rudder propeller as a sail drive, which works as a hydro generator to generate power while being pushed in the convoy. These barges are designed to sail the "last mile" fully electrically to and from the end destination by itself powered by battery packs which are not easily exchangeable and permanently installed, but they are still accessible and can be removed or exchanged if repairs are needed. The battery packs are placed midships above the rudder propeller dependent on the barges purpose, either below the crane or in the first layer of containers which involves capacity loss of about two TEU, or between the cargo rooms on the bulk barge (see Figure 14). Further, integration of solar panels on deck of the barges supports the power supply on board ([17], [18], [19]), although the space is very limited and exposed to a high risk of damage.

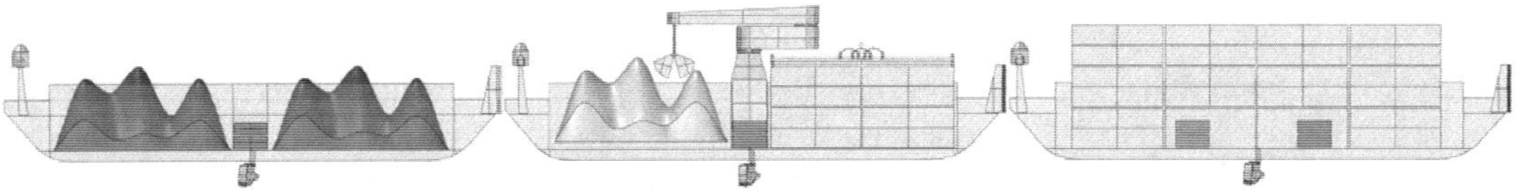

Figure 14 Possible battery (green) storage inside the different barge concepts.

The cargo handling equipment for the modular designed barges depend on the transported cargo. In case of mixed cargo, it is required to handle different cargo with the same crane. Thus, equipment for handling both cargo units is needed and should be easily and preferably autonomously exchangeable. All cargo handling equipment will be carried on the barge, easily accessible for the installed crane.

The barges are designed to be fully autonomous (level 4). Equipment for positioning, navigation, and communication is installed on board. Furthermore, automated mooring and automated coupling in and out of the convoy is supported by the on-board equipment.

Push boat

This vessel (see Figure 15) is developed to be able to push a maximum of nine barges (described above) at an operational speed of 8 kn. Resultant from these assumptions the main dimensions are listed in Table .

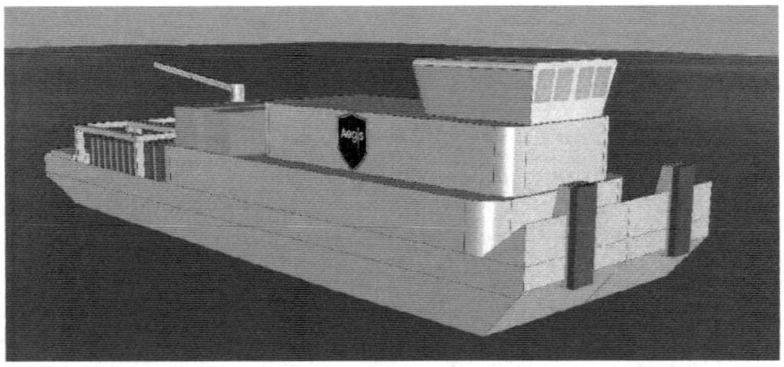

Figure 15 Push boat for convoy concept.

Table 5 Main dimensions for push boat.

L_{OA}[m]	B[m]	T_{Design}[m]
33.4	10.8	2.0

The coupling system at the bow of the vessel is similar to the one designed for the barge concepts. Such coupling system allows the coupled units movements in vertical direction while it suppresses any movements in horizontal direction. Furthermore, no payload is stored on the push boat but only on the barges.

However, some storage areas on the push boat are considered for fuels. The vessel is designed as hybrid solution with fuel cells driven by hydrogen, and battery packs. As machinery two Azimuth thruster are considered to ensure high manoeuvrability of the vessel. At the very end of the vessel the hydrogen (red colour) is stored in pressure bottles at 200–250 bar which are combined to three packs, each in the size of a 20 ft container. This storage solution is designed to be standardized, and therefore easy to exchange which saves time for loading and refuelling. Furthermore, permanently installed battery packs inside the vessel ensure a homogeneous power supply.

As mentioned above, no cargo capacity is considered on the vessel and therefore no cargo-handling equipment per se is needed. Nevertheless, the vessel is equipped with a small, stationary, and automated crane to load and unload the hydrogen packs when exchanging empty by fuelled ones.

The autonomy level of this vessel is considered to grow from 2 up to the level 4, hence, a bridge and crew accommodations are needed at the beginning. The sensors for positioning, navigation, and communication as well as supporting software and systems are also on board installed. Route and fuel consumption optimization are currently available and required for this push boat concept. Step by step, the vessel could be operated by a reduced crew with many supporting systems (level 2). Later the operation via a remote-control centre (level 3) on shore with no crew on board is possible. The last step is a fully autonomous operation (level 4) monitored in a control centre.

4.2. IWW vessel designs for Use-case B

In this subsection, the current status of the developed vessel designs for use-case B is described. Note, that these designs are in concept state and not fully developed or detailed.

As mentioned in subsection 2.2, these RoRo vessels are required to sail on inland water ways (IWW) in the region of Netherland and Belgium to reduce congestion of the roads. Therefore, the vessel design is strictly limited to the available waterway dimensions along the expected route. These dimensions are classified within the CEMT and provide basis for the development of green and advanced vessel designs.

In Figure 16, three different sizes of RoRo vessels are shown. The two upper designs are suitable for CEMT class II (top left) and CEMT class IV (top right) where the bottom design is a new combination of the CEMT classes IV with an adapted breadth according to CEMT class VIb.

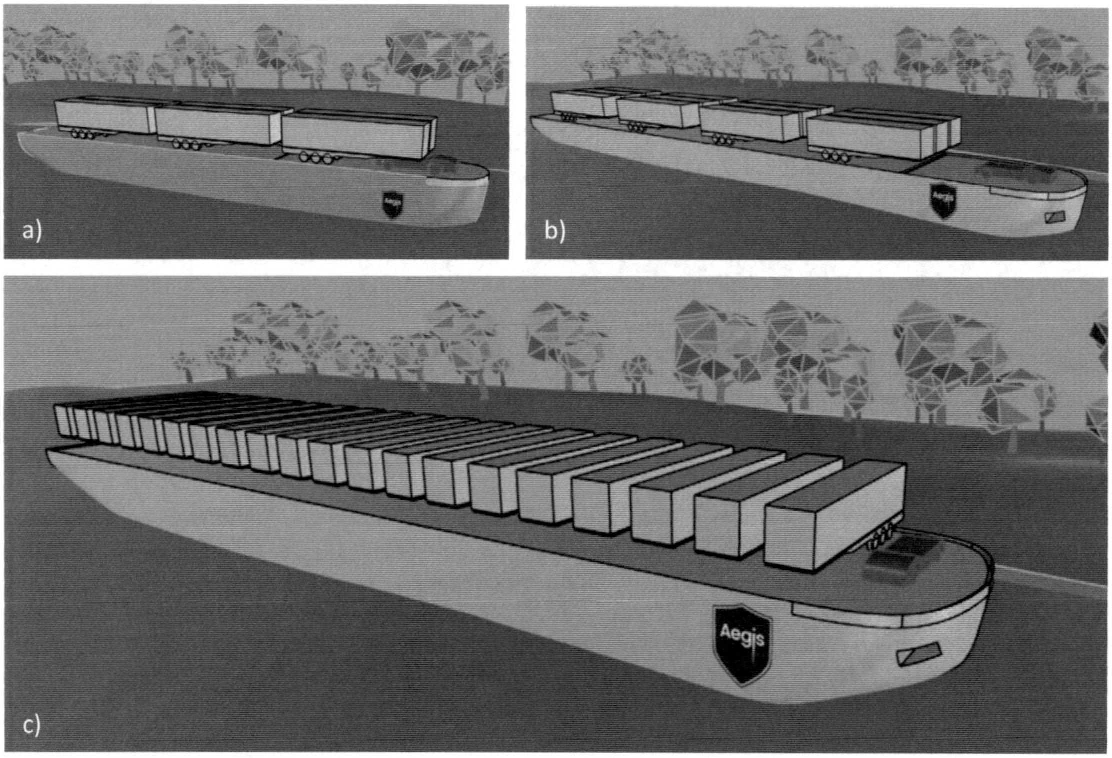

Figure 16 IWW RoRo vessel designs; suitable for CEMT II (a), CEMT IV (b) and CEMT IV+ (c).

These approaches yield to main dimensions, which are listed in Table . In general, the propulsion and cargo handling systems are the same for each vessel, however they are adopted to the vessel size. The capacity of the vessels are 12 trailers for CEMT II, 21 trailers for CEMT IV and 38 trailers for CEMT IV+.

Table 6 Main dimensions for IWW vessels.

Vessel	L_{OA}[m]	B[m]	T_{Design}[m]
CEMT II (a)	55.0	6.6	2.3
CEMT IV (b)	85.0	9.5	2.5
CEMT IV+ (c)	85.0	15.0	2.5

The developed vessels are designed to drive fully electrically powered by batteries. These battery packs are located at the end of the vessels and due to fire protection and on-board safety [20], it is required to surround the batteries with transverse bulkheads. Further, the vessel is equipped with two azimuth thrusters at the stern, and on rotatable bow thruster [21] to maximise the manoeuvrability and flexibility of the developed vessel designs.

For loading and unloading, it is required for the vessels to have a ramp build into it (not shown in Figure 16) which is located at the bow, automated and connected with an automated mooring system. Furthermore, two layers of cargo are planned which requires a lift system for the trucks or trailers where the distribution into the two layers is possible. In addition, the vessels will also have an own AGV (automated guided vehicle) on board for loading and unloading the trailers in smaller ports along the route. The AGV will be stored on deck when the vessel is sailing. All loading systems are developed to be fed by the vessels power supply or by green shore power, if available.

Loading the trailers transversal to the sailing direction requires a different type of cargo handling. Every pair of trailers stored on top of each other uses its own lift combined with a moveable side ramp to load and unload over portside. Therefore, an automated stability system for loading and unloading is required. This system is considered to continuously pump water in and out the ballast water tanks.

Not depending on the loading direction of the trailers, the vessels are designed to be highly autonomous. Regarding the level of autonomy, it is either fully autonomous (level 4) or remotely controlled from a remote-control centre on shore without crew on board (level 3), leading research projects are FernBin [22] and AutoBin [23]. Furthermore, autonomous mooring systems are used and digital twins will be implemented in operating the vessels.

4.3. Combined concepts for Use-case C

This subsection describes development approaches for use case C. The developed vessel designs are deducted from the two other use cases. As mentioned in subsection 2.3, LoLo as well as RoRo vessels are required to cover the possibilities of the defined regions in Denmark.

First approaches are shown in Figure 17 and Figure 18. In the top picture, a combined SSS and IWW container vessel with a capacity of about 128 TEU is depicted. The lower picture shows also a combined SSS and IWW vessel but for mixed cargo transportation.

Figure 17 Container vessel design.

Table 7 Main dimensions for container vessel.

L_{OA}[m]	B[m]	T_{Design}[m]
71.0	10.9	2.8

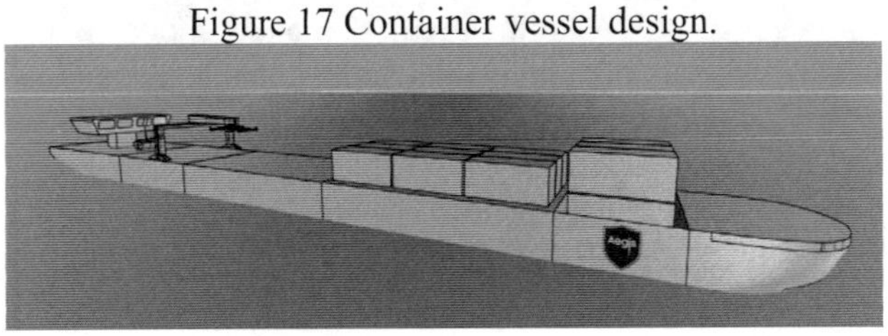

Figure 18 Mixed cargo vessel design.

Table 8 Main dimensions for mixed cargo vessel.

L_{OA}[m]	B[m]	T_{Design}[m]
85.0	9.5	2.5

For both vessel designs, a hybrid propulsion solution is considered. Batteries are required to support the power supply during peak performance and when entering and mooring in ports. For the container vessel, the main propulsion system is considered to be fuel cell technology driven by hydrogen which requires high pressure tanks and extra space for bunkering. On the other hand, the mixed cargo vessel is considered to use methanol or ammonia as the main power source for the vessel. The concept design with the arrangement of cargo, fuel tank (pink), and battery packs (orange) is shown schematically in Figure 19.

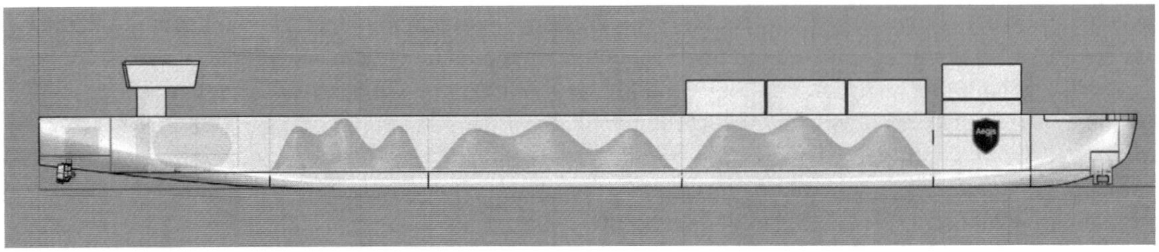

Figure 19 Mixed cargo vessel schematically; fuel tank (pink), battery packs (orange).

This hybrid solution is selected due to the sailing distances expected for the vessel and due to the future infrastructure of methanol built by Maersk ([24], [25]). Ammonia might also be available due to its usage in fertilizer production. The infrastructure for handling of ammonia is established but not yet enough extended to cover the consumption needs for worldwide shipping.

The applied cargo handling equipment on the developed vessel designs are deduced from use case A vessel designs. One design on the container vessel design is a triple joint crane from project Partner MacGregor. These kinds of cranes are self-stabilising and therefore very suitable for the concept seen in Figure 17. Due to the location of the crane on the vessel, all cargo can easily be reached and loaded/unloaded without requiring any further equipment from the portside. The other design is a moveable gantry crane similar to the one described in subsection 4.2. For the concept seen in Figure 18, the width of the crane is adapted to the vessel breadth and exchangeable equipment for different cargo units are required. When the ship is sailing the crane is secured in front of the bridge as close to the deck as possible.

These first approaches tend to align only one part of use case C. Therefore, further concepts need to be developed to fit all requirements. Especially a RoRo vessel serving along the south of Scandinavia will be designed. For this purpose, the CEMT class IV+ RoRo design from subsection 4.2 will be adopted to the requirements and conditions of this use case.

These vessels are designed to be operated medium to highly autonomous. During operation, the autonomy level will increase from level 2 up to level 4. Therefore, planned captain's bridges and crew accommodation will be modal and easy to deconstruct. Additionally, automated mooring and automated cargo handling as well as the implementation of a digital twin is considered for this developed design.

5. Conclusions and future steps

The central objective of AEGIS is to develop a new waterborne transport system for Europe that leverages the benefits of ships and barges while overcoming the conventional problems like dependence on large terminals, high transhipment costs, low speed and frequency and low automation in information processing.

To reach this goal, new advanced, efficient, green and smaller vessel designs are developed besides other factors like terminal automation and digital connectivity, especially the communication between all the participants of the future transport chain. These vessel designs, as well as the whole transport chain including all transport modes and terminal activities are developed regarding three different real life use cases in Europe with respect to the environmental and regional conditions.

One main part of the vessel development is the implementation of new technology for autonomous vessel operations and transhipments. Therefore, new technology approaches from the project partners MacGregor and Kalmar for autonomous transhipment are investigated, simulated and built in terms of a demonstrator.

Investigations within the use cases leads to the development of a logistics analysis tool. This tool is required to be able to simulate and compare costs of shipping routes with a current vessel design or a comparable autonomous vessel design. This is going to be just one aspect of the logistics analysis tool.

Acknowledgements
The AEGIS project has received funding from the European Union's Horizon 2020 research and innovation program under Grant Agreement N°859992. Partners in AEGIS are Port of Aalborg (DK), Aalborg University (DK), Kalmar (FI), DFDS (DK), DTU (DK), Grieg Connect (NO), Institut für Strukturleichtbau und Energieeffizienz (DE), MacGregor (FI), NCL (NO), SINTEF Ocean (NO), Port of Trondheim (NO) and Port of Vordingborg (DK).

Appendix A

Table A 1 Summary Table KPIs for AEGIS.

KPI Level	KPI Sublevel	KPI Name	KPI Measurement	KPI Description
Economic	Cost	CAPEX	€	Capital expense
Economic	Cost	OPEX	€	Operating expense
Economic	Cost	Maintenance costs	€	All expenses to ensure the c orrect operation of an asset a nd keep reliability high
Economic	Cost	Port charges	€	Fees paid to port authorities for the use of its facilities and services
Economic	Cost	Fuel cost	€/NM	Total amount of money spent in fuel
Economic	Cost	Wages	€	Total amount of money spent on salaries
Economic	Cost	Cargo unit cost	OPEX/TEUs	OPEX divided by the numbe r of cargo units
Economic	Time	Loading time	h	Duration of the loading proc ess
Economic	Time	Unloading time	h	Duration of the discharging process
Economic	Time	Sailing time	h	Duration of the vessel voyag e
Economic	Time	Waiting time	h	Time during which cargo is i dle or delayed
Economic	Time	Drive time	h	Duration of the trip of the ca rgo from to its final inland d estination, and vice versa
Economic	Time	Punctuality rate	% of port calls	Mean deviation from expect ed arrival/departing time.
Economic	Time	Recovery time	h	Time from the detection of a disruption to when full level of performance is restored
Economic	Time	Cargo handling time	TEUs/h	Time to move goods on and off ships plus the terminal handling time
Economic	Others	Energy consumption	KWh	Total energy needed
Economic	Others	Cargo carried	TEUs/ship	Cargo carried from loading t o discharging
Economic	Others	Percentage of load	Cargo car/max cap.	Actual cargo carried compar ed to vessel maximum loadi ng capacity
Economic	Others	Cargo lost	% total cargo	Cargo unable to be found
Economic	Others	Number of Cyber-attacks	#	Quantity of cyber-attacks suffered
Economic	Others	Restored level of	%	How fully performance is

		performance		restored after a disruption occurs
Economic	Others	Autonomy level	levels	Degree of autonomy
Economic	Others	Frequency of service	Shipments/week	Number of available sailings per week
Economic	Others	Energy efficiency	%	Energy-expenditure required to achieve a target
Economic	Others	Number of container moves	#TEU/route	Amount of goods shipped per route
Environmental	Emissions	CO_2	Kg of CO_2/tkm	CO_2 emissions
Environmental	Emissions	NO_x	Kg of NO_x/tkm	NO_x emissions
Environmental	Emissions	SO_x	Kg of SO_x/tkm	SO_x emissions
Environmental	Emissions	Particulate matter	Kg of PM_{10}/tkm	PM_{10} emissions
Environmental	Emissions	Acoustic emissions - Noise	dB	Noise emitted
Environmental	Others	Use of renewable energy sources	%	Percentage of energy consu med that comes from enviro nmental-friendly energy sour ces
Social	Security /Safety	Accident rate	#	Amount of unfortunate incidents resulting in damage or injury
Social	Security /Safety	Fatality rate	#	Amount of occurrences of de ath by accident
Social	Security /Safety	Fire incidents	#	Amount of incidents involvi ng smoke, heat and flames c ausing damage
Social	Work-life	Labor conditions	Work-life-balance	Quality of working environ ment
Social	Work-life	Employment	% of change	Influence on the occupationa l rate
Social	Work-life	Income	% of change	Influence on earnings
Social	Work-life	Training	Time/worker	Time invested in teaching an employee a particular working skill

References

[1] 2009 *Stratetic goals and recommendations for EU's maritime transport policy until 2018* /COM/2009/0008 final/Brussels

[2] 2017 *National Transport Plan 2018-2029* Meld. St. 33 (2016-2017) https://www.regjeringen.no/no/dokumenter/meld.-st.-33-20162017/id2546287/

[3] Auditor General Norway 2018 *Transport of cargo* Retrieved June 2020 from: https://www.riksrevisjonen.no/rapporter/Sider/Godstransport.aspx

[4] IMO 2021 *Outcome of the regulatory Scoping Exercise for the use of Maritime Autonomous Surface Ships (MASS)* MSC.1/Circ.1638

[5] Eurostat 2020 *Modal split of inland freight transport, EU-27, 2013-2018 (% share in tonne-kilometres)* Retrieved from ec.europa.eu: https://ec.europa.eu/eurostat/statistics-explained/index.php?title=File:Modal_split_of_inland_freight_transport,_EU-27,_2013-2018_(%25_share_in_tonne-kilometres).png

[6] Netherlands Ministry of Infrastructure and Water Management *Key Transport Figures 2018*

[7] *Filezwaarte* Retrieved from: statistiekvlaanderen.be: https://www.statistiekvlaanderen.be/filezwaarte#kilometeruren

[8] AEGIS 2021 *Potential transfer from road transport to short-sea-shipping in Denmark*

Retrieved Oktober 2021 from: https://aegis.autonomous-ship.org/resources/reports-and-articles/

[9] IMO *Autonomous Shipping* Retrieved January 2022 from: https://www.imo.org/en/MediaCentre/HotTopics/Pages/Autonomous-shipping.aspx

[10] *Autonomous Ships* Retrieved January 2022 from: https://www.standard-club.com/fileadmin/uploads/standardclub/Documents/Import/publications/bulletins/split-articles/2018/2767846-autonomous-ships.pdf

[11] Rødseth Ø J and Nordahl H 2017 *Definition for Autonomous Merchant Ships* Retrieved January 2022 from: https://nfas.autonomous-ship.org/wp-content/uploads/2020/09/autonom-defs.pdf

[12] Bureau Veritas 2018 *Autonomous Ships* Retrieved January 2022 from: https://marine-offshore.bureauveritas.com/insight/autonomous-ships

[13] Kretschmann L 2015 *D9.3: Quantitative assessment* Retrieved January 2022 from: http://www.unmanned-ship.org/munin/wp-content/uploads/2015/10/MUNIN-D9-3-Quantitative-assessment-CML-final.pdf

[14] Molland A F 2008 *The Maritime Engineering Reference Book* (Butterworth-Heinemann) pp. 344-482

[15] Hansa 2021 *Vale bringt Bulker mit Flettner-Rotoren in Fahrt* Retrieved November 2021 from: https://hansa-online.de/2021/05/featured/175454/vale-bringt-bulker-mit-flettner-rotoren-in-fahrt/?utm_source=CleverReach&utm_medium=email&utm_campaign=14-05-2021+HANSADaily+14.05.2021&utm_content=Mailing_12749131

[16] Port of Helsinki 2019 *Second automated mooring system to West Harbour for faster, safer and cleaner operations* Retrieved January 2022 from: https://www.portofhelsinki.fi/en/port-helsinki/whats-new/news/second-automated-mooring-system-west-harbour-faster-safer-and-cleaner

[17] Energy Observer Retrieved January 2021 from: https://www.energy-observer.org/

[18] Trends der Zukunft 2021 *Autonomes Segelboot kartiert den Meeresboden: Jungfernfahrt von San Francisco nach Hawaii* Retrieved November 2021 from: https://www.trendsderzukunft.de/autonomes-segelboot-kariert-den-meeresboden-jungfernfahrt-von-san-francisco-nach-hawaii/

[19] Lewis N 2021 *Greener and cheaper: Crewless vessels are mapping the ocean thanks to an Irish startup* Retrieved January 2022 from: https://edition.cnn.com/2021/09/07/business/xocean-uncrewed-surface-vessels-ireland-spc-intl/index.html

[20] CESNI 2016 Retrieved July 2021 from: https://www.cesni.eu/wp-content/uploads/2016/06/ES_TRIN_de.pdf

[21] Schottel Retrieved November 2021 from: https://www.schottel.de/portfolio/produkte/produkt-detail/spj-schottel-pumpjet

[22] Kracht F E 2020 *FernBin Ferngesteuertes, koordiniertes Fahren in der Binnenschifffahrt* Retrieved January 2022: https://www.uni-due.de/mechatronik/forschung/fernbin

[23] DST *Autonomous Inland Vessel – Simulation and demonstration of automated inland navigation* Retrieved January 2022 from: https://www.autobin.de/en/

[24] Hansa 2021 *Maersk sichert sich E-Methanol aus neuem Werk für Feeder-Neubau* Retrieved November 2021 from: https://hansa-online.de/2021/08/featured/180711/maersk-sichert-sich-e-methanol-aus-neuem-werk-fuer-feeder-neubau/?utm_source=CleverReach&utm_medium=email&utm_campaign=19-08-2021+HANSADaily+19.08.2021&utm_content=Mailing_12956788

[25] Frangoul A 2021 *Maersk links deal to secure supply of "green" e-methanol for vessel, targets 2023 launch* Retrieved November 2021 from: https://www.cnbc.com/2021/08/19/maersk-inks-deal-to-secure-supply-of-green-e-methanol-for-vessel.html

Intelligent Operator Support Concepts for Shore Control Centres

H. van den Broek[1], J. van der Waa[1]

[1] TNO, Kampweg 55, 3769 DE Soesterberg, The Netherlands

Hans.vandenbroek@tno.nl

Abstract. This paper introduces three intelligent operator support functions to allow multiple operators to effectively and efficiently supervise multiple autonomous operations. The many-to-many concept assumes a stage in human-automation collaboration design where supervision of maritime autonomous surface ships is not permanently required anymore. Only in extreme and very rare situations the human may need to intervene. One of the challenges is balancing the task assignments and support functions over the operators to ensure the cognitive task load matches the operator's mental capacity. For this purpose we introduce and described a dynamic task allocation algorithm. Also, human attention is limited and operators therefore must constantly shift attention resulting in moment-to-moment fluctuations in situation awareness. To overcome these reductions in situation awareness, operators must reassess the environment to recover situation awareness. We introduce the concept of continuous risk assessment to initiate the process of situation awareness recovery. Furthermore, the many-to-many ratio between supervising operators and autonomous ships implies that operators will not be able to supervise all ships in parallel. This makes current supervisory control interfaces less suitable. Instead we opt to apply the idea of progressive disclosure in the operator's interface and interactions. The work described in this paper is directed towards developing an intelligent operator support system with which the operator support functions will be demonstrated as part of a Robotic Container Handling System, an innovation of the European MOSES research project.

1. Introduction

It is expected that in the maritime domain, automation will advance to levels where supervision of maritime autonomous surface ships is not permanently required anymore, as systems are able to work fully autonomously in almost every condition. Only in extreme and very rare situations the human may need to intervene. This implies that the human operator will be out-of-the-loop by definition and will not have the skills to take measures in these, probably, difficult conditions. This has led to different opinions and strategies on how to deal with remote operator support in the transition to fully autonomous systems.

Van den Broek et. al [1], distinguish three different stages for highly automated human-automation collaboration that require different operator support designs (see figure 1). In the supervision stage, the first distinguished phase, the human operator monitors the system 100 percent of his time and is not involved in other tasks. Supervision of dynamic positioning systems by onboard operators is an example of this (e.g., [2], [3]). Such one-to-one supervision duties require continuous high vigilance levels and are prone to loss of situation awareness (SA), out-of-the-loop problems [4], and complacency [5].

Stage	Support functions	References
1. Supervision stage. The human operator is 100% of his time supervising the system(s) and is not involved in other tasks	Common (SCADA) interface support	[2], [3]
2. Partial supervision/autonomy stage. Self-directness is higher. The human operator spends part of his or her time conducting secondary tasks.	Supervisory displays; SA recovery support after returning to the main task; Increase the reaction window by detecting early signals; Situation awareness recovery.	[6], [8], [9], [10], [13]
3. Intervener/full autonomy stage. Both self- sufficiency and self-directedness are high. The system is working autonomously almost 100%	Dynamic task allocation; Continuous Risk Assessment; Progressive Disclosure Interface Design.	This paper, [17], [21]

Figure 1. Overview of three different stages for highly automated human-automation collaboration that require different operator support designs.

SA is defined as the conscious knowledge of the immediate operational environment and the events that are occurring in it. It involves the perception of elements in the environment, comprehension of what they mean and how they relate, and projection of their future state. It refers to the operator's awareness of the current operational status and the anticipated future status of a system, necessary to intervene in an effective and timely manner.

To address these operator related issues (i.e., human factors issues), a new operator support concept has been developed and demonstrated in previous research by TNO to enable partial supervision [6]. In the partial supervision/autonomy stage, the second distinguished human-automation collaboration stage, the system self-directness is higher. This allows the human operator to spend time performing secondary tasks and thereby leave the control station on the bridge. In case a critical situation is building, the operator gets informed, via a smart watch and tablet computer. It is the operator's decision whether or not it is necessary to return to the bridge and to intervene. The main challenge for stage two is to keep the operator posted of the status of critical tasks and to enable him or her to resume control effectively when required. Both the human operator and the automation develop SA relevant for the primary task [7].

The operator support for the partial supervision/autonomy stage consists of the following functions [8]:

1. Support the upkeep of operator SA using supervisory displays [9] to help the operator decide whether involvement in the primary task is required;
2. Provide SA recovery support after returning to the main task;
3. Increase the reaction window by detecting early signals and providing on-time alerts for the primary task;
4. Situation awareness recovery [10]: provide, change detection and option awareness support for quick decision making when a critical event has occurred at the primary task.

However, it is expected that future shore control centres will be staffed in such a way to allow for the supervision of multiple operations in parallel, because it will be more cost efficient compared to the one-to-one set-up (a single human supervises a single operation) described above. Supervisory control by a single operator over multiple operations could match the efficiency requirement. When a human supervises a set of multiple operations (one-to-many) care must be taken to ensure that the operator has the capacity to give adequate attention to each operation [11]. Neerincx [12] developed a cognitive load model, distinguishing three load factors that have a substantial effect on task performance and mental effort: percentage time occupied, level of information processing, and task-set switching. The combination of the three load factors determines the cognitive task load. The challenge is to balance the task load in such a way that it matches the operator's mental capacity in a certain task setting. In addition, unexpected situations can arise in which autonomous systems require human assistance. It can be assumed that these will be safety-critical and difficult situations that will require full attention. As a result, the operator span of facilitation will be limited to one system, with the result that other operations under the responsibility of the operator are forced to left unattended [13], providing a potential risk. For reasons of operational safety, therefore, a transfer of supervision responsibility to another operator is necessary under these circumstances. So, for reasons of cognitive load balancing and for guaranteeing the overall span of facilitation, it is necessary to have some sort of task allocation over multiple operators. For this reason, we propose a many-to-many remote control centre concept, in which several operators supervising a set of multiple operation (several one-to-many instantiations) allowing for dynamic task allocation over operators.

For a many-to-many shore operation concept to work requires that both self-sufficiency and self-directedness of the autonomous systems are high and the systems are working autonomously nearly 100 percent of the time. This allows the operator to work on other tasks, i.e., to work on other operations. But even fully autonomous systems fail sometimes, for different reasons, and in these exceptional cases the operator needs to intervene. This third and highest level of human-automation collaboration is the so-called intervener/full autonomy stage.

In line with the above, this paper advocates the necessity of additional operator support concepts to enable this high level human-automation collaboration.

The additional operator support concepts that will be discussed in this paper are:

1. *Dynamic task allocation.*
   Assign ships (operations) to operators based on user and task profile.
2. *Continuous Risk Assessment.*
   Assess in real-time potential risks and warn, inform, explain and help solve them.
3. *Progressive Disclosure Interface Design.*
   Show information and offer control on different abstraction levels.

The research outlined in this paper is carried out as part of the European MOSES[1] research project. As a system demonstration is part of the MOSES scope, the concepts will be integrated into a single AI-based Intelligent Operator Support System (IOSS).

The use-case within MOSES is to enable multiple operators to oversee and track multiple loading and offloading operations, executed by robotic container handling systems mounted on autonomous feeder vessels, and to take action and intervene when necessary. The MOSES research scope and the robotic container handling systems innovation will be described in the next section.

The three additional operator support concepts are described in the following paragraphs. The article ends with a conclusion section.

2. The MOSES research context

The European MOSES research project [14] is one of three research and innovation projects within the Horizon 2020 program that contributes to more automation and autonomy in Europe's short sea logistics. The other two projects are AUTOSHIP and AEGIS. The main focus of AUTOSHIP is on vessel technology. The AEGIS consortium, on the other hand will design Europe's next generation sustainable and highly competitive waterborne logistics system comprising more autonomous ships and automated cargo handling.

The innovations within the MOSES project, aim to significantly enhance the Short Sea Shipping (SSS) component of the European container supply chain by implementing a constellation of innovations including innovative vessels and the optimization of logistics operations.

The work described in this paper is directed towards developing and demonstrating a Robotic Container Handling System (RCHS) that, when mounted upon a hybrid electric and autonomous feeder

[1] AutoMated Vessels and Supply Chain Optimization for Sustainable Short SEa Shipping.

vessel, will stimulate and support the use of short sea container services for small ports that have limited or no (un)loading infrastructure.

Normally, a crane operator controls the crane based on his/her eye-hand coordination, e.g. knowing which container to pick-up, he/she looks for the position of the container, estimates the distance between spreader and container, reduces speed if needed, locks, etc. Furthermore, the safety of the operation is guaranteed with a relatively high degree of oversight from other people as well as the creation of a secured and sealed-off area. The RCHS research challenge[2] is to bridge the large gap between the current way of working and the envisioned concept where the RCHS can perform all these tasks on its own, i.e. autonomously.

The RCHS is approached as a systems-of-systems. In fact, the RCHS consists of five major components, each consisting of multiple subcomponents.

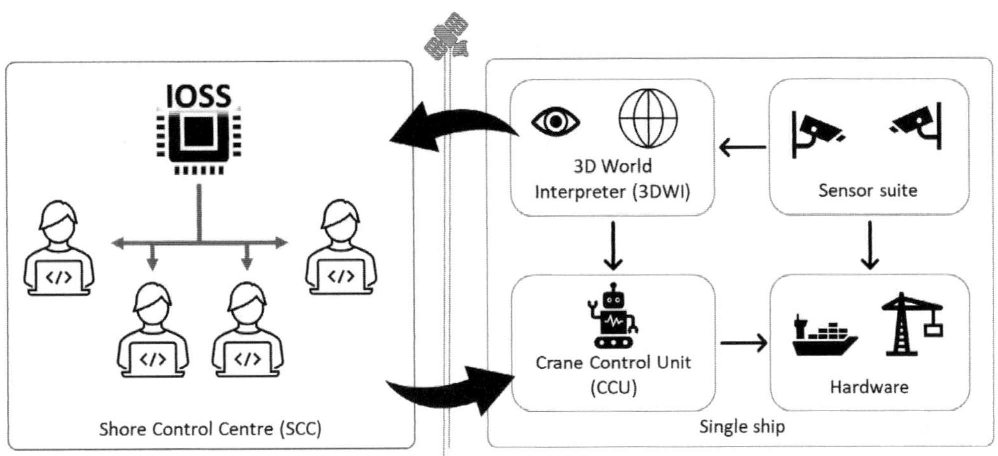

Figure 2. As a systems-of-systems, the robotic container handling system consists of five major components.

1. The crane, spreader, and automatic swivel control system.
2. The Crane Control Unit (CCU), responsible for both the control and high-level decision-making needed to load and unload containers safely in a public area at a small port. The control functionality of CCU is responsible to drive the crane and all of its sub-components. The decision-making function is responsible to decide which container to load or unload when and where.
3. The 3D World Interpreter (3DWI), is responsible for the perception (detection) of elements in the environment, comprehension of what they mean (identification) and how they relate. This SA is required by the CCU for controlling the crane and high-level decision-making. Likewise, it provides the local information needed both for operator SA as for the functions the operators support. Combined, the CCU and the 3DWI is the core of a self-sufficient autonomously operating crane system.
4. Sensor suite, a combination of (stereo) cameras and lidars to provide the 3DWI with images of the operational area.
5. The Intelligent Operator Support System (IOSS) is responsible for safely managing a large amount of autonomous container handling operations occurring in parallel and their supervision from a remote shore control centre. This system ensures that every ongoing operation is performed effectively and safely. It does so by providing support to enable only a few operators

[2] Within the RCHS research and development TNO closely collaborate with MacGregor Finland Oy, the company that provides the physical crane system and crane control software.

to supervise many operations. Examples are the allocation of operations, SA support, risk assessment, control suggestions and more.

3. Dynamic task allocation

As indicated, within the concept it is envisioned that several remote operators within the Shore Control Centre will be responsible for several loading and unloading operations at the same time.

To deal with workload fluctuations in a dynamic way, the dynamic task allocation algorithm connects to the voyage and operational stage planning on fleet level and estimates the operator workload level associated with the number of ships under supervision and their voyage stage [13]. When workload thresholds threaten to be exceeded, the span of control, i.e. the number of ships under control, can be enlarged or reduced through task reallocation.

The algorithm takes five stages into account: sea passage (transit), port departure/approach, leaving/approaching berth, and container handling. For each stage the autonomy mode changes for which the workload demands can be planned in advance with exception of the stage in which exceptional circumstances occur.

We face two challenges when allocating ships to operators based on expected workload per stage. From a human factors perspective we face the challenge of modelling workload. Second, this model needs to be incorporated into a feasible optimization algorithm to compute an responsible allocation.

To model workload we opted to define a cost function. This function takes a given ship-to-operator allocation and returns how costly it is in terms of workload. Workload is defined as a percentage of an operator's total capacity (e.g., 100%). This abstraction allows us to circumvent the issue of formally defining the cognitive processes that determine workload. For instance, a task with an expected load of 10% implies that the operator can handle ten of such tasks in parallel given a capacity of 100%.

Given a ship and a certain loading/offloading at a port, we assign such a workload to each of the stages. For instances, for a particular combination of a ship, port and containers the sea passage might require 5%, the port approach 10%, and the container handling stage 15%. These percentages can vary given the complexity of the tasks. For instance, autonomously berthing to a crowded port might be more difficult and thus potentially require more supervision. To determine these relative percentages an approach such as by Wilson et. al [15] can be used that utilizes physiological measurements to assess the cognitive strain a task has on people. Such an approach should typically be performed in the design phase of the shore control centre and can be repeated when necessary.

We also account for the fact that these workloads per stage depend on an operator's expertise. As such, we increase these expected workloads as operator expertise decreases. For instance, in the example above, the stages might require double the workload for a junior operator resulting in an actual workload of 10%, 20% and 30% respectively. Operator expertise can be simply defined by their years of experience with a fixed factor affecting the workload of various tasks. However, it is more optimal to perform a more in-depth task analysis on what expertise each task and difficulty in an operation requires combined with assessments how skilled an operator is in such an expertise. Although this remains an open challenge, works such as by Allais et. al [16] illustrate how this can be done in other domains.

This provides us with a basic model of workload that is feasible to work with. Next we define the cost function. The purpose of this function is to compute how "bad" any given allocation of ships to operators is. The higher the cost, the worse the allocation in terms of workload distribution. We define this function as a simple linear additive function, where several components model an aspect of human factors knowledge.

The components we defined are as follows:

1) The average and variance of workload over time and all operators. Minimizing this ensures that the workload is equally distributed over all operators.
2) The average time in a critical workload for all operators. A critical workload is defined as either too high of a workload (e.g., above 80%) or too low of a workload (e.g., below 20%). Minimizing this ensures that operators are not cognitive over- or underloaded.

3) The average time operators still have work after a certain amount of time. This models the requirement to take breaks. For instance, every four hours operators should have time with no workload. Minimizing this ensures operators have sufficient breaks.

These components are each weighted in linear fashion. The selection of weights should be considered in the design phase of a shore control centre as it is dependent on the models of workload, operator expertise, and tasks as well as the desired working environment within the centre. The summation of the weighted components result in single combined cost. For instance, an allocation with a cost of 0.5 is a worse fit given these than one with a cost of 0.25. See figure 3 for an illustration how workload is modelled over time for three operators and six operations. For a more detailed overview and evaluation of this cost function we refer to Brug et. al [17].

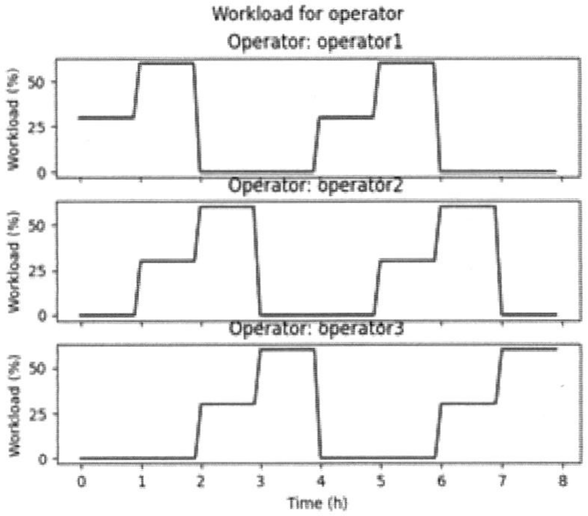

Figure 3. Three graphs depicting how we model workload for three operators given the allocated ships for him/her to supervise. The x-axis depicts workload capacity and the y-axis the time.

Given this cost function we can know apply common optimization strategies to find the best possible allocation of ships to operators based on our workload model. We opted for the Iterative Deepening Depth First Search (IDDFS) [18]. This is not necessarily the only available algorithm as we defined the challenge of dynamic task allocation as a straightforward optimisation problem. The IDDFS algorithm however is a relatively straightforward approach that is reasonably efficient and performant, and it lends itself to reallocate tasks by reusing previous computations. Which is a beneficial property to dynamically reallocate tasks based on events (i.e., a new operation becomes known or the difficulty of an operation changes suddenly). Alternative approaches could include classical approaches such as A* [19] and derivatives, or more recent algorithms based on linear programming [20]. The only additional requirement is that the selected algorithm can efficiently handle the reallocation of already assigned operations. Otherwise, the algorithm will be uncapable of handling the dynamic reallocation of operations effectively when a new operation is introduced.

The IDDFS algorithm functions by iteratively selecting an unallocated ship and assigns it to each operator. It computes the cost of each potential allocation and selects the best fit. Then it continues with the next unallocated ship, until all are allocated. Since the order of ships matters, several samples must be taken to improve the allocations. This is repeated for a number of times with the operations being allocated in different orders to ensure enough of the search space is covered. If all orders are addressed, the optimal allocation will be found. This is still more time and memory efficient (both linear) then a brute force search (exponential) since computations can be reused and only a fraction of all potential allocations are actually computed. For a more detailed discussion of this we refer to Brug et al. [17].See figure 4 for an overview how this algorithm functions in one iteration given three operators and three ships to allocate.

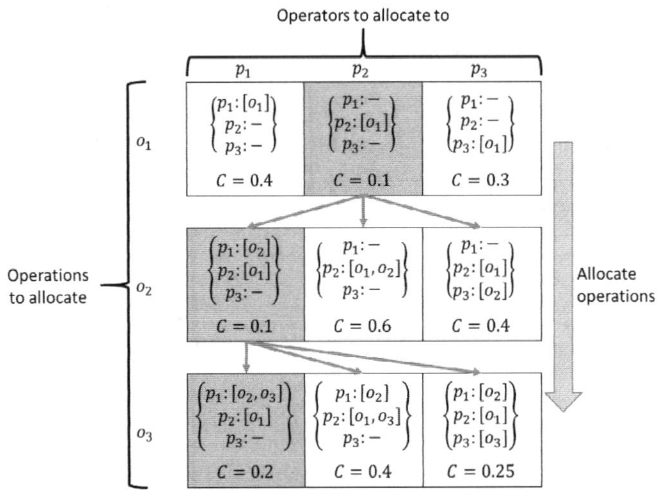

Figure 4. An illustration of the IDDFS algorithm used to allocate three autonomous ship operations (rows; o1, o2 and o3) to three operators (columns; p1, p2 and p3). At each allocation iteration the cost of allocating a ship to any operator is assessed (depicted as C), and the best one is selected (green highlight) until all ships are allocated. Such a search tree is repeated with the operations in various orders to ensure convergence to the optimal allocation.

This approach fulfils three major requirements;

1) It models workload in an abstract sense that is easily extendable with more notions from the human factors literature.
2) It is guaranteed to find the optimal allocation given sufficient time, which increases only linear given the number of operators and ships.
3) The allocation is dynamic.

Especially the latter is a major advantage. As this approach allows for the introduction of new ships requiring supervision as well as operators to flag ships they do not feel they are suited to supervise. In both cases a new allocation needs to be found. This can be done by allocating that ship to the operator with the lowest workload at the time of the operation. However, there might be a better allocation if additional ships are reallocated. For instance, in cases when the new operation's difficulty requires a more experienced operator then is available.

With the IDDFS algorithms we can efficiently reallocate operators by reverting several steps of the computation. Each reverted step signifies an operation that might potentially be reallocated. For instance, we could undo the last five allocated operations. This would give the IDDFS algorithm more freedom to allocate the new operation to a suitable operator without having to reallocate all operations. To determine how many operations should be reallocated we introduce a fourth component in the cost function that penalizes each operation that is being reallocated (e.g., reallocating five operations might introduce an additional cost of 0.1 whereas three operation an additional cost of 0.06). This allows us to incorporate the decision how many operations should be reallocated into the optimisation process itself. In other words, the IDDFS algorithm can revert steps incrementally and reallocate those operations until the gain in reallocating operations balances the cost of doing so. See for more details on this the work by Brug et. al [17] that describes this process in more detail and demonstrates it on several simulated cases.

Since we provide operators to flag operations they feel should not be allocated to them for some reason, it might occur that a single operator does not accept a necessary reallocation. In this case the IDDFS can fall back to a less optimal allocation. However, if operators are reluctant and flag operations too often, we put the computed plans up for review for a team leader to allocate operations manually by presenting the expected cost and workload over time.

To conclude, this dynamic task allocation method allows for the efficient allocation of supervisory tasks to operators under changing conditions while accounting for human factors knowledge on the ideal workload for operators.

4. Continues risk assessment

Because human attention is limited, operators in this dynamic environment must constantly shift attention between different autonomous ships and operations. These shifts in attention result in moment-to-moment fluctuations in situation awareness (SA). For example, when one ship needs assistance, the operator's attention is mainly focused on a single ship and the operator cannot know or forgets about other ships, situations and critical information. Such reductions in SA are a constant, given the fact that attention has to be divided over several operations. To overcome these reductions in SA, a skilled operator must reassess the environment to recover SA; Gartenberg et al. [10], call this process situation awareness recovery (SAR).

It is vital to the SAR process that operators are informed in a timely manner about a critical situation or receive information about a situation that could develop into one. The IOSS has therefore been given the functionality to continuously assess risks for each ship and to adequately bring a risk to the attention of the operator. This raises the question of what is 'adequate' and what this means for the assessment of such risks.

In the case of our operators, every intervention requires SAR. Risk assessment can provide the means for IOSS to evaluate whether an intervention is, or will be, required. For instance, if the weather worsens the risk of an unsuccessful autonomous berthing might increase and thus offer a trigger for IOSS to bring this to the attention of the allocated operator.

This raises the question however, what needs to be communicated to provide this awareness. A risk can be defined by its expected impact, which is a combination of its likelihood of occurring and severity of its consequences if it occurs. While this seems to be an adequate definition, it lacks real world complexity, and in particular the complexity seen in maritime industry. It is common for errors, malfunctions, calamities and other events to be interrelated, resulting in a potential long list of risks with various likelihood and severity. Such a long list does not improve an operator's SAR process.

Instead we opt to also include the connectivity and velocity of risks in its expected impact. A highly connected risk, is a risk that – if it occurs – affects the likelihood of many other risks. The velocity of a risk dictates how much more likely it makes those other risks it is connected with. This makes the expected impact of a risk much richer as it also includes the relations between risks. For instance, a malfunctioning sensor and a busy port might not be that severe on their own, but combined make the risk of human injury much more likely.

Without modelling the connections between risks, operators might not be involved in a sufficiently timely manner. In addition, modelling such relations allows for a straightforward root cause analysis – and even tracing causes and effects – by analysing the network of risks and their effects on each other given a situation. This kind of information can be visualized and communicated to an operator to facilitate a quick grasp of the situation and how to effectively intervene.

The challenge is to find a way to model these risks. We opted for the use of graphs, where a single node represents a certain risk. Within that node is a computational model of its likelihood given a current situation. This can be as straightforward as a slowly increasing likelihood of a sensor malfunctioning as time progresses, or include a much more sophisticated predictive maintenance model using AI methods. See figure 5 for an example of such a graph.

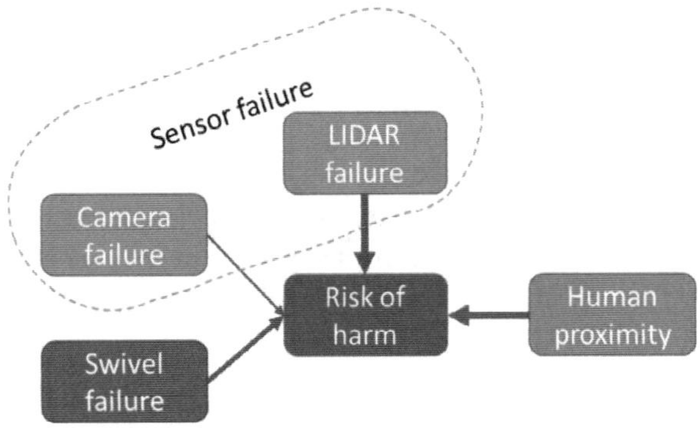

Figure 5. An example on how we propose to model risks and their relation through the use of graphs. It illustrates four risks (rectangles) and their causal connections (red arrows, strength depicted by thickness). A single high-level risk category is depicted for sensor failures. This small toy example shows how the risk of harm is significantly more likely if the LIDAR fails and people are detected nearby.

Also contained in that node is a similar model of its severity. Again, this can be as straightforward as a relative number dictating that a risk such as a busy port is of medium severity. Instead, it might again be a much more complex model where the severity depends on the situation as well. For instance, that a port busy with commercial traffic is less severe than a port with much recreational boating.

The edges between the nodes dictate the relations between the risks. On each edge there might be a simple weight to model its connectivity; the larger the weight, the more it impacts the likelihood of a connected risk. Again, it might also be much more complex where connectivity depends on the situation. For instance, the sensor failure and busy port can be both connected to the risk of human injury. However, if the sensor has another as backup, then its velocity – its weight on the edge between itself and the risk of human injury – might be much lower.

The impact of each risk can then be computed using so-called "spreading activation algorithms". These algorithms propagate the effects, a node's "activation", of risk to its connected risks. This "activation" of a node is equal its likelihood, which in turn is affected by the likelihood of the nodes connected to this it. Ideally, such graphs are acyclic, meaning that risks are not connected to themselves through other risks. As this would introduce a feedback-loop, resulting in a cascade of spreading activations that end with all risks being entirely likely to occur. If, for some reason, the graph needs to be acyclic this effect can be easily prevented by introducing a diminishing effect of one risk on another based on how many risks are between them.

As stated, this approach allows for a straightforward approach to root cause analysis, which can be used to explain to an operator why a certain risk has such a high likelihood. What is needed is only to trace back how much each linked risk affected that risk's likelihood. This can also be visualized in various ways. The most interestingly is presenting the graph itself as an overview of all current risks. If such a graph contains many nodes, a (hierarchical) classification of risks can be defined. For instance, a class "malfunctions" can be made that in turn can contain a class "sensor failures" which contains a singular risk such as "camera #1 failure". Not only can such a classification be used to visualize a comprehensible graph, it can also offer a way to organize risks. For instance, a risk of "malfunctions" can be defined which is simply the highest likelihood of one of the risks contained in this class.

To conclude, a graph-based risk assessment approach enables the use of state of the art risk models to provide an operator with timely notification of a potential risk occurring and a relevant causal analysis to support his/her SAR process.

5. Progressive disclosure interaction design

With IOSS we aim to achieve this many-to-many ratio between supervising operators and autonomous ships. By definition this implies that operators will not be able to supervise all ships in parallel under their responsibility. This makes current supervisory control interfaces less suitable. Instead we opt to apply the idea of progressive disclosure in the operator's interface and interactions.

Progressive disclosure is an interaction design pattern that provides a user with increasingly more and detailed interactions about a topic [21]. The pattern entails the idea that interactions (or control signals) should start simple, and progressively become more complex and intricate based on previous interactions and the situation. The reason to do so is to prevent overload, reduce the risk of human-error, and prevent micro-management.

The idea of progressive disclosure for operators is not novel [22]. We argue however that this idea should be much pervasively present to enable a many-to-many ratio. Furthermore, progressively disclosing interaction designs tend to be entirely user-driven where the user determines the pace and next steps. With IOSS functions as dynamic task allocation and continuous risk assessment this does not necessarily need to be the case.

We developed for IOSS a three-layer interface that progressively discloses more information about the operations an operator supervises and enables increasingly more complex and involved control actions. These three layers are; 1) a global view providing an overview of the entire fleet an operator is responsible of, 2) a localized view providing an overview of the status of a single ship, and 3) a situational view providing an entirely immersive view using virtual reality on a single situational perspective of a ship. See figure 6 for an overview of each level.

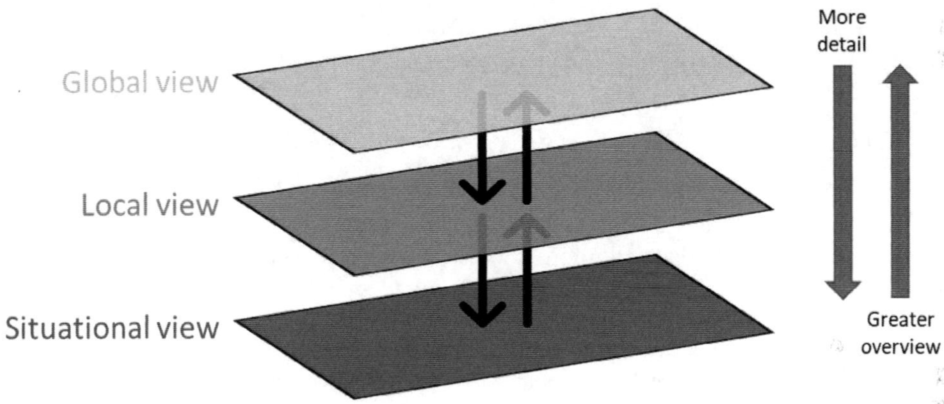

Figure 6. An overview of the three interface layers for the operator. This follows the design pattern of progressively disclosing more information and allowing more complex control actions.

The global view lists all ships that were allocated and accepted by an operator. For each ship it provides general information, such as tasks (e.g., berthing checklist), operation phase (e.g., ship approaching port), and issues (e.g., low risk level). In addition we opted to illustrated the expected workload over time of the operator, as well as give an indication of the workload of his/her close colleagues. Control actions are limited here to accepting responsibility of newly allocated ships or flagging a ship to allocate to a different operator, potentially accounting for his/her own workload compared to that of any colleagues. From here the operator can decide to move towards a localized view of a single ship, or IOSS can proactively suggest this based on an upcoming activity for the operator or a risk that is increasing. See figure 7 for an initial interface design of this global view.

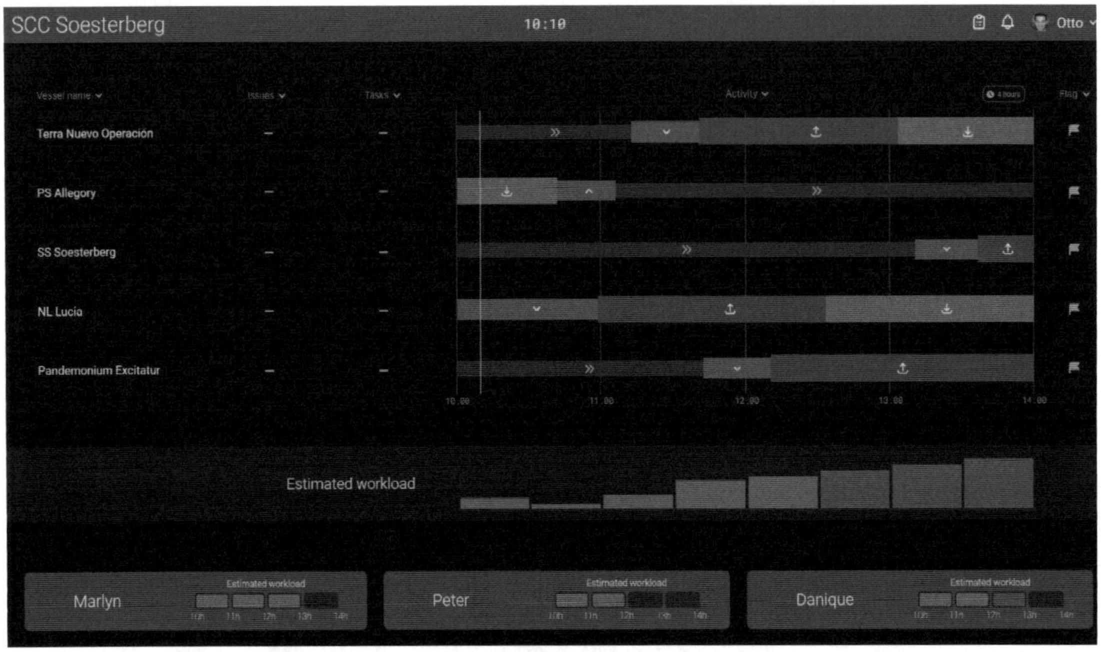

Figure 7. Our concept for the global view of an operator's interface. It lists all ships the operator is currently responsible for and detailed information such as open tasks, ship phases, and the expected workload of him-/herself and direct colleagues.

The local view shows the detailed status of a single ship. Here the operator can review the voyage plan and -stages as well as various tabs with detailed information, such as assessed risks, open tasks, operation progression, cargo status and manifest, and the ship's history (e.g., sensor malfunctions, ports visited, etc.). Typical control actions are here to complete open tasks, understand the assessed risks and act upon them. Acting on such risks, whether current or expected, is done through contacting relevant parties (e.g., terminal operator or cargo owner). This is done through a message-based interaction if low priority. Where – on the operator's or IOSS' initiative – a conversation can be started. Given the topic, IOSS will automate any information transfer. For example, if a sensor is malfunctioning and needs to be checked in the next port, the terminal operator can receive the details from IOSS on the sensor's make, location and malfunctioning without involvement from the operator. See figure 8 for an initial interface design of this local view.

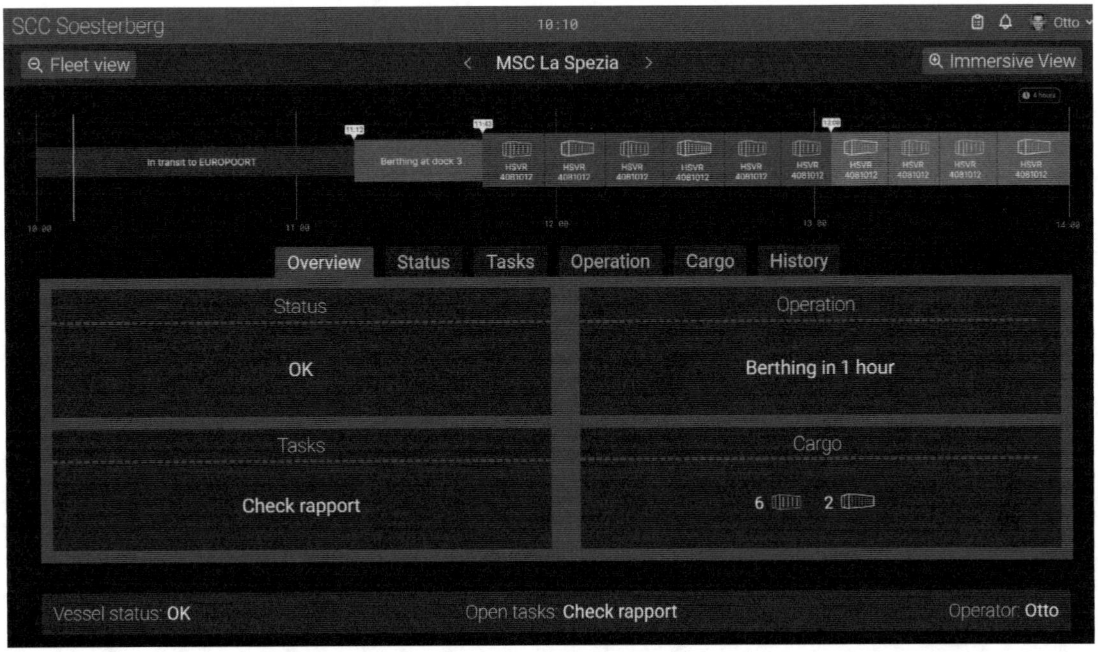

Figure 8. Our concept for the local view of an operator's interface. It presents a single ship's voyage plan as well as various tabs with detailed information. Typical control actions are contacting relevant people on risks and completing verification tasks.

The situational view aims to immerse the operator entirely in the ship's current situation. Here we rely on the real-time reconstruction of a ship's situation into a digital twin. This digital twin is used as a virtual reality environment in which the operator can freely explore the sensor data the ship acquires. Outside the ship's sensor range either a default ocean view is presented when the ship is in transit or a 3D reconstruction of a port is rendered based on predefined models. The role of IOSS is to retrieve the high-level sensor data and create this digital twin (e.g., position and classification of objects), and where needed or requested provide the actual raw data (e.g., a camera feed). This ensure this immerse view can be maintained with a regular 4G connection. Control actions in this view may vary depending on the level of autonomy of the ship. For instance, control can be restricted to a (safe) stop signal and informing relevant parties or might support an IOSS assisted tele-operator. We opted for the former, as we assume a high degree of autonomy. The situational view is currently in development, see figure 9 for an impression.

Figure 9. An indication how the situational view could look like. It is a real-time digital twin rendered by IOSS to enable an operator to view an autonomous ship's current situation through virtual reality.

Note that the idea of progressive disclosure can and should be implemented throughout. For instance, the local view presents a tab of assessed risks, which can also first present only the high level and summarized risks and through interaction or necessity further details can be presented.

6. Conclusions

The work described in this paper is directed towards developing an intelligent operator support system (IOSS) with which the operator support functions will be demonstrated as part of a robotic container handling system, an innovation of the European MOSES research project. For this purpose, we introduced three intelligent operator support functions to allow multiple operators to effectively and efficiently supervise multiple autonomous operations.

Despite the fact that we have developed IOSS to fit in the operational context of the robotic container handling system, we believe that the operator support functions introduced and described above should be considered as generic and necessary support functions for many-to-many remote control centre concepts, in which several operators supervising a set of multiple operation.

1) Due to the dynamics in the task environment, the many dependencies in the overall planning and the unpredictability of system failures and unsafe situations, the cognitive task load will not be completely predictable and will fluctuate as a result. To mitigate these fluctuation and to balance cognitive workload distributions a dynamic task allocation mechanism should be in place. One can argue whether it should be an AI algorithm or human supervisor to execute the load balancing is yet another discussion.

2) Human attention is limited and operators therefore must constantly shift attention resulting in moment-to-moment fluctuations in situation awareness. Such reductions in SA are a constant, given the fact that attention has to be divided over several automated operations. The more reliable and robust that automation is, the less likely that operators overseeing the automation will be aware of critical information and are able to take action when needed [23]. The concept of continuous risk assessment support the process of situation awareness recovery by presenting operators information over critical situation or development that could develop into a critical situation.

3) The many-to-many ratio between supervising operators and autonomous ships implies also that operators will not be able to supervise all ships in parallel nor in full detail. Because operators want to

have on overview of all the operation (the broader picture) but also want to be able to zoom in to a higher level of detail, a three-layer interface that progressively discloses more information about the operations an operator supervises and enables increasingly more complex and involved control actions is eminent.

Although generic in nature, to be able to deploy the support functions for effective and efficient human automation collaboration in other contexts, adaptation is necessary. It is the domain, i.e., task context in which IOSS is applied that determines what critical information is, which risks can arise and how the layers of a progressive disclosure interface should be designed.

7. Acknowledgements

This research has been conducted as part of MOSES project, which has received funding from the European Union's Horizon 2020 Research and Innovation Program under Grant agreement No. 861678.

References

[1] Van den Broek, J., Schraagen, J. M., te Brake, G. and Van Diggelen, J. 2017 Approaching full autonomy in the maritime domain: paradigm choices and Human Factors challenges. Proceedings of MTEC2017, 375–389.

[2] Van der Kleij, R., Te Brake, G. and Van den Broek, J. 2015 Enabling swifter operator response in the event of a fault initiation through adaptive automation. Dynamic positioning conference, October 13-14, Houston.

[3] Tjallema, A., Van der Nat, C., Grimmelius, H. and Stapersma, D. 2007 The road to eliminating operator related dynamic positioning incidents. Dynamic Positioning Conference (pp. 1-17), Houston, Texas.

[4] Kaber, D. B. and Endsley, M. R. 1997 Out-of-the-loop performance problems and the use of intermediate levels of automation for improved control system functioning and safety. Process Safety Progress, 16(3), 126-131.

[5] Parasuraman, R. and Manzey, D. H. 2010 Complacency and bias in human use of automation: An attentional integration. Human Factors, 52(3), 381-410.

[6] Van Diggelen, J., van den Broek, J., Schraagen, J. M. and van der Waa, J. 2018 An intelligent operator support system for dynamic positioning. Advances in Intelligent Systems and Computing, 599, 48–59.

[7] Stanton N. A., Stewart R, Harris D, Houghton R. J., Baber C, McMaster, R, Salmon P. M., Hoyle G., Walker G. H., Young M. S., Linsell M., Dymott R. and Green, D. 2006 Distributed situation awareness in dynamic systems: theoretical development and application of an ergonomics methodology. Ergonomics 49:1288–1311

[8] Van den Broek, J., van Diggelen, J., van der Kleij, R., Hueting, T. F., van der Waa, J., van Schendel, J. A. and Langefeld, J. J. 2018 Adaptive Maritime Automation: Final Report.

[9] St. John, M. F. 2013 Janus Design Principles and the Acquisition Process for a Supervisory Situation Awareness Display. In Proceedings of the Human Factors and Ergonomics Society Annual Meeting (Vol. 57, No. 1, pp. 2002- 2006). SAGE Publications.

[10] Gartenberg, D., Breslow, L., McCurry, J. M. and Trafton, J. G. 2014. Situation Awareness Recovery. Human Factors: The Journal of the Human Factors and Ergonomics Society 56 (4), 710–727. https://doi.org/10.1177/0018720813506223.

[11] Crandall, J. W. and Cummings, M. L. 2007. Developing performance metrics for the supervisory control of multiple robots. In Proceedings of the ACM/IEEE international conference on Human-robot interaction (pp. 33-40).

[12] Neerincx, M.A. 2003. Cognitive task load design: model, methods and examples. In: E. Hollnagel (ed.), Handbook of Cognitive Task Design. Chapter 13. 283-305. Mahwah, NJ: Lawrence Erlbaum Associates.

[13] Van den Broek, J., Griffioen, J. R. and Van Der Drift, M. 2020 Meaningful Human Control in Autonomous Shipping: An Overview. IOP Conference Series: Materials Science and Engineering, 929(1). https://doi.org/10.1088/1757-899X/929/1/012008.

[14] MOSES. n.d. Automated Vessels and Supply Chain Optimization for Sustainable Short Sea Shipping. Retrieved from https://moses-h2020.eu/
[15] Wilson, G. F., & Eggemeier, F. T. (2020). Psychophysiological assessment of workload in multi-task environments. *Multiple-task performance*, 329-360.
[16] Allais, I., Perrot, N., Curt, C., & Trystram, G. (2007). Modelling the operator know-how to control sensory quality in traditional processes. *Journal of food engineering*, *83*(2), 156-166.
[17] Brug, T., van der Waa, J., Maccatrozzo, V., & van den Broek, H. (2022). Dynamic task allocation algorithms within intelligent operator support concepts for shore control centres. In proceedings of the conference on Computer Applications and Information Technology in the Maritime Industries
[18] Korf, R. E. (1985). Depth-first iterative-deepening: An optimal admissible tree search. Artificial intelligence, 27(1), 97-109.
[19] Hart, P. E., Nilsson, N. J. & Raphael, B. (1968). A Formal Basis for the Heuristic Determination of Minimum Cost Paths. IEEE Transactions on Systems Science and Cybernetics. 4 (2): 100–107.
[20] Fleischer, L., Goemans, M.X., Mirrokni, V.S. and Sviridenko, M. (2006). Tight approximation algorithms for maximum general assignment problems.
[21] Johnson, J., Roberts, T. L., Verplank, W., Smith, D. C., Irby, C. H., Beard, M. and Mackey, K. 1989. The xerox star: A retrospective. Computer, 22(9), 11-26.
[22] Cochran, E. L. 1992. Control Room User Interface Technology in the Year 2000: Evolution or Revolution? In Proceedings of the Human Factors Society Annual Meeting (Vol. 36, No. 4, pp. 460-464). Sage CA: Los Angeles, CA: SAGE Publications.
[23] Endsley, M. R. 2017. From Here to Autonomy. Human Factors: The Journal of the Human Factors and Ergonomics Society,59(1), 5–27. https://doi.org/10.1177/0018720816681350/

Journal of Physics: Conference Series **2311** (2022) 012033 doi:10.1088/1742-6596/2311/1/012033

On the implementation of a rule-based system to perform assessment of COLREGs onboard maritime autonomous surface ships

M Constapel[1*], P Koch[1] and H-C Burmeister[1]

[1]Fraunhofer Center for Maritime Logistics and Services, Hamburg, Germany

manfred.constapel@cml.fraunhofer.de

Abstract. Complying to the Convention on the International Regulations for Preventing Collisions at Sea (COLREGs) is one of the key challenges for Maritime Autonomous Surface Ships (MASS) to perform autonomous navigation. We propose a rule-based approach utilizing operationalized situation statements to assess vessel encounters with regards to COLREGs. The statements contain basic informational pieces to describe the relation between two vessels. For processing with a rule-based inference system, the statements are grouped into rule sets to determine applicable COLREGs rules for an encounter situation. Implementation results are shown in the context of the B ZERO project, outlining how the approach can be utilized onboard a seagoing ship.

1. Introduction

Maritime autonomous surface ships (MASS) are vessels to be controlled by automation without human assistance [1]. Control hereby includes the task to perform navigation. Developments of such systems have gained considerable importance over the past years as demand for automated navigation grows. Enabling safe, reliable, and efficient autonomous operation and navigation is an active area of research. The task of autonomous navigation is driven by a complex set of procedures, decisions, and obligations [2], e.g. the Convention on the International Regulations for Preventing Collisions at Sea (COLREGs) [3], Standing Orders, and regulations set by the International Maritime Organization (IMO). Such tasks, typically performed by a trained navigational watch officer, must be transformed into mechanisms that can be performed by an autonomous system without the need for human supervision or intervention.

1.1. Situation Assessment

Observation and assessment of the surrounding, identification of vessels and hazards, and derivation of obligations is a key responsibility of any MASS [4]. Therefore, the translation of human-readable rules (e.g. COLREGs, but also from other regulatory texts) into evaluable rules to derive obligations is among other challenges an important task. Providing explicit and machine-interpretable rules to a MASS comes in conjunction with better maintainability of the rules as they are more human-relatable, and ease interpretability of the decision-making process by doing the rule inference. The challenge remains in identifying the underlying building blocks, i.e. operationalization of regulatory texts, particularly the COLREGs. Rules shall be understandable to both, machine and human beings, to allow for the reasoning of decisions done by either to the other.

1.2. Rule-based Systems
A rule-based system stores, interprets and manipulates knowledge (i.e. facts) to change or build up new knowledge in a process called inference. The inference engine of a rule-based system uses predefined rules to conduct and execute the inference process by continuously looking at facts and activating applicable rules as long as the inference process finds active rules applicable to the facts. Any rule can be considered being a conditional statement (IF-THEN) with a left-hand-side (LHS) and right-hand-side (RHS), which allows building up a flat and declarative hierarchy of rules, avoiding complex and nested trees of conditional statements. In common, rule-based approaches can be split into forward and backward chaining, either starting from the data available to deduce conclusions or performing the inference backwards from a goal state back to the preconditions necessary to meet that goal. Our approach uses the forward chaining mechanism, hence performing inference on the data (i.e. information from the perceived environment) to infer the set of rights and obligations which need to be considered in the decision-making process.

1.3. Practical Considerations
Any COLREGs-compliant system needs to be extensible so that new regulations (e.g., federal ones, or standing orders) can be added, resulting in the need for an architecture that allows existing rules to be enforced, restricted or annulled. In a rule-based system that requirement is considered for any implementation off-the-shelf. Besides the translation of COLREGs knowledge (i.e. rights and obligations applicable in certain situations) to evaluable rules as a result of the operationalization, the process of inference along such rules creates a decision path in the rule-based system. Hence, the operation of the decision-making component of a MASS remains deterministic and can be easily monitored and evaluated. Being predictable in behavior may support the acceptance of MASS as it becomes able to clearly describe its decision-making process. The most significant drawback of a rule-based approach is the inability to model uncertainty as the rules (i.e. IF-THEN statements) are conditionals which need to be either evaluated true or false. This raises the requirement for an accurate representation of the perceived environment with facts even in noisy environments. However, we deal with the uncertainty issue in the operationalization of COLREGs by introducing the rule-dependent design parameter $\varepsilon \geq 0$ to model the situation uncertainty.

1.4. Our Contribution
In this work, we focus on the operationalization and subsequent implementation of the core rules of COLREGs to be suitable for being processed with a rule-based system: Overtaking (§13), head-on situations (§14), crossing situations (§15), give-way vessel (§16), stand-on vessel (§17), and responsibilities between vessels (§18). Therefore, the paper is structured as follows: After a brief introduction to the core concept of COLREGs, related work regarding COLREGs-compliant and rule-based systems is outlined in Section 2. Section 3 describes the operationalization of the aforementioned COLREGs to evaluable statements. Subsequently, Section 4 provides an overview about the most important implementation details of the rule-based system utilizing the statements from the operationalization, and presents an application developed in the B ZERO project as a result of this work. Section 5 concludes the paper and Section 6 points out future developments of our approach.

2. Background

2.1. COLREGs in a Nutshell
COLREGs is a regulatory text split into parts and sections which describe ship-to-ship encounters and their resolutions at open sea. In Part B the steering and sailing rules are covered, which is subdivided into three sections: Section 1 with rules 4 to 10 (conduct of vessels in any condition of visibility), Section 2 with rules 11 to 18 (conduct of vessels in sight of one another), and Section 3 with rule 19 (conduct of vessels in restricted visibility). COLREGs Part B exclusively considers encounter situations with two vessels, whereas the overall scheme is to firstly analyse and assess the traffic

situation from the ownships point of view (observer) whether a risk of collision exists, and secondly, based on the analysis, to determine appropriate counter measures to avoid collision with the other vessel (target) [5]. The surrounding of the vessel that is assessing an encounter situation (i.e. observer) is split into sectors with reference to the vessels own heading in order to describe the relative positions of other vessels. Figure 1 illustrates the four sectors (ahead, starboard, astern, port) of a vessel and five typical encounter situations (being crossed, headon, crossing, being overtaken, overtaking). The starboard and port sectors do have an angle of 112.5°, and astern is 135° in angle. Navigator training and practical seamanship often considers angles of up to 10° to each side from the centerline as the sector ahead, even though this cannot be derived directly from COLREGs, it is a compromise based on international decisions made [6]. The encounter situations are covered by rules §13 (overtaking), §14 (headon situations), and §15 (crossing situations) as we primarily focus on these in this work.

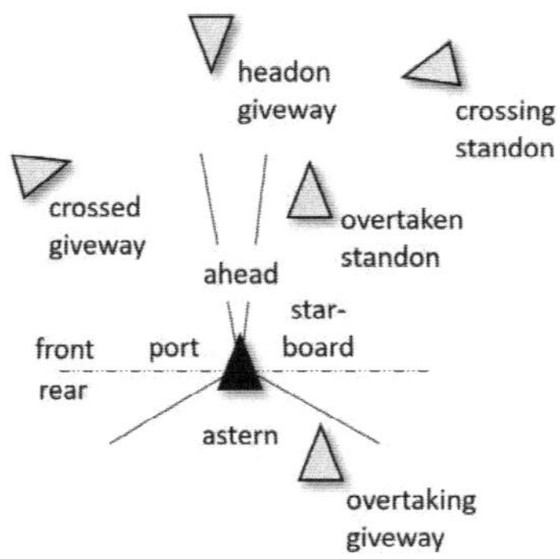

Figure 1. COLREGs sectors and encounter situations (the black shape is the observer heading north which assesses the situation, gray shapes are target vessels)

2.2. COLREGs-compliant Systems

Implementing the rules formulated in COLREGs can be achieved by several approaches. [7] gives an overview about a variety of actual approaches including which parts of COLREGs are covered, even though not detailing the criteria for the traffic analyses in detail. Most approaches are concentrating solely on rule 13 to 15 during the traffic assessment, which is often based on the Closest Point of Approach and derived metrics. Information uncertainty during COLREGs determination is often neglected, however exceptions to that are [5] [8]. Approaches using Fuzzy logic mainly concentrate on overcoming the vaguely defined nature of COLREGs, and approaches shown in [9] aim at providing a guideline to the mariner and serve as a decision support during various types of encounter situations. Due to the nature of COLREGs, fuzzy sets are often highly debatable making verifiability and testability of these systems challenging. Some approaches use the concept of "survival of the fittest" as shown in [10] to iterate on a planned path under the governance of a genetic cost function. Whilst these approaches can yield promising results, due to the dynamic nature of these algorithms, a solution may be just locally optimal. Some developments combine the task of rule application and path planning into a single system. Such systems aim at enhancing existing path planning techniques with constraints given by the COLREGs rules [11].

2.3. Rule-based Approaches

Knowledge-based approaches as presented in [12] can be considered to be at the core of COLREGs-compliant approaches. Developments as shown in [13] devise a question-and-answer style architecture, prompting the user (officer) to enter the observed state of the surroundings into the system. Results are obtained following a nested IF-THEN-ELSE structure, however, such a system can appear convoluted, and dependencies between different rules can potentially complicate debugging. In [14] the architecture of the rule-based system is shifted to an agent-based approach. Ships are treated as entities which can, due to movement or state change, activate rules. Like in our approach, relations between two vessels are determined. However, there are no connections between calculated values, instead all decisions are mainly based on CPA (closest point of approach) metrics like DCPA (distance to CPA) and TCPA (time to CPA). Whilst showing conceptual aspects, and in accordance to our approach, the developments shown in [15] state the importance of supplementing COLREGSs with additional rules representing federal regulations or practices on good seamanship.

3. Operationalization

This section describes how a set of situation statements which is necessary to assess situations are derived (operationalized) from COLREGs. An operationalization scheme is proposed, and the required mathematical background is introduced. Despite the lack of an agreed maritime perceived data set, the operationalization assumes the existence of it as outlined in [16].

3.1. Situation Statements

A key part for the operationalization of the COLREGs with regards to vessel encounters is the sector model as shown in Figure 1 as the assessment is highly dependent on where the other vessel is located relative to the observers heading. The aspects for assessing a situation regarding the sector model is addressed in Table 1 with sector statements. Another criterion for correctly assessing a situation is the orientation (or pose) of vessels to one another, which are covered with the so-called scene statements. The COLREGs type or the circumstances of operation of a vessel, e.g. power-driven, sailing, fishing, and its state, e.g. moored, at anchor or underway, are important for the situation assessment as well and are considered by the type statements and status statements. The need to introduce these could arise from a vessel that is not underway which can therefore be considered an obstacle. Furthermore, motion-related attributes of vessels are considered for the situation assessment as illustrated in Figure 2 and do not involve the often-used CPA. The motion statements describe how vessels move relative to one another, while the passage statements expand the motion statements and describe which sector the other vessel will pass. All situation statements from our operationalization of COLREGs encounter situations are categorized and briefly described in Table 1.

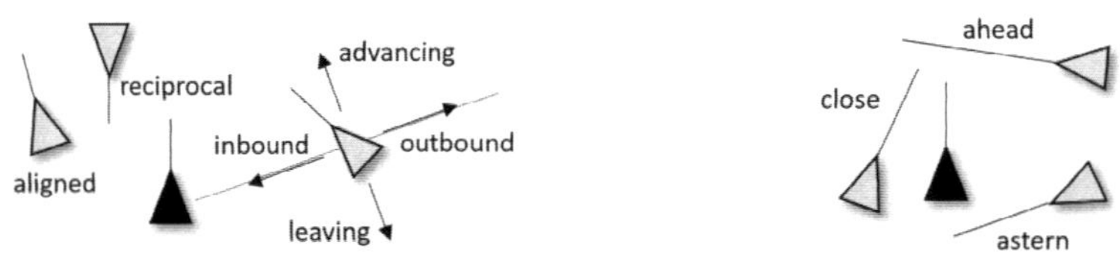

Figure 2. Illustration of situation statements as being used from an observer (on the left subsets of scene and motion statements, on the right passage statements)

3.2. Definitions

In subsequent sections as well as in Table 1, X and Y denote two different vessels as COLREGs encounter situation consists of two vessels always, either being the observer, and the other being the target vessel. The orientation and direction of absolute angles (bearings, courses, headings) are referenced to the common true north and given in degrees (0° to 359°). Relative angles (difference of two absolute angles) can be negative (to port side) or positive (to starboard side). The observation uncertainty of the perceived environment can be incorporated and specified with the design parameter $\varepsilon > 0$ on a per-statement basis. This parameter can be either a scalar value gathered from sensor calibration or obtained by other methods and models (e.g. type2 fuzzy sets).

For the sake of readability, in the statements listed in Table 1 the observer (or the ownship) is denoted as O, any other vessel (or a target) which is not the observer is denoted as V. Statements with the expression *n/a* cannot be evaluated nor inferred directly, thus their values need to be provided from an external information source from the perceived environment. Furthermore, statements must be aligned with respect to time, thus, it is assumed that the actual (system) time matches the timestamps of both, the states of O and V, ensuring to evaluate statements based on recent and synchronized vessel state information.

Table 1. Operationalization of COLREGS vessel encounter rules to evaluable situation statements

	Statement	Description	Expression
Sector	*V ahead O*	*V* is straight ahead of *O*	$\lvert brg(O,V) - hdg(O)\rvert < 10° + \varepsilon$
	V astern O	*V* is astern of *O*	$\lvert brg(O,V) - hdg(O)\rvert > 112{,}5° - \varepsilon$
	V starboard O	*V* is on the starboard side of *O*	$10° - \varepsilon \leq brg(O,V) - hdg(O) \leq 112{,}5° + \varepsilon$
	V port O	*V* is on the port side of *O*	$-112{,}5° - \varepsilon \leq brg(O,V) - hdg(O) \leq -10° + \varepsilon$
Scene	*V visible O*	*V* can be seen at *O* with optical means	*n/a*
	V facing O	*V* is showing a small aspect angle to *O*	$\lvert brg(V,O) - hdg(V)\rvert < 5° + \varepsilon \wedge V\ visible\ O$
	V front O	*V* is in front of *O*	$\lvert brg(O,V) - hdg(O)\rvert < 90° + \varepsilon$
	V rear O	*V* is rear of *O*	$\lvert brg(O,V) - hdg(O)\rvert \geq 90° - \varepsilon$
	V reciprocal O	*V* and *O* move contrary to each other	$\lvert inv(cog(V)) - cog(O)\rvert < 10° + \varepsilon$
	V aligned O	*V* is aligned to *O*	$max(\lvert brg(O,V) - cog(V)\rvert, \lvert cog(V) - cog(O)\rvert) < 67{,}5° + \varepsilon$
Type	*V SP*	*V* is a seaplane	*n/a*
	V PD	*V* is a power-driven vessel	*n/a*
	V S	*V* is a sailing vessel	*n/a*
	V EIF	*V* is engaged in fishing	*n/a*
	V EIT	*V* is engaged in trawling	*n/a*
	V F	*V* is a fishing vessel	$V\ EIF \vee V\ EIT$
	V CBD	*V* is constrained by her draught	*n/a*
	V RAM	*V* is restricted in her ability to manoeuvre	*n/a*
	V NUC	*V* is not under command	*n/a*
Status	*V handling*	*V* is towing or pushing	*n/a*
	V anchored	*V* is at anchor	*n/a*
	V moored	*V* is tied to shore or a stationary object	*n/a*
	V aground	*V* is aground	*n/a*
	V underway	*V* is underway or adrift	$\neg(V\ anchored \vee V\ moored \vee V\ aground)$
Motion	*V across O*	*V* and *O* proceed the same way across	$sgn(sac(O,V)) \neq sgn(sac(V,O))$
	V advancing O	*V* proceeds faster across than *O*	$(V\ across\ O) \wedge \lvert sac(V,O) - \varepsilon\rvert > \lvert sac(O,V) + \varepsilon\rvert$
	V leaving O	*V* proceeds slower across than *O*	$\neg(V\ across\ O) \vee \lvert sac(V,O) + \varepsilon\rvert < \lvert sac(O,V) - \varepsilon\rvert$
	V steady O	*V* and *O* proceed across equally (steady bearing)	$\lvert sac(O,V) + sac(V,O)\rvert \leq \varepsilon$
	V inbound O	*V* and *O* are getting closer to each other	$sal(O,V) + sal(V,O) > \varepsilon$
	V outbound O	*V* and *O* are moving away from each other	$sal(O,V) + sal(V,O) < -\varepsilon$
	V fellow O	*V* and *O* not closing in nor moving away	$\lvert sal(O,V) + sal(V,O)\rvert \leq \varepsilon$
Passage	*V pass ahead O*	*V* will pass ahead of *O* (bow crossing)	$(V\ advancing\ O) \wedge (V\ inbound\ O)$
	V pass astern O	*V* will pass astern of *O*	$(V\ leaving\ O) \wedge (V\ inbound\ O)$
	V pass close O	*V* and *O* will pass close (close quarter)	$(V\ steady\ O) \wedge (V\ inbound\ O)$

For the evaluation of sector and scene statements the sector model as introduced in Figure 1 has to be considered, thus the calculation of relative bearings and aspects is necessary to determine a sector:

$brg(X,Y)$	returns the true bearing from *X* to *Y*,
$hdg(X)$	returns the heading of *X*,
$brg(X,Y) - hdg(X)$	thus, returns the relative bearing from *X* to *Y*.

The passage sector of a vessel to one another can be determined from the decomposition of the relative motion into a tangential (across) and a radial (along) component with respect to the bearing.

The speed across *sac* of *X* to *Y* can be obtained with

$$sac(X,Y) = \sin\big(brg(X,Y) - hdg(X)\big) \cdot spd(X),$$

whereas the speed along *sal* of *X* to *Y* is expressed by

$$sal(X,Y) = \cos\big(brg(X,Y) - hdg(X)\big) \cdot \mathrm{spd}(X).$$

For relative motion decomposition and scene statements the state function

$spd(X)$	returns the speed over ground of *X*,
$cog(X)$	returns the course over ground of *X*.

Finally, the value function

$sgn(a)$	returns 1 $if a \geq 0\ else - 1$,
$inv(b)$	flips the absolute bearing *b* to the opposite on a compass (*b* +/- 180°),
$max(m,n)$	returns $m\ if m \geq n\ else\ n$.

4. Rule-Based System

The operationalization of COLREGs provides evaluable situation statements and forms the basis for the assessment of encounter situations with regards to COLREGs. In this section the most relevant details of the rule-based system to perform such assessments is described: The overall shape of the fact base to store acquired and inferred knowledge as fact objects in the fact memory is outlined, and the situation statements as shown in Table 1 are linked to the rule sets as shown in Ruleset 1 to Ruleset 8 forming the rule base.

4.1. Shape of the Fact Base

The UML diagram as shown in Figure 3 illustrates the overall architecture of the fact base with three fact classes: Vessel, Situation, and Encounter. The inference engine of the rule-based system applies rules from the rule base during the inference process to all fact objects stored in the fact memory. The inference process can be triggered by external means or the rule-based system itself, and evaluates properties (i.e. attribute values) and types (i.e. class memberships) of fact objects to create, remove, or change fact objects in the fact memory. Vessel facts are inserted and maintained from an external system (e.g. sensor fusion component). Situation facts are created for any ownship-vessel pair in the fact memory during the inference process, and appropriate situation tags are determined from the situation statements as listed in Table 1. Subsequently, encounter facts are inferred and annotated by applying the rules from Ruleset 1 to Ruleset 8 to the situation facts. Encounter facts are pushed back to externally attached systems (e.g. collision avoidance component) for decision-making, decision support or assistance.

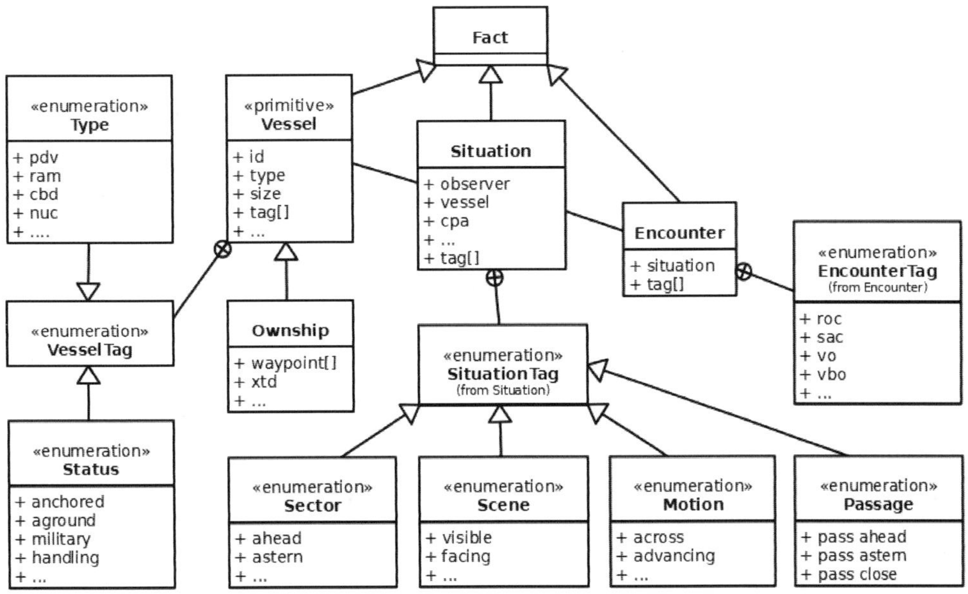

Figure 3. UML class diagram of the fact base

4.2. Rule Base for Vessel Encounters

The rules declared in the rule base are specified with the Rulesets 1 to 8. The observer (ownship) is denoted as *O*, the target (vessel observed from the ownship) is denoted as *V*. The implemented statements *safe*, *overtaking*, *(being) overtaken*, *headon*, *crossing*, and *(being) crossed* fully describe all possible encounters from an observer point of view, thus cover all possible situations as defined in COLREG Part B Section 2 except for §9 (Narrow Channels) and §10 (Traffic Separation Schemes).

ROC describes the risk of collision as some vessel is going to come close to the observer, and SAC applies when the vessel is visibly moving away from the ownship, thus considered safe and clear.

ROC	V threatening $O =$ $(V \text{ pass close } O) \wedge \neg(V \text{ aligned } O)$
SAC	V safe $O =$ $(V \text{ outbound } O) \wedge (V \text{ visible } O)$

Ruleset 1. Risk of collision (§7)

A precondition for the application of rules of subsequent rulesets is B2, which is assumed to be evaluated true when the distance between two vessels is less than 6 nautical miles (considered an implementation detail).

B2	V sight $O =$ $(V \text{ visible } O) \wedge (O \text{ visible } V)$

Ruleset 2. In sight of one another (§11)

Overtaking situations are covered from both views, the overtaking vessel with VO and the vessel being overtaken due VBO.

VO	V overtaking $O =$ $(V \text{ astern } O) \wedge \neg(V \text{ outbound } O) \wedge \neg(V\ \mathit{leaving}\ O)$
VBO	V overtaken $O =$ $(V \text{ aligned } O) \wedge (O \text{ inbound } V)$

Ruleset 3. Overtaking situation (§13)

Vessels in head-on situations are described by VH.

VH	V headon $O =$ $(V \text{ PD}) \wedge (O \text{ PD}) \wedge (V \text{ threatening } O) \wedge$ $(V \text{ ahead } O) \wedge (V \text{ facing } O) \wedge (V \text{ reciprocal } O)$

Ruleset 4. Head-on situation (§14)

Both cases of crossing situations are covered: A vessel is crossing according to VC, and VBC describes when a vessel is being crossed. If the ownship has been overtaken by a vessel previously, that vessel cannot become a crossing vessel for the ownship anymore.

VC	V crossing $O =$ $(V \text{ PD}) \wedge (O \text{ PD}) \wedge (V \text{ threatening } O) \wedge$ $(V \text{ starboard } O) \wedge \neg(O \text{ overtaken } V)$
VBC	V crossed $O =$ $(V \text{ PD} \wedge O \text{ PD}) \wedge (V \text{ threatening } O) \wedge (V \text{ port } O)$

Ruleset 5. Crossing situation (§15)

SOV applies when a vessel is being overtaken or being the crossing vessel or privileged by its navigational status. Therefore, the vessel shall keep course and speed (stand-on obligation).

SOV	V standon $O =$ $(V \text{ crossing } O) \vee (V \text{ overtaken } O) \vee$ $((V \text{ privileged } O) \wedge (V \text{ threatening } O))$

Ruleset 6. Stand-on vessel (§17)

The give-way obligations are included in GWV: A vessel has to give-way to the ownship if it is either the overtaking one, or in a head-on situation, or being crossed, but neither the stand-on vessel nor safe and clear.

GWV	V giveway $O =$ $(V \text{ headon } O) \vee (V \text{ crossed } O) \vee (V \text{ overtaking } O) \wedge$ $\neg(V \text{ standon } O) \wedge \neg(V \text{ safe } O)$

Ruleset 7. Give-way vessel (§16)

Responsibilities between vessels are expressed with RBV. The function $\text{rank}(X)$ returns the associated numerical value of the vessel *X* depending on its operationalized navigational type: $4 \coloneqq \text{rank}(NUC) = \text{rank}(RAM) = \text{rank}(CBD)$, $3 \coloneqq \text{rank}(F)$, $2 \coloneqq \text{rank}(S)$, $1 \coloneqq \text{rank}(PD)$.

RBV	V privileged $O =$ $\text{rank}(V) > \text{rank}(O) \wedge \neg(V \text{ overtaking } O)$

Ruleset 8. Responsibilities (§18)

4.3. B ZERO Assist

The B ZERO project develops a bridge for unmanned operation for a period of up to 8 hours [17]. B ZERO Assist is one software module in the whole B ZERO system responsible for performing navigation-related tasks. The approach presented in this paper has been implemented for building and maintaining a robust navigational situation picture for B ZERO Assist to aid the situational awareness of the system. The COLREGs obligations yielded by the rule-based system are used by subsequent decision-making modules, i.e. collision avoidance and path planning. Figure 4 shows an example of an assessed situation perceived by B ZERO Assist onboard the observer (round symbol) for three vessels (triangular shapes) along with the inferred labels from the rule inference; green shapes indicate give-way vessels, whereas red shapes indicate stand-on vessels.

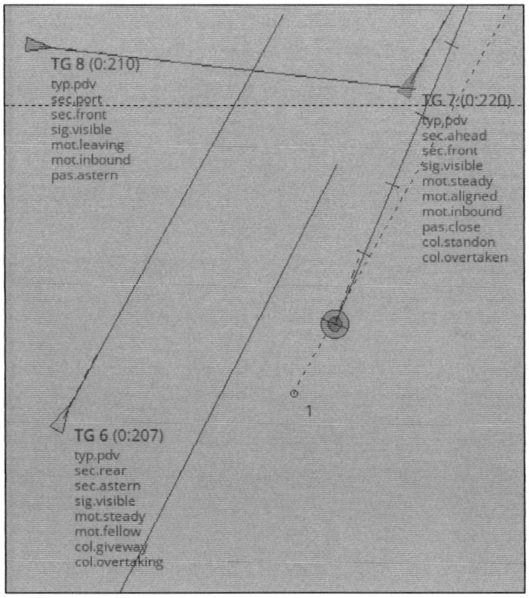

Figure 4. A compiled situation picture in B ZERO Assist with labels from the ruleset inference shown on an Electronic Navigational Chart (ENC). Similar encounter situations have been extracted from a AIS database consisting of 11 years of data from North sea and Baltic sea and assessed as proposed for validation purposes.

5. Conclusion

In this paper, we propose an extensible approach utilizing situation statements to assess vessel encounters with regards to COLREGs rules 13 (overtaking), 14 (head-on situations), 15 (crossing situations), and 18 (responsibilities between vessels). The situation statements are the result of the operationalization of the related COLREGs rules which are described in detail. We have shown our implementation of a rule-based system, in particular the shape of the fact base and the underlying rule sets for the rule base. As a proof-of-concept, we show the implementation of the approach to assess encounter situations in the B ZERO system for subsequent decision-making processes.

6. Future work

The operationalization of (all) COLREGs rules to comprehensively assess situations is still a field of research as such an endeavor raises the necessity to robustly perceive the own ship's state and the environment. This includes (and is not limited to) estimation of visibility, recognition, and identification of vessels, buoys, lights, shapes, and noise signals. Some COLREGs (e.g. traffic separation schemes and narrow channels) can probably be covered with contextual information from the electronic navigation chart (ENC), which is currently a work in progress. Moreover, the ENC can help to put spatial constraints to the assessment of a situation by taking depth contours, coastlines, shallow waters, no-go areas, fairways, and routes into account. Furthermore, it is assumed that a complete set of COLREGs statements from the operationalization similar to the proposed one in this work can serve as a set of labels to fully describe an environment suitable for Reinforcement Learning (RL): Agents can be trained to find collision avoidance strategies - optimal to some criteria - just by looking at the relatively sparse state space given by the labels.

Acknowledgements

The B ZERO project has received funding by the German Federal Ministry of Economic Affairs and Energy under support code 03SX500.

References

[1] International Organization for Standardisation, ISO/PRF TS 23860 - Ships and marine technology — Vocabulary related to autonomous ship systems, vol. 1.I. 8. S. a. m. technology, Ed.

[2] S. N. MacKinnon, R. Weber, R. Olindersson and M. Lundh, "Artificial Intelligence in Maritime Navigation: A Human Factors Perspective," in *Advances in Human Aspects of Transportation*, Division of Maritime Studies, Chalmers University, 2020.

[3] International Maritime Organization, COLREGS : Convention on the International Regulations for Preventing Collisions at Sea, London: IMO, 1972.

[4] International Mairitime Organization (IMO), "IMO's Maritime Safety Committee finalizes its analysis of ship safety treaties, to assess next steps for regulating Maritime Autonomous Surface Ships (MASS).," 2021.

[5] H.-C. Burmeister und W. Bruhn, „Designing an Autonomous Collision Avoidance Controller Respecting Colreg," in *Maritime-Port Technology and Development : Proceedings of the Conference on Maritime-Port Technology, MTEC 2014, Trondheim, Norway, 27-29 October 2014*, B. E. Asbjornslett, O. J. Rodseth und T. E. Berg, Hrsg., CRC Press, 2015, p. 83–88.

[6] H. Hilgert und R. Schilling, (ger.) Ein Kommentar der internationalen Kollisionsverhütungsregeln, Dt. Kommunal-Verlag Naujoks und Behrendt, 1997.

[7] H.-C. Burmeister und M. Constapel, „Autonomous Collision Avoidance at Sea: A Survey," *Frontiers in Robotics and AI,* Bd. 8, 2021.

[8] D. K. M. Kufoalor, E. Wilthil, I. B. Hagen, E. F. Brekke und T. A. Johansen, „Autonomous Colregs-Compliant Decision Making Using Maritime Radar Tracking and Model Predictive Control," *18th European Control Conference, ECC 2019, Naples, Italy,* p. 2536–2542, 2019.

[9] M. K. James, "Modelling the Decision Process in Computer Simulation of Ship Navigation," James Cook University of North Queensland, 1986.

[10] R. Smierzchalski und Z. Michalewicz, „Modeling of ship trajectory in collision situations by an evolutionary algorithm," *IEEE Transactions on Evolutionary Computation,* pp. 227-241, September 2000.

[11] Y. Kuwata, M. Wolf, D. Zarzhitsky and T. Huntsberger, "Safe Maritime Autonomous Navigation with COLREGS, Using Velocity Obstacles," in *Oceanic Engineering, IEEE Journal of. 39. 110-119.*, 2014.

[12] M. Salous, A. Hahn and C. Denker, "COLREGs-Coverage in Collision Avoidance Approaches: Review and Identification of Solutions," OFFIS e.V. Institute for Information Technology, Oldenburg; University of Oldenburg, Department of Informatics.

[13] T. W. Lewallen, "A COLLISION-PREVENTION EXPERT SYSTEM FOR A NAVY OFFICER OF THE DECK," Naval Postgraduate School, Monterey, California, 1987.

[14] P. Sosnin, "Question-Answer Expert System for Ship Collision Avoidance," Ulyianovsk State Technical University, 2009.

[15] P. Banas and M. Breitsprecher, "Knowledge Base in the Interpretation Process," in *International Journal on Marine Navigation and Safety of Sea Transportation*, Institute of Marine Technology, Maritime University of Szczecin, 2011.

[16] H.-C. Burmeister, M. Constapel, C. Uge and C. Jahn, "From Sensors to MASS: Digital Representation of the Perceived Environment enabling Ship Navigation From Sensors to MASS: Digital Representation of the Perceived Environment enabling Ship Navigation," *IOP Conference Series Materials Science and Engineering,* November 2020.

[17] C. Uge und S. Hochgeschurz, „Learning to Swim - How Operational Design Parameters Determine the Grade of Autonomy of Ships," *TransNav, the International Journal on Marine Navigation and Safety of Sea Transportation,* Bd. 15, Nr. 3, pp. 501-509, 2021.

Journal of Physics: Conference Series **2311** (2022) 012034 doi:10.1088/1742-6596/2311/1/012034

On maritime electrification – electrification technologies, charging infrastructure and energy management strategies

X Feng[1,2], E B Zainudin[1], K J Tseng[1] and K C A Low[1].

[1] Singapore Institute of Technology
[2] Elsaaafeng@outlook.com

Abstract. This paper aims to give a landscape introduction and review on maritime electrification technologies and related applications, ranging from onboard vessel electrification technologies, standards, onshore charging infrastructure to onboard and onshore energy management strategies. Summaries and references given in this review paper can serve as a guidance with regard to the aforementioned aspects on marine electrification. The paper starts with an introduction on various vessel electrification technologies which include incorporation of renewable energy resources as part of the onboard energy supply and also configurations that incorporate these electrification technologies. Onshore charging infrastructure is essential to support the operation of electric vessels and they will be reviewed along with relevant standards. In addition to electrification technologies and onshore charging infrastructure, another important aspect of marine electrification lies with effective energy management strategies that ensure optimal operations. Energy management strategies with various objectives will be reviewed.

1. Introduction

The International Maritime Organization (IMO) has a goal to reduce the CO2 emissions from international shipping by at least 40% by the year 2030 [1]. In line with these goals, Singapore has detailed a roadmap to steer the marine sector to a more sustainable and eco-friendly direction. Due to Singapore's position as a maritime trade hub, it would be vital to have infrastructure and regulations that can accommodate these alternative sources of energy, as the world has an increasing demand for green energy. To achieve this goal, the electrification of sea vessels as well as the port technology to charge new types of vessels will need to be developed.

The first step of the vessel electrification involves incorporating alternative sources, which include solar, battery storage and hydrogen fuel cells [2 - 4]. Vessel electrification can be achieved through incorporating components such as electric motors, photovoltaic panels, battery storage and hydrogen fuel cells and the combination of them. Industry standards to be taken into consideration when introducing these technologies will also be reviewed. Based on the types of electrification technologies adopted, electric vessels can be further categorized into five different types including Diesel-Electric, Hybrid, Battery Hybrid, Battery Plug in Hybrid and Full Electric [3]. Each configuration comes with their own characteristics and will be detailed in this paper. Typical system architecture of vessels based on AC or DC propulsion will be reviewed.

Onshore charging infrastructure is then required to effectively support these vessels, namely shore to ship charging technologies [5] as well as the grid that supplies the power to the onshore charging facilities [6 – 7]. Firstly, four types of charger-to-vessel configurations, including AC charger to AC propulsion system, AC charger to DC propulsion system, DC charger to AC propulsion system and DC charger to DC propulsion system, as well as the associated power conversion system design will be discussed. Different types of charging interface technologies which include contact charging, inductive charging, and battery swapping will then be elaborated. Lastly, typical charging control systems, which include two main types that are low-level control (Power Converter control) and high-level control (System Control) will be discussed. Standards to take into consideration when designing onboard and onshore charging technologies [8] will also be reviewed wherever available.

Energy management strategies are essential to optimise the operation of electric vessels. Optimal component sizing sets the energy saving on a correct path, upon which effective energy management strategies can further be developed. Component sizing methods for vessel electrification will be reviewed in which details about how to determine the optimal power/energy ratings of components will be discussed. Energy management strategies with various objectives including cost minimization, fuel consumption optimisation and carbon emission optimization with case studies of various vessels, will then be introduced.

The rest of the paper is organized as below. Section 2 reviews onboard electrification technologies. Section 3 follows up with onshore charging technologies that support operations of electric vessels. Section 4 reviews energy management strategies for both onboard and onshore applications. Section 5 concludes the paper.

2. Electrification Technologies and Vessel Configurations

2.1 Electrification Technologies

To achieve the goal of reducing emissions of GHG's and reducing the dependency on fossil fuels, vessels will need to tap into electric power. Often this will include the installation of an electric motor to drive the propulsion system of the vessel in replacement of the traditional combustion engine-based drivetrain system. The electric motor can be powered by diesel generators, however, to achieve long term sustainability, alternative sources of renewable energy will need to be used. In this section, integration of photovoltaic (PV) cells, energy storage systems and fuel cells onto vessels will be detailed as alternate sources of energy.

2.1.1 Photovoltaic cells

The effectiveness of PV cells when used on maritime vessels is dependent on a variety of factors such as ambient temperature, irradiance and harsh weather conditions out at sea. In [9], a mathematical model of a cruise vessel fitted with an onboard renewable power system which consists of a PV array, a PEM fuel cell, a diesel generator, an electrolyser, and a DC/AC inverter shows a reduction of GHG by 9.84%. It also shows that 13.83% of the total energy needed on the vessel was produced by the onboard renewable power system. This study also noted that in countries that experience higher solar irradiance, ferries outfitted with the above-mentioned technologies will produce better results reflected in lower emissions and cost. A case study showed a vessel operated between Sweden and the Aland Islands, in which authors noted that with a solar irradiance of 2.87 $kWh/m^2/day$ it was sufficient to meet the power demands of the vessel. In comparison, regions such as Sharjah, United Arab Emirates receives a solar irradiance of 8.30 $kWh/m^2/day$ and hence provide great potential for the integration of such onboard renewable energy systems.

Authors in [10] detailed a cost-benefit analysis of using a solar hybrid power system on merchant vessels. It focused on vessels between 500 to 10,000m^2 as 90% of the world's bulk carriers are less than 10,000m^2. The authors concluded that the cost benefit of installing PV panels on merchant ships is closely related to the annual price increase of fuel. The higher the increase in the fuel price is, the shorter the payback period is for the installation of PV panels with the minimum possible period of being 10 years. Furthermore, vessels that are exposed to more solar radiation (vessels near the equator) stand to benefit more due to the increased efficiency and more extensive utilisation of the PV panels.

Environmental effects on the PV panels should also be taken into account in maritime applications. Authors in [11] detail the impacts on PV panels when they are exposed to rainwater, seawater and salt particles. The experiment aims to "obtain the variation trend under maritime environment". The authors concluded that rainwater, saltwater and salt particles will increase the maximum voltage (Vm) and open circuit voltage (Voc) of the panel while at the same time the short circuit current and maximum current decreases. This causes a small reduction in efficiency. Rainwater and salt water can increase heat dissipation thus reducing the temperature of the panel. Due to this increased heat dissipation, rainwater may result in increased efficiency as compared to a PV panel with salt layered on top. The salt layer on the panel will increase the temperature which impacts its efficiency as well as reduce the solar radiation exposed to the panel. Seawater at higher concentrations of salt causes a smaller reduction to the efficiency of the panel as compared to seawater with lower salt content.

2.1.2 Energy Storage System

The availability of an onboard energy storage system allows flexibility in the optimization of the energy dispatch on a vessel, especially considering the presence of intermittent renewable energy. In the case of [12], a ferry is able to benefit from the fluctuating grid prices when charging from an onshore charging station through energy arbitrage. Having the vessel charge more from the grid during off-peak periods further increases the cost effectiveness of an electrified system. Authors in [12] also aim to optimize the energy dispatch of a hybrid ferry taking into account the degradation of the onboard energy storage system. This degradation, as described in the paper, can be attributed to the corrosion/oxidation of electrolytes, thermal and electrical stress when cycling, environmental factors and the depth of discharge, some of which are unique to maritime applications.

The inclusion of onboard batteries involves significant investment. Cargo ships which require larger power demands amount to larger capacities of battery storage required and larger space onboard the vessel. In [13], a battery was size according to an electrical machine rated at 875kW. It was sized to provide power for the longest usage with a depth of discharge of 60%, which leads to an energy rating of 3000 kWh. A battery of this capacity would demand significant vessel space onboard with a "2% reduction in the lane-metre capacity of the vessel".

Relevant Standard relating to energy storage systems can be found in IEC 60092-305 – Accumulator (Storage Batteries). This standard governs the requirements when permanently equipping a maritime vessel with storage batteries. It does not cover batteries that are portable or temporary. It covers topics such as types of batteries, construction, assembly, charging facilities, etc.

2.1.3 Fuel Cells

Fuel cells are seen as an attractive source of energy as it boasts many advantages over conventional maritime propulsions systems. These advantages include lower emissions, higher fuel efficiency, lower noise while operating, efficient low load operation and lower operating and maintenance costs [14].

A typical fuel cell consists of an anode, cathode and electrolyte shown in Figure 1.

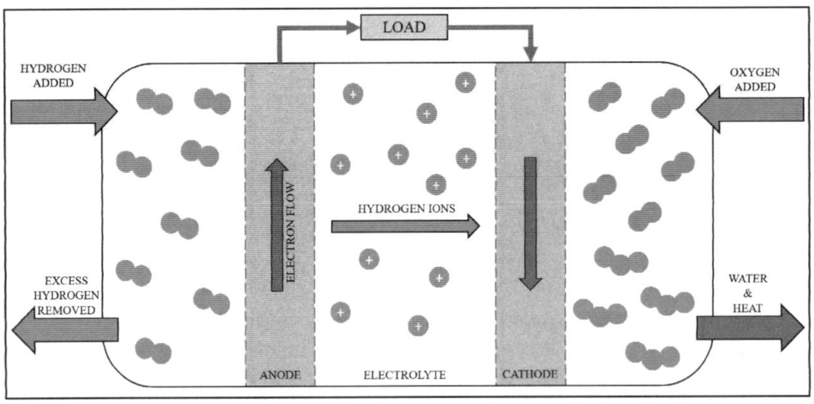

Figure 1. Typical Fuel Cell construction [14]

In the case of a polymer electrolyte membrane fuel cell (PEMFC), the anode and cathode are separated by an electrolyte membrane. Hydrogen is fed to the anode while oxygen is fed to the cathode. The hydrogen atoms split into protons and electrons by a catalyst with the protons being able to pass through the membrane. In order to achieve an equal charge, the electrons will move through the anode causing an electric current to run between the anode, load and cathode. At the cathode, the hydrogen protons, electrons and oxygen molecules combine to form H2O as a by-product. Table 1 gives an overview of different types of fuel cells using various types of electrolytes and their associated parameters.

Despite various advantages offered by fuel cells, the cost that is needed to implement such a system remains a hurdle to its wider adoption. Authors in [10] note that integration of fuel cells in an onboard PV system decreases the profitability of the system. As the solar energy directed to the operation of the fuel cell increases, the cost-benefit of the system decreases. Authors in [9] note that the cost per MWh for a diesel engine vessel was $223 while a vessel with a PV array, a PEMFC and a diesel generator was $260. This highlights that the installation of a fuel cell, might incur greater financial cost, despite the reduction of emissions.

Table 1. Types of Fuel Cells [14]

Type of Fuel Cell	Electrolyte	Fuel	Efficiency (%)	Power Rating per module
PEMFCs	MEA	Hydrogen	50-60	Up to 120 kW
PAFCs	Silicon carbide matrix saturated with liquid phosphoric acid	Hydrogen, LNG, Methanol, Diesel	40	100 to 400 kW
MCFCs	Combination of alkali carbonates	Hydrogen, CO, LNG, Methanol, Diesel	45-60	Up to 500 kW
SOFCs	Metal Oxide	Hydrogen, CO, LNG, Methanol, Diesel	45-60	20 to 60 kW
DMFCs	MEA	Methanol	35-80	Up to 5 kW

2.2 Electric Vessel Configurations

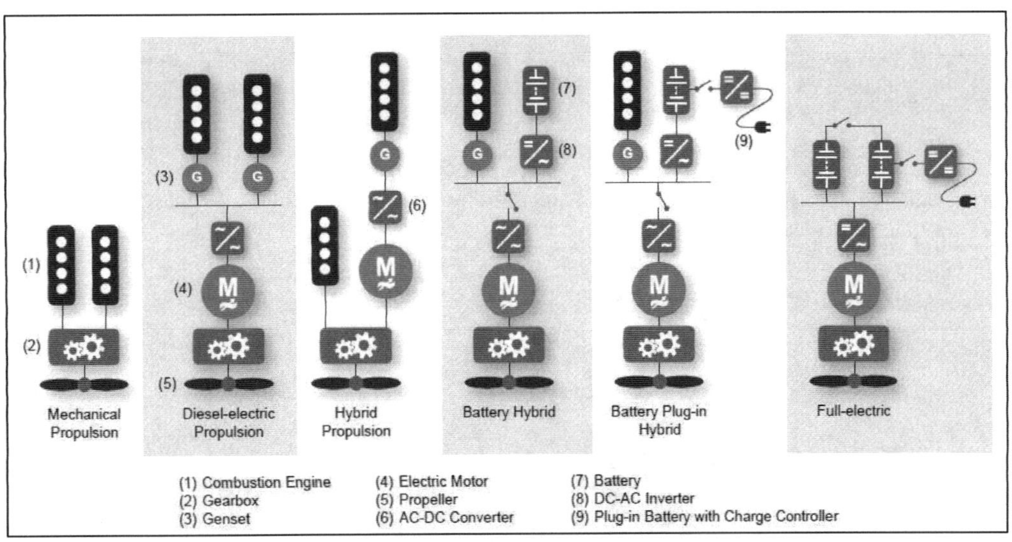

Figure 2. A Classification of Different Vessel Configurations [3]

This section details different vessel propulsion configurations for integration of alternative sources. A general classification of different vessel configurations is shown in Figure 2 [3].

Relevant standards are listed below:

- IEC 60092-301 – Generators and Motors. This standard applies to all rotating electrical machines rated for 750W and above. This includes all excitation machines and primes movers. It details topics such as general requirements, voltage regulation, parallel operation, running speed, protection etc.
- IEC 60092-501 – Electric Propulsion Plant. This standard will detail the design, installation and testing for most components that are within an electric propulsion plant. These components include generators, prime movers, transformer/reactors, semiconductor convertors, propulsions motors, excitation systems, control, monitoring and safety systems, wires, busbars, trunking and harmonic filters. The topics detailed in these standards include system design, propulsion switchboards/transformers, propulsion motors, control systems etc.

2.2.1 Diesel-Mechanical

This type of propulsion is the most widely used in marine ports all around the world which consists of engines that burn fossil fuels to rotate propeller blades. An alternator is also attached to the engine to supply electrical power to the onboard ship systems.

2.2.2 Diesel-Electric

In this configuration, fossil fuel is still used as the fuel source, however the diesel generators are connected to a common electrical bus which then supplies power to an electric motor. Electric motors can achieve a much higher efficiency as compared to a diesel engine.

The advantages of diesel-electric configuration are as below:

- Propulsion system can be decoupled [5][15].
- Better load sharing of generators [5].
- Gearbox optional due to full torque at start-up [5].
- Variable speed control electric motors have high efficiency between 5% and 100% rated power. Compared to combustions engine that have an optimum operating range of 85% to 90% [15].
- Some of the vessels that use this config are cruise ships, LNG Carriers, drill ships, etc [15].

2.2.3 Hybrid

As the name suggests, hybrid is a combination of both diesel and electric. In this configuration, a combustion engine can be connected directly to the propeller to provide high efficiency when the vessel is travelling at a consistent high speed while the electric motor is used at lower speeds. Alternatively, the two can be separated, with each motor having its own propeller. Depending on the operating mode, the electric motor can be used as a generator to supply onboard appliances as well [15].

Other configurations classified as hybrids will solely use the electric motor to drive the propulsion system. In these cases, the common electrical bus that provides power to the motor is connected to either a genset or battery.

There are many ways the combustions engine and the electric motors can work together depending on the use case, these are some examples [15]:

1) Full-electric: Can be used in case of combustion engine failure, zero-emission operation and low speed operation.

2) Combined propulsion: Both the engine and electric motor supply power to the propeller, initiating a boost of speed.

2.2.4 Full-Electric

In a full electric vessel, the power is supplied solely by batteries. There are a variety of ways these batteries can be charged including being charged by onboard PV panels, battery swapping and onshore charging stations. Large vessels engaged in ocean-going activities, station keeping and icebreaking have implemented electric propulsion, however it is unlikely that these vessels can be solely supported by batteries, as the wattage requirements for the propulsion system can range between 20 to 25MW [15]. Normally these large vessels have generators to supply the power.

However, it is possible to reduce the scale of these vessels to the point that batteries alone can be used to supply the vessel fully. The Yara Birkeland is a cargo vessel from Norway that is able to carry 120TEU (Twenty Foot Equivalent) with a battery capacity of 6.8MWh and is designed for a relatively short route of 12 nautical miles near the coastline [16]. As a comparison, typical cargo ships can have a cargo capacity of 10,000 to 21,000 TEU. The vessel will be charged by onshore charging stations that utilize renewable energy thus further reducing the carbon footprint and GHG Emissions.

3. Onboard and onshore charging infrastructure

3.1 Typical Charging System

Onshore charging systems and vessel propulsion systems can be categorized into AC or DC [5] based on the type of electricity adopted on the busbar. For example, if the main bus on the vessel is of a DC type, the vessel is categorized as a DC propulsions system.

Relevant Standards are listed below:

- IEC 60092-201 – General. This standard details general system design elements found in a standard electrical system These include AC/DC distribution systems, earthing, protection, motors circuits, etc.
- IEC 60092-503 – AC Supply System. This standard is specific to AC supply systems that are 1kV to 36kV(inclusive). It details topics such as insulation, electrical protection, high voltage equipment, transformers, reactors, converters, etc.
- IEC/IEEE 8005-1 – High Voltage Shore Connection Systems (HVSC). This standard comprises of the design, testing, and installation of HVSC systems for both onboard and onshore. High voltage is defined as vessels that require 1MVA and above. The topics include, high voltage distribution systems, shore to ship connection, transformer/reactors, control, monitoring, interlocking and power management systems.
- IEC PAS 8005-2 – Low Voltage Shore Connection Systems (LVSC). Comprising the same topics as the HVSC systems, however this standard, covers vessels that require up to 1MVA.

3.1.1 AC Charging System to AC/DC Propulsion System

Shown in Figure 3 is an example of an AC Charging station connected to DC and AC propulsion systems.

For an AC charging station connected to a DC propulsion system, the AC power from the charging station is converted to DC power onboard the ship through the converter onboard, seen in Figure 3a as Y. If the converter (Y) is able to regulate the voltages to match the requirements of the onboard batteries, a buck and boost converter (seen in Fig 3a as X) will not be required. This can be employed as a cost-saving measure.

When an AC charging station is connected to an AC-based propulsions system, the main concern is with the synchronization of voltage, phase and frequency between the onboard power system to the onshore charging station. The accurate synchronization is time consuming which renders time sensitive operations unsuitable for this kind of configuration. This configuration is useful for cold ironing where the time constraints are not as critical.

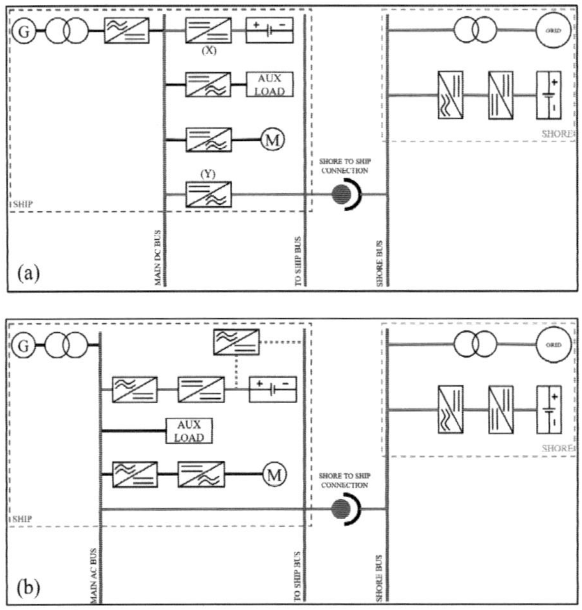

Figure 3. AC Charging station connected to DC(a) and AC(b) propulsion system [5]

3.1.2 DC Charging System to AC/DC Propulsion System

Shown in Figure 4 is an example of a DC Charging station connected to DC and AC propulsion systems.

A DC charging station can be conveniently and relatively straightforwardly connected to a DC propulsion system if the power requirements are met. In this configuration, the onshore charging stations take charge of regulating/controlling the incoming AC power supply which removes the need for onboard inverters/converters, leading to savings in vessel weight and cost.

For a DC charging station to AC propulsion system configuration, energy can be transferred to the onboard batteries without the need to go through the AC bus bar. In figure 4b, the onshore charging station transfers power to the buck/boost converter onboard the vessel. By bypassing the onboard AC busbar, it effectively reduces one step in the conversion process.

 2311 (2022) 012034 doi:10.1088/1742-6596/2311/1/012034

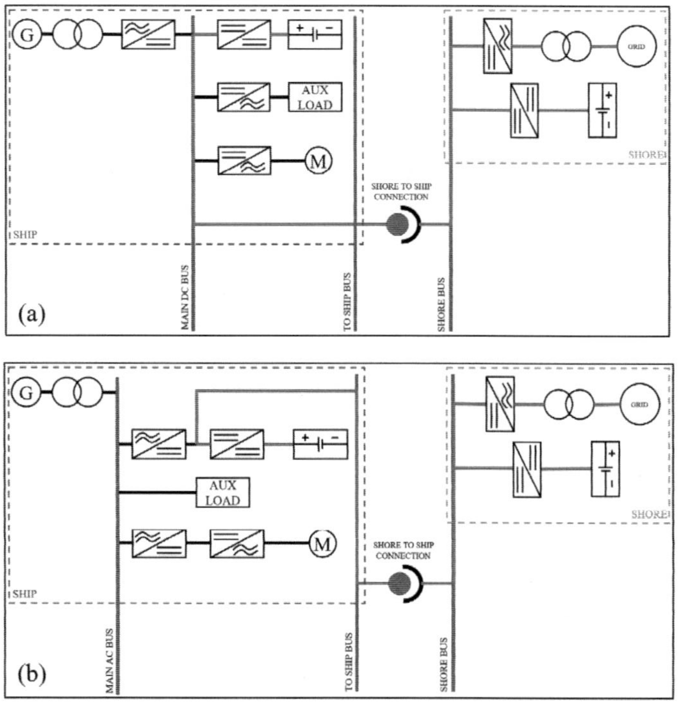

Figure 4. DC Charging station connected to DC(a) and AC(b) propulsion system [5]

3.1.3 Efficiency Comparison

Authors in [17] evaluated the efficiency of the different configurations using a short-distanced ferry as the case study. Efficiency is defined as the "utilization of the aforementioned power loss models for evaluating the total energy drawn from the grid, directly and indirectly through the onshore battery, for supplying the required energy to the battery" [17]. It was concluded that for a ferry with an AC-based propulsion system, charging from a DC charging station is more efficient than charging from an AC charging station. This is because in an AC propulsion to AC charging station configuration, the energy from the onshore battery has to go through multiple conversions before reaching the onboard battery. From DC (onshore batteries) to AC (ship bus) and then back to DC (Onboard battery). Each conversion introduces a power loss. In comparison, a DC charging station does not go through the conversions, unless there is a requirement for a buck/boost converter connected to the onboard battery. Even so, the power loss from one DC to DC converter is far smaller than the power losses through multiple conversions. For a DC-based propulsion system, either AC or DC charging could be the more efficient depending on the charging parameters depending on the charging parameters. on the charging parameters.

3.2 Shore to Ship Charging Interfaces

Shore to ship interface refers to the point at which the shore charging station and the electrical power system on the ship interact with each other. It is through this interface that charging power flow from the source to the ship. These may include different components such as electrical contacts, mechanical systems and monitoring systems [5]. The transfer of electrical power can be done via contact charging, wireless charging and battery swapping. Each method has advantages and disadvantages and therefore should be chosen based on specific applications.

3.2.1 Contact Charging

Contact charging is the simplest way for the electrical connection to be established between the ship and charging station. The most common way is using a plug. It simply involves plugging a cable into the charging dock at the charging station. At lower voltages, this connection can be done by personnel manually plugging the cable shown in Figure 5. At higher voltages, the process can be automated using a mechanical arm to engage the plug out of safety considerations shown in Figure 6. Another method is a tower that drops the charging cable from a height onto the vessel shown in Figure 7.

The main issue of contact charging may face is the fact that a vessel is not stationary when docked due to the varying water levels. The connection can be disrupted should a particularly big wave hit the vessel. To overcome this issue, a vacuum mooring system need to be employed to maintain the ship stationary for a stable charging connection.

Table 2. Advantages and disadvantages of various contact charging methods

Method	Advantages	Disadvantages
Manual Plug	• Simple	• Limited to low voltages
Automated plug	• Safe	• Expensive • Slow
APS Tower	• Flexible for high and low water levels	• Expensive • Slow
Pantograph	• Fast to connect	• Expensive

Table 2 summarizes the advantages and disadvantages of the various shore to ship connection methods:

Figure 5. Automated Arm [5]

Figure 6. Plug in system tower (APS) [5]

Figure 7. Manual Plug [5]

3.2.2 Wireless charging

Wireless charging involves the transfer of electrical energy using inductive coils with an example given in Figure 8. The absence of contact allows greater flexibility when it comes to the bobbing of the vessel. However, this comes at the cost of efficiency. It can be noted in [5] that inductive charging is not significantly less efficient than contact charging as zero voltage switching in inductive charging can reduce the switching losses. Table 3 summarizes the advantages and disadvantages of inductive charging.

Table 3. Advantages and disadvantages of inductive charging

Method	Advantages	Disadvantages
Inductive	• Less prone to mechanical wear & Corrosion. • Safety due to lack of exposed contacts. • Reduces need for attaching plug, therefore less time wasted.	• Expensive. • Heavy, to increase efficiency, larger coils need to be installed. • Slightly Lower Efficiency.

Figure 8. Wartsilla Inductive charging system. [5]

3.2.3 Battery Swapping

The process of battery swapping involves swapping out discharged battery on the vessels with freshly charged ones from the battery swapping stations. Due to the speed at which this can be done, time-sensitive operations can utilize this method to the greatest effect. In the case of marine vessels, ferries that are on strict schedules and who operate near to shore are prime candidates.

Batteries in the battery swapping stations can be charged in a multitude of ways depending on the requirements. With the reduction of GHG and green energy in mind, renewable energy sources such as wind turbines and solar panels can be considered. Alternatively, the grid can also be used.

Table 4 gives a summary of advantages and disadvantages of battery swapping.

Table 4. Advantages and disadvantages of battery swapping method

Method	Advantages	Disadvantages
Battery swap	• Speed. • Able to make use of off-peak grid prices. • Incorporation of renewable energy sources.	• Initial capital investment. • Requires robotic arm to swap the batteries. • Extra battery packs.

3.3 Charging control systems

Charging control systems can be categorised into low-level and high-level [5]. Low-level control includes power and voltage control at the bottom of the control chain. High-level control involves power management systems (PMS) and energy management system (EMS) which send commands to low-level control. These systems regulate the power and voltage limits to ensure the efficiency and stability of a system while charging. The high-level control system may also include the battery management system of the onboard/onshore batteries. The battery management systems are responsible for monitoring the state of charge of the battery, state of health, cell balancing and temperature regulation [5].

3.3.1 Low-level Control

An example of a low-level control system within a charging path is the converter control.

Figure 9 shows a DC Charging station with a DC-based propulsion vessel. In this configuration, there are two converter controllers each with their own objectives. Converters X and Y are responsible for controlling the power coming from the DC bus. Likely they will have to buck/boost the DC power appropriately based on the charging specifications of the battery supplier. In the case of X, there will be two operating modes, 1) Charging from the charge station or 2) Discharging from the ship. Y on the other hand will either 1) Charge from the grid or 2) Discharge to the ship.

Converter Z is responsible for controlling the AC power incoming from the grid. Z includes a sync block (phase-locked loop), current controller (PI Controller) and PWM generator. This converter has two operating modes, it either 1) Charges to the onshore battery banks or 2) Maintaining the correct DC level for the bus to charge the on-ship battery.

3.3.2 High-level Control

Figure 10 shows the typical onshore and onboard controller systems. This figure illustrates the how the individual control systems coordinate within the onshore and onboard energy/power systems respectively. The onboard systems will communicate to the onshore charging system the requirements based on specific operation tasks. The information includes voltage and current requirements, state of charge, operating mode, etc. The onshore charging station will then execute as required.

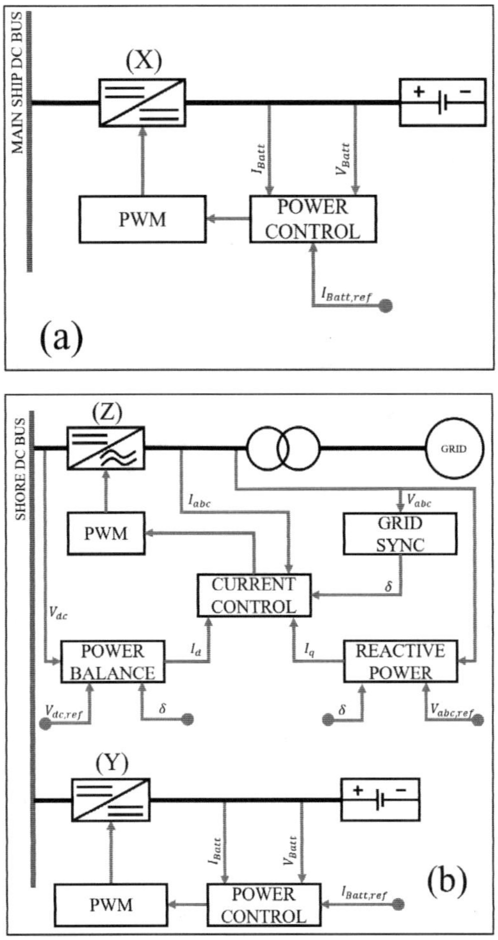

Figure 9. Converter Level Control for DC Charging, Onboard Battery(a), Onshore Grid & Battery [5]

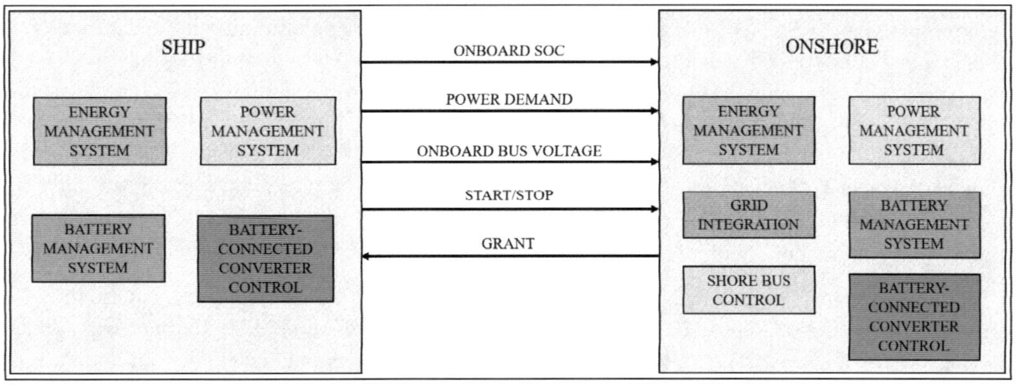

Figure 10. System Level Control for Shore to Ship Charging [5]

3.4 Onshore Charging System Operation

There are four operating modes for a charging station [5]:

1) Charging onshore batteries while no vessel is docked.
2) Charging onshore batteries from the grid when vessel is docked.
3) Supplying power to the vessel only from the onshore batteries.
4) Supplying power to the vessel from both the grid and onshore batteries.

The onshore EMS is responsible for the dispatch of energy to ensure cost savings and efficiency where possible. When a vessel is at berth, the onboard system sends the power demand to the onshore charging station. The EMS controls the share of the grid power and the battery banks to supply the power needed. An example of optimization using the onshore EMS would be prioritizing the use of the grid power when the tariff is low, to charge the vessel. The PMS is responsible for maintaining the limits of the system to ensure reliable operation. An example would be maintaining the power demand of the batteries while it is being charged from the grid. The shore bus controller regulates the voltage of the shore bus according to the operating mode. The battery management system performs the monitoring and cell balancing of the battery banks. It works in tandem with the EMS to ensure safety and stability.

3.5 Onboard power system operation

There are three main operating mode for an onshore power system [5]:

1) All-Electric mode.
2) Hybrid mode.
3) Charging mode.

Similar to the EMS on the onshore charging system, an onboard EMS manages the dispatch of the possible sources on board. These are primarily the diesel generator and onboard battery banks. The PMS will dictate the power demand from the sources according to the operating mode it is set in. The battery management system will do the same function as mentioned above.

4. Energy management strategies

Energy management and optimization generally comes in two stages. Firstly, each component in the concerned system would need to be sized optimally for the best possible optimization outcomes. Optimal component sizing would typically consider demand, supply and other constraints with an aim to achieve highest cost-effectiveness. Other factors such as available space, safety and standards can also affect the component sizing. One example would be the size of a PV array. On a vessel, there is space limitation as well as weight limitation. This could mean that when optimizing for the size of a PV array, the optimized size in theory might exceed the available space on the vessel. Therefore, various constraints linked to practical applications need to be carefully included. The remaining content will look into optimal sizing and

4.1 Optimal component sizing

Authors in [18] aim to optimize the size and energy dispatch of a hybrid vessel using a multi-objective, double-layer optimization procedure. The components being optimized are a diesel generator, an energy storage system and a hydrogen fuel cell. The paper describes an inner and outer layer for optimization. The inner layer focuses on the energy dispatch of the different components at each interval in time using MILP (Mixed Integer Linear Programming) and minimizing OPEX as the objective function. The outer layer focuses on the optimized power rating of each component. This layer uses NASGA-II as the optimization algorithm while using the minimization of CAPEX, OPEX and Emissions as the objective functions.

4.2 Energy management and optimization

Energy management and optimization is required to enhance the overall system efficiency. Electric vessels, considering their structure. Integration of renewable energy systems come with them an aspect of uncertainty. In the context of electrical vessels and shore charging installations, the power output of the PV panels is directly related to real time weather conditions which can be intermittent and cause fluctuations. Other uncertainties include tidal fluctuations, voyage uncertainties and so on. In order to meet the operational requirements while achieving high-cost effectiveness, energy management strategies should be carefully designed to optimally manage the controllable resources in the overall system while taking into account various uncertainties and constraints.

Energy management strategies for marine electrification can be designed to meet different purposes through varying the objective functions of the optimisation problem formulation. They generally include minimizing cost (OPEX, CAPEX), fuel consumption, genset running hours, genset start/stop, and weight of diesel fuel consumed.

4.2.1 Minimizing Operating Cost

To optimize the operating cost of a vessel while on a voyage, the authors in [19] proposed a coordination model for a multi-energy vessel to optimize the energy dispatch in a voyage. The optimization considers uncertainties arising from load demand, ambient temperature and solar radiations. This coordination model is then converted to a mixed integer linear programming problem which can be solved using commercial solvers such as Cplex or Gurobi. The inputs of this optimization model include the forecasted PV generation of the journey, onboard load demand, ambient temperature, and ship voyage information. The outputs of the optimization determine the voyage scheduling and energy dispatch decision. In more detail, a two-stage optimization process is proposed. The 1st stage is called the "Voyage Ahead Operation". This stage predetermines the cruising speed and distance, charging/discharging of the onboard batteries and switching on/off of the diesel engines. These decisions are set a day to half a day in advanced and will be strictly followed. The 2nd stage is optimised every 30 minutes in real time which controls the power outputs from the combined cooling heating and power (CCHP) unit, diesel generator, power to thermal unit, power transfer of the bidirectional converter, cold ironing as well as the indoor temperature. This is the stage that makes up for the deficits caused by the decisions in the 1st stage, thus ensuring a more reliable system.

In [19], the optimisation problem aims to minimize the operating cost which include the fuel cost, maintenance cost, start-up cost of the diesel generator, grid power purchasing cost and general emission cost.

4.2.2 Minimize Fuel Consumption and GHG Emissions

Another type of optimisation is to minimize the fuel usage. Authors in [20] proposed a method that optimizes the usage of diesel generators and ESS onboard a vessel considering ESS degradation. The optimisation results control states of the diesel generator, power delivered by each generator and power delivered by the energy storage system at each time step. This problem was formulated as a mixed integer linear programming and solved using Cplex in a GAMS environment. The case studies used in [20] include a ferry and a platform supply vessel. Both these vessels only have generators and energy storage systems as the power source. The authors concluded with 6% savings for the ferry while the platform support vehicle observed a 32% to 42% net savings considering the vessel's lifetime. Savings were defined as the fuel cost per mission in addition to the installation cost of the power inverter and ESS and replacement cost of the ESS.

Authors in [21] propose a way to reduce the GHG emissions by optimizing the scheduling of fixed speed generators. For the GHG emissions and fuel consumption to be kept to a minimal, the generators need be kept at optimal loading. This loading is dictated by the "specific fuel oil consumption curve" SFOC of the generators. Optimal loading is defined as 90% of the rated power of the generators. Three configurations were looked into including 1) Four Fixed Speed Generators 2) Three Fixed Speed Generators with One Variable Speed Generator and 3) Three Fixed Speed Generators with ESS. Every configuration has a different algorithm that manages the energy dispatch. The case studies evaluated in this paper were a ferry, a platform supply vessel and seismic survey vessel. The results showed that in the case of the ferry, the highest fuel savings was observed for the 3rd configuration. The platform survey vessel showed similar results with the ferry with a fuel savings of 10.3%. The seismic survey vessel showed the most fuel savings of 10.3% in the 2nd configuration. It was noted that this could be because of the stable nature of the load profile. The seismic survey vessel does not have dramatic shifts in the load profile as compared to the ferry and platform survey vessel. It was concluded that adding a variable speed generator or an ESS to a system can significantly reduce the GHG and fuel consumption of a vessel as it allows for the system to be more reactive to any load fluctuations in the vessel. Low loading operations can be allocated to the variable generators or ESS more heavily as compared to the fixed speed generators which are best run close to their max rating.

5. Conclusion

This paper gives a landscape review on maritime electrification, ranging from electrification technologies, charging infrastructure to energy management strategies. Electrification technologies are the first step to transform the traditional vessels to ones with electrification potential through incorporation of electric motors and alternative means of energy sources including photovoltaic cells, ESS and fuel cells. Possible impacts from the special environment at sea to these electrification technologies also need to be carefully considered in actual design and operation. Charging infrastructure is an integral part of electrification and essential to ensure safe and efficient operation of the electrified vessels. Various types of charging system architecture, charging interfaces and control systems are available when it comes to actual implementation. It is worth mentioning that land-based EV charging systems and standards are more mature than the marine counterparts, it would be a good approach to develop marine charging systems by modification of existing land-based systems. The last part of the equation is energy optimization for the overall shore to ship charging ecosystem ranging from design to actual operation. Energy optimisation can be designed for specific purposes for different applications based on actual needs.

6. References

[1] "Greenhouse Gas Emissions." https://www.imo.org/en/OurWork/Environment/Pages/GHG Emissions.aspx (accessed Sep. 20, 2021).

[2] https://ebook.ntu.edu.sg/a-study-on-the-future-energy-options-of-singapore-harbour-craft.html

[3] https://ebook.ntu.edu.sg/electrification-of-singapore-harbour-craft.html

[4] "Singapore R&D Roadmap 2030: Maritime Transformation." https://www.maritimeinstitute.sg/roadmap2030/ (accessed Sep. 20, 2021).

[5] Karimi S, Zadeh M and Suul J A 2020 Shore charging for plug-in battery-powered ships: Power System Architecture, infrastructure, and Control *IEEE Electrification Magazine* **8** 47–61

[6] Kumar J, Kumpulainen L and Kauhaniemi K 2019 Technical design aspects of harbour area grid for shore to ship power: State of the art and future solutions *International Journal of Electrical Power & Energy Systems* **104** 840–52

[7] Kumar J, Palizban O and Kauhaniemi K 2017 Designing and analysis of innovative solutions for Harbour Area Smart Grid *2017 IEEE Manchester PowerTech*

[8] Anon IEC/IEEE International Standard - utility connections in port -- part 1: High voltage shore connection (HVSC) systems -- general requirements

[9] Ghenai C, Bettayeb M, Brdjanin B and Hamid A K 2019 Hybrid solar PV/PEM fuel cell/diesel generator power system for cruise ship: A case study in Stockholm, Sweden *Case Studies in Thermal Engineering* **14** 100497

[10] Glykas A, Papaioannou G and Perissakis S 2010 Application and cost–benefit analysis of solar hybrid power installation on merchant marine vessels *Ocean Engineering* **37** 592–602

[11] Sun Y, Yan X, Yuan C, Luo H and Jiang Q 2015 The I–V characteristics of solar cell under the Marine Environment: Experimental Research *2015 International Conference on Renewable Energy Research and Applications (ICRERA)*

[12] Hein K, Yan X and Wilson G 2020 Multi-objective optimal scheduling of a hybrid ferry with shore-to-ship power supply considering energy storage degradation *Electronics* **9** 849

[13] Sciberras E A, Zahawi B and Atkinson D J 2017 Reducing shipboard emissions – assessment of the role of Electrical Technologies *Transportation Research Part D: Transport and Environment* **51** 227–39

[14] Shakeri N, Zadeh M and Bremnes Nielsen J 2020 Hydrogen fuel cells for ship electric propulsion: Moving toward greener ships *IEEE Electrification Magazine* **8** 27–43

[15] Hansen J F and Wendt F 2015 History and state of the art in commercial electric ship propulsion, Integrated Power Systems, and future trends *Proceedings of the IEEE* **103** 2229–42

[16] https://www.yara.com/news-and-media/press-kits/yara-birkeland-press-kit/

[17] Karimi S, Zadeh M and Suul J A 2020 Evaluation of energy transfer efficiency for shore-to-ship fast charging systems *2020 IEEE 29th International Symposium on Industrial Electronics (ISIE)*

[18] Wang X, Shipurkar U, Haseltalab A, Polinder H, Claeys F and Negenborn R R 2021 Sizing and control of a hybrid ship propulsion system using multi-objective double-layer optimization *IEEE Access* **9** 72587–601

[19] Li Z, Xu Y, Fang S, Zheng X and Feng X 2020 Robust coordination of a hybrid AC/DC multi-energy ship microgrid with flexible voyage and thermal loads *IEEE Transactions on Smart Grid* **11** 2782–93

[20] Boveri A, Silvestro F, Molinas M and Skjong E 2019 Optimal sizing of energy storage systems for shipboard applications *IEEE Transactions on Energy Conversion* **34** 801–11

[21] Skjong E, Johansen T A, Molinas M and Sorensen A J 2017 Approaches to economic energy management in Diesel–electric marine vessels *IEEE Transactions on Transportation Electrification* **3** 22–35

Biofuel Adoption Pathways for Cargo Vessels under Carbon Tax

Zengqi XIAO

Maritime Energy & Sustainable Development Centre of Excellence (MESD)
Nanyang Technological University, Singapore

Jasmine Siu Lee LAM*
Maritime Energy & Sustainable Development Centre of Excellence (MESD)
School of Civil and Environmental Engineering, Nanyang Technological University, Singapore

E-mail address: sllam@ntu.edu.sg (Corresponding author)

Prapisala THEPSITHAR

Maritime Energy & Sustainable Development Centre of Excellence (MESD)
Nanyang Technological University, Singapore

Kevin MILLA

Maritime Energy & Sustainable Development Centre of Excellence (MESD)
Nanyang Technological University, Singapore

Abstract. According to the International Maritime Organization (IMO) Greenhouse Gas (GHG) Strategy, the shipping industry by 2050 aims to reduce by at least 50% of GHG emission level in 2008. Among various alternative energy sources, biofuels illustrate potential as a drop-in fuel using existing shipboard technologies, bunkering infrastructure, and a new fuel replacing conventional fossil-based fuels. However, the decision to adopt biofuel in vessels depends on various factors. Biofuels may be derived from different feedstock and production routes and thus demonstrate different economic and environmental performances. The compatible type of engines also differs across biofuels, implying differences in capital cost. In addition, a high carbon tax may push the usage of a more expensive biofuel, while a low carbon tax may not drive the switch from fossil fuel. This study aims to analyse biofuel adoption pathways under carbon tax to achieve a net 50% GHG emission reduction for cargo vessels, including bulk carriers, container vessels and tankers. The study focuses on the following types of fuels: low sulphur fuel oil (LSFO), marine gas oil (MGO), Liquefied natural gas (LNG), biodiesel in the form of palm methyl ester (PME), hydrotreated vegetable oil (HVO), and bio-LNG.

Keywords: Ship emission, Greenhouse gas, Biofuel adoption, Carbon tax

1. Introduction

Maritime shipping is indispensable to global trade and the economic development of countries. Over 80% of world trade is transported by sea. The cost of shipping operation involves heavily in cost of energy required for ship powering. During the past two centuries, ocean-going vessels for commercial trade have evolved from using wind till the early 19th century to engines fuelled by coal, and later to diesel engines fuelled by oil from the second half of the 20th century. Among various oil products, heavy fuel oil (HFO) has become the dominant choice of fuel in the shipping industry due to its cost efficiency, ease of handling, and applicability. It is estimated that about 98% of vessels are driven by diesel engines nowadays [1].

Currently, more than 62,000 vessels in the world trading fleet are operating using fossil fuels. These ships account for about 250 to 300 million tonnes of fossil usage per year, about 3% of the world's fossil fuel consumption. Currently, international shipping is responsible for about 3% of total anthropogenic CO2 emission or 11% of global CO2 emission in the transport industry [1]. If no measures were taken, the emission of CO2 might increase by about 50% to 250% by 2050. To further enable the reduction of GHG emissions from ships and better align with the Paris Agreement, MEPC 72 established the "Initial IMO Strategy on Reduction of GHG Emissions from Ships" in 2018 [2]. In this strategy, various levels of ambitions are set as follows:

1. To implement further phases of the energy efficiency design index (EEDI) for new ships with the percentage improvement for each phase to be determined for each ship type, as appropriate;
2. To reduce CO2 emissions per transport work by at least 40% by 2030, pursuing efforts towards 70% by 2050, compared to 2008; and
3. To peak GHG emissions from international shipping as soon as possible and to reduce the total annual GHG emissions by at least 50% by 2050 compared to 2008.

It is most unlikely to meet IMO's ambitious goal by 2050 without the help of alternative energy sources. Examples of such energy sources include fuels containing lower carbons, biogenic carbons, recycled carbons from carbon capture technology, and ultimately zero carbons. Among various alternative energy sources, biofuels illustrate potential as a drop-in fuel using existing shipboard technologies, bunkering infrastructure, and a new fuel replacing conventional fossil-based fuels. Yet, according to IEA's latest tracking report on international shipping [3], almost no low-carbon fuels are used today, with biofuel accounting for only 0.1% of the final energy consumption in shipping. Based on the current progress, IEA estimated that low- or zero-carbon fuels could only account for 5% of the final energy consumption by 2050, falling severely short of the expected 83% to meet the IEA's Net Zero Scenario.

A plausible explanation for the projected low adoption of biofuel in shipping is that the decision to adopt biofuel in vessels depends on various factors that are complex and uncertain. Biofuels may be derived from different feedstock and production routes and thus demonstrate different economic and environmental performances. The compatible type of engines also differs across biofuels, implying differences in capital cost. In addition, a high carbon tax may push the usage of a more expensive biofuel, while a low carbon tax may not drive the switch from fossil fuel. This study aims to analyse biofuel adoption pathways under cargo tax to achieve a net 50% GHG emission reduction for cargo vessels, including bulk carriers, container vessels and tankers. The study focuses on the following types of fuels: low sulphur fuel oil (LSFO), marine gas oil (MGO), Liquefied natural gas (LNG), biodiesel in the form of palm methyl ester (PME), hydrotreated vegetable oil (HVO), and bio-LNG.

2. Biofuels for Shipping Sectors

In the search for energy sources alternative to using conventional fossil-based fuels for the shipping industry, a potential option is to adopt biofuels. In this context, the term "biofuel" refers to liquid or gaseous fuels for the transportation sector that is predominantly produced from biomass [4]. Biomass is renewable organic material that comes from plants and animals [5][6]. The biomass feedstock can be categorised based on their sources into four major groups [7]. Through thermal, chemical and biochemical conversion processes, biomass can be converted into a variety of biofuel products. Biofuels provide several advantages over fossil fuels especially in the nature of the feedstock (i.e. biomass). Biomass tends to be more renewable, carbon-neutral, and evenly distributed on Earth than fossil fuel resources [4]. As such, biofuels tend to be associated with lower life-cycle GHG emissions. After coal and oil, biomass stands as the third-largest energy source in the world [8]. There is a wide range of biofuel products produced from different sources of feedstock (as described in Table 1) through several conversion routes. The biofuel products include (but are not limited to) bio-oil, biodiesel (Fatty Acid Methyl Ester, FAME), synthetic diesel (hydrotreated vegetable oil), Fischer-Tropsch diesel (F-T diesel or BTL), bio-alcohol and biohydrogen. Given the varieties of biofuels and in view of IMO's 50% emission reduction target by 2050, a research question that remained unanswered in the existing literature would be which biofuel adoption pathway is more feasible at certain conditions. This study focuses on the first-generation biofuels as a start, namely: biodiesel in the form of palm methyl ester (PME) and hydrotreated vegetable oil (HVO).

Table 1. Biomass as feedstock for biofuel production

Type	Source	Biofuel Products	Note
First generation (1G)	Vegetable oils & animal fats and plants containing sugar & starch	Biodiesel (Fatty acid methyl ester, FAME), synthetic diesel (hydrotreated vegetable oil) and bioethanol	- Conflict of 1G biofuels with food supply - The use of only 2% of the planet's arable land to produce the 1G biofuels had a significant impact on food and animal feed price.
Second generation (2G)	Lignocellulosic biomass	A wide range of biofuels, i.e. bio-oil, synthetic diesel, FT diesel (or BTL), biomethane, bio-ethanol, bio-methanol and biohydrogen	- Costly pre-treatment and advanced technologies are needed for biomass conversion to biofuels.
Third generation (3G)	Algae (macro and microalgae)	A wide range of biofuels, i.e. bio-oil, synthetic diesel, FT diesel (or BTL), biomethane, bioethanol, bio-methanol and biohydrogen	Macroalgae: areas for their cultivation Microalgae: Oil content in microalgae, harvesting, oil extraction
Fourth generation (4G)	Wastes and genetically modified algae (Waste includes municipal waste, food processing waste, sewage, plastic, organic waste, etc.)	A wide range of biofuels, i.e. bio-oil, synthetic diesel, FT diesel (or BTL), biomethane, bio-ethanol, bio-methanol and biohydrogen	To eradicate the constraints in 2G biofuel production

3. Methodology

The evaluation of the adoption of alternative fuels is performed mainly based on economic criteria, Relative Net Present Value, designated as NPV, to justify whether ship owners will adopt biofuels instead of continuing using conventional fuel.

$$NPV = \sum_{t=0}^{T} \frac{Fuel\ cost_{FO} - Fuel\ cost_{Biofuel}}{(1+r)^t} - CAPEX$$

Where,

T = Remaining service year (years). The remaining service year of a ship is derived from the lifespan of a ship (typically 25 years) with subtraction of ship age.

CAPEX = Conversion cost (USD) from using conventional fuels to adopt alternative fuel. It includes engine and its system with their installation.

Fuel $Cost_{FO}$ = Annual fuel cost for ship operation when using conventional fuel (USD/year). It can be calculated from annual fuel consumption (tonnes) multiplied by fuel cost per tonne (USD/tonne). Annual fuel consumption (tonnes) is the product of operational power of the ship (kW), operational hours (hours) and specific fuel consumption (tonnes/kWh).

Fuel $cost_{Biofuel}$ = Annual fuel cost for ship operation when adopting biofuel (USD/year). Its calculation can be performed in a similar approach to that of fuel oil.

r = Discount rate (%). The discount rate of investment used in this study is based on the weighted average cost of capital (WACC) of the Global Shipbuilding and Marine Industry.

Table 2. Pathways for the adoption of biofuels for the shipping industry

Pathway	Description	Input		Results
1 Fossil dominance	New-built vessels continue using fossil fuels (i.e., LSFO/MGO and LNG) [1.1 Conventional marine fuel; and 1.2 LNG]	Cost:	CAPEX, fuel cost, and space loss	1. CO_2 monetary value to enable the adoption 2. Fuel requirement: types and amounts 3. Emission of CO_2 (onboard and LCA)
		Fuel:	Characteristics including energy density, density as well as onboard and LCA emission factors	
		Ships:	Specifications, operating profiles, SFOCs	
2. Biodiesel (in the form of PME)	Consideration of the adoption of B30	Cost:	CAPEX, fuel cost, and space loss	1. CO_2 monetary value to enable the adoption 2. Requirement of B30 3. Emission of CO_2 (onboard and LCA)
		Fuel:	Characteristics including energy density, density as well as onboard and LCA emission factors	
		Ships:	Specifications, operating profiles, SFOCs	
3. HVO	Consideration to substitute fossil fuels using HVO	Cost:	CAPEX, fuel cost, and space loss	1. CO_2 monetary value and desirable HVO cost to enable the adoption 2. Requirement of HVO 3. Emission of CO_2 (onboard and LCA)
		Fuel:	Characteristics including energy density, density as well as onboard and LCA emission factors	

		Ships: Specifications, operating profiles, SFOCs	
		Aim: Targeted amount of CO2 to be reduced to meet IMO target by 2050	

4. Results and Discussion

Table 3 presents the results of pathways for the adoption of biofuels in the shipping industry with CO2 monetary value. Pathway 1.1 is the fossil LSFO pathway. This pathway involves ships' operation without implementing CO2 monetary value, i.e., (cost of CO2, expressed in the unit of USD/ton CO2). The shipping industry continues using fossil fuels in this pathway, including LSFO, MGO and LNG. Although the adoption of fossil-LNG can be observed, a majority still continue using conventional fuel oils (LSFO) due to the economic decision. If following this pathway, the shipping industry will not achieve IMO GHG emission reduction target by 2050. Therefore, the pathway is considered not sustainable.

Pathway 1.2 is the fossil LNG pathway. The monetary value of CO2 is introduced to this pathway now. It could be seen that the shipping industry will adopt fossil-LNG when CO2 monetary value is set at 70 USD/tons. The existing ships will continue using conventional marine fuels, i.e., LSFO. In contrast, the adoption rate of fossil LNG increases over the period of time due to the implementation of CO2 monetary value. If following this pathway, the shipping industry will not achieve IMO GHG emission reduction target by 2050 for both SSPs. Therefore, the pathway is also considered not sustainable.

Pathway 2 is the biodiesel B30 pathway. The monetary value of CO2 is introduced now in order to enable the adoption of B30. The adoption of B30 can be seen as early as 2022. It could be seen that the shipping industry will adopt B30 when CO2 monetary value is set at 200 USD/tons. In addition to B30, the shipping industry will also adopt fossil LNG in this pathway. This is purely based on the economic decision. With reference to the current production of biodiesel (40 million tons per year), the requirement of B30 will be around 0.9 million tons and, consequently, B100 of 0.28 million tons. The supply should be sufficient to support the shipping industry. However, the pathway cannot facilitate the industry to meet IMO target by 2050. Therefore, the pathway is considered not sustainable.

Pathway 3 is the HVO pathway. The monetary value of CO2 is introduced with immediate effect. The CO2 monetary level of 262 USD/ton is required in order to enable the adoption of HVO. With the implementation of CO2 monetary value, the first adoption of HVO can be observed in 2025. In addition, the adoption of fossil LNG and bio-LNG can also be observed. Although the pathway has the potential for the shipping industry to meet IMO GHG emission reduction target by 2050, there will be an issue with the supply of HVO with limited feedstock. In order to fulfil this pathway, the industry requires HVO of around 160 million tons/year. Without 4th generation feedstock, i.e., algae, this pathway is likely to be "unrealistic".

Table 3. Results of pathways for the adoption of biofuels in the shipping industry with CO_2 monetary value

Description	1.1. Fossil LSFO Pathway			1.2. Fossil LNG Pathway			2. Biodiesel B30 Pathway			3. HVO Pathway		
	2020	2030	2050	2020	2030	2050	2020	2030	2050	2020	2030	2050
Fossil-fuel demand (million tons/year)												
LSFO requirement	222.76	258.9	345.9	222.76	204.6	115.6	222.76	180.7	72.6	222.8	180.7	72.6
LNG requirement	Negligible	1	4.8	Negligible	43.7	185.8	0	61.7	218.8	0	44.2	44.2
Biofuel requirement (million tons/year)												
B30	0	0	0	0	0	0	0	0.9	0.9	0	1.5	1.5
HVO	0	0	0	0	0	0	0	0	0	0	9	160.6
Bio-LNG	0	0	0	0	0	0	0	0	0	0	8.8	28.8
CO_2 emission (million tons/year)												
LCA CO_2	805.7	943	1,272.60	805.7	901.7	1,097.40	805.7	883.4	1064.5	805.7	840.1	696.2
Onboard CO_2 emission	693.4	808.9	1,090.20	693.4	757.1	870.9	693.4	734.3	829.8	693.4	687.5	350.8
CO_2 emission target by 2050 (50% of 2008 emission level)	NA	NA	354	NA	NA	354	NA	NA	354	NA	NA	354
Difference in CO_2 to meet target by 2050	NA	NA	**736.2**	NA	NA	**516.9**	NA	NA	**475.8**	NA	NA	**-3.2**
Required CO_2 monetary value (USD/ton)	**0**	**0**	**0**	**0**	**70**	**70**	**0**	**200**	**200**	**0**	**262**	**262**
Cost of biofuels (USD/ton)												
B30	702	702	702	702	702	702	702	702	702	702	702	702
HVO	1,474	1,357	1,122	1,474	1,357	1,122	1,474	1,357	1,122	1,474	1,357	1,122
Bio-LNG	1,437	1,369	1,234	1,437	1,369	1,234	1,437	1,369	1,234	1,437	1,369	1,234

5. Conclusion

This study performs an economic analysis on biofuel adoption pathways under carbon tax to achieve a net 50% GHG emission reduction for cargo vessels, including bulk carriers, container vessels and tankers. The study presents three pathways with a focus on the following types of fuels: low sulphur fuel oil (LSFO), marine gas oil (MGO), Liquefied natural gas (LNG), biodiesel in the form of palm methyl ester (PME), hydrotreated vegetable oil (HVO), and bio-LNG. The respective CO2 monetary value to enable the adoption of individual fuel is highlighted in this study. A key finding is that a minimum carbon value of USD 262 should be implemented as soon as possible to achieve IMO's 50% emission reduction target. A carbon value lower than that would either promote biofuels that could not meet IMO's emission target (e.g. biodiesel) or would not be enough to encourage the use of biofuels. Future research could apply the adoption model to other types of alternative fuels such as bio-methanol, BTL diesel, renewable ammonia, renewable hydrogen etc.

Acknowledgements

The research is supported by the Singapore Maritime Institute under Maritime Energy and Sustainable Development Centre of Excellence.

References

[1] International Maritime Organization (IMO) 2020 *Fourth IMO Greenhouse Gas Study* (London: International Maritime Organization)
[2] International Maritime Organization (IMO) 2018 *Resolution MEPC.304(72), Initial IMO Strategy on Reduction of GHG Emissions from Ships* (London: International Maritime Organization)
[3] IEA 2021 *International Shipping – Analysis* IEA Retrieved February 13, 2022, from https://www.iea.org/reports/international-shipping
[4] Florentinus A, Hamelinck C, Bos A, Winkel R and Cuijpers M 2012 *Potential of Biofuels for Shipping – Final Report* (Prepared by ECOFYS for European Maritime Safety Agency)
[5] Houghton R, Hall F and Goetz S 2009 Importance of biomass in the global carbon cycle. *Journal of Geophysical Research: Biogeosciences* **114** G2
[6] US Energy Information Administration (EIA) 2021 *Biomass Explained* Retrieved February 13, 2022, from https://www.eia.gov/energyexplained/biomass/
[7] Darda S, Papalas T and Zabaniotou A 2019 Biofuels Journey in Europe: Currently the way to low carbon economy sustainability is still a challenge *Journal of Cleaner Production* **208** 575-588
[8] Tumuluru J, Wright C, Kenny K and Hess J 2010 *A Review on Biomass Densification Technologies for Energy Application* (U.S. Department of Energy: Office of Biomass Program)

Abatement of GHG emissions onboard mobile offshore units

Chunwee Ng[1,3], **Jizhao Li**[1], **Rakesh Mishra**[2], **Conn Fagan**[2]

[1] DNV Singapore, Centre of Excellence for Maritime Decarbonisation & Autonomy, 16 Science Park Drive Singapore 118227

[2]DNV, Offshore Classification, Veritasveien 1 1363 Oslo Norway

[3] Author to whom any correspondence should be addressed

Abstract. The International Maritime Organisation (IMO) is setting clear and ambitious goals for the decarbonisation of the shipping industry through the introduction of new regulations like the Energy Efficiency Design Index (EEDI), Energy Efficiency Existing Ship Index (EEXI) and Carbon Intensity Indicator (CII). However, these new regulations do not cover mobile offshore units like the Floating Production Storage and Offloading (FPSO) units that fall under the same maritime regime. To embrace sustainability in their corporate strategies and improve their public image, it is in the interest of existing oil and gas companies and operators to embark on a program of abating Greenhouse Gas (GHG) emissions through careful measurement and analysis. This paper is aimed at providing an overview of the typical FPSO's GHG emissions that originates from sources such as power generation, flaring, leakages and tank venting. The paper goes in-depth concerning some potential abatement measures that can be taken to reduce such emissions. This is followed by a case study of an FPSO assessed for the voluntary DNV class notation, Abate, to discuss how a competent third-party like DNV can play a significant role in helping offshore oil and gas companies demonstrate and achieve abatement in GHG emissions from their assets.

1. Introduction

To control escalating acquisition costs in the offshore oil and gas industry, use of the maritime approach, (i.e. Classification and Flag State compliance) for design and construction has achieved wide acceptance in recent years, when applied to floating offshore installations.

Supporting this approach, the maritime classification principles [1] will help offshore asset owners to be assured that an adequate level of safety of life, property and environment has been achieved in the design, construction and operation of these offshore units. Each of these project phases can take place in different locations across the world. Classification Societies, being multi-national organisations will have resources in different countries to independently verify this process.

Under the maritime execution model for mobile offshore units, environmental protection requirements will be mostly dictated by international shipping regulations developed by the International Maritime Organisation (IMO). One of these international regulations for marine pollution from is the International Convention for the Prevention of Pollution from Ships (MARPOL) which covers pollution by oil, noxious liquids, packages, sewage, garbage and air pollution. Over the past few years, several new regulations on Greenhouse Gas (GHG) reduction like the Energy Efficiency Design Index (EEDI), Energy Efficiency Existing Ship Index (EEXI) and Carbon Intensity Indicator (CII) that mandate a quantified reduction in CO_2 emission intensity over time have been introduced by IMO to the MARPOL regulations [2,3]. However, these new IMO regulations do not cover floating offshore installations as the basic carbon intensity calculation they describe involves dividing the quantity of CO_2 emitted by the transport work which is then dependent on ship capacity and speed. For an offshore installation operating at an offshore oil and gas field, such a metric will not be appropriate.

Responsibility for regulating emissions from the oil and gas processes usually falls upon the country or shelf state with its continental shelf being the location where the installation is operating. While most of them have implemented requirements on marine pollution from ships that could occur in waters under their jurisdiction, most have not instituted regulations on the management of GHG emissions from floating offshore installations.

In absence of more definitive statutory requirements, the job of reducing GHG emissions falls upon the oil and gas companies and operators, some of whom may adopt more forward-thinking abatement measures as part of their corporate strategy and to make them more attractive to funding sources and improve their public image.

This paper will look into the regulatory compliance issue regarding GHG emissions amongst the major stakeholders including Shelf States, oil companies and operators and Classification Societies. A study of the various sources of GHG emissions will be presented together with some proposed abatement measures. Lastly, a recent DNV development to introduce a new classification notation, Abate, for a Floating Production Storage and Offloading (FPSO) will be examined and discussed.

The proposed notation will give oil companies and operators a structured way to demonstrate that they are deeply committed to reducing and managing the GHG emissions from their offshore MOUs.

2. Emissions management strategies adopted by various stakeholders

2.1. *Shelf states*

Under the United Nations Convention on the Law of the Sea (UNCLOS), Coastal States and in the case of oil and gas exploration, Shelf States are obliged to protect and preserve the marine environment and adopt and enforce laws and regulations for offshore oil and gas activities within their jurisdiction. While the IMO provides a framework for international regulations on marine pollution, only ship-related marine emissions are regulated. For MOUs, the emissions from the specialised offshore oil and gas exploration and exploitation activities are excluded. This is illustrated in the coverage of IMO's MARPOL regulations for oily discharges from fixed or floating platforms process systems as shown in Figure 1 [4]. The same principles apply to other parts of MARPOL.

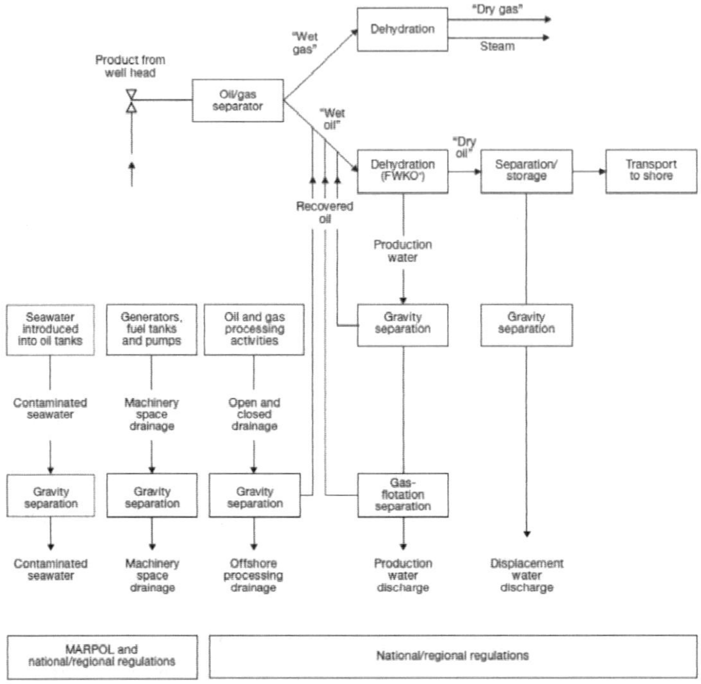

Figure 1. Coverage of offshore oily discharge by MARPOL Annex I (Source: IMO)

Hence, for other emissions originating from the hydrocarbon process activities, the Shelf State will need to develop the requirements which it does under the United Nations Environment Programme (UNEP) Regional Seas Programme (RSP) which involves multilateral environmental agreements signed by governments to address the degradation of the oceans and seas at a regional level [5].

Despite their importance, Watson identified regulatory inconsistencies and gaps under the various RSPs concerning GHG emissions from their offshore oil and gas activities [6]. Due to the fragmented nature of these RSPs, there are notable inconsistencies in the standards of regulation regarding the prevention of pollution from GHG emissions from offshore oil and gas activities. The regulatory gap is due to the non-binding nature of some RSP agreements and how compliance is ensured.

While the RSPs work to improve the situation, an alternative voluntary compliance mechanism needs to be in place so that impacts to emissions reduction can be made with immediate impact.

2.2. Oil companies and operators

Oil companies and operators of offshore oil and gas installations face intense public scrutiny for their contribution to the climate change problem due to their close connection to fossil fuels, and their connection to global warming. For instance, an oil company has to suspend its development of an oilfield offshore Shetland Islands after facing fierce opposition from environmentalist groups in 2021.

Some oil companies are required to pay carbon tax which is a fixed tax rate levied on CO_2 emissions or participate in Emission Trading Scheme (ETS) to sell and buy emission credits between companies. This is expected to have a significant influence on the profitability of some of their asset where at US$200/tonne, Wood Mackenzie found that a third of all assets would have at least 50% of their remaining value lost due to carbon charges [7].

Being capital intensive assets, offshore oil and gas installations require substantial funding from banks and financial institutions who are increasingly looking at minimising their exposure to the fossil fuel industry and moving into green renewable ones.

With the major factors above, oil companies and operators who are installation owners need to demonstrate that their installations are as environmentally friendly as possible.

2.3. Classification Societies

With increasing demand from the shipping market for sustainability, Classification Societies have started offering class notations that show that a ship design has achieved a certain level of environmental friendliness. A listing of some examples of these class notations is shown in Table 1. The foundation of the requirements of these class notations is the prevailing international regulations with the addition of more stringent class requirements. Only when all the stated requirements are met, the additional class notation will be awarded to the ship that will help the shipowner in the marketing of their products.

Table 1. Class notations for environmental ship designs

Class society	Class notation
America Bureau of Shipping (ABS)	ENVIRO
Bureau Veritas (BV)	CLEANSHIP
DNV	CLEAN
Lloyd's Register (LR)	ECO

Since the end of the year 2020, some classification societies have offered further class notations like the SUSTAIN notation by ABS and SUSTAINABLESHIP by BV to help shipowners document that their ship design will meet parts of the United Nations' Sustainable Development Goals (SDG). These class notations can be described as combination notations in that they require the ship design to meet the requirements of several basic class notations. For example, in BV's SUSTAINABLESHIP-1 notation, it requires the basic CLEANSHIP, BWT for ballast water treatment, SEEMP for ship energy efficiency management plan and GREEN PASSPORT for ship recycling to be met before the overall SUSTAINABLESHIP-1 notation can be awarded.

While the above class notations will encourage ship designers to develop ships that will reduce the risks of marine pollution and improve their sustainability, it may not directly lead to a quantifiable reduction of GHG emissions. IMO's EEDI regulations will help to drive down the CO_2 emissions from ships but we have seen from earlier that this does not apply to MOUs. Further work needs to be be done for encouraging clean and sustainable design for MOUs too.

To plug this gap for the MOUs, Classification Society DNV has introduced the new Abate notation in July of 2021 to assist the owners and operators of MOUs to identify, assess and implement measures to reduce GHG emissions. Unlike earlier conventional class notations which are typically prescriptive, the ABATE notation is based on the management system principle which is used in the Quality, Environmental and Energy Management Systems, combined with certain prescriptive requirements.

Before the Abate notation is discussed in detail with the application to an FPSO, the following section will first cover the sources of GHG emissions and some ways they can be mitigated by various technologies.

3. Sources of emissions and abatement measures

According to a report by UK Oil and Gas Industry Association [8], they found that emissions from the sector reduced by 10% in the year 2020 compared to the baseline in the year 2018. They estimated that around half of the emission reduction is attributed to a fall in production and maintenance activities due to the COVID-19 pandemic. The other half of the emission reduction is due to the operator's actions like reduced flaring and venting, streamlining operations and investing in targeted plant modifications which improved the UK Continental Shelf (UKCS) average carbon intensity from 23 kg CO_2 per barrel of oil equivalent (boe) to 21 kg/boe in the year 2020. This example shows that abatement measures taken by the operator are important to lead to a further reduction in GHG emissions.

In terms of the breakdown of GHG emission gases and sources, the Oil and Gas Authority of UK [9] presented (as shown in Figure 2) that 88% of total emissions is by CO_2 followed by CH_4 at 10% and a smaller amount of 2% by N_2O. Whereas the majority of CH_4 originates from ventingand methane release during flaring, fuel combustion and flaring contributes to most of the CO_2 and N_2O emissions.

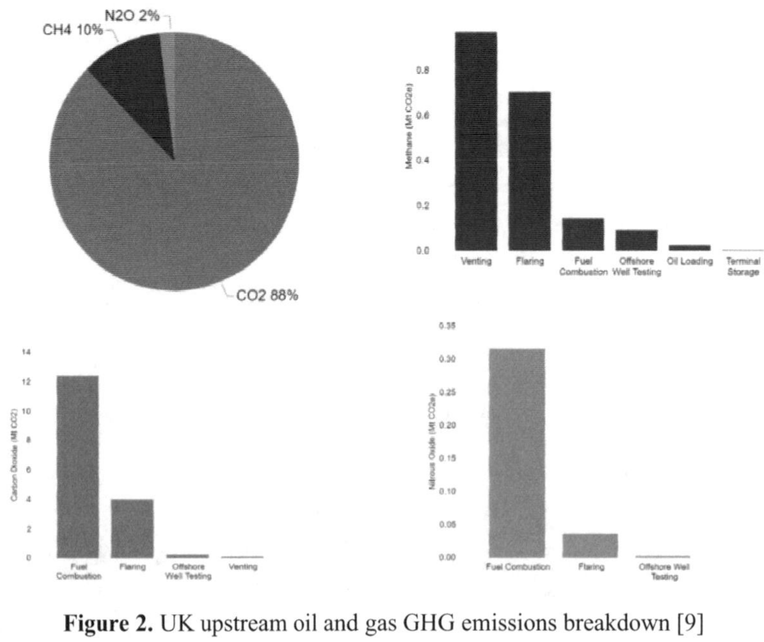

Figure 2. UK upstream oil and gas GHG emissions breakdown [9]

The following section will discuss the main sources of emissions highlighted earlier and some corresponding abatement measures that can be applied.

3.1. CO_2 emissions

CO_2 is a major by-product of combustion and hence originates mainly from fuel combustion from engines, particularly the gas turbines in the offshore oil & gas industry. The flaring process is the next contributor to CO_2 emissions.

There are several ways to reduce CO_2 emissions, as follow:

- Reduce energy demand onboard and/or improve energy efficiency during power/heat generation.
- Reduce flaring with optimised production and flaring process.
- Reduce emissions from power and/heat generation.

3.1.1. Improve energy efficiency

Improving the energy efficiency of MOUs can lead to significant energy savings and CO_2 emissions reduction by 15 to 20% as found by Nguyen [10] through the use of advanced machinery and systems, process integration, waste heat recovery, downsizing and replacement of gas turbines, et cetera.

One of the most energy-intensive systems on an MOU is the gas turbine system, which usually consists of a generation skid, pumps, heat exchanger known as Waste Heat Recovery Unit (WHRU), and associated piping arrangements as shown in Figure 3.

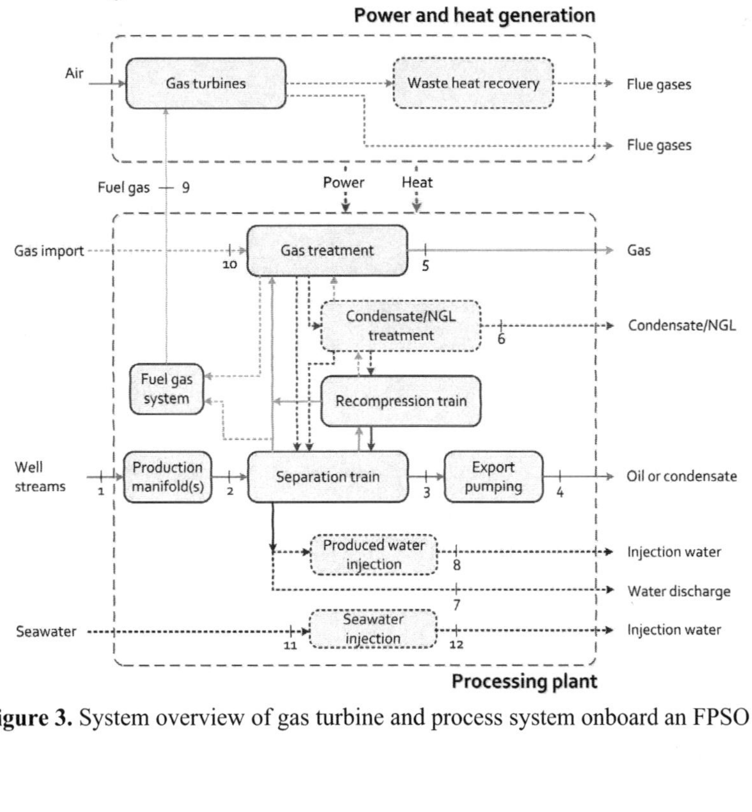

Figure 3. System overview of gas turbine and process system onboard an FPSO [10]

Heat is recovered from the exhausts of the gas turbines and transferred to a heated medium like thermal oil or hot water loop which can be used for heating purposes. If the heated medium like steam is used to power a turbine-generator, electrical power can then be extracted in a combined cycle with steam Rankine cycle as shown in Figure 4.

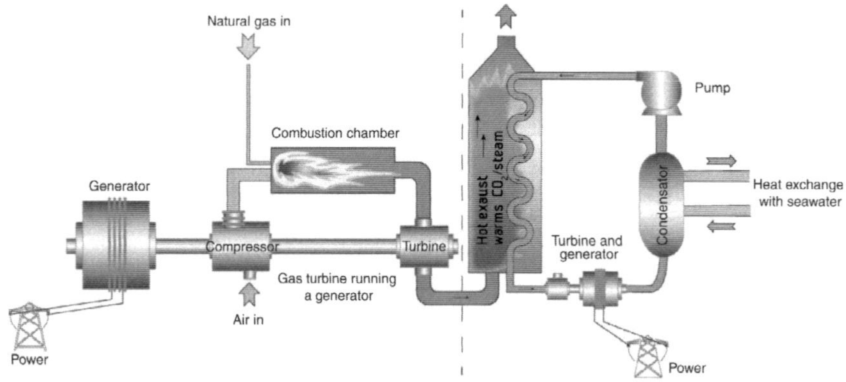

Figure 4. Combined cycle with gas turbine and Steam Rankine cycle [11]

The process integration of waste heat recovery cycle is complex in practice with a large number of factors to consider, such as operating parameters (e.g. pressure, temperature) and strategy (e.g. maximise power, cost, efficiency), as well as the selection of the utilities (e.g. seawater). By system optimisation, the dual goals of maximising power production and minimising CO_2 emissions can be achieved.

Other abatement measures that can be considered include the following:

- Fuel choice and treatment
- Advanced process control system
- Good design and maintenance of combustion equipment

3.2. Emissions from flaring

A report published by IEA on Flaring Emissions [12] estimates average global combustion efficiency (including both normal operating and extinguished flares) to be around 92%, which means that 8% of natural gas and NGLs directed into a flare stack were not combusted.

To reduce flaring, productive uses for the associated gas must be developed to minimise start-up and shutdown of process systems, or associated gas must be safely injected into the reservoir. An alternative solution is to integrate a flare gas recovery (FGR) system with a liquid ring compressors and separators into the installation (Figure 5). Gas that would otherwise be flared is captured and turned into other useable products or electrical power.

3.3. *Emissions from venting and leakages*

Tanks, vents and leakages release methane and Volatile Organic Compounds (VOC), which all contributes to methane emissions. Specific measures can be adopted to achieve the reduction of methane and VOC emissions. For example, methane emissions can be reduced by minimising venting from processes, improving leakage management and applying optimised maintenance procedures. On the other hand, VOC emissions shall be treated differently due to the nature and location of the compounds. Stricter tank inspection regimes can be applied to reduce the number of tank inspection activities, and thus reduce the chances of tank openings to the atmosphere. Newer processes are being utilised on VOC recovery systems. There is a growing trend in the industry to increase the reliability of equipment and systems and to eliminate shutdowns employing intelligent and predictive maintenance.

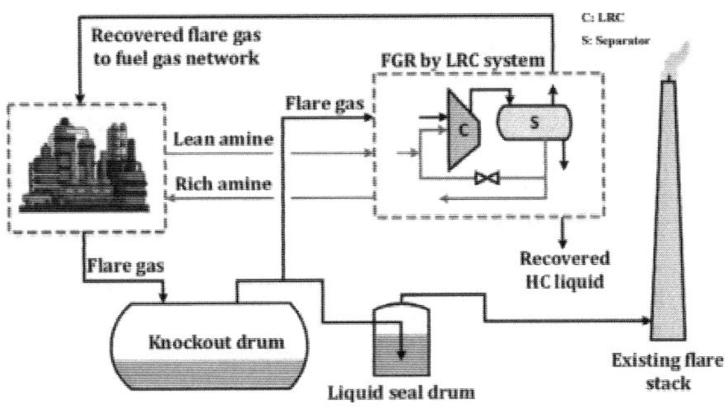

Figure 5. Flare Gas Recovery System with an existing plant [13]

3.3.1. *Reduce emissions from venting*

The emission of volatile organic compounds can be reduced by introducing tank vent gas into the VOC recovery system, followed by a condensation process to separate the heavier hydrocarbon fractions of the gas.

3.3.2. *Reduce process leakages*

A large number of offshore platforms around the globe make careful measurement and analysis of leakage critical for global emissions inventories. For the Norwegian Continental Shelf (NCS) oil and gas installations, an estimation of the methane venting and NMVOC emissions has been performed as shown in Table 2.

In 2016, the U.S. Environmental Protection Agency (EPA) announced a set of rules to reinforce the reduction of methane emissions through the management of process leaks [14]. The following points can be taken as good practices:

- A fixed schedule can be set for monitoring leaks.
- Adopt alternative approaches for finding leaks, such as the use of portable monitoring instruments rather than conventional optical gas imaging.

3.3.3. *Carbon capture storage*

As power generation using fossil fuels on both production and drilling units represents the largest contribution to CO_2 emissions on such units, capturing such emissions and preventing them from entering the atmosphere would be very positive regarding overall environmental footprint.

The most commonly suggested technology is to remove CO_2 from the flue gas by contact with an amine solvent and then regenerate the solvent and release the CO_2 for disposal. Disposal of course can be a challenge, however, there are several options including injection in a reservoir (pure CO_2 injection, comingling CO_2 with existing natural gas re-injection, and comingling CO_2 with typical water injection), pipeline export, and temporary storage and offloading. The latter option might be appropriate for drilling units and is at present being considered for shipping.

Table 2. Estimated release of methane and VOC in NCS [15]

Main Source	Methane	NMVOC	VOC	% Total
Dry compressor seal	2500	1100	3600	18%
Common vent (measured)	1950	1350	3300	17%
Produced water treatment	2300	600	2900	15%
HC blanketing gas and flushing	1100	1300	2400	12%
Gas leakages	1250	950	2200	11%
Flare gas not burned	1500	600	2100	11%
Glycol re-generation	550	1000	1550	8%
Compressor sealing oil	900	300	1200	6%
Other sources	300	250	550	3%
Total	12350	7450	19800	100%

4. Application of a GHG emission abatement class notation (Abate) for an FPSO

The earlier sections have presented the importance of GHG emission abatement for the main stakeholders: shelf state, oil companies and operators and how some emission abatement technologies are available in the market to be adopted.

This section will focus on how one of the classification societies, DNV can offer a new service to help the asset owner demonstrate that the MOUs have incorporated efficient emission abatement measures and processes to meet their own Company emission goals and for better public image and easier access to financial capital.

4.1. *Demonstrating emission reduction along unit's lifecycle with Abate notation*

In July 2021, DNV became the first classification society to publish the rules for the new class notation, Abate, which helps to identify, assess and implement the measures taken by rig owners for GHG emission abatement [16]. Shortly after that, the first notation was awarded to FPSO owner, Altera Infrastructure's Petrojarl Knarr FPSO [17] shown in Figure 6.

Figure 6. The Petrojarl Knarr FPSO (Image care of Altera Infrastructure)

The notation is voluntary and comprises a basic scope which is associated with the installation owner's management system of emissions and assessment of their potential abatement measures, and then a detailed scope, including some prescriptive requirements, associated with specific features of a floating offshore installation through five different qualifiers that include power generation (P), carbon capture (CC), flaring (F), process leaks (Pr) and venting from storage (S). An overview of this is shown in Figure 7.

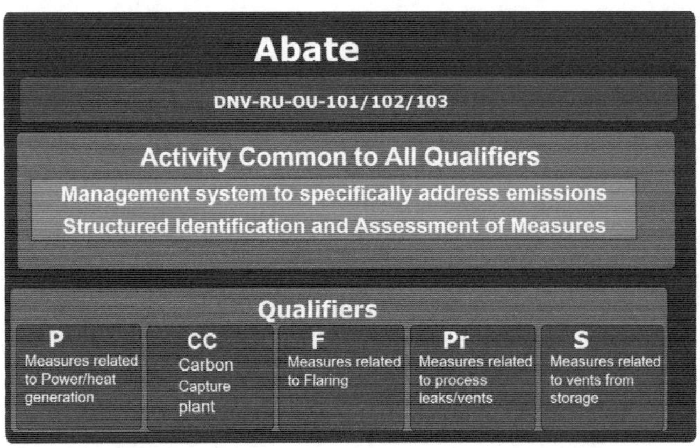

Figure 7. Overview of DNV's Abate notation

The Abate notation can be awarded as part of a newbuilding project or for an existing installation. Existing units will have some additional constraints with regards to the potential for physical modifications to the existing plant and the capability of installing additional physical measures. These constraints would be reflected in the required assessment of the measures identified by the required studies. The basic minimum prescriptive requirements would however still need to be complied with.

The notation can be maintained in the in-service phase by regular review of how the emission management philosophy is being followed up, confirmation that implemented measures function satisfactorily, and that there is a continuous review of future potential abatement measures.

4.2. General requirements for the Abate notation

4.2.1. Management system

As the basic scope, the requirements for a management system and the need to identify and assess abatement measures are a prerequisite to obtaining any of the Abate qualifiers. The emission management system shall follow similar principles as those documented in ISO 50001 which addresses energy management [18] as shown in Figure 8. Essentially the term "energy" can be replaced by the term "emission".

Within the framework of the emission management system, the management level needs to delegate the responsibility to a dedicated person or team and direct a policy for emission abatement on how the targets and pathways will be developed to meet the policy.

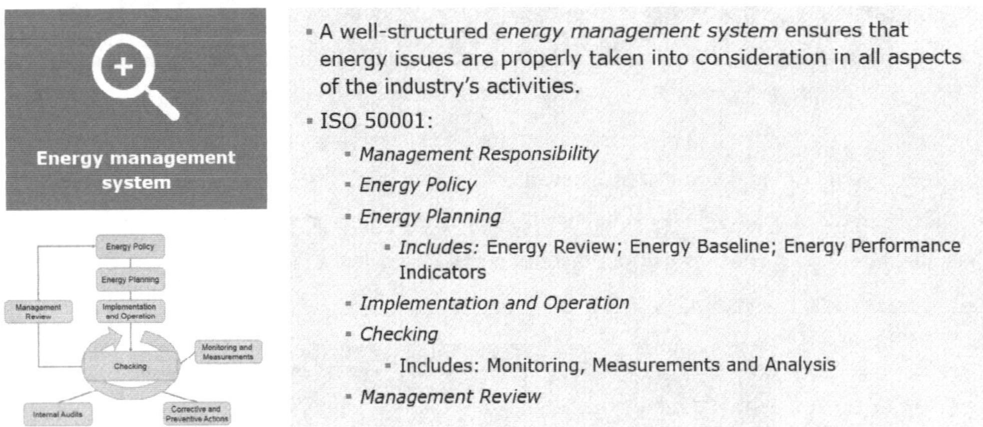

Figure 8. Emission management following principles in ISO 50001

The initial management system assessment is aimed to verify and document that the client has carried out a review of the status of their GHG emissions and has carried out studies to identify potential measures which can be reasonably implemented. The review should confirm that the installation owner has a suitable management system to address the issue of emissions, has set emission goals, will continuously monitor progress, and will update their system as appropriate.

The management system should describe what emissions are being recorded, how they are being recorded and the accuracy of the recorded values. A baseline emission level should be established and targets specified for abatement relative to this reference level at specified time intervals. Monitoring and assessment of progress should be addressed and documented. There should also be a procedure to review how well the management policy has worked with a feedback mechanism to continually improve emission management.

The second key part of this initial assessment is to confirm that abatement measures based on current Best Available Technology (BAT) have been identified, and consideration of their applicability and potential for implementation has been documented. One structured approach for assessment of abatement measures for this is known as the DNV Marginal Abatement Cost Curve (MACC). Identified measures are also assessed for suitability of application based on a defined set of parameters that typically include reliability, weight/space demands, technology maturity, suitability for location and CAPEX/OPEX. This process is not stagnant after the award of the Abate notation and needs to be carried out at least every 5 years as part of the renewal survey for classification.

The requirements to the management system and the need to identify and assess abatement measures are a prerequisite to obtaining any of the Abate qualifiers. These qualifiers will be explained in the following sections.

4.2.2. Power system emission abatement (P)

As power generation accounts for a large proportion of GHG emissions, this is the area where it is possible to achieve the most emission abatement. This qualifier is intended to permit us to give credit for physical measures which are implemented, and which will reduce GHG emissions.

As emissions from power generation are closely linked to energy consumption, several energy assessments shall be made to reduce and optimise energy demand as below:

- Energy demand optimisation study that may have the following findings-
  - Running a fewer number of units at higher capacity might reduce overall reliability/redundancy but reduce overall usage and fuel consumption
  - Assessing the number of major consuming units necessary for operation (e.g pumps, compressors)
  - Applying closed bus technology to reduce the number of units needed (e.g Dynamic Positioning units)
- Energy source study
  - External electrification by using shore power
  - Hybrid systems using batteries or fuel cells to supplement fossil fuel power generation
  - Future consideration of use of alternative fuels with lower GHG footprint (H_2, NH_3, LNG rather than fossil fuels)
- Energy efficiency study
  - Applying combined cycle power generation by adding a steam turbine to use waste heat (most appropriate for FPSOs)
  - Using heat from a WHRU to provide process heat (assume most heat is recovered and that this is as efficient as a combined cycle power generation) thereby avoiding using additional power for that purpose
  - Selection of equipment with consideration of power consumption

Several prescriptive requirements include implementing measures identified in the energy assessments above, having a system to monitor energy consumption, implementing a combined cycle power generation or waste heat recovery if supplying process heat is more relevant or efficient. Also, if other power sources with lower GHG emissions are shown to be feasible, they should be implemented.

In addition to technical requirements, the procurement process should also consider the selection of equipment based on their efficiency and emission level, following the declared procurement philosophy which will be reviewed in the management system.

In general, it is expected that items related to power and heat generation identified as feasible and "reasonably practicable" in the review phase are implemented onboard.

4.2.3. Carbon capture of power generation emissions (CC)

Modular carbon capture technology is currently being developed and is expected to be relevant for application already in the year 2022 and 2023. The technology needs to capture most of produced CO_2. The CC plant needs to be linked to technology to store and/or appropriately dispose of the captured CO2. Given the relative novelty of current CC technology it would be necessary to document that the technology had undergone some form of technology qualification procedure.

4.2.4. Flare emission abatement (F)

Flaring contributes to the next major proportion of offshore GHG emissions. The amount of flaring depends on many factors including reservoir properties, local legislation, carbon taxation among others.

Most appropriate for production units engaging in non-emergency flaring (e.g. at start-up, shut down, maintenance), however, could potentially be relevant for drilling units that can address abatement during well test flaring. The following assessments would need to be performed:

- Flare scenario identification and assessment study

To address flare emission reduction, the rig owner needs to document under what circumstances flaring is being carried out. The study needs to address whether it is possible to eliminate or reduce the occurrence of some of these scenarios.

- Flare gas capture study

In a process system, gas at different locations will be directed to the flare during normal operation. The study should address the possibility, and technology needed, to re-direct such flows to other parts of the process plant.

- Flare efficiency study

The prime concern with flaring is the production and emission of CO_2, however, the release of methane due to incomplete combustion will also have a significant GHG effect. The study should address how efficient the flare arrangement is, both concerning the actual flare burner design and also the availability of the flare when required.

It is required that a flare gas recovery system be installed to capture gas that would otherwise be flared. In many locations a philosophy of zero operational flaring is is required and adopted, in areas where non-emergency flaring is permitted other gas usage or disposal technologies that include gas export, gas injection, gas-to-liquid or gas-to-wire applications should also be considered.

4.2.5. Process system vent and fugitive emissions abatement (Pr)

The process plant will normally be the source of some leakages and intentional venting to the atmosphere that occurs due to the design of some equipment (e.g. sealing arrangement on some compressors), leakage due to failure of equipment or seals in the system and releases due to normal vent design in the process. To reduce such emissions, the usual steps taken are as follow:

- selection of equipment with lower leakage potential compared to any alternatives
- reducing the frequency of failure by improving initial reliability
- reducing the frequency of failure by implementing a maintenance system that involves planned shutdowns rather than being based on action on failure/leakage. Shutdowns may result in blowdown/purging of equipment, process train, or an entire plant, so the limitation of the need to shut down will reduce emissions.
- reducing the volume of leakage by early detection of any leaks (e.g. additional sensors and monitoring)
- collection of normally vented releases rather than release to the atmosphere
- special consideration should be given to the design of purge gas system with a preference for use of inert gas rather than hydrocarbon gases

The following assessments would normally be required to ensure the above are achieved:

- Vent and leakage study to identify sources
- Reliability, Availability and Maintainability (RAM) study
- Vent capture study
- Detection and prevention of leakages study (e.g. gas detection, procurement, maintenance)

Furthermore, the procurement process should be explicit in minimising emissions in the selection of equipment.

In some locations, there may be a very large concentration of CO_2 in the well stream which should be considered in the above assessments.

4.2.6. Storage tank emission abatement (S)

Venting from hydrocarbon storage tanks either during normal operation, loading or unloading and maintenance can release GHGs like methane or non-methane Volatile Organic Compound (VOC) to the atmosphere.

The following assessments would need to be carried out:

- Storage tank release study that identifies release scenarios and assesses whether they can be reduced e.g. frequency of tank entry
- Identification and assessment of both technical and operational abatement measures. This should include the provision of a Vapour Recovery Unit where VOC emissions cannot be avoided

5. Conclusions

The offshore oil and gas industry is not currently regulated in the same way as the international shipping industry for GHG emission reduction. However, the same issue is highly relevant to the offshore industry, even though many of the characteristics of these industries differ.

It is discussed in this paper that many oil companies now have green ambitions and will want to document this by independent verification. They are increasingly looking to see emission reductions and some are including such requirements in tenders. The rewards include easier access to lending institutions that will want to see that they are investing in greener projects and responding positively to social and political pressure on them to reduce emissions as best as they can.

Classification Societies such as DNV have an important role to play when there are uncertainties in regulatory frameworks for offshore emission reduction. Offering specialised classification notations like Abate can help their client demonstrate their commitment to GHG emissions abatement in a structured independently verifiable way.

6. References

[1] C. Ng and R. Jiang, *Classification Principles for Very Large Floating Structures*, vol. 41. Singapore: Springer Singapore, 2020.

[2] IMO, "Guidelines on the method of calculation of the attained energy efficiency design index (EEDI) for new ships.," London, 2018.

[3] IMO, "Report of the Marine Environment Protection Committee on its seventy-sixth session," London, 2021.

[4] IMO, "Resolutions from the fifty-second session of the Marine Environment Protection Committee in October 2004. MEPC.117(52)," London, 2004.

[5] UNEP, "Regional Seas Programme," 2022. [Online]. Available: https://www.unep.org/explore-topics/oceans-seas/what-we-do/regional-seas-programme. [Accessed: 17-Feb-2022].

[6] S. M. Watson, "Greenhouse gas emissions from offshore oil and gas activities — Relevance of the Paris Agreement, Law of the Sea, and Regional Seas Programmes," *Ocean Coast. Manag.*, vol. 185, no. March 2019, p. 104942, 2020.

[7] G. Kellas, "How carbon pricing could reshape upstream oil and gas economics," *Wood Mackenzie Report*, 2021. [Online]. Available: https://www.woodmac.com/news/opinion/how-carbon-pricing-could-reshape-upstream-oil-and-gas-economics/. [Accessed: 08-Feb-2022].

[8] OGUK, "Energy Transition Outlook 2021," 2021.

[9] OGA, "Emissions Monitoring Report," London, 2021.

[10] T. Van Nguyen, M. Voldsund, P. Breuhaus, and B. Elmegaard, "Energy efficiency measures for offshore oil and gas platforms," *Energy*, vol. 117, pp. 325–340, 2016.

[11] M. J. Mazzetti, P. Nekså, H. T. Walnum, and A. K. T. Hemmingsen, "Novel energy efficiency technologies for reduction of offshore CO2 emissions," *Offshore Technol. Conf. Proc.*, vol. 2, no. February, pp. 1339–1350, 2013.

[12] IEA, "Flaring emissions," 2021. [Online]. Available: https://www.iea.org/reports/flaring-emissions. [Accessed: 16-Feb-2022].

[13] J. Asadi, E. Yazdani, Y. Hosseinzadeh Dehaghani, and P. Kazempoor, "Technical evaluation and optimization of a flare gas recovery system for improving energy efficiency and reducing emissions," *Energy Convers. Manag.*, vol. 236, p. 114076, 2021.

[14] EPA, "New Source Performance Standards and Permitting Requirements," 2016. [Online]. Available: https://www.epa.gov/controlling-air-pollution-oil-and-natural-gas-industry/new-source-performance-standards-and. [Accessed: 17-Feb-2022].

[15] G. Husdal, L. Osenbroch, Ö. Yetkinoglu, and A. Østebrøt, "Cold venting and fugitive emissions from Norwegian offshore oil and gas activities," 2016.

[16] DNV, "Rules for classification: Offshore units — DNV-RU-OU-0102. Edition July 2021.," Hovik, 2021.

[17] DNV, "DNV and Altera Infrastructure pilot new Abate notation to reduce GHG emissions on offshore installations," 2021. [Online]. Available: https://www.dnv.com/news/dnv-and-altera-infrastructure-pilot-new-abate-notation-to-reduce-ghg-emissions-on-offshore-installations-207513. [Accessed: 10-Feb-2022].

[18] ISO, "ISO 50001:2018: Energy management systems - Requirements with guidance for use."

Acknowledgements

The authors would like to thank the Singapore Economic Development Board (EDB) for their grant for the Centre of Excellence for Maritime Decarbonisation and Autonomy which funded the preparation of this paper.

Ship route optimization using hybrid physics-guided machine learning

Ulrik Jørgensen[1], Pauline Røstum Belingmo[1], Brian Murray[1], Svein Peder Berge[1] and Armin Pobitzer[2]

[1]SINTEF Ocean AS, Postbox 4762 Torgarden, NO-7465 Trondheim, Norway
[2]SINTEF Ålesund, Norsk Maritimt Kompetansesenter, Borgundvegen 340, NO-6009 Ålesund, Norway

E-mail: ulrik.jorgensen@sintef.no

Abstract. This paper presents a method for energy efficient weather routing of a ferry in Norway. Historical operational data from the ferry and environmental data are used to develop two models that predict the energy consumption. The first is a purely data-driven linear regression energy model, while the second is as a hybrid model, combining physical models with data-driven models using machine learning techniques. With an established energy model, it is possible to develop a route optimization that proposes efficient routes with less energy usage compared to fixed speed and heading control.

1. Introduction

To support the UN's sustainable development goal of combatting climate change, the International Maritime Organization (IMO) has adopted mandatory measures to reduce ship emissions through the Ship Energy Efficiency Management Plan (SEEMP) (IMO, 2019). The initial IMO greenhouse gas (GHG) strategy is to reduce average CO2 emissions across the international shipping fleet by at least 40 % by 2030, and aiming towards 70 % by 2050, compared to 2008 values. The total annual GHG emissions from international shipping should be reduced by at least 50 % by 2050, compared to 2008. One manner to support this goal is through efficient voyage planning. To do so, the energy consumption of a vessel under various operational and environmental conditions must be estimated. By optimizing the route of the vessel with respect to the prevailing conditions, the energy consumption of the voyage can be minimized. Route optimization has the potential for significant reductions in energy consumption, where the associated cost and emission savings are estimated to be around 7 % for the global fleet [1].

1.1. Related work

As originally presented in [2], an overview of the state-of-the-art optimization methods used in weather routing has been presented in [3], while a weather routing system based on travel time, added resistance, and safety is developed in [4]. Many of the related works for weather routing revolve around long-distance route optimization where a large reduction in energy consumption can be achieved by avoiding the roughest environmental conditions. However, a considerable reduction of energy consumption can also be achieved for short-sea shipping, due to the high frequency of voyages. An example is presented in [5] where weather routing system for short-sea shipping is based on a pathfinding algorithm with the use of met-oceanographic predictions.

With large amounts of operational ship data and environmental data available, there has been an increased interest in using machine learning methods for estimating the energy consumption [6, 7, 8]. Often, regression models are used to predict the energy consumption, using engine RPM, ship velocity and weather data as inputs.

Machine Learning (ML) is being applied in a wide variety of domains due to its state-off-the-art performance in many classification and regression tasks. ML techniques are data-driven, i.e. they fit a model to a data set without explicit knowledge of underlying relationships. Applying such techniques in safety-critical systems can therefore be a challenge since ML models are often used as "black boxes", where the underlying causality is unknown. Facilitating trust in such methods is, therefore, a challenge, as the interpretability of ML models is limited. As a result, there is a growing consensus that ML techniques should be coupled with physical knowledge to strengthen and overcome some of the drawbacks with ML. This technique is called hybrid physics-guided ML (also referred to as knowledge-guided ML, science-guided ML, physics-guided ML, physics-informed ML, physics-aware AI, and theory-guided data science) and is gaining popularity within both the ML and scientific communities [9, 10]. Recently, several works have applied hybrid physics-guided ML, i.e., within the fields of air-foil aerodynamics [11] and energy fusion [12] with promising results.

1.2. Objective

The objective of this paper is to develop a weather routing system for a ferry that will propose the most energy efficient route. The accuracy of the weather routing system depends on the accuracy of the model used to predict the energy consumption (hereby referred to as energy model). In this paper, we seek to investigate if hybrid physics-guided ML techniques can prove useful when applied within the maritime domain. More specifically, we investigate if it is possible to estimate the energy consumption of a vessel more accurately using hybrid physics-guided ML techniques than using pure ML techniques.

1.3. Assumptions

When referring to energy consumption, we focus on the propulsion energy alone i.e., hotel loads are not considered. Moreover, passenger comfort is not included in the optimization schema, but low energy consumption is often correlated with low environmental disturbances, which in turn increase the comfort level. For the ferry route used in this work, transit is the longest and most energy consuming phase (75 % to 80 % of the overall propulsion energy used for an individual trip) with the most intricate interplay between vessel and environmental forces [2]. Consequently, we only focus on the transit phase in this paper. Finally, as this is a route suggestion optimizer, we do not consider dynamic traffic, since the captain will always be in direct control of the vessel.

2. Methodology

From the ship we gather GPS positions, speed over ground (SOG), course over ground (COG), vessel heading, power consumed by the electric motors driving the propulsors, and local wind measurements on the ship. In addition, vessel trim angle, and ship draft are measured, which often are not easily available. Draft measurements are gathered by mounting two displacement radars at the side of the ferry, measuring the distance between the sensors and the sea surface. Knowing the distance between the sensors, distance measurements from the two sensors are used to calculate the average draft measurement as well as vessel trim angle.

The ship data from about 1000 half hour ferry crossings is logged at a frequency of 1 Hz. Waves and current values are estimated by meteorological forecasting models gathered from the open-source weather service provided by MET Norway [13]. From the waves model we collect the wave period and the wave direction, measured from true north. From the current model we collect the speed and direction, relative true north. The weather models have a temporal resolution of one hour and a spatial resolution of 800 m × 800 m.

The waves and current models are interpolated in time and space to relate them to the ship data. Weather data are first interpolated spatially using inverse distance weighted interpolation with the four grid points closest to the current position. Then, the values are interpolated linearly in time. Directional variables, e.g. current direction, are interpolated in the space of complex numbers, mapping angles between 0° and 360° to the unit circle before interpolation.

To remove unphysical data, all trips are filtered by discarding trips that do not have a reasonable duration or speed profile. The stored data is used to develop the energy models, as illustrated in Figure 1.

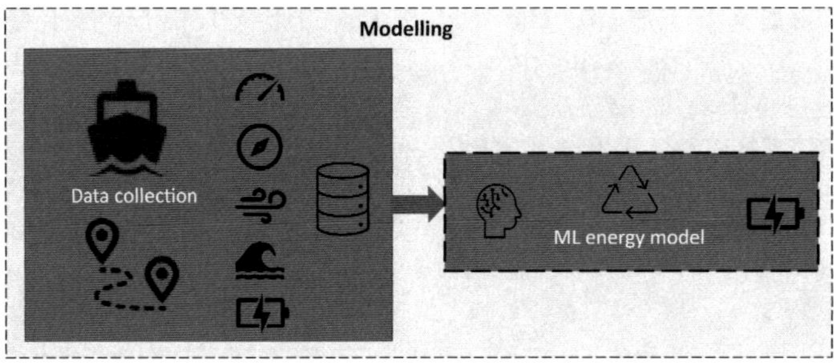

Figure 1: Collected data is used as input to the energy models.

2.1. Data analysis

This paper is a continuation of the work done in [2]. In this study, draft, trim and wind measurements are collected from the ship in addition to the other ship data. An analysis of the three newly added data is given in the following.

In this study, the static draft and trim are utilized. These values are estimated as the mean of the prior 30 s of data for each departure when the vessel is free floating and not affected by engine forces. In this way measurements are not affected by dynamic variations from external forces during the voyage. To remove measurement bias, all used measurements are scaled by subtracting the mean draft value for each sensor.

Having two draft measurements, one located at the front, and one located at the back of the vessel, the trim can be calculated using the draft measurements. Additionally, the pitch angle is measured by a vessel motion reference unit (MRU). The calculated trim is compared with the pitch angle. As can be seen in Figure 2, there is a high correlation (Pearson correlation coefficient $P = 0.97$) between calculated trim angle and the pitch angle. Moreover, since there is a high correlation between the pitch angle and the trim angle calculated using the draft measurements, we conclude that both signals are applicable as model input. For simplicity by keeping the number of inputs to a minimum, we choose to use the draft measurement to estimate trim angles.

The wind measurements from the vessel are compared with wind models from the weather service provided by MET Norway. Wind speed and direction analysis are shown in Figure 3 and Figure 4 with Pearson correlation coefficient $P = 0.76$ and $P = 0.49$, respectively. The level of correlation is as expected since wind measurements are often highly varying since it is affected by the vessel itself as well as measuring a combination of average and wind gusts.

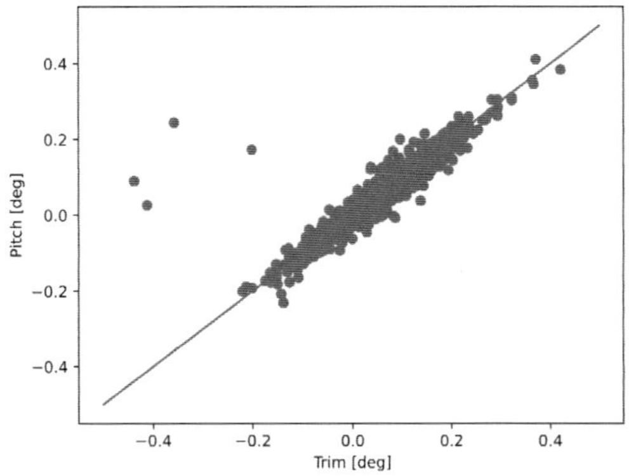

Figure 2: Correlation between calculated trim angle from the displacement radars and pitch angle from the vessel MRU.

Data from MET Norway on the other hand, is model based and does not capturing the complex dynamics around the vessel in the local area. Moreover, the measurement height is not the same as the MET Norway model. Nevertheless, since there are clear relationships between the wind measurements and the MET Norway model, we conclude that the vessel wind measurements are reliable and will therefore use these measurements in the energy model to capture the complex wind dynamics.

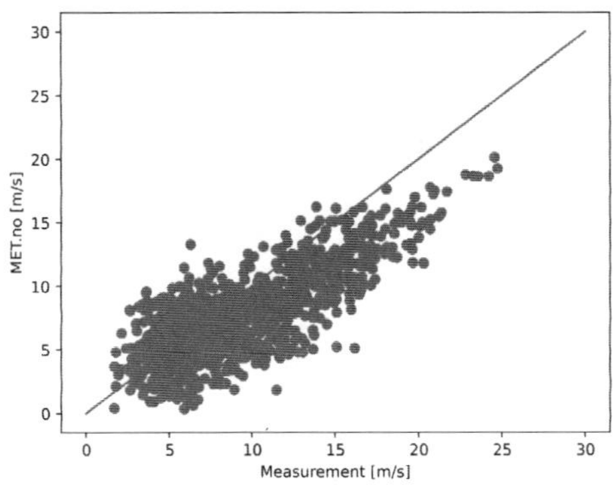

Figure 3: Correlation between measured wind speed and wind speed from MET Norway model.

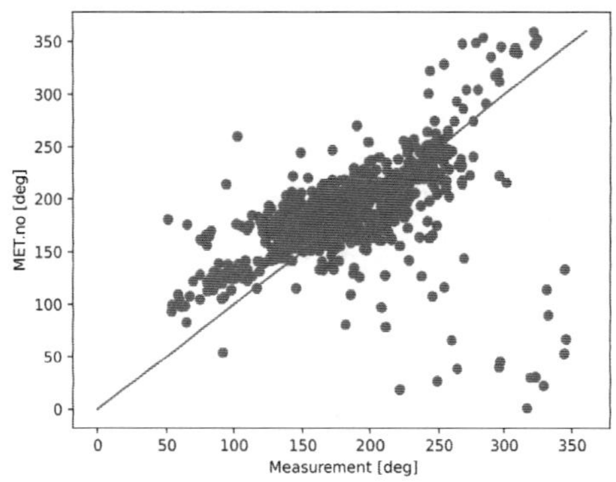

Figure 4: Correlation between measured wind direction and wind direction from MET Norway model.

2.2. Energy models
Using the collected weather and ship data, two energy models are developed and evaluated.

2.2.1. Linear model The first energy model is a linear regression model based on selected parameters derived from the collected data. These parameters are selected based on their correlation with the energy consumption. This model is purely data driven, where the correlation between the input parameters and the energy consumption are identified via ML.

To improve the linear model proposed in [2], vessel draft and trim measurements have been added to the energy model. To further improve the energy model, model parameters are expressed such that the correlation between them are minimized. For example, instead of using wave peak period ω_0 and the wave angle of attack β separately, the wave encounter frequency ω_e is used instead and given by

$$\omega_e = \left| \omega_0 - \frac{\omega_0^2}{g} U \cos(\beta) \right|, \tag{1}$$

where g is the acceleration of gravity and U is the speed of the vessel. Moreover, the current velocity $v_{c,rel}^n$ relative the vessel operating at velocity v_v^b, is given by

$$v_{c,rel}^b = R_n^b(\theta) v_{c,true}^n - v_v^b, \tag{2}$$

where $v_{c,true}^n$ is the current velocity relative true north, θ is the vessel heading, $R_n^b(\theta)$ is the rotation matrix between the vessel body attached frame $\{b\}$ and the North-East-Down (NED) frame $\{n\}$. Superscript $\{b, n\}$ denotes which frame the vector is expressed in. See (Fossen, 2021) for details.

Based on the relationships above, a regression model for the transit phase is developed. The model estimates energy per distance used during transit based on average SOG, longitudinal and transverse current and wind speed, wave encounter frequency, significant wave height H_s, trim and draft.

2.2.2. Hybrid physics-guided model As an extension to the linear model, a hybrid physics-guided model combining physical models with data driven ML models is developed (hereby referred to as hybrid energy model). The model utilizes known physical relations between the selected parameters and the energy consumption. Machine learning methods are then applied to find the unknown relations.

The transverse (denoted t) and longitudinal (denoted l) wind (denoted w) and current (denoted c) forces $F_{i,j}$ acting on a vessel are estimated as

$$F_{i,j} = \frac{1}{2} \rho_i A_{i,j} v_{i,j} \left| v_{i,j} \right|, \tag{3}$$

where $i \in \{w, c\}$, $j \in \{t, l\}$, ρ_i is the air and water density, $A_{i,j}$ is the transverse and longitudinal air and underwater cross-section areas and v_i is the relative wind and current velocities. For modelling simplicity, the static cross-section areas are given by the vessel drawings, hence we do not consider dynamic cross-section areas due to time varying trim and draft. The drag coefficients in Equation 3 are not included, as they are assumed to be estimated as part of the ML-based regression.

In addition to the linear model parameters presented in Section 2.2.1, squared SOG, wind and current forces given in Equation 3 have been added to the physics-guided ML model.

3. Optimization

Once the energy models have been sufficiently trained and evaluated, they can be used to optimize planned routes. This is illustrated in Figure 5. Given the trained energy model, an initial route, weather forecast, and ferry schedule, the goal of the Route Optimizer (RO) is to return an optimized route that minimizes the energy consumption. The RO optimizes the route by modifying the geography and speed profile of the initial route.

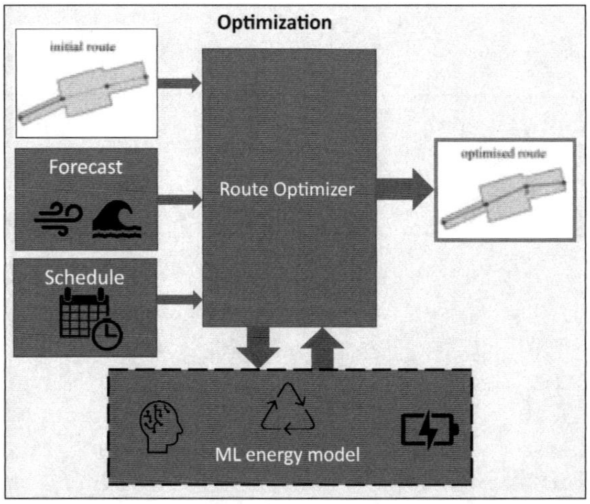

Figure 5: Optimization flow diagram.

The initial route, shown in the upper left corner of Figure 5, is typically defined as a collection of waypoints from the start position to the desired end position of the voyage. The legs between two consecutive waypoints are then often associated with a desired speed value and a cross track distance where it is safe for the vessel to operate. In fact, this is so common that there exists a standard route plan exchange format (RTZ) specifically for this purpose [14, 15]. The cross-track distances for all the legs in a route constitutes a safety zone the vessel can safely operate in, illustrated as the grey area around the routes in Figure 5. The safety zone will typically be configured to avoid shallow water and fixed obstacles. This area also set an absolute limit for where the route optimizer can manipulate the voyage route.

Another element that constrains the route optimizer is the ferry time schedule. The ferry typically needs to be at the destination within a required time. The schedule sets an upper bound for the duration of the trip, and the route optimizer must balance the arrival time with the energy consumption since these are in general competing parameters.

The final input to the RO is the trained energy model presented in Section 2.2. The energy model will be used to estimate the energy consumption for the current proposed route by the RO. To score the current route, the RO proposes an objective function for the current route as a sum of the energy estimation, penalty terms, and regularisation terms. Given the objective function score, the RO can iteratively update the route until an optimal route is given. For details about the optimization, please see [2].

4. Result and discussion

Energy consumption is a sensitive parameter for operating companies, hence the presented data has been anonymised by removing explicit reference to location and energy consumption is not reported in absolute numbers, but relative to reference values, e.g. the average energy consumption for the route or the energy consumption before optimisation.

To evaluate model performance, the data set was split into training and test sets. In this manner, the models can be evaluated on unseen data, i.e. the test set. The test error of both models is given in Table 1, where the mean error of the predictions is evaluated as a percentage of the true values. Furthermore, the root mean squared error (RMSE) and coefficient of determination (R^2) are evaluated.

In [2], the mean error of the energy model was found to be 7 %. By utilizing the improved linear model in this study, the error has decreased to 5.44 %. As such, it appears that the

Table 1: Model accuracy metrics.

	Training error [%]	Test error [%]	Test RMSE $[\frac{\mathrm{kW\,h}}{\mathrm{m}}]$	Test R^2
Linear model	4.86	5.44	0.0024	0.79
Hybrid model	4.26	4.70	0.0020	0.86

outlined parameter selection and transformation yields a more accurate energy model.

The second model investigated was a hybrid model that leverages known physical relationships. In this manner, classical physics-based models can be fused with data-driven ML-based models. Furthermore, non-linearity is added to the model via such relationships. The hybrid model was found to have superior performance, with a mean error of 4.70 % for the test set. Furthermore, the RMSE was found to be 0.002 kW h/m, with an R^2 value of 0.86. This indicates that the model can describe most of the variation in the data with high accuracy. It should be noted that the accuracy and generalizability of ML-models will increase for larger data sets. As such, the accuracy of the models can be expected to increase further given more data.

Figure 6 shows the predicted energy consumption (y-axis) plotted against the true energy consumption (x-axis) of the linear model (blue dots) and hybrid model (orange dots). If a model is zero error, the dots will appear along the diagonal line, indicating that the predicted energy equals the true energy.

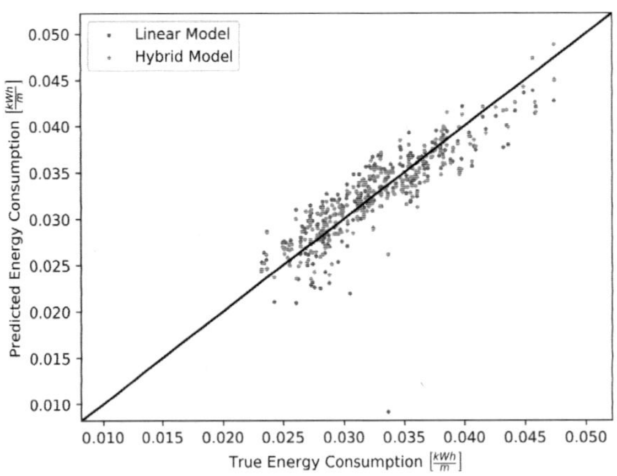

Figure 6: Energy model predictions on test set.

The results indicate that the hybrid model is more robust with respect to predicting energy consumption, as the data are more concentrated about the line representing zero error. Furthermore, the figure shows that the linear model has several outliers, indicating that the linear model was not able to capture the true relationships in the data set. This may be due to the inherent inability of a linear model to capture nonlinear physical relationships. In addition, the model may be overly sensitive to certain parameters, where anomalous values in the input parameter space result in erroneous predictions. It appears that the hybrid model could handle such data in a better manner. This may be due to its ability to leverage the known physical relationships between parameters in a better manner, making it less sensitive to outliers, and thus more generalizable. Hence, it is of interest to investigate the sensitivity of the models to various input parameters, to better understand what the respective models are focusing on.

4.1. Model interpretation

Generally, machine learning models are viewed as black boxes, where end-users have little to no insight into the reasoning behind a prediction. By providing a form of explainability to the model, users may gain increased confidence in the models [16].

Figure 7 and Figure 8 illustrate the scaled mean energy contributions of the input parameters for the linear model and hybrid model respectively. These contributions are calculated as the mean of all energy contributions derived from each parameter irrespective of sign. In this manner, the relative energy contribution of each parameter can be compared, yielding insight into the reasoning behind the models.

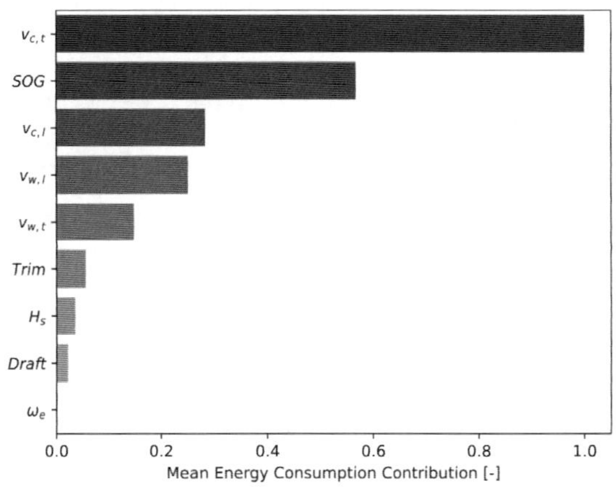

Figure 7: Relative mean energy contributions of the linear model.

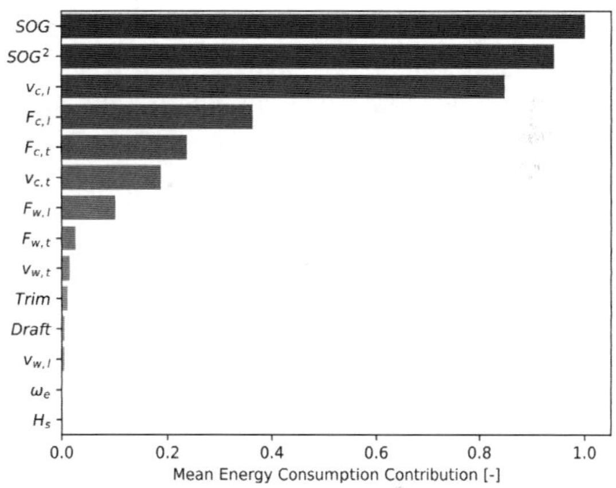

Figure 8: Relative mean energy contributions of the hybrid model.

From Figure 7 it is evident that the linear model has focused primarily on the relative transverse current speed, $v_{c,t}$, as well as the speed over ground. From a physical standpoint, the relationship between SOG and energy is sound. However, the focus of the model on the relative transverse current speed does not seem physically meaningful. It may be attributed to the dominating current direction in the region in which the ferry is operating. As such, the model may be indirectly able to identify the direction in which the ferry is sailing. Each direction the ferry sails will constitute a specific operational mode, due to the varying characteristics of each route. As such, the model may focus on differentiating between these modes, as this minimizes the prediction error. This, however, may come at the expense of the generalizability of the model. This is supported by Table 1, which illustrates the training and test error of the investigated models. The difference between the performance of the linear model on the training and test data sets are higher than that found for hybrid model, which may indicate overfitting to the training data.

Figure 8 shows that the primary focus of the hybrid model is on speed related parameters, with current forces as secondary contributions, and wind contributions as tertiary contributions. Furthermore, the contributions of the longitudinal components of the respective external forces appear to have a higher magnitude than their transverse components. This is conducive with physical phenomena related to ship resistance. As such, the model appears to be physically meaningful. The relative longitudinal current speed, however, has a more significant contribution than expected. This may be due to correlation with the speed over ground.

In both the linear and hybrid models, the trim, draft, significant wave height and wave encounter frequency parameters provide limited to no contribution to the predicted energy consumption. Such results are unexpected, as these parameters are known to affect ship resistance. However, when inspecting the data, it was found that the draft varies little between trips. Such low variation in the data set likely results in the inability of the ML model to learn meaningful correlations. Furthermore, as seen in Figure 2, the observed trim angles in the data set are small. From a physical standpoint, it follows that there should be little to no energy contribution from such angles. Similarly, the wave related parameters have little effect on the energy models, despite wave resistance constituting a significant portion of the energy consumption of sea-going vessels. This can be explained by the region and season the data were gathered in, where the wave height was generally low with little variation, i.e. less than 1 % of the vessel length. As such, the model is unable to learn any meaningful relationship between the wave related parameters and energy consumption, as other effects are dominant.

In general, it appears that the use of a hybrid model provides multiple advantages compared to a purely data-driven linear model. In addition to achieving better accuracy, the hybrid model behaves in a manner that conforms to known physical relationships. As such, the causation of the model is more apparent, increasing confidence in the predictions. The linear model, being purely data-driven, appears to focus more on correlations in the data than physical relationships founded in causation. Furthermore, the hybrid model introduces known non-linear relationships that allow the model to learn more complex relationships than a linear model.

It should, however, be noted that the developed models are specific for this ferry and will need to be re-trained for another vessel and/or region. The focus of the models may also shift depending on the use case. For instance, draft, trim and wave related parameters may play a larger role for vessels that experience greater variation. Also note that by specifying the wave encounter frequency as in (1), one do not get any information how the wave direction affects the energy. For cases with larger H_s this could influence the model performance and should be investigated further. Nonetheless, hybrid ML-based energy models should generally outperform linear ML models due to their ability to leverage more complex causal relationships, as well as be more generalizable.

4.2. Optimization

Once an energy model has been trained on the operational data of the vessel, it can be utilized for route optimization, as illustrated in Figure 5. In this study, the hybrid energy model was chosen for use due to its enhanced performance. To investigate the performance of the route optimization, a historical voyage was randomly selected from the data set. This test voyage can then be optimized, and the relative energy consumption of the respective routes compared.

Using the test voyage, the operational data were extracted and input to the route optimization function. The test voyage is input as the initial route, along with the prevailing met-ocean conditions for the voyage. The original, unoptimized route is illustrated in blue in Figure 9. Using the safety zone, illustrated in green, as a constraint, the route optimization outputs an optimized route, illustrated in orange. Note that the optimized route is slightly longer than the original, but provides an energy reduction of 3.7 %. The optimization algorithm therefore finds a longer path which is advantageous for the current met-ocean conditions. The energy reduction is achieved solely via minor alterations to the route and speed profile prior to departure. If implemented, such route optimizations can lead to a reduction in fuel consumption and associated operational costs.

5. Conclusion

In this paper, a linear regression, and a hybrid physics-guided ML energy model have been developed to facilitate route optimization. The models were developed to predict the energy consumption of a ferry based on operational data. Compared to the data-driven linear regression model, the hybrid model was found to have superior results. Furthermore, by attempting to

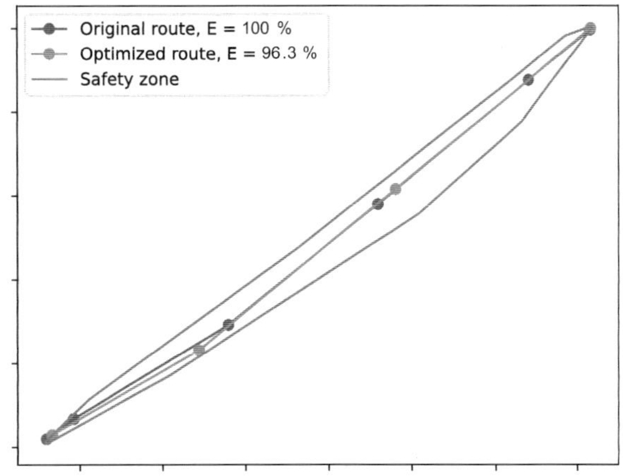

Figure 9: Example figure showing optimized route, energy savings and safety zone.

interpret the reasoning behind the models, it appeared that the hybrid model discovered causal relationships grounded in known physical relationships. The linear model, however, appeared to discover correlations without direct causality.

Based on the findings, the hybrid energy model was tested in an optimization schema showing an energy reduction of 3.7 % compared to the actual consumption, simply by applying minor route and speed profile alterations. Such route optimizations prior to departure can lead to reduced fuel consumption and operational costs.

Acknowledgments

The work presented here is in part funded by the Norwegian Research Council (NRC) through the project Self-learning route optimization for ships (SMARTSHIPROUTING), NRC project no. 295763. The authors would like to thank the project owner Norwegian Electric Systems and the project partners HAV Design and Havila Kystruten. Furthermore, we would like to thank Fjord1 for allowing us to collect operational data from one of their ferries.

References

[1] Lindstad H 2013 Strategies and measures for reducing maritime CO2 emissions

[2] Bellingmo P R, Pobitzer A, Jørgensen U and Berge S P 2021 Energy efficient and safe ship routing using machine learning techniques on operational and weather data (Hamburg University of Technology)

[3] Walther L, Rizvanolli A, Wendebourg M and Jahn C 2016 *International Journal of e-Navigation and Maritime Economy* **4** ISSN 24055352

[4] Fabbri T and Vicen-Bueno R 2019 *Journal of Marine Science and Engineering* **7**(5) 127 ISSN 2077-1312 URL `https://www.mdpi.com/2077-1312/7/5/127`

[5] Grifoll M, de Osés F X M and Castells M 2018 *WMU Journal of Maritime Affairs* **17**(2) 195–211 ISSN 16541642 URL `https://doi.org/10.1007/s13437-018-0143-6`

[6] Bui-Duy L and Vu-Thi-Minh N 2021 *Asian Journal of Shipping and Logistics* **37**(1) 1–11 ISSN 20925212

[7] Parkes A I, Sobey A J and Hudson D A 2018 *Ocean Engineering* **166** 92–104 ISSN 00298018

[8] Petersen J P, Jacobsen D J and Winther O 2012 *Journal of Marine Science and Technology* **17**(1) 30–39 ISSN 09484280

[9] Karpatne A, Atluri G, Faghmous J, Steinbach M, Banerjee A, Ganguly A, Shekhar S, Samatova N and Kumar V 2016 *IEEE Transactions on Knowledge and Data Engineering* **29**(10) 2318–2331 URL `http://arxiv.org/abs/1612.08544` `http://dx.doi.org/10.1109/TKDE.2017.2720168`

[10] Rai R and Sahu C K 2020 *IEEE Access* **8** 71050–71073 ISSN 21693536

[11] Pawar S, San O, Aksoylu B, Rasheed A and Kvamsdal T 2021 *Physics of Fluids* **33**(1) 011701 ISSN 1070-6631 URL `https://aip.scitation.org/doi/abs/10.1063/5.0038929`

[12] Piccione A, Berkery J W, Sabbagh S A and Andreopoulos Y 2020 *Nuclear Fusion* **60**(4) 46033 URL https://doi.org/10.1088/1741-4326/ab7597
[13] The Norwegian Meteorological Institute 2022 Met norway URL https://www.met.no
[14] IEC 2015 IEC 61174:2015: Maritime navigation and radiocommunication equipment and systems - electronic chart display and information system (ecdis) - operational and performance requirements, methods of testing and required test results
[15] IEC 2021 IEC PAS 61174-1:2021 maritime navigation and radiocommunication equipment and systems – part 1: Route plan exchange format (rtz) – general requirements, methods of testing and required test results
[16] Shin D 2021 *International Journal of Human-Computer Studies* **146** 102551 ISSN 1071-5819

AUTHOR INDEX